Smart Innovation, Systems and Technologies

Volume 56

About this Series

The Smart Innovation, Systems and Technologies book series encompasses the topics of knowledge, intelligence, innovation and sustainability. The aim of the series is to make available a platform for the publication of books on all aspects of single and multi-disciplinary research on these themes in order to make the latest results available in a readily-accessible form. Volumes on interdisciplinary research combining two or more of these areas is particularly sought.

The series covers systems and paradigms that employ knowledge and intelligence in a broad sense. Its scope is systems having embedded knowledge and intelligence, which may be applied to the solution of world problems in industry, the environment and the community. It also focusses on the knowledge-transfer methodologies and innovation strategies employed to make this happen effectively. The combination of intelligent systems tools and a broad range of applications introduces a need for a synergy of disciplines from science, technology, business and the humanities. The series will include conference proceedings, edited collections, monographs, handbooks, reference books, and other relevant types of book in areas of science and technology where smart systems and technologies can offer innovative solutions.

High quality content is an essential feature for all book proposals accepted for the series. It is expected that editors of all accepted volumes will ensure that contributions are subjected to an appropriate level of reviewing process and adhere to KES quality principles.

More information about this series at http://www.springer.com/series/8767

Ireneusz Czarnowski · Alfonso Mateos Caballero
Robert J. Howlett · Lakhmi C. Jain
Editors

Intelligent Decision Technologies 2016

Proceedings of the 8th KES International Conference on Intelligent Decision Technologies (KES-IDT 2016) – Part I

Editors
Ireneusz Czarnowski
Gdynia Maritime University
Gdynia
Poland

Alfonso Mateos Caballero
Artificial Intelligence Department
Universidad Politécnica de Madrid
Madrid
Spain

Robert J. Howlett
KES International
Shoreham-by-sea
UK

Lakhmi C. Jain
University of Canberra
Canberra
Australia

and

Bournemouth University
Poole
UK

and

KES International
Shoreham-by-sea
UK

ISSN 2190-3018 ISSN 2190-3026 (electronic)
Smart Innovation, Systems and Technologies
ISBN 978-3-319-81927-3 ISBN 978-3-319-39630-9 (eBook)
DOI 10.1007/978-3-319-39630-9

Softcover reprint of the hardcover 1st edition 2016

Printed on acid-free paper

This Springer imprint is published by Springer Nature
The registered company is Springer International Publishing AG Switzerland

Preface

This volume contains the proceedings (Part I) of the 8th International KES Conference on Intelligent Decision Technologies (KES-IDT 2016) held in Puerto de la Cruz, Tenerife, Spain, in June 15–17, 2016.

The conference was organized by KES International. The KES-IDT Conference series is a subseries of the KES Conference series.

KES-IDT is a well-established international annual conference, an interdisciplinary conference in nature. It provides excellent opportunities for the presentation of interesting new research results and discussion about them, leading to knowledge transfer and generation of new ideas.

This edition, KES-IDT 2016, attracted a substantial number of researchers and practitioners from all over the world, who submitted their papers for the main track and 12 special sessions. All papers have been reviewed by 2–3 members of the International Program Committee and International Reviewer Board. Following a rigorous review process, only the highest quality submissions were accepted for inclusion in the conference. The 80 best papers have been selected for oral presentation and publication in the two volumes of the KES-IDT 2016 proceedings.

We are very satisfied with the quality of the program and would like to thank the authors for choosing KES-IDT as the forum for the presentation of their work. Also, we gratefully acknowledge the hard work of the KES-IDT international program committee members and of the additional reviewers for taking the time to review the submitted papers and selecting the best among them for the presentation at the conference and inclusion in its proceedings.

We hope and intend that KES-IDT 2016 significantly contributes to the fulfillment of the academic excellence and leads to even greater successes of KES-IDT events in the future.

June 2016

Ireneusz Czarnowski
Alfonso Mateos Caballero
Robert J. Howlett
Lakhmi C. Jain

KES-IDT 2016 Conference Organization

Honorary Chairs

Lakhmi C. Jain, University of Canberra, Australia and Bournemouth University, UK
Gloria Wren-Phillips, Loyola University, USA
Junzo Watada, Waseda University, Japan

General Chair

Ireneusz Czarnowski, Gdynia Maritime University, Poland

Executive Chair

Robert J. Howlett, KES International and Bournemouth University, UK

Program Chair

Alfonso Mateos Caballero, Universidad Politécnica de Madrid, Spain

Publicity Chair

Izabela Wierzbowska, Gdynia Maritime University, Poland

Special Sessions

Specialized Decision Techniques for Data Mining, Transportation and Project Management
Piotr Jędrzejowicz, Gdynia Maritime University, Poland
Ireneusz Czarnowski, Gdynia Maritime University, Poland

Intelligent Decision Technologies for Water Resources Management
Wojciech Froelich, University of Silesia, Sosnowiec, Poland
Ewa Magiera, University of Silesia, Sosnowiec, Poland

Intelligent Methods for Eye Movement Data Processing and Analysis
Katarzyna Harezlak, Silesian University of Technology, Poland
Paweł Kasprowski, Silesian University of Technology, Poland

Intelligent Data Analysis and Applications
Urszula Stańczyk, Silesian University of Technology, Gliwice, Poland
Beata Zielosko, University of Silesia, Katowice, Poland

Intelligent Decision Making for Uncertain Unstructured Big Data
Bharat Singh, Big Data Analyst, Hildesheim, Germany
Neel Mani, ADAPT Centre for Digital Content Technology, Dublin City University, Dublin, Ireland
Pandian Vasant, PETRONAS University of Technology, Malaysia
Junzo Watada, Waseda University, Japan

Decision Making Theory for Economics
Eizo Kinoshita, Meijo University, Japan

New Advances of Soft Computing in Industrial and Management Engineering
Shing Chiang Tan, Multimedia University, Malaysia
Chee Peng Lim, Deakin University, Australia
Junzo Watada, Waseda University, Japan

Interdisciplinary Approaches in Business Intelligence Research and Practice
Ivan Luković, University of Novi Sad, Serbia

Recent Advances in Fuzzy Systems
Jose L. Salmeron, University Pablo de Olavide, Spain
Wojciech Froelich, University of Silesia, Sosnowiec, Poland
Elpiniki Papageorgiou, Technological Education Institute (TEI) of Central Greece, Lamia, Greece

Pattern Recognition for Decision Making Systems
Paolo Crippa, Università Politecnica delle Marche
Claudio Turchetti, Università Politecnica delle Marche

Pattern Recognition in Audio and Speech Processing
Claudio Turchetti, Università Politecnica delle Marche
Paolo Crippa, Università Politecnica delle Marche

Reasoning-Based Intelligent Systems
Kazumi Nakamatsu, University of Hyogo, Japan
Jair M. Abe, Paulista University, Brazil

International Program Committee

Mohamed Arezki Mellal, M'Hamed Bougara University, Boumerdes, Algeria
Dariusz Barbucha, Gdynia Maritime University, Poland
Alina Barbulescu, Ovidius University of Constanta, Romania
Mokhtar Beldjehem, University of Ottawa, Ontario
Monica Bianchini, Department of Information Engineering and Mathematics, Italy
Gloria Bordogna, CNR—National Research Council, Italy
Janos Botzheim, Tokyo Metropolitan University, Japan
Wei Cao, School of Economics, HeFei University of Technology, China
Michele Ceccarelli, Qatar Computing Research Institute, HBKU
Shing Chiang Tan, Multimedia University, Malaysia
Marco Cococcioni, University of Pisa, Italy
Paolo Crippa, Università Politecnica delle Marche, Ancona, Italy
Alfredo Cuzzocrea, University of Trieste, Italy
Ireneusz Czarnowski, Gdynia Maritime University, Poland
Ana de Almeida, ISCTE Instituto Universitário de Lisboa, Portugal
Dawn E. Holmes, University of California, USA
Margarita Favorskaya, Siberian State Aerospace University, Russian Federation, Russia
Antonio Fernández-Caballero, Universidad de Castilla-La Mancha, Spain
Wojciech Froelich, University of Silesia, Sosnowiec, Poland
Marcos G. Quiles, Federal University of São Paulo (UNIFESP), Brazil
Mauro Gaggero, National Research Council of Italy
Daniela Godoy, Unicen University, Argentina
Raffaele Gravina, DIMES, University of Calabria, Italy
Christos Grecos, Sohar University, Oman
Katarzyna Harezlak, Silesian University of Technology, Gliwice, Poland
Ioannis Hatzilygeroudis, University of Patras, Department of Computer Engineering and Informatics, Greece
Katsuhiro Honda, Osaka Prefecture University, Japan
Daocheng Hong, Fudan University, China
Tzung-Pei Hong, National University of Kaohsiung, Taiwan
Yuh-Jong Hu, National Chengchi University, Taipei, Taiwan
Yuji Iwahori, Chubu University, Japan

Dilhan J. Thilakarathne, VU University Amsterdam, The Netherlands
Piotr Jędrzejowicz, Gdynia Maritime University, Poland
Vyacheslav Kalashnikov, Tecnologico de Monterrey, Campus Monterrey, Mexico
Nikos Karacapilidis, University of Patras, Greece
Pawel Kasprowski, Silesian University of Technology, Poland
Hideki Katagiri, Hiroshima University, Japan
Eizo Kinoshita, Meijo University, Japan
Petia Koprinkova-Hristova, Bulgarian Academy of Science
Boris Kovalerchuk, Central Washington University, USA
Marek Kretowski, Bialystok University of Technology, Poland
Vladimir Kurbalija, University of Novi Sad, Serbia
Kazuhiro Kuwabara, Ritsumeikan University, Japan
Jose L. Salmeron, University Pablo de Olavide, Spain
Mihaela Luca (Costin), Romanian Academy, Iasi Branch, Romania
Ivan Luković, University of Novi Sad, Serbia
Ewa Magiera, University of Silesia, Sosnowiec, Poland
Neel Mani, ADAPT Centre for Digital Content Technology, Dublin City University, Dublin, Ireland
Alfonso Mateos Caballero, Universidad Politécnica de Madrid, Spain
Shimpei Matsumoto, Hiroshima Institute of Technology, Japan
Lyudmila Mihaylova, University of Sheffield, UK
Toshiro Minami, Kyushu Institute of Information Sciences and Kyushu University Library, Japan
Jair Minoro Abe, Paulista University and University of Sao Paulo, Brazil
Daniel Moldt, University of Hamburg, Germany
Stefania Montani, Universita' del Piemonte Orientale, Italy
Mikhail Moshkov, King Abdullah University of Science and Technology, Saudi Arabia
Fionn Murtagh, University of Derby and Goldsmiths University of London, UK
Kazumi Nakamatsu, University of Hyogo, Japan
Marek Ogiela, AGH University of Science and Technology, Krakow, Poland
Eugénio Oliveira, LIACC/DEI, Faculty of Engineering, University of Porto, Portugal
Elpiniki Papageorgiou, Technological Education Institute (TEI) of Central Greece, Lamia, Greece
Mario Pavone, University of Catania, Italy
Chee Peng Lim, Deakin University, Australia
Petra Perner, Institute of Computer Vision and Applied Computer Sciences, IBaI
Georg Peters, Munich University of Applied Sciences, Germany
Camelia Pintea, UT Cluj-Napoca Romania
Clara Pizzuti, Institute for High Performance Computing and Networking (ICAR) National Research Council of Italy (CNR)
Bhanu Prasad, Florida A&M University, USA
Radu-Emil Precup, Politehnica University of Timisoara, Romania
Jim Prentzas, Democritus University of Thrace, Greece

Milos Radovanovic, University of Novi Sad, Faculty of Sciences, Russia
Sheela Ramanna, University of Winnipeg, Canada
Ewa Ratajczak-Ropel, Gdynia Maritime University, Poland
Ana Respício, Universidade de Lisboa, Portugal
John Ronczka, SCOTTYNCC Independent Research Scientist
Helene S. Wallach, University of Haifa, Israel
Mika Sato-Ilic, University of Tsukuba, Japan
Miloš Savić, University of Novi Sad, Serbia
Rafał Scherer, Czestochowa University of Technology, Poland
Hirosato Seki, Osaka University, Japan
Bharat Singh, Big Data Analyst, Hildesheim, Germany
Urszula Stańczyk, Silesian University of Technology, Gliwice, Poland
Catalin Stoean, University of Craiova, Romania
Ruxandra Stoean, University of Craiova, Romania
Mika Sulkava, Natural Resources institute, Finland
Claudio Turchetti, Università Politecnica delle Marche, Italy
Pandian Vasant, PETRONAS University of Technology, Malaysia
Jerzy W. Grzymala-Busse, Computer Science University of Kansa, USA
Fen Wang, Central Washington University, USA
Junzo Watada, Waseda University, Japan
Snejana Yordanova, Technical University of Sofia, Bulgaria
Dmitry Zaitsev, International Humanitarian University, Odessa, Ukraine
Danuta Zakrzewska, Lodz University of Technology, Poland
Beata Zielosko, University of Silesia, Katowice, Poland
Alfred Zimmermann, Reutlingen University, Germany

International Referee Board

Alireza Ahrary, Barbara Marszał-Paszek
Richard Anderson, Alina Momot
Nureize Arbaiy, Fionn Murtagh
Piotr Artemjew, Thanh Long Ngo
Giorgio Biagetti, Michał Niezabitowski
Krisztian Buza, Agnieszka Nowak-Brzezińska
Tan Choo Jun, Shih Yin Ooi
Giuseppe De Maso Gentile, Simone Orcioni
Vladimir Dimitrieski, Tomasz Orzechowski
Laura Falaschetti, Małgorzata Pałys
Bogdan Gliwa, Piotr Paszek
Aleksandra Gruca, Adam Piorkowski
Kenneth Holmqvist, Agnieszka Piotrowska
Noriko Horibe, Małgorzata Przybyła-Kasperek
Hung-Hsuan Huang, Azizul Azhar Ramli

Dragan Ivanović, Shohel Sayeed
Tomasz Jach, Ralf Seepold
Nikita Jain, Roman Simiński
Przemyslaw Juszczuk, Krzysztof Simiński
Tomasz Kajdanowicz, Aleksander Skakovski
Jarosław Koźlak, Rafał Skinderowicz
Robert Koprowski, Fred Stefan
Aleksandar Kovačević, Choo Jun Tan
Jan Kozak, Kay Meng Tay
Michał Kozielski, Magdalena Tkacz
Agnieszka Lijewska, Katarzyna Trynda
Way Soong Lim, Marco Vannucci
Chee Lim, Robert Wójcicki
Pei-Chun Lin, Berlin Wu
Agnieszka Lisowska, Tomasz Xięski
Vannucci Marco, Zhenyuan Xu
Joanna Marnik, Raimondas Zemblys

Contents

Main track

Grouping Like-Minded Users for Ratings' Prediction 3
Soufiene Jaffali, Salma Jamoussi, Abdelmajid Ben Hamadou and Kamel Smaili

An Approach for Designing Order Size Dependent Lead Time Models for Use in Inventory and Supply Chain Management 15
Peter Nielsen and Zbigniew Michna

Clique Editing to Support Case Versus Control Discrimination 27
Riccardo Dondi, Giancarlo Mauri and Italo Zoppis

Detecting Value-Added Tax Evasion by Business Entities of Kazakhstan 37
Zhenisbek Assylbekov, Igor Melnykov, Rustam Bekishev, Assel Baltabayeva, Dariya Bissengaliyeva and Eldar Mamlin

On Classification of Linguistic Data—Case Study: Post-operative 51
Kalle Saastamoinen

Solving Technician and Task Scheduling Problems with an Intelligent Decision Heuristic 63
Amy Khalfay, Alan Crispin and Keeley Crockett

Scheduling System for Multiple Unmanned Aerial Vehicles in Indoor Environments Using the CSP Approach 77
Youngsoo Park, Yohanes Khosiawan, Ilkyeong Moon, Mukund Nilakantan Janardhanan and Izabela Nielsen

An Ontology Supporting Multiple-Criteria Decision Analysis Method Selection 89
Jarosław Wątróbski

A Hybrid Approach to Decision Support for Resource-Constrained Scheduling Problems 101
Paweł Sitek, Izabela Nielsen, Jarosław Wikarek and Peter Nielsen

Prediction of Length of Hospital Stay in Preterm Infants a Case-Based Reasoning View 115
Ana Coimbra, Henrique Vicente, António Abelha, M. Filipe Santos, José Machado, João Neves and José Neves

The Shapley Value on a Class of Cooperative Games Under Incomplete Information 129
Satoshi Masuya

Modeling and Property Analysis of E-Commerce Logistics Supernetwork 141
Chuanmin Mi, Yinchuan Wang and Yetian Chen

Maximum Lifetime Problem in Sensor Networks with Limited Channel Capacity 151
Zbigniew Lipiński

Statistical Method for the Problem of Bronchopulmonary Dysplasia Classification in Pre-mature Infants 165
Wiesław Wajs, Hubert Wojtowicz, Piotr Wais and Marcin Ochab

The Rank Reversals Paradox in Management Decisions: The Comparison of the AHP and COMET Methods 181
Wojciech Sałabun, Paweł Ziemba and Jarosław Wątróbski

A New Approach to a Derivation of a Priority Vector from an Interval Comparison Matrix in a Group AHP Framework 193
Jiri Mazurek

Toward a Conversation Partner Agent for People with Aphasia: Assisting Word Retrieval 203
Kazuhiro Kuwabara, Takayuki Iwamae, Yudai Wada, Hung-Hsuan Huang and Keisuke Takenaka

Intelligent Monitoring of Complex Discrete-Event Systems 215
Gianfranco Lamperti and Giulio Quarenghi

Anticipation Based on a Bi-Level Bi-Objective Modeling for the Decision-Making in the Car-Following Behavior 231
Anouer Bennajeh, Fahem Kebair, Lamjed Ben Said and Samir Aknine

Probabilistic Ontology Definition Meta-Model 243
Hlel Emna, Jamoussi Salma, Turki Mohamed
and Ben Hamadou Abdelmajid

Development Aid Decision Making Framework Based on Hybrid MCDM 255
Eric Afful-Dadzie, Zuzana Komínková Oplatková, Stephen Nabareseh
and Michael Adu-Kwarteng

Specialized Decision Techniques for Data Mining, Transportation and Project Management

Measuring Quality of Decision Rules Through Ranking of Conditional Attributes 269
Urszula Stańczyk

Greedy Algorithm for Optimization of Association Rules Relative to Length 281
Beata Zielosko and Marek Robaszkiewicz

Decision Rules with Collinearity Models 293
Leon Bobrowski

PLA Based Strategy for Solving MRCPSP by a Team of Agents 305
Piotr Jędrzejowicz and Ewa Ratajczak-Ropel

Apache Spark Implementation of the Distance-Based Kernel-Based Fuzzy C-Means Clustering Classifier 317
Joanna Jędrzejowicz, Piotr Jędrzejowicz and Izabela Wierzbowska

Ant Clustering Algorithm with Information Theoretic Learning 325
Urszula Boryczka and Mariusz Boryczka

Kernel-Based Fuzzy C-Means Clustering Algorithm for RBF Network Initialization 337
Ireneusz Czarnowski and Piotr Jędrzejowicz

Properties of the Island-Based and Single Population Differential Evolution Algorithms Applied to Discrete-Continuous Scheduling 349
Piotr Jędrzejowicz and Aleksander Skakovski

An Improved Agent-Based Approach to the Dynamic Vehicle Routing Problem 361
Dariusz Barbucha

Pattern Recognition for Decision Making Systems

Predictive Strength of Bayesian Networks for Diagnosis of Depressive Disorders 373
Blessing Ojeme and Audrey Mbogho

Automatic Human Activity Segmentation and Labeling in RGBD Videos 383
David Jardim, Luís Nunes and Miguel Sales Dias

Smart Under-Sampling for the Detection of Rare Patterns in Unbalanced Datasets 395
Marco Vannucci and Valentina Colla

Personal Recommendation System for Improving Sleep Quality 405
Patrick Datko, Wilhelm Daniel Scherz, Oana Ramona Velicu, Ralf Seepold and Natividad Martínez Madrid

Multivariate Direction Scoring for Dimensionality Reduction in Classification Problems 413
Giorgio Biagetti, Paolo Crippa, Laura Falaschetti, Simone Orcioni and Claudio Turchetti

An Efficient Technique for Real-Time Human Activity Classification Using Accelerometer Data 425
Giorgio Biagetti, Paolo Crippa, Laura Falaschetti, Simone Orcioni and Claudio Turchetti

New Advances of Soft Computing in Industrial and Management Engineering

A Double Layer Neural Network Based on Artificial Bee Colony Algorithm for Solving Quadratic Bi-Level Programming Problem 437
Junzo Watada and Haochen Ding

A Memetic Fuzzy ARTMAP by a Grammatical Evolution Approach 447
Shing Chiang Tan, Chee Peng Lim and Junzo Watada

Particle Swarm Optimization Based Support Vector Machine for Human Tracking 457
Zhenyuan Xu, Chao Xu and Junzo Watada

Author Index 471

Main Track

Grouping Like-Minded Users for Ratings' Prediction

Soufiene Jaffali, Salma Jamoussi, Abdelmajid Ben Hamadou and Kamel Smaili

Abstract Regarding the huge amount of products, sites, information, etc., finding the appropriate need of a user is a very important task. Recommendation Systems (RS) guide users in a personalized way to objects of interest within a large space of possible options. This paper presents an algorithm for recommending movies. We break the recommendation task into two steps: (1) Grouping Like-Minded users, and (2) create model for each group to predict user-movie ratings. In the first step we use the Principal Component Analysis to retrieve latent groups of similar users. In the second step, we employ three different regression algorithms to build models and predict ratings. We evaluate our results against the SVD++ algorithm and validate the results by employing the MAE and RMSE measures. The obtained results show that the algorithm presented gives an improvement in the MAE and the RMSE of about 0.42 and 0.5201 respectively.

Keywords Rating prediction · Social recommendation · Grouping like-minded users

S. Jaffali (✉) · S. Jamoussi · A.B. Hamadou
MIRACL Laboratory, Higher Institute of Computer Science and Multimedia,
University of Sfax, BP 1030 Sfax, Tunisia
e-mail: jaffali.soufiene@gmail.com

S. Jamoussi
e-mail: jamoussi@gmail.com

A.B. Hamadou
e-mail: abdelmajid.benhamadou@gmail.com

K. Smaili
Campus Scientifique LORIA, Nancy, France
e-mail: smaili@loria.fr

I. Czarnowski et al. (eds.), *Intelligent Decision Technologies 2016*,
Smart Innovation, Systems and Technologies 56,
DOI 10.1007/978-3-319-39630-9_1

1 Introduction

Given the huge amount of products to purchase or information to browse on the web, matching users with the most appropriate items becomes a very important task. Recently, the interest to the Recommendation Systems (RS) has increased considerably. Indeed, RS play an important role in the well noted web sites such as Netflix,[1] Tripadvisor,[2] etc. The RS aim to provide suggestions for items to be of use to a user. An item refers to any object, it can be a music to listen, product to buy, film to watch, etc. In this scope, many algorithms and methods are proposed. Burke [3] distinguishes six families of RS approaches (Content-Based, Collaborative Filtering, Demographic, Knowledge-Based, Community-Based and Hybrid Recommendation Systems). As one of the most representative categories, collaborative filtering (CF) is popular for its high performance and simple requirements. The principle of CF is to create patterns based on user preferences and use them to produce suggestions.

One of the emergent tasks in RS field is *Movie Recommendation*. The Netflix and MovieLens are the most popular *Movie Recommendation* tasks [14, 17]. Those tasks attract many researchers and many algorithms are proposed in this topic. Most of the CF algorithms used for *Movie Recommendation*, handle the problem as a *Partial Matrix Factorization*, and aim to find the missing values in the matrix $users \times movies$ containing user-movie ratings [13]. The major problem with these solutions is the sparsity of data (given the huge number of users and movies, in one hand and the lack of stored users-movies ratings, in other hand). Furthermore, we cannot predict recommendations for a new movie or user that does not appear in the training data.

We propose a new algorithm to recommend movies based on the Like-Minded user groups (groups of users having similar cinimatographic tastes). The main idea of our algorithm is to infer user u rating on a movie m based on the stored ratings of the user u and users having similar interests. Such prediction can improve the recommendation accuracy and decrease the complexity of the prediction model, as it builds patterns for groups of homogeneous users. To retrieve Like-Minded users, we use the Principal Components Analysis (PCA). Thus, we reduce data dimensions and solve the sparsity problem that characterizes RS data. Also, the Like-Minded user groups are generated based on the tastes of users (genres of rated movies), without any knowledge about the social network relation between users. Thus, we remedy the limitation of *community-Based RS*.

The remaining of this paper is organized as follows. Section 2 gives an overview of the related work. In Sect. 3, we present our approach to estimate ratings. Then, we describe the data set and the baseline in Sect. 4. The experimental results are presented in Sect. 5, and we conclude in Sect. 6 by pointing out some future works.

[1]https://www.netflix.com/.

[2]https://www.tripadvisor.com/.

2 Related Work

In this section, we review a small set of Collaborative Filtering (CF) researches. We can find two major groups of CF methods, namely: *Neighborhood* and *Latent-Model-Based* methods. In the case of *Neighborhood* methods, the user-item ratings stored in the system are used to predict ratings for other items directly. The prediction is done either by the *user-based* or the *item-based* way. In the first case, for a given user u and an item i, Heckmann et al. [7] use the stored ratings of i, given by users which are similar to u (called neighbors) to infer R_{ui} (the interest of the user u to the item i). In the *item-based* case, Khabbaz and Lakshmanan [12] use the ratings of u for items similar to i to predict R_{ui}. The methods used in *Neighborhood* CF can be grouped into two main approaches:

1. *Dimensionality reduction* such as Latent Semantic Indexation (LSI) [22] and Singular Value Decomposition (SVD) [13].
2. *Graph-based* such as Random Walk [4].

The major problem of *Neighborhood* methods is the limited space of predicted items. As the system recommends only items similar to those rated by the user or by his neighbors, it is hard to cover the huge number of items in the system, such as products in the *Amazon.com*[3] or films on *IMDB*.[4]

The *Latent-Model-Based* methods use the stored ratings to learn a model. The main idea is to model the user-item interactions using factors representing latent characteristics of both users and items within the system. This model is learnt using available data and then utilized to predict user ratings on new items. In this context, many technics are used, such as Bayesian clustering [2], probabilistic Latent Semantic Analysis (pLSA) [8], Latent Dirichlet Allocation (LDA) [1], SVD [15], etc.

Thanks to its accuracy and scalability, the SVD technic is deeply used in CF [18, 21]. Most of the recent researches and recommendation tools (*LingPipe*,[5] *pyrsvd*[6]) are based on the basic model or improved models of SVD like SVD++ [16] and timeSVD++ [15].

Although it is a very prevalent Matrix Factorization (MF) method, just a few works use Principal Component Analysis (PCA) in RS field. Goldberg et al. [6] propose *Eigentaste*, a CF algorithm to recommend jokes. The proposed algorithm starts by creating M, an $n \times m$ matrix, containing the ratings given by n users to m items. In the second step, the algorithm extracts the two eigenvectors having the highest eigenvalues of the correlation matrix of M. Then, it projects data in the eigenvectors plan. Finally, it uses *Recursive Rectangular Clustering* to create clusters of neighbors. Those clusters are used to predict users' preferences.

---

[3] http://www.amazon.fr.

[4] http://www.imdb.com.

[5] http://alias-i.com/lingpipe/index.html.

[6] https://code.google.com/p/pyrsvd/.

With the growth of social networks, a new kind of RS referred to as *Social Recommender Systems* (also known as *Community Recommender Systems*) gained popularity. The main idea is that users tend to trust more the recommendations from their friends than an anonymous user or seller [25]. FilmTrust [5] is the ideal example, it combines a movie rating and review system with a trust network (A social network expressing how much the members of the community trust each others). Yang et al. [25] present a deep review of collaborative filtering based social recommender systems. As the *Community RS* are based on link information only, they are limited to the user friends recommendations. Furthermore, given the big amount of items to recommend, and the fact that the *egocentric network*[7] is very small compared to the whole social network, by looking only in the egocentric network, a considerable part of items are ignored. Also, the social network is not available in all RS.

3 The GLER Algorithm

This work is a continuation of our previous work [10], where we cluster users in social networks based on their interest centers and their opinions. In this work, we apply grouping of like-minded users on the ratings prediction task in recommendation systems.

In this section, we describe our algorithm GLER (**G**rouping **L**ike-minded users to **E**stimate **R**atings). GLER algorithm aims to create groups of Like-Minded people, in order to estimate future user-movie ratings, based on the stored ratings of that user and those having similar cinimatic tastes. Such a prediction leads to triple improvements:

1. Enhancing the training base of the predictor, as we exploit Like-Minded users' experiences.
2. Facilitating the deployment of the recommendation algorithm, by creating groups based on the interest centers and not on the information about social network links, which are not available in all RS.
3. Improving the accuracy of the predictor by dividing the training base into sub-bases containing homogeneous entries.

To create groups of Like-Minded users, we use PCA, which proved its effectiveness in this area [10]. Once created, we use these groups to train the RS, and so, to build a model for each group. To estimate a new user-movie rating for an existing user, we use the model corresponding to the group that the user belongs to. For a new user, we calculate the distance between this user and the users of each group, and we use the model of the closest group. Figure 1 presents the proposed method. In the remainder of this section, we detail the steps of our algorithm.

[7]The network around a single node (ego).

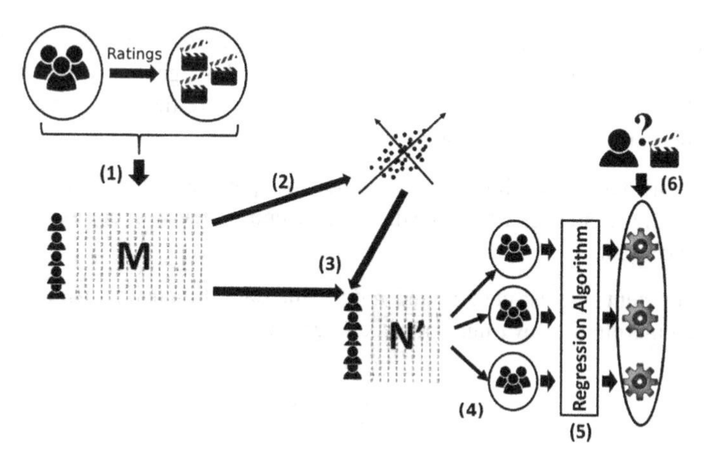

Fig. 1 The proposed method steps

3.1 Grouping Similar Users

Create the Matrix of Users' Interests Most of the RS algorithms use a $n \times m$ matrix of raw ratings from n users and m items. The matrix is partial (several rating are missed). Those algorithms aim to infer the missing ratings using a matrix factorization algorithm, such as SVD or LSI. In our work, we create a matrix representing users interests, and then use this matrix to retrieve the latent common interest centers. We start by creating a vector U for each user. The vector U contains the averages of user ratings per movie genre (Sect. 6). Next, we group users' vectors into a matrix M. Then, we normalize the matrix M to produce the matrix N. For the normalization, we use two methods. The first consists of normalizing all values in the matrix M as follows:

$$N_{ij} = \frac{max_j - m_{ij}}{max_j - min_j} \quad (1)$$

where max_j and min_j present respectively the maximum and minimum values in the column j of the matrix M. m_{ij} is the average of ratings by the user i of the movie genre j. In the second normalization method, we divide each value by the standard deviation of each column.

$$N_{ij} = \frac{m_{ij}}{\sigma_{m_{kj}}} \text{ ; } with\ k = 1 \dots n \quad (2)$$

With n is the number of users. In the remainder of this article, we use $Norm_1$ and $Norm_2$ to refer to the first and the second normalization methods respectively.

Retrieve Latent Factors In this step, we seek the latent factors to model users using PCA which is a popular matrix factorization method, generally used to reduce the data dimensions or to retrieve the latent centers where the data is concentrated [11]. In our work, we use PCA for both reasons:

1. Reduce the data dimensions, and so reduce the sparseness and data noise.
2. Retrieve the latent axes where the data is concentrated which represent the users' interest centers.

To find the latent interest centers and reduce the data dimensions using PCA, we start by calculating the covariance matrix C, given by:

$$C = \frac{1}{n-1} N^T N \tag{3}$$

With N is the normalized form of the Users' Interests matrix, created in the previous step. Then, we retrieve the matrices Λ and E by solving the Eqs. (4) and (5):

$$C = E^T \Lambda E \tag{4}$$

$$ECE^T = \Lambda \tag{5}$$

The columns of the matrix E correspond to the eigenvectors of C. Λ is a diagonal matrix whose elements correspond to the eigenvalues of C.

As the principal components correspond to the axis around which the data is concentrated, we can say that those axis present the groups of users where the correlation is maximal. We use the retrieved eigenvalues to determine the number of eigenvectors to retain, and the number of user groups [10].

To retrieve the number of eigenvectors to retain and user groups from eigenvalues, we adopt the Scree Test Acceleration Factor proposed in [20]. Finally, we project our data in the space of the retained eigenvectors:

$$N' = NE_v^T \tag{6}$$

where v is the number of eigenvectors to retain, E_v is the matrix formed by the retained eigenvectors, and N' is the reduced form of the matrix N.

Create Groups Many algorithms can be used to cluster users. In this work, we adopt the K-Means algorithm to group the similar users, for its simplicity and effectiveness. As inputs, K-Means takes the set of data N' and the number of groups to be identified k. In our work, we set $k = v$, as the number of eigenvectors to retain reflects the number of interest centers within the input data. At the end of this step, we obtain v clusters of users. Each cluster presents a group of users having similar proprieties and tastes.

3.2 *Predict Ratings*

Predicting a user rating could be treated as both a regression and a classification problem. If the ratings are in the form of classes or labels, the prediction is treated as a classification problem. In average, the prediction is considered as a regression issue, if the ratings are discrete numerical values (such as the five star evaluation in the MovieLens RS). In this work, we adopt three regression algorithms using Weka.[8] The adopted regression algorithms are:

M5P M5P is a reconstruction of Quinlan's M5 algorithm [19] for inducing trees of regression models. It builds regression trees whose nodes are chosen over the attribute that maximizes the expected error reduction, and the node leaves are composed of multivariate linear models.

Multi-Layer Perceptron (MLP) An MLP can be viewed as a logistic regression classifier where the input is first transformed using a learnt non-linear transformation Φ. This transformation projects the input data into a space where it becomes linearly separable. Hornik et al. [9] prove that three-layer perceptron networks are theoretically universal approximators.

Support Vector Regression (SVR) Support Vector Machines (SVM) can also be used as a regression method, maintaining all the main features that characterize the algorithm (maximal margin). The SVR uses the same principles as the SVM for classification, with the introduction of an alternative loss function. The loss function must be modified to include a distance measure. In Weka, there are two alternatives of SVR, namely: ε-SVR [24] and ν-SVR [23].

For each user group produced in the previous step, we use the stored users ratings to learn the regression algorithm. Once learnt, the system is able to infer user ratings in this group. For a new user—who is not included in the learn set—we assign him to the nearest group, based on the cosine distance between the new user vector and the average of users vectors in each group.

4 Dataset and Baseline

4.1 *Dataset*

To evaluate the accuracy of the proposed algorithm, we use the *MovieLens-100K data sets*.[9] The data set was collected by the GroupLens Research Project, through the MovieLens web site[10] during the 7 month period from September 19th, 1997 through April 22nd, 1998. This data has been cleaned up—users who had less than 20 ratings or did not have complete demographic information were removed from

[8]http://www.cs.waikato.ac.nz/ml/weka/.

[9]http://grouplens.org/datasets/movielens/.

[10]https://movielens.org.

this data set. This data set consists of 100,000 ratings [1–5] from 943 users on 1,682 movies. Demographic information about the users are: age, gender, occupation and zip code. The information about a movie in the dataset are: movie title, release date, video release date, *IMDB* URL and 19 binary values corresponding to the 19 movie genre. Those values indicate if the movie is or is not of the corresponding genre.

The data is distributed over 5 dataset. Each dataset contains a train and test data. The train and test sets contain respectively 80,000 and 20,000 entries. Each entry is in the form: "*user id*" | "*item id*" | "*rating*" | "*timestamp*".

4.2 Baseline

We use SVD++ to evaluate and position our new algorithm. SVD++ is a popular Matrix Factorization RS. Matrix factorization models map both users and items to a joint latent factor space of dimensionality f, such that user-item interactions are modeled as inner products in that space. The latent space tries to explain the ratings by characterizing both products and users on factors automatically inferred from user feedback [15].

Accordingly, each item i is associated to a vector $q_i \in R^f$, and each user u is associated to a vector $p_u \in R^f$. For a given item i, the elements of q_i measure the extent to which the item possesses those factors. For a given user u, the elements of p_u measure the extent of interest the user has in items that are high on the corresponding factors. In the movie RS context, the elements of q_i express the extent to which the movie i belongs to movie genres, and the elements of p_u measure the extent of interest the user u has in movies of corresponding genres. The resulting dot product, $q_i^T p_u$ captures the interaction between the user u and the movie i.

Let b_i and b_u denote the observed deviations of user u and item i, respectively, from the average μ (the overall average rating). For example, if the average of the observed ratings $\mu = 2.5$, and the average of the stored ratings on the item i is 3, and the average of the ratings of the user u is 2, then $b_i = 3 - \mu = 0.5$ and $b_u = 2 - \mu = -0.5$.

The prediction rule of the basic SVD recommender is given by adding the product $q_i^T p_u$ to the sum of the three parameters μ, b_i and b_u:

$$\hat{r}_{ui} = \mu + b_i + b_u + q_i^T p_u \tag{7}$$

To improve the basic SVD recommender by adding implicit feedback, Koren [13] adds a second set of item factors, relating each item i to a factor vector $y_i \in R^f$. Those new item factors are used to characterize the users based on the set of items that they rated. The new rule to predict ratings is given by:

$$\hat{r}_{ui} = \mu + b_i + b_u + q_i^T \left(p_u + |R(u)|^{-\frac{1}{2}} \sum_{j \in R(u)} y_j \right) \tag{8}$$

where $R(u)$ contains the items rated by the user u. Thus, the user u is represented by $p_u + |R(u)|^{-\frac{1}{2}} \sum_{j \in R(u)} y_j$, instead of p_u. Thus, the user representation in the Eq. 7 is enhanced, by adding the perspective of implicit feedback.

5 Accuracy Measures

Typically, the rating dataset is split into a training set R_{train} and a test set R_{test}. R_{train} is used to generate models and tune the recommender system parameters. R_{test} is used to evaluate the recommender system. Let $\hat{R}_{u,i}$ the predicted ratings, and $R_{u,i}$ the recorded ratings. To assess the recommender system accuracy, we use two prevalent measures of accuracy, namely: *Root Mean Squared Error* and *Mean Absolute Error*.

Root Mean Squared Error (RMSE)

RMSE is one of the most popular metric used in evaluating the accuracy of predicted ratings. RMSE between known and predicted ratings is given by:

$$RMSE = \sqrt{\frac{\sum_{(u,i) \in R_{test}} (\hat{r}_{ui} - r_{ui})^2}{|R_{test}|}}$$

RMSE disproportionately penalizes large errors. For example, given a test set with four hidden items, RMSE would prefer a system that makes an error of 2 on three ratings and 0 on the fourth to one that makes an error of 3 on one rating and 0 on all three others.

Mean Absolute Error (MAE)

MAE is a popular alternative of RMSE, given by:

$$MAE = \frac{\sum_{(u,i) \in R_{test}} (\hat{r}_{ui} - r_{ui})}{|R_{test}|}$$

Unlike RMSE, MAE prefers systems that make the least number of errors. In the previous example, MAE would prefer the system that makes an error of 3 on one rating and 0 on all three others.

6 Results and Discussion

To evaluate our algorithm on the the MovieLens-100k dataset, we start by creating the input matrix, as mentioned in the Sect. 3.1. Thus, we create a user vector $U_x = \{age, gend, occ, zip, year, rg_1, \ldots, rg_{19}\}$, where "*age*", "*gend*", "*occ*", and "*zip*" represent respectively the age, gender, occupation and zipcode of the user x.

Table 1 Comparison with the baseline accuracy

	Baseline	$GLER_{M5P}$		$GLER_{MLP}$		$GLER_{\varepsilon\text{-}SVR}$		$GLER_{\nu\text{-}SVR}$	
		Norm1	Norm2	Norm1	Norm2	Norm1	Norm2	Norm1	Norm2
MAE	0.7135	0.3083	0.3057	0.2969	0.2902	0.306	0.30102	**0.2869**	**0.2809**
RMSE	0.9065	0.4312	0.4322	0.4198	0.4228	**0.3909**	**0.3864**	0.3972	0.3928

"*year*" is the average of rated movies release years. "rg_i" is the average of ratings that the user x assigns to movies belonging to the genre i. Thus, the vector U presents a user not only by his personal information (age, gender, etc.), but also by his interest to movies genre and release year. Then, we gather user vectors to create the matrix M. Thereafter, we normalize M to obtain the matrix N, and retrieve the eigenvalues and eigenvectors of the covariance matrix of N. The obtained eigenvectors correspond to the latent centers where the users interests are concentrated. Using those centers, we form groups of like-minded users. Finally, we divide the training base into several dataset according to the users' groups, and learn the regression algorithm to build a model for each group. Once learnt, we use these models to estimate ratings for users of the corresponding group.

As we use three different regression algorithms, we designate by $GLER_{M5P}$ (respectively $GLER_{MLP}$, $GLER_{\varepsilon\text{-}SVR}$ and $GLER_{\nu\text{-}SVR}$) the *GLER* using M5P algorithm (respectively MLP, ε-SVR and ν-SVR). We applied the proposed method and the baseline algorithm (SVD++) on the MovieLens-100k dataset, and we calculate the accuracy using MAE and RMSE. Table 1 illustrates the accuracy of the baseline algorithm and the four alternatives of GLER on the MovieLens-100k five bases.

We remark that the GLER algorithm provides MAE values lower than 0.31 and RMSE values lower than 0.44 for all the tests. Using the SVR regression algorithms ($GLER_{\nu\text{-}SVR}$ and $GLER_{\varepsilon\text{-}SVR}$), we obtain the best predictions, compared to $GLER_{M5P}$ and $GLER_{MLP}$. We can conclude that the predicted ratings using $GLER_{\varepsilon\text{-}SVR}$ are most of the time close from the right ones, and $GLER_{\nu\text{-}SVR}$ provides more exact ratings but the errors are larger than those in ε-*SVR* case. This is can be explained by the fact that the ε-*SVR* tends to control the amount of error in the model and go for the best performance without control of the support vectors in the resulting model, unlike the ν-*SVR* algorithm.

The results show that the performances of our algorithm GLER exceed those of the baseline, and we obtain an improvement in the MAE of about 0.42 and a drop from 0.9065 to 0.3864 in the RMSE values. Thanks to the PCA used in our algorithm, we overcome the sparsity problem which characterizes RS data, and reduce the data dimension, which is very important and helpful for building the prediction model. Creating Like-Minded user groups based on their tastes leads to a better performance. Indeed, the prediction will be based, not only on the user's rating history, but also on users having the same cinematographic tastes. Furthermore, following the rule of "*divide and conquer*", creating similar-user groups facilitates the task of the regression algorithms. In fact, creating a model for a smaller number of homogeneous users is more easier and improves the prediction accuracy.

7 Conclusion

In this work we presented the GLER algorithm for recommending movies. Our algorithm creates Like-minded user groups based on the user-movie ratings. Then, it uses those groups to build patterns and predict user-movie ratings. Experimenting on the MovieLens-100k dataset, we showed that GLER achieves better results than the SVD++ algorithm. This improvement manifests in the decrease of MAE and RMSE values of about 0.42 and 0.5201 respectively.

In our algorithm we used three regression algorithms to build recommendation model and infer ratings. The experimentations show that the $GLER_{\nu\text{-}SVR}$ and $GLER_{\varepsilon\text{-}SVR}$ have the highest accuracy.

Our algorithm uses explicit feedback (user ratings) to build the recommendation model. Also, the training and test experimentation are done offline. In future research, we plan to enhance our algorithm by including users' implicit feedback, and handle the online recommendation.

References

1. Blei, D.M., Andrew, Y., Ng., Jordan, M.I., Lafferty, J.: Latent dirichlet allocation. J. Mach. Learn. Res. **3**, 2003 (2003)
2. Breese, J.S., Heckerman, D., Kadie, C.: Empirical analysis of predictive algorithm for collaborative filtering. In: Proceedings of the 14th Conference on UAI, pp. 43–52 (1998)
3. Burke, R.: The Adaptive Web, pp. 377–408. Springer, Heidelberg (2007)
4. Fouss, F., Pirotte, A., Renders, J.M., Saerens, M.: Random-walk computation of similarities between nodes of a graph with application to collaborative recommendation. IEEE Trans. Knowl. Data Eng. **19**(3), 355–369 (2007)
5. Golbeck, J., Hendler, J.: Filmtrust: movie recommendations using trust in web-based social networks. In: CCNC 2006. 3rd IEEE, vol. 1, pp. 282–286
6. Goldberg, K., Roeder, T., Gupta, D., Perkins, C.: Eigentaste: a constant time collaborative filtering algorithm. Inf. Retrieval **4**(2), 133–151 (2001)
7. Heckmann, D., Schwartz, T., Brandherm, B., Schmitz, M., von Wilamowitz-Moellendorff, M.: Gumo -the general user model ontology. In: User Modeling (2005)
8. Hofmann, T.: Latent semantic models for collaborative filtering. ACM Trans. Inf. Syst. 89–115 (2004)
9. Hornik, K., Stinchcombe, M., White, H.: Multilayer feedforward networks are universal approximators. Neural Netw. **2**(5), 359–366 (1989). July
10. Jaffali, S., Ameur, H., Jamoussi, S., Ben Hamadou, A.: Glio: a new method for grouping like-minded users. In: Transactions on Computational Collective Intelligence XVIII. LNCS, vol. 9240, pp. 44–66. Springer, Heidelberg (2015)
11. Jaffali, S., Jamoussi, S.: Principal component analysis neural network for textual document categorization and dimension reduction. In: 6th International Conference on SETIT, pp. 835–839 (2012)

12. Khabbaz, M., Lakshmanan, L.V.S.: Toprecs: top-k algorithms for item-based collaborative filtering. In: Proceedings of the 14th International Conference on Extending Database Technology, EDBT/ICDT '11, pp. 213–224. ACM (2011)
13. Koren, Y.: Factorization meets the neighborhood: a multifaceted collaborative filtering model. In: Proceedings of the 14th ACM SIGKDD, pp. 426–434. ACM (2008)
14. Koren, Y.: The bellkor solution to the netflix grand prize. Netflix prize documentation (2009)
15. Koren, Y., Bell, R.: Advances in collaborative filtering. In: Recommender Systems Handbook, pp. 77–118. Springer, US (2015)
16. Kumar, R., Verma, B.K., Rastogi, S.S.: Social popularity based SVD++ recommender system. Int. J. Comput. Appl. **33–37** (2014)
17. Lu, Z., Shen, H.: A security-assured accuracy-maximised privacy preserving collaborative filtering recommendation algorithm. In: Proceedings of the 19th International Database Engineering and Applications Symposium, Japan, pp. 72–80 (2015)
18. Paterek, A.: Improving regularized singular value decomposition for collaborative filtering. In: Proceedings KDD Cup Workshop at SIGKDD'07, pp. 39–42 (2007)
19. Quinlan, J.R.: Learning with continuous classes. In: Proceedings of the Australian Joint Conference on Artificial Intelligence, pp. 343–348. World Scientific (1992)
20. Raîche, G., Walls, T.A., Magis, D., Riopel, M., Blais, J.: Non-graphical solutions for cattells scree test. Methodol.: Eur. J. Res. Methods Behav. Soc. Sci. **9**(1), 23–29 (2013)
21. Salakhutdinov, R., Mnih, A., Hinton, G.: Restricted boltzmann machines for collaborative filtering. In: Proceedings of the 24th International Conference on Machine Learning, ICML '07, pp. 791–798, New York, NY, USA. ACM (2007)
22. Sarwar, B.M., Karypis, G., Konstan, J.A., Riedl, J.T.: Application of dimensionality reduction in recommender system—a case study. In: ACM WebKDD Workshop (2000)
23. Schölkopf, B., Smola, Williamson, A.J., R.C., Bartlett, P.L.: New support vector algorithms. Neural Comput. **12**(5), 1207–1245 (2000)
24. Vapnik, V.N.: Statistical Learning Theory. Wiley (1998)
25. Yang, X., Liu, Y., Guo, Y., Steck, H.: A survey of collaborative filtering based social recommender systems. Comput. Commun. (2013)

An Approach for Designing Order Size Dependent Lead Time Models for Use in Inventory and Supply Chain Management

Peter Nielsen and Zbigniew Michna

Abstract This paper addresses the issue of lead time behavior in supply chains. In supply chains without information sharing a supply chain member can only use the information they observe; orders/demand and their lead times. Using this information four different scenarios of lead time behavior are suggested and discussed. Based on this discussion an analytical approach is proposed that investigates the link between order quantities and lead times. This approach is then demonstrated on data from a company. In the particular case it is determined that there seems to be a link between order quantities and lead times, indicating that a complex lead time model may be necessary. It is also concluded that current state of supply chain management does not offer any methods to address this link between order quantities and lead times and that therefore further research is warranted.

Keywords Supply chain management · Lead times · Bullwhip effect · Stochastics

1 Introduction

The bullwhip effect is a term that covers the tendency for replenishment orders to increase in variability as one moves up-stream in a supply chain. The term is also often referred to as demand amplification from its technical definition as the variance of orders divided with the variance of the observed demands. In the current state of research typically five main causes of the bullwhip effect are considered

P. Nielsen (✉)
Department of Mechanical and Manufacturing Engineering,
Aalborg University, Fibigerstraede 16, 9220 Aalborg Oest, Denmark
e-mail: peter@m-tech.aau.dk

Z. Michna
Department of Mathematics and Cybernetics, Wroclaw University
of Economics, Wroclaw, Poland
e-mail: zbigniew.michna@ue.wroc.pl

I. Czarnowski et al. (eds.), *Intelligent Decision Technologies 2016*,
Smart Innovation, Systems and Technologies 56,
DOI 10.1007/978-3-319-39630-9_2

(see e.g. Lee et al. [9]): demand forecasting, non-zero lead time, supply shortage, order batching and price fluctuation. Recently Michna et al. [10] has added lead time forecasting and variability of lead times as a sixth main cause of the bullwhip effect. Forecasting of lead times is necessary when a member of the supply chain places an order and the signal processing of lead times in a similar manner as signal processing of demands causes bullwhip effect. While demand forecasting is a well-known challenge in planning [15], lead time forecasting is a phenomenon that is of particular interest for the management of supply chains. A number of approaches to manage supply chains have been proposed ranging from simulation, to optimization and control theory (see e.g. Sitek [18]). This work focuses on lead times and their behavior. Despite it being well established that lead times are critical in terms of both supply chain management and bullwhip effect they have received surprisingly limited attention in literature. This work outlines a step in remedying this through proposing an approach for how a supply chain member using observations of their up-stream orders (lead times and order quantities) can develop a model of lead time behavior. Firstly we want to list the main problems arising in supply chains when lead times are not deterministic. Secondly we propose an approach for how a supply chain member using paired observations of their up-stream orders (lead times and order quantities) can develop a model of lead time behavior depending on orders. The aim is to use this model to improve inventory management and decision making. It is also the aim to establish whether complex models of lead time behavior should be studied further.

The remainder of the paper is structured as follows. First, the theoretical background is established and the relevant literature is reviewed. Second, an approach to investigate the link between lead times and order quantities is proposed. Third, a test [of the approach is conducted using data from a manufacturing company. Finally conclusions and further potential avenues of research are presented.

2 Theoretical Background and Literature Review

This research focuses on supply chains that do not use information sharing, but where each echelon acts solely based on the information it can observe. In the simplified case each member of a supply chain echelon can only observe (1) the demand received from the previous echelon and the orders it itself places in the next echelon (Q) (2) the lead time (LT) for the orders it places and the lead time it itself gives its customers. If we limit the scope of the research to using information that any echelon should be able to gather by observing its suppliers' behavior we arrive at four different cases of lead time models that are plausible.

In the first case LT's are deterministic. This case is trivial to establish and will not be subject to further study. However, it is worth noting that this particular scenario has received significant attention in literature. Chen et al. [3] conclude that the mean lead time and the information used to estimate demand are the single two most important factors in determining the bullwhip effect in a supply chain.

In the second case we assume that LT's are mutually independent identically distributed independent of demands and orders, i.e. lead times are exogenous. This case has received some attention in literature in the context of the bullwhip effect, see recent papers by Disney et al. [4], Do et al. [5], Duc et al. [6], Kim et al. [8], Michna et al. [10] and Nielsen et al. [13]. From Michna et al. [10] we know that the consequence of having i.i.d. lead times is a significant increase in the bullwhip effect and they indicate that lead time forecasting is another important cause of the bullwhip effect. This is seen in the equation below. Where the demand amplification (bullwhip measure BM) is expressed as the ratio $\frac{Var(q_t)}{Var(D_t)}$:

$$BM = \frac{Var(q_t)}{Var(D_t)} = \frac{2\sigma_L^2(m+n-1)}{m^2n^2} + \frac{2\sigma_L^2\mu_D^2}{m^2\sigma_D^2} + \frac{2\mu_L^2}{n^2} + \frac{2\mu_L}{n} + 1$$

where $\sigma_L^2, \sigma_D^2, \mu_L^2, \mu_D^2$ are the observed variance and means of the lead times and demand distributions and *m* and *n* are respectively the number of observations used to estimate the lead time and demand. It is a relatively trivial matter to develop an appropriate lead time model if in fact lead times are i.i.d. It is simply a matter of having sufficient observations available to estimate a distribution [13]. However, under the assumption of i.i.d. lead times there is a problem with the so-called crossovers which happens when replenishments are received in a different sequence than they were ordered see e.g. Bischak at el. [1, 2], Disney at el. [4] and Wang and Disney [19]. The crossover (see Fig. 1) effect is especially severe when a member of the supply chain forecasts lead times to place an appropriate order.

In the third case LT depends on Q i.e. the distribution of LT depends on the parameter m = Q

$$F = F(x, Q)$$

Consider the situation where a number of thresholds of order sizes exist i.e. several intervals $Q \in [Q_i; Q_j]$ exist for each of which there is a corresponding distribution of LT. In practice this seems like a potentially likely relationship between Q and LT. It also seems reasonable that LT unidirectionally depends on Q. From the perspective of inventory and supply chain management it is highly complicated if LT depends on Q in any form as the estimate of LT is used to determine an appropriate Q. The work presented by Nielsen et al. [12] indicates that such relationships may in certain cases be appropriate. However, to the best of the authors' knowledge no supply chain models exist that take this into account. In information sharing supply chains it could well be that Q or total demand depends

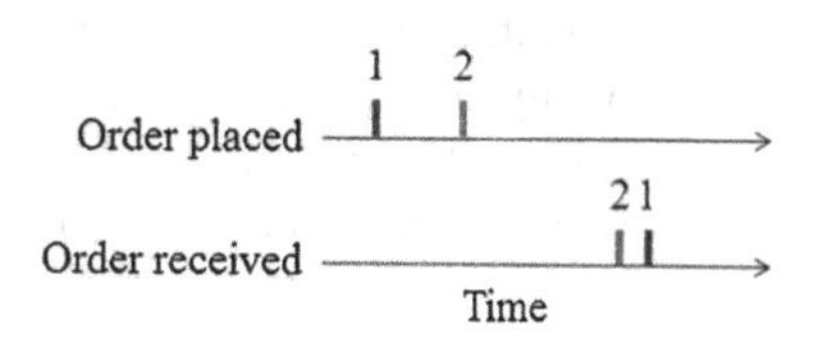

Fig. 1 Order crossover where order 1 is placed prior to order 2, but received after order 2

on LT. However, as only non-information sharing supply chains are considered this case is not investigated in this research.

In the final case the joint Q-LT distributions is not the product of the marginal distributions of Q and LT, that is we observe a stochastic dependence of lead times and orders which means that

$$H(x,y) = C(F(x), G(y))$$

where H is a joint distribution of lead times and orders, C is a copula function, F is a marginal distribution of lead time and G is a marginal distribution of orders see e.g. Joe [7] and Nelsen [11]. This case is by far the most complicated of the four, both to establish and to address. However, this scenario seems rather unlikely as it implies that Q is in fact a random variable (although jointly distributed with LT). From practice this is unlikely to be the case for any echelon of a supply chain baring the final echelon before the end-customer. It would also be potentially very complicated to manage inventory if such a relationship did in fact exist.

There is of course any number of other appropriate models for LT behavior. However, the models above have the benefit that they depend on information that a supply chain member can acquire through observation, i.e. observing actual lead times achieved for orders. Order quantities may be an important factor in determining lead times, but this is likely due to capacity constraints at up-stream echelons where the order quantity acts as a proxy estimator for capacity utilization. Pahl et al. [14] reviews a large body of literature for situations where lead times depend on capacity utilization. However, without information sharing, capacity utilization at the supplier is not known by the customer at the time of order placement. So the above proposed LT models seem easy to implement and appropriate to use in the general case.

3 Approach

The following approach to determine an appropriate lead time model relies on having a number of pair observations (Q_x, LT_x) of order quantities and their corresponding lead times. This information can be observed by any member of a supply chain by monitoring its orders to suppliers. It also assumes that it has been established that LT can be considered to be random variables, i.e. not constant.

The proposed approach contains three main steps as seen in Fig. 2. The first step is data cleaning, where any incorrect data entries are removed (negative values, non-integer values etc.). Following this each pair of observations (Q_x, LT_x) is removed if either the LT or the Q observation can be considered a statistical outlier. For simplicity in this research any observation of either Q or LT that is larger than the observed mean ($\bar{\mu}_Q$ and $\bar{\mu}_{LT}$) and four standard deviations is considered an outlier. Also for simplicity all outliers are removed in one step rather than iteratively.

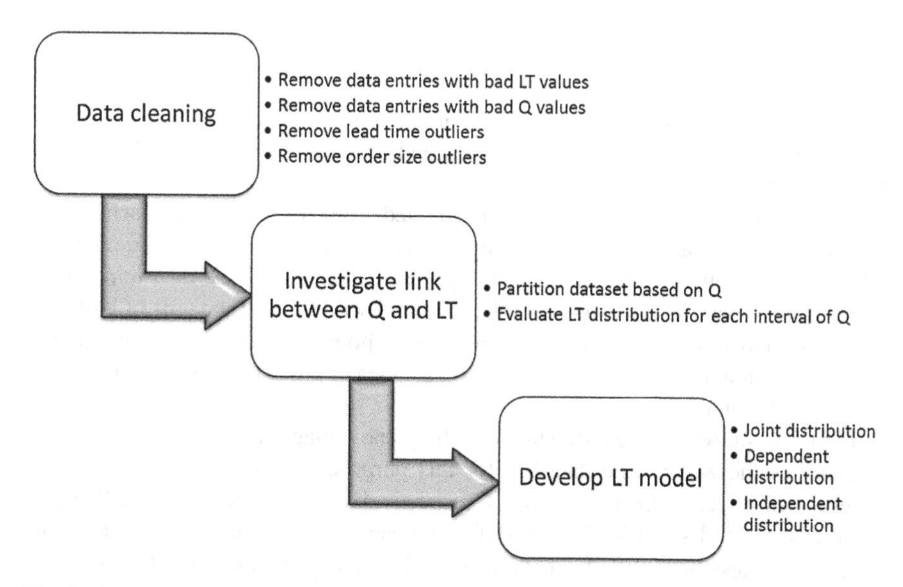

Fig. 2 A three step approach to determine an appropriate LT model

In the second step each pair of observations is portioned into buckets using Q to determine into which bucket the pair is placed. The buckets are derived such that the total range of Q is partitioned into a reasonable number of buckets covering an equal range. Subsequently the distribution of LT for each of these buckets is evaluated. The evaluation is a simple comparison of the main shape parameters of the LT distribution; mean, median, standard deviation and skewness. The mean and standard deviation of the LT distributions are used because we know from Duc et al. [6], Kim et al. [8] and Michna et al. [10] that they are critical for the size of the bullwhip effect incurred due to stochastic lead times. The median and skewness are included as they are critical for the time it takes for a supply chain to reach steady state. In this case a simple benchmark is used to determine if the different set of observations behave in conflicting manners. The benchmark is calculated in the same manner for all four shape parameters. Exemplified by the mean LT: $\bar{\mu}_{LT,set}/\bar{\mu}_{LT,all}$, where $\bar{\mu}_{LT,set}$ is the mean for the particular set of orders partitioned on Q and $\bar{\mu}_{LT,all}$ is the calculated mean LT for all observations. Values above 1 indicate that the particular subset has a higher mean than the whole data set. Values close to 1 indicate a similar mean on the particular parameter.

The third step involves evaluating the output from step two to determine an appropriate model of LT. Here the benchmarks are needed to determine whether there is any significant difference in LT behavior for the different ranges of Q, whether they cover a significant amount of observations or total demand to warrant modeling separately or a simple i.i.d. model for LT can be used instead.

4 Test Case

In the following data from four products from the same product family and a total of 29,897 observations of orders with corresponding order sizes (Q) and lead times (LT) are analyzed using R [16].

Figure 3 shows the cumulative data removed from the samples in each of the four steps and as can be seen, removing outliers, using a four standard deviations criteria, reduces the available data with between 2 and 3 %. This data removal is however necessary as there are several extreme observations in the tails of both the order size and lead time distributions that could potentially skew any subsequent analysis. In an application of the approach data cleaning must always be conducted in the given context.

Figure 4 shows a three dimensional frequency diagram with the frequency intensity as the z-axis. The data has for this purpose been cut into five intervals covering an equal range of Q and LT respectively. As can be seen, small order quantities (Q) and small lead times (LT) dominate the frequency plots for all four products. X^2-tests using the buckets depicted in the Fig. 4 shows that LT depends on Q on a better than 0.001 level in all four cases.

To evaluate the buckets' suitability in representing the actual data we investigate the number of observations in each bucket of Q and their contribution to total demand. The left hand side of Fig. 5 shows that for the particular data sets used in the analysis small orders (order size interval 1) very much dominate the data in the sense of observations as also seen on Fig. 4.

From the right hand side of Fig. 5 it should be noted that the total volume of demand is actually very dispersed between the order size intervals with no clear pattern between the four products. In the present analysis the observations have simply been partitioned into five groups of equal range for Q and LT. Another

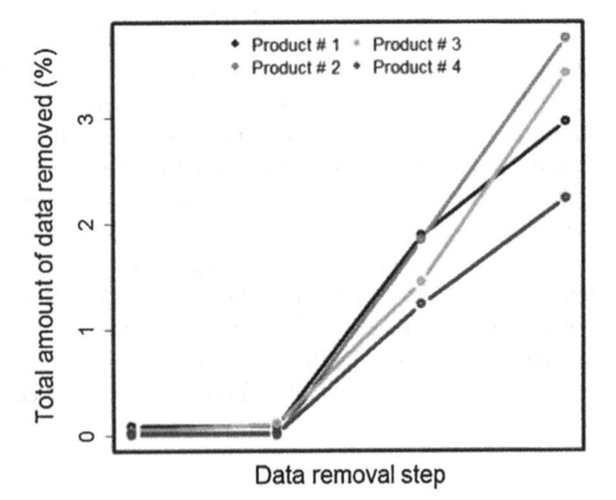

Fig. 3 Data removed in each of the four data cleaning steps

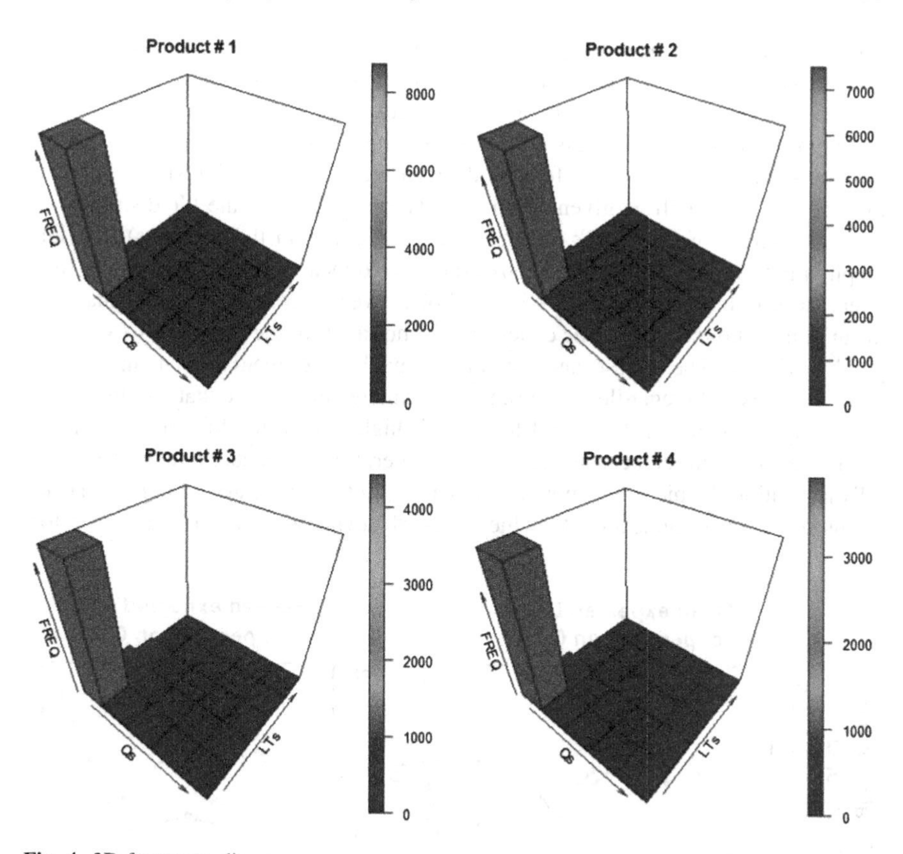

Fig. 4 3D frequency diagrams

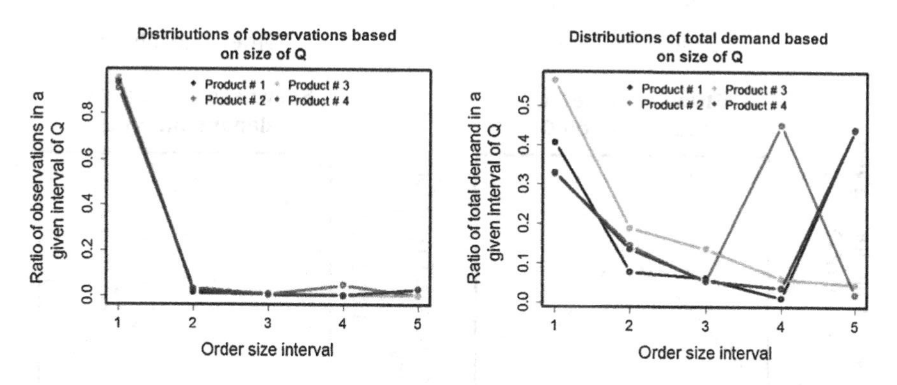

Fig. 5 Ratio of observations and the ratio of total demand in a given order interval for each of the four products

approach could have been to split the data into groups based on equal volume of sales or equal number of observations. However, these approaches would potentially be misleading with regards to the link between Q and LT.

Figure 6 shows the comparison across the range of Q for all four products and the four shape parameters. The four products appear to be behaving similarly so an overall evaluation will be given. With regards to the mean of the LT distributions the first interval of Q (i.e. the smallest order quantities in the data sets) (top left graph in Fig. 6) have uniformly the lowest expected mean where the remaining four intervals of Q have an expected mean LT of between 1.5 and 2.5 higher than the mean of the whole data set. The conclusion is the same for the median LT (top right graph in Fig. 6). The lowest range of Q has a much lower median LT than the four other intervals of Q regardless of product. It is interesting to note that the difference here is upwards to a factor 8 and thus much higher than the difference in mean, where the largest differences are a factor 2.5 larger. For the standard deviation of the LT distribution the picture is more complicated (bottom left graph in Fig. 6). There seems to be some indication that the standard deviation of the LT is lower for

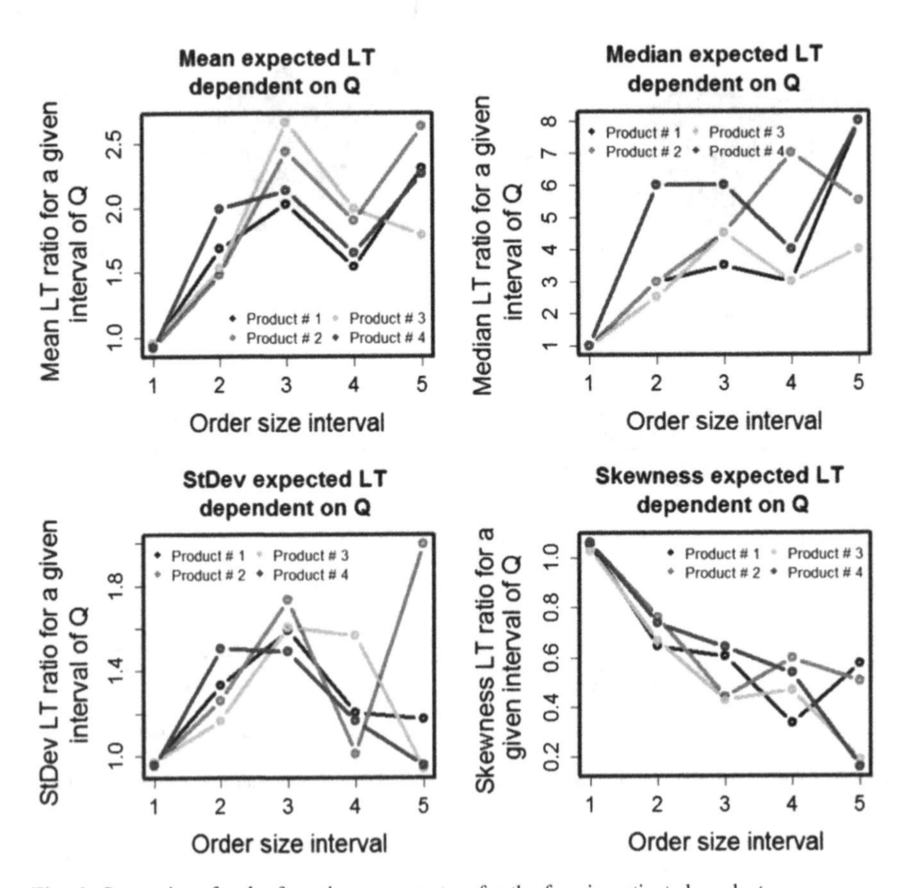

Fig. 6 Comparison for the four shape parameters for the four investigated products

smaller Q. However, the evidence is not conclusive. The final parameter is the skewness (bottom right graph in Fig. 6). Here the results are the same for all four products. The skewness of the distribution is lower for larger ranges of Q and although they are significantly less all LT distributions are skewed in the same direction regardless of which interval of Q is considered.

The overall conclusion is that there seems to be ample evidence that in the particular case lead times can be considered to depend to some extend on the order quantities. From the number of observations available in each order size interval (regardless of product) one could consider to merge the four order size intervals that cover the larger orders into one interval. In all four cases they behave similar and doing this would mean that 59.4, 67.4, 43.6 and 67.0 % of the total volume would be in one order size interval (i.e. large orders). This would allow for simplifying the LT model to two functions:

$$F_1(x, Q)\, for\, Q \in [Q_i; Q_j]$$
$$F_2(x, Q)\, for\, Q \in [Q_j; Q_h]$$

where $Q_i; Q_j$ limits the range of the first order interval for a given product, and Q_h is the largest order size included in the final range and F_1 and F_2 are density functions for each of the respective intervals of Q. In the data preparation values of Q larger than Q_h have been removed, so in practice some approximation of expectations for LT for $Q > Q_h$ must be included. For illustrative purposes the corresponding density functions have been numerically approximated [17] for product 1 for all intervals of Q and for the last four intervals of Q combined. The results are illustrated in Fig. 7.

While the presented example is just one case, there is no reason to believe that this case is unique and it underlines the need to conduct research into supply chains with order quantity dependent lead time distributions. No doubt in practice this type

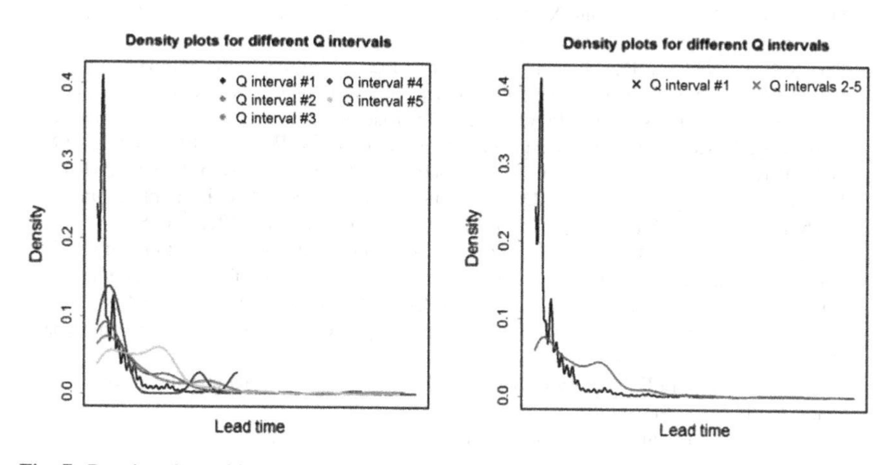

Fig. 7 Density plots of lead times for product 1 for different intervals of Q

of supply chain without information sharing is more difficult to manage than one where lead times are independent of order quantities or even more ideally one where lead times can be considered to be deterministic.

5 Conclusions and Future Research

From literature it is established that lead times and their behavior are a major source of bullwhip effect in supply chains. It is also established that there are limited studies of how actual lead times behave. To remedy this, a simple approach for analyzing lead times in supply chains with no information sharing is proposed. The approach utilizes data that any company should have readily available to establish if there is a link between order quantities and lead times. The approach is tested on data from a company and it is concluded that the lead times for four products appear to depend on the order quantity. Future research will focus on analyzing the impact of quantity dependent lead times on supply chains.

References

1. Bischak, D.P., Robb, D.J., Silver, E.A., Blackburn, J.D.: Analysis and management of periodic review order-up-to level inventory systems with order crossover. Prod. Oper. Manag. **23**(5), 762–772 (2014)
2. Bradley, J.R., Robinson, L.W.: Improved base-stock approximations for independent stochastic lead times with order crossover. Manuf. Serv. Oper. Manag. **7**(4), 319–329 (2005)
3. Chen, F., Drezner, Z., Ryan, J.K., Simchi-Levi, D.: Quantifying the bullwhip effect in a simple supply chain: the impact of forecasting, lead times, and information. Manag. Sci. **46**(3), 436–443 (2000)
4. Disney, S.M., Maltz, A., Wang, X., Warburton, R.D.H.: Inventory management for stochastic lead times with order crossovers. Eur. J. Oper. Res. **248**, 473–486 (2016)
5. Do, N.A.D., Nielsen, P., Michna, Z., Nielsen, I.E.: Quantifying the bullwhip effect of multi-echelon system with stochastic dependent lead time, advances in production management systems. In: Proceedings International Conference Innovative and Knowledge-Based Production Management in a Global-Local World: IFIP WG 5.7, APMS 2014, Part I, 20–24 Sept 2014, vol. 438, pp. 419–426. Springer, Ajaccio, France (2014)
6. Duc, T.T.H., Luong, H.T., Kim, Y.D.: A measure of the bullwhip effect in supply chains with stochastic lead time. Int. J. Adv. Manuf. Technol. **38**(11–12), 1201–1212 (2008)
7. Joe, H.: Multivariate Models and Dependence Concepts. Chapman & Hall, London (1997)
8. Kim, J.G., Chatfield, D., Harrison, T.P., Hayya, J.C.: Quantifying the bullwhip effect in a supply chain with stochastic lead time. Eur. J. Oper. Res. **173**(2), 617–636 (2006)
9. Lee, H.L., Padmanabhan, V., Whang, S.: The bullwhip effect in supply chains. Sloan Manag. Rev. **38**(3), 93–102 (1997)
10. Michna, Z., Nielsen I.E., Nielsen P.: The bullwhip effect in supply chains with stochastic lead times. Math. Econ. **9**(16) (2013)
11. Nelsen, R.: An Introduction to Copulas. Springer, New York (1999)

12. Nielsen P., Michna Z., Do, N.A.D., Sørensen B.B.: Lead times and order sizes—a not so simple relationship. In: 36th International Conference Information Systems Architecture and Technology 2015, Karpacz, Part II, pp. 65–75 (2016). doi:10.1007/978-3-319-28555-9_6
13. Nielsen, P., Michna, Z., Do, N.A.D.: An empirical investigation of lead time distributions, advances in production management systems. In: Proceedings International Conference Innovative and Knowledge-Based Production Management in a Global-Local World: IFIP WG 5.7, APMS 2014, Part I, Sept 20–24 2014, vol. 438. Springer, Ajaccio, France, pp. 435–442 (2014)
14. Pahl, J., Voß, S., Woodruff, D.L.: Production planning with load dependent lead times. 4OR **3** (4), 257–302 (2005)
15. Relich, M., Witkowski, K., Saniuk, S., Šujanová, J.: Material demand forecasting: an ERP system perspective. Appl. Mech. Mater. **527**, 311–314 (2014)
16. R-Project.org: www.r-project.org (2016)
17. Scott, D.W.: Multivariate Density Estimation. Theory Practice Vision Wiley, New York (1992)
18. Sitek P.: A hybrid CP/MP approach to supply chain modelling, optimization and analysis. In: Federated Conference on Computer Science and Information Systems (FedCSIS), pp. 1345–1352 (2014). doi:10.15439/2014F89
19. Wang, X., Disney, S.M.: Mitigating variance amplifications under stochastic lead-time: the proportional control approach. Eur. J. Oper. Res. (2015)

Clique Editing to Support Case Versus Control Discrimination

Riccardo Dondi, Giancarlo Mauri and Italo Zoppis

Abstract We present a graph-based approach to support case vs control discrimination problems. The goal is to partition a given input graph in two sets, a clique and an independent set, such that there is no edge connecting a vertex of the clique with a vertex of the independent set. Following a parsimonious principle, we consider the problem that aims to modify the input graph into a most similar output graph that consists of a clique and an independent set (with no edge between the two sets). First, we present a theoretical result showing that the problem admits a polynomial-time approximation scheme. Then, motivated by the complexity of such an algorithm, we propose a genetic algorithm and we present an experimental analysis on simulated data.

1 Introduction

Graph modification has been widely applied in different contexts to provide sets of homologous instances from a set of different elements. For example, correlation clustering [3, 15], a well-known problem with applications in machine learning and bioinformatics, asks for the partition of the vertices of a graph in cliques, so that the similarity (expressed as the number of common edges between the input graph and

R. Dondi (✉)
Dipartimento di Scienze Umane e Sociali, Università Degli Studi di Bergamo, Bergamo, Italy
e-mail: riccardo.dondi@unibg.it

G. Mauri · I. Zoppis
Dipartimento di Informatica, Sistemistica e Comunicazione, Università Degli Studi di Milano-Bicocca, Milano, Italy
e-mail: mauri@disco.unimib.it

I. Zoppis
e-mail: zoppis@disco.unimib.it

I. Czarnowski et al. (eds.), *Intelligent Decision Technologies 2016*,
Smart Innovation, Systems and Technologies 56,
DOI 10.1007/978-3-319-39630-9_3

the output graph, plus the number of pairs of vertices not connected in both input and output graphs) is maximized.[1]

Here, we focus on a similar problem known as *Sparse Split Graph* problem in the context of case vs control experimental design. In this situation, an important goal is to distinguish healthy subjects (controls) from patients (cases). For instance, consider a set of individuals that could be either affected by some disease or randomly sampled from a healthy population in some clinical trial. We can represent the observed sample as the set of vertex in a graph by connecting two vertices whenever two patients share similar symptoms (for example when correlation between their blood pressure is statistically significant). Following this process, a clique might be ideally used to provide a set of homologous patients. Similarly, an independent set of vertices, well separated from that clique, could ideally be used to instantiate healthy subjects. Clearly, the results of the trial may be motivated by many reasons, and not all patients (i.e. cases) affected by the expressed disease share similar symptoms; as a result the graph we obtain is not necessarily partitioned into ill (clique) and healthy subjects (independent set). Nevertheless, optimizing such structure could benefits further classification tasks. In fact, most real domains are best described by structured data where instances of particular types are related to each other in complex ways. In these contexts, relationships (e.g. expressing similarities or differences between instances) can provide suitable source of information to inference mechanism in data mining or machine learning problems. Such an approach has been proved useful for both classification and clustering in many situations, see e.g. [6, 7] for molecular data analysis, [11] for network data, or [12] for a broader view of fundamentals in cluster analysis.

The goal of our approach is then to obtain a suitable partition of the given input graphs in two sets, a clique (cases) and an independent set (controls), such that no edge connecting vertices of the clique with vertices of the independent set are allowed. Importantly, following a parsimonious principle, we expect that such process preserves the maximum number of original relations (edges or missing edges) from the graph observed as an input instance, i.e. preserving the original structure as well as possible. At this point, it is not difficult to imagine how this model can be applied for future discrimination, i.e. when new subjects have to be classified either as cases or controls. For example a simple approach could be to evaluate whether new instances correlate mostly with the ("case") clique or with the independent set of healthy controls. Please notice that, here we do not focus on the classification question. Instead we analyze the combinatorial problem described above, i.e. the Maximum Sparse Split Graph Problem ($\mathcal{MAX}$-$\mathcal{SSG}$).

$\mathcal{MAX}$-$\mathcal{SSG}$ has the same input of the Minimum Sparse Split Graph Problem [9], but with a complementary objective function. Given a graph, Minimum Sparse Split Graph and $\mathcal{MAX}$-$\mathcal{SSG}$ aim to partition the graph in two sets, a clique and an independent set, such that there is no edge connecting a vertex of the clique with a vertex of the independent set; the objective function of Minimum Sparse Split Graph is the

[1]Different variants of the problem have been investigated, including the weighted and constrained variants [3, 8].

minimization of the number of modified relations (edges removed or inserted with respect to the input graph), the objective function of $\mathcal{MAX}$-$\mathcal{SSG}$ is the maximization of the number of unchanged relations.

Since Minimum Sparse Split Graph is NP-hard [9], the same complexity result holds also for the Maximum Sparse Split Graph problem. Indeed a graph G' which is an optimal solution of Minimum Sparse Split Graph is also an optimal solution of $\mathcal{MAX}$-$\mathcal{SSG}$: since it modifies the minimum number of inserted or deleted edges, it maximizes the number of relations of the input graph that are not modified. Clearly, also the converse is true: a graph G' which is an optimal solution of $\mathcal{MAX}$-$\mathcal{SSG}$ is also an optimal solution of Minimum Sparse Split Graph.

We thus consider two possible directions of research, namely approximation algorithms and genetic algorithms. For the first direction, we present in Sect. 3, a theoretical result, that is we prove that the Maximum Sparse Split Graph problem admits a Polynomial Time Approximations Scheme (PTAS), which means that the problem can be solved within any factor $1 + \varepsilon$ in time $O(n^{1/\varepsilon})$, for any constant ε, where n is the number of vertices of the input graph. While this result shows that solutions close to the optimal can be computed in polynomial time, such a result is only of theoretical interest, due to the time complexity of the proposed PTAS. Hence, we consider genetic algorithms as a possible direction to compute optimal or near optimal solution for the Maximum Sparse Split Graph problem. We give in Sect. 4 a natural genetic algorithm and we present some experimental results on a simulated data set, that show that the genetic algorithm converges fast, in particular when the input data has a moderate similarity to Sparse Split Graph Problem.

2 Preliminaries

In this section we introduce some basic definitions that will be useful in the rest of the paper and we give the formal definition of the Maximum Sparse Split Graph problem. We recall that a graph is a clique if it is a complete graph, that is a graph where each pair of vertices is connected by an edge. A graph is an independent set if no pair of vertices is connected by an edge.

Given a graph $G = (V, E)$ and a set V' of vertices, with $V' \subseteq V$, we denote by $G[V']$ the graph induced by V', that is $G[V'] = (V', E')$, where $E' = \{\{v_i, v_j\} : v_i, v_j \in V' \wedge \{v_i, v_j\} \in E\}$.

Now, we give the formal definition of *Sparse Split Graph*, since starting from a given graph, we aim to compute a sparse split graph that can be obtain by preserving the maximum number of relations represented by the input graph.

Definition 1 Given a graph $G = (V, E)$, G is a *Sparse Split Graph* if V can be partitioned in two sets V_1 and V_2 such that $G[V_1]$ is a clique, $G[V_2]$ is an independent set and, for each $v_i \in V_1$ and $v_j \in V_2$, $\{v_i, v_j\} \notin E$.

Before giving the formal definition of the Maximum Sparse Split Graph Problem, we introduce the definition of agreement between two graphs, as it is useful to describe the objective function we aim to maximize.

Definition 2 Given a graph $G = (V, E)$ and a graph $G' = (V, E')$, we define the *agreement* of G and G', denoted by $A(G, G')$, as the number of edges that belong to both E and E' plus the number of edges that do not belong to both E and E'.

Now, we are ready to give the definition of the Maximum Sparse Split Graph Problem ($\mathcal{MAX}$-$\mathcal{SSG}$).

Problem 1 Maximum Sparse Split Graph Problem ($\mathcal{MAX}$-$\mathcal{SSG}$)
Input: a graph $G = (V, E)$.
Output: a graph $G' = (V, E')$, such that G' is a sparse split graph.
Objective function: $A(G, G')$ is maximized.

In the remaining part of the paper, we denote by n the number of vertices in V, that is $n = |V|$.

3 A PTAS for $\mathcal{MAX}$-$\mathcal{SSG}$

In this section, we present a PTAS for the $\mathcal{MAX}$-$\mathcal{SSG}$ problem. The algorithm is based on the smooth polynomial programming technique of [1], a technique that has been previously applied to design PTAS for graph modification problems (for example a variant of Correlation Clustering [5]) and problems in phylogenetics (Maximum Quartet Inconsistency [10]).

We now briefly present the smooth polynomial programming technique. A c-*smooth polynomial integer program* over variables $x_1, \dots, x_m$ is a problem having the following form:

$$\begin{aligned} &\text{maximize } p(x_1, \dots, x_m) \\ &\text{subject to } z_j \le q_j(x_1, \dots, x_m) \le u_j \\ &\qquad\qquad x_i \in \{0, 1\} \text{ for } 1 \le i \le m \end{aligned} \tag{1}$$

where each $q_j(x_1, \dots, x_m)$ is a polynomial on variables $x_1, \dots, x_m$ that has maximum degree d, and the coefficients of each degree-ℓ monomial are in the interval $[-cm^{d-\ell}, cm^{d-\ell}]$. Denote by OPT the optimal value of a c-smooth polynomial integer program. In [1] it is shown that, for each constant $\delta > 0$, there exists an approximation algorithm that, in time $O(m^{\delta^{-2}})$, computes a 0/1 assignment $\langle a_1, \dots, a_m \rangle$ to the variables $\langle x_1, \dots, x_m \rangle$ of a c-smooth polynomial integer program, such that, by assigning $x_1 = a_1, \dots, x_m = a_m$, $p(x_1, \dots, x_m)$, has value at least $OPT - \delta m^d$.

Before presenting our formulation of $\mathcal{MAX}$-$\mathcal{SSG}$ as a c-smooth polynomial integer program, we prove a property of the $\mathcal{MAX}$-$\mathcal{SSG}$ problem, namely we give a lower bound on the value of an optimal solution of $\mathcal{MAX}$-$\mathcal{SSG}$ on any instance G.

Lemma 1 *Given an instance $G = (V, E)$ of $\mathcal{MAX}$-$\mathcal{SSG}$, an optimal solution of $\mathcal{MAX}$-$\mathcal{SSG}$ on instance G has value at least $\frac{n(n-1)}{4}$.*

Proof Given a graph $G = (V, E)$, consider the following solutions of $\mathcal{MAX}$-$\mathcal{SSG}$ on instance $G = (V, E)$:

- $G' = (V, \emptyset)$, that is G' contains no edge;
- $G'' = (V, V \times V)$, that is G'' is a clique.

Consider an edge $\{v_i, v_j\} \in E$, then by construction $\{v_i, v_j\}$ is an edge of G''. Consider two vertices v_i, v_j of V such that $\{v_i, v_j\} \notin E$; it follows by construction that $\{v_i, v_j\}$ are not connected by an edge in G'. Hence denote by $A(G', G)$ ($A(G'', G)$, respectively) the number of agreement between G' and G (between G'' and G, respectively). Then

$$A(G', G) + A(G'', G) = \frac{n(n-1)}{2}$$

hence one of $A(G', G)$ or $A(G'', G)$ is at least $\frac{n(n-1)}{4}$. Since an optimal solution has agreement value greater or equal than $A(G', G)$ and $A(G'', G)$, the lemma follows. □

From Lemma 1, it follows that an approximation algorithm with additive error δn^2 is a $(1 + \frac{\delta}{\alpha})$ approximation algorithm, for some fixed constant α, which proves the existence of a PTAS for the problem.

Now we present a formulation of $\mathcal{MAX}$-$\mathcal{SSG}$ as a c-smooth polynomial integer program. Consider the variables $x_{i,j}$, with $1 \leq i \leq n$ and $1 \leq j \leq 2$; $x_{i,1}$ has value is 1 if and only if the v_i is assigned to the clique of a solution, otherwise is 0; $x_{i,2}$ has value is 1 if and only if the v_i is assigned to the independent set of a solution, otherwise is 0. Denote by $e_{i,j}$ a constant equal to 1 if $\{v_i, v_j\} \in E$ (else $e_{i,j}$ is equal to 0), and denote by $n_{i,j}$ a constant equal to 1 if $\{v_i, v_j\} \notin E$ (else $n_{i,j}$ is equal to 0). We can given a formulation of $\mathcal{MAX}$-$\mathcal{SSG}$ as follows:

$$\begin{aligned}
\text{maximize} \sum_{i,j} & e_{i,j} x_{i,1} x_{j,1} + n_{i,j}(x_{i,2}x_{j,2} + x_{i,1}x_{j,2} + x_{i,2}x_{j,1}) \\
\text{subject to } & x_{i,1} + x_{i,2} = 1 \qquad \text{for } 1 \leq i \leq n \\
& x_{i,t} \in \{0, 1\}
\end{aligned} \tag{2}$$

The formulation given for $\mathcal{MAX}$-$\mathcal{SSG}$ is a c-smooth polynomial integer program with degree $d = 2$. Following the approach in [10], we design a PTAS by first computing in polynomial time a fractional solution for the c-smooth polynomial integer and then rounding this solution as follows: the variable among $x_{i,1}$ and $x_{i,2}$ having the maximum value is rounded to 1, the other variable is rounded to 0. Such a rounding procedure has an additive error δn^2 (for some constant $\delta > 0$, see the proof of Theorem 1) with respect to an optimal solution of the c-smooth polynomial integer program.

Now, we can conclude this section by proving the that we have indeed designed a PTAS.

Theorem 1 *$\mathcal{MAX}$-$\mathcal{SSG}$ admits a PTAS.*

Proof Consider the value OPT of an optimal solution of $\mathcal{MAX}$-$\mathcal{SSG}$ on instance G and let OPT_c be the value of an optimal solution to the corresponding c-smooth polynomial integer program. It follows that $OPT \leq OPT_c$. Let A_c be a solution of the rounding procedure of the c-smooth polynomial integer program of $\mathcal{MAX}$-$\mathcal{SSG}$. By [1, 10], A_c can be computed in time $O(n^{\delta^{-2}})$, for each constant $\delta > 0$, so that $A_c \geq OPT_c - \delta n^2 \geq OPT - \delta n^2$.

Since, by Lemma 1, $OPT \geq \alpha n^2$, for some constant $\alpha > 0$, it follows that

$$A_c \geq \alpha n^2 - \delta n^2 = \alpha n^2(1 - \frac{\delta}{\alpha}) = OPT(1 - \varepsilon)$$

where $\varepsilon = \frac{\delta}{\alpha}$. □

4 A Genetic Algorithm for $\mathcal{MAX}$-$\mathcal{SSG}$

The algorithm presented in Sect. 3 is mainly of theoretical interests due to its computational complexity. Therefore to provide an efficient heuristic for $\mathcal{MAX}$-$\mathcal{SSG}$, we consider an approach based on genetic algorithms [13].

Given an input graph $G = (V, E)$, the genetic algorithm represents a solution (that is a bipartition (V_1, V_2) of the vertex-set V, where V_1 is a clique and V_2 an independent set), as a chromosome that represents the vertex membership properties. Given a chromosome c, we have $c[i] = 1$ if vertex i is part of a clique; 0 otherwise. As usual for genetic algorithms, we consider standard operators such as *mutations* and *crossover*. More specifically, we adapt the mutation operator as follows. The mutation process applied here aims to modify the set that a vertex belongs to, that is if a vertex belong to the independent set is moved to the clique, whereas a vertex that belongs to the clique is moved to the independent set. Selection of elements in the population for modification follows a fitness proportionate scheme (see e.g. [13]).

Next, we give more details about the mutation and crossover operators.

- Mutation. Each individual from current population at time i is modified with probability 0.1 as follows:
  - A random sample of vertices with low degree in the original graph takes value 0 over the chromosome (that is it is moved to the independent set);
  - A random sample of vertices from cliques of size three of the original graph takes value 1 over the chromosome (that is the three vertices belong to the clique).
- Crossover is applied with probability 0.8.

We consider cliques of size three of the original graph, as they can be computed efficiently (in time $O(n^3)$), and they may represent subsets of larger cliques.

Genetic algorithm uses a fitness function to evaluate a solution. The fitness function for our genetic algorithm is defined as the negation of the Hamming distance between the adjacency matrix of the solution graph and the adjacency matrix of the input graph (we omit here the constant value $\binom{n}{2}$ used in the objective function of $\mathcal{MAX}$-$\mathcal{SSG}$). Finally we also consider elitist selection (or elitism) when constructing a new population from an existing one, to allow the best individuals to be part of the next generation. It is known that such an approach can be used to guarantee that the solution quality does not decrease from one generation to the next [2].

The genetic algorithm was coded in R using the Genetic Algorithm package [14] downloadable at https://cran.r-project.org/web/packages/GA/index.html.

4.1 Experimental Results

We apply the genetic algorithm to synthetic data. We sample input graphs from two classical random models (see e.g. [4]), Erdos-Renyi and Barabasi.[2] The first model, *Erdos-Renyi* $ER(n, p)$ is used to generate graphs with n vertices (every possible edge is created with the same constant probability p).

The second model, *Barabasi* $BA(n)$, uses a simple random mechanism to generate graphs with n vertices through *preferential attachment*. A vertex is added for each step to the current growing graph, starting with a single vertex and no edges. Then, in the next steps, a vertex v is added to the graph by defining edges that connects the new vertex with vertices already part of the graph. The probability that a vertex i already in the graph is chosen is given by $Pr(i) \sim k_i^{\alpha} + a$, where k_i is the degree of vertex i at the current time and a and α are parameters. In our simulation, we use $\alpha, a = 1$.

From each random model we sample 10 graphs for each choice of n (number of vertices of the graph) in $\{10, 50, 100, 200\}$. We increase the size of the population accordingly (see Table 1). The results are given in Table 2. We stop the execution of the genetic algorithm when the fitness value is not increased in more than 15 generations. As shown in the same table, the final fitness value (as expected), decreases as the number of vertices increases. Moreover, we observe that for graphs from *ER* this value is smaller than for graphs from *BA*. Accordingly, the required CPU time for convergence is greater for graphs sampled from *ER* than for graphs from *BA*.

Our results[3] show that the genetic algorithm has good performances on simulated data, even for graph with 200 of vertices. In this case the algorithm was able to converge in a reasonable time (approximately 1428 s for samples from *ER* and approximately 239 s for samples from *BA*).

In the analysis we consider also the *relative mutation rate*, that is the percentage of the number of modifications (edge additions and edge deletions) in a graph among all the potential mutations in the original graph. Interestingly, this value is significantly

---

[2]We use Random Graphs (RGs) as generative models to simulate observations.

[3]The tests were performed on a machine with 8 GB RAM, Intel Core i7 2.30 GHz.

Table 1 Models and parameters

Type	Pop. size	Generat.	Elitism	Cross pr	Mut pr
ER(10, 1/2)	100	17.7	10	0.8	0.1
ER(50, 1/2)	500	64.3	50	0.8	0.1
ER(100, 1/2)	1000	106.6	100	0.8	0.1
ER(200, 1/2)	2000	176.2	200	0.8	0.1
BA(10)	100	15.1	10	0.8	0.1
BA(50)	500	41.1	50	0.8	0.1
BA(100)	1000	53.9	100	0.8	0.1
BA(200)	2000	46.4	200	0.8	0.1

ER Erdos-Renyi model; *BA* Preferential attachment (*Barabasi*). *GA parameters* Population size, Generation, Elitism, Crossover Probability and Mutation probability

Table 2 Models and results

Type	Final fitness value	CPU			Relative mut. rate
		User	Sys	Elaps.	
ER(10, 1/2)	−13.0	1.777	0.00	1.778	28.9
ER(50, 1/2)	−508.0	29.918	0.470	30.442	41.6
ER(100, 1/2)	−2196.1	153.196	0.360	153.727	44.7
ER(200, 1/2)	−9182.5	1365.656	62.717	1428.373	46.4
BA(10)	−8.0	1.512	0.000	1.517	17.8
BA(50)	−54.0	25.207	0.040	26.378	4.1
BA(100)	−130.1	73.594	0.112	272.099	2.5
BA(200)	−372.0	231.403	6.769	239.079	2.2

ER Erdos-Renyi model; *BA* Preferential attachment (*Barabasi*). *Results* Fitness value, *CPU time* user, system, elapsed (in seconds) and Relative mutation rate, computed as the percentage of the number of mutations (edge addition and edge deletions) in a graph among all the potential mutations in the original graph

lower in graphs from *BA* than in graphs from *ER*. Moreover, in graphs from *ER* this value increases as the number of vertices increases, while in graphs from *BA* we observe the opposite behavior. This fact suggests that graphs from *BA* have a structure that is closer to that of a Sparse Split Graph with respect to the graphs from *ER*; furthermore, this property becomes stronger as the number of vertices increases.

We present the performance results for graphs with 50, 100, 200 vertices in Figs. 1, 2 and 3, respectively. As the results shown, the GA algorithm converges before 54 generations for graphs from *BA* even with 200 vertices. The convergence is much slower for graphs from *ER*. For example, for a graph with 200 of vertices, the GA algorithm requires more than 170 generation to converge.

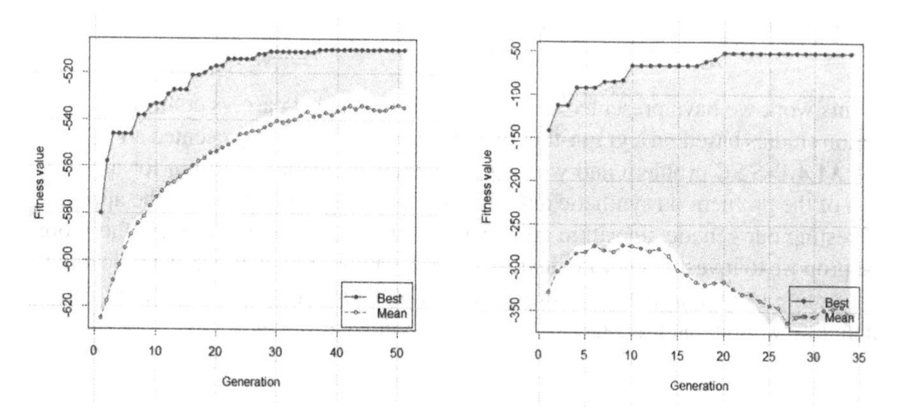

Fig. 1 Experimental results of the GA for $ER(50, 1/2)$ (*left diagram*) and $BA(50)$ (*right diagram*)

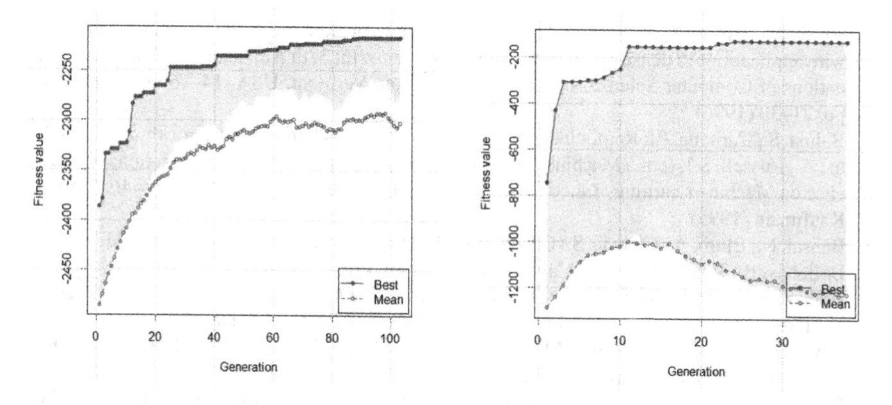

Fig. 2 Experimental results of the GA for $ER(100, 1/2)$ (*left diagram*) and $BA(100)$ (*right diagram*)

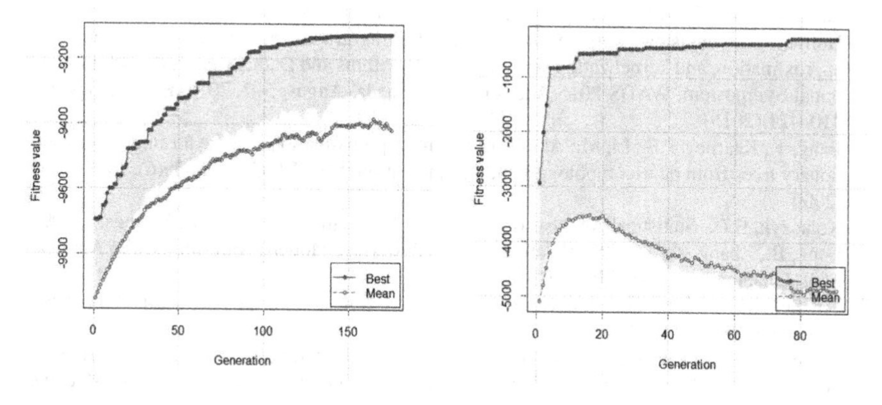

Fig. 3 Experimental results of the GA for $ER(200, 1/2)$ (*left diagram*) and $BA(200)$ (*right diagram*)

5 Conclusion

In this work we have presented a methodology to support case vs control discrimination studies based on a graph-theoretical approach. We have presented a PTAS for the $\mathcal{MAX}$-$\mathcal{SSG}$ problem and we have presented a genetic algorithm for the solution of the problem on synthetic data. In the future, we aim to validate the approach by testing our genetic algorithm on a real case study and on real data. Furthermore, we propose to investigate more flexible variants of our approach, for example allowing a more general structure of the output graph than that of Sparse Split Graph or including weights on the edges to represent confidence values.

References

1. Arora, S., Frieze, A.M., Kaplan, H.: A new rounding procedure for the assignment problem with applications to dense graph arrangement problems. In: 37th Annual Symposium on Foundations of Computer Science, FOCS '96, Burlington, Vermont, USA, 14–16 October, 1996, pp. 21–30 (1996)
2. Baluja, S., Caruana, R.: Removing the genetics from the standard genetic algorithm. In: Prieditis, A., Russell, S.J. (eds.) Machine Learning, Proceedings of the Twelfth International Conference on Machine Learning, Tahoe City, California, USA, July 9–12, 1995, pp. 38–46. Morgan Kaufmann (1995)
3. Bansal, N., Blum, A., Chawla, S.: Correlation clustering. Mach. Learn. **56**(1–3), 89–113 (2004)
4. Bollobas, B.: Random Graphs. Cambridge University Press (2001)
5. Bonizzoni, P., Della Vedova, G., Dondi, R., Jiang, T.: On the approximation of correlation clustering and consensus clustering. J. Comput. Syst. Sci. **74**(5), 671–696 (2008)
6. Cava, C., Zoppis, I., Gariboldi, M., Castiglioni, I., Mauri, G., Antoniotti, M.: Copy–number alterations for tumor progression inference. Artif. Intell. Med. 104–109 (2013)
7. Cava, C., Zoppis, I., Gariboldi, M., Castiglioni, I., Mauri, G., Antoniotti, M.: Combined analysis of chromosomal instabilities and gene expression for colon cancer progression inference. J. Clin. Bioinform. **4**, 2 (2014)
8. Giotis, I., Guruswami, V.: Correlation clustering with a fixed number of clusters. Theory Comput. **2**(13), 249–266 (2006). http://dx.doi.org/10.4086/toc.2006.v002a013
9. Hüffner, F., Komusiewicz, C., Nichterlein, A.: Editing graphs into few cliques: complexity, approximation, and kernelization schemes. In: Algorithms and Data Structures - 14th International Symposium, WADS 2015, Victoria, BC, Canada, August 5–7, 2015. Proceedings, pp. 410–421 (2015)
10. Jiang, T., Kearney, P.E., Li, M.: A polynomial time approximation scheme for inferring evolutionary trees from quartet topologies and its application. SIAM J. Comput. **30**(6), 1942–1961 (2000)
11. Kolaczyk, E.D.: Statistical Analysis of Network Data: Methods and Models. Springer (2009)
12. Long, B., Zhang, Z., Yu, P.S.: Relational Data Clustering: Models, Algorithms, and Applications. Chapman and Hall/CRC (2010)
13. Mitchell, M.: An Introduction to Genetic Algorithms. Complex Adaptive Systems. MIT Press, Cambridge (Mass.) (1996)
14. Scrucca, L.: GA: a package for genetic algorithms in R. J. Stat. Softw. **53**(4), 1–37 (2013)
15. Wirth, A.: Correlation clustering. In: Sammut, C., Webb, G.I. (eds.) Encyclopedia of Machine Learning, pp. 227–231. Springer (2010)

Detecting Value-Added Tax Evasion by Business Entities of Kazakhstan

Zhenisbek Assylbekov, Igor Melnykov, Rustam Bekishev, Assel Baltabayeva, Dariya Bissengaliyeva and Eldar Mamlin

Abstract This paper presents a statistics-based method for detecting value-added tax evasion by Kazakhstani legal entities. Starting from features selection we perform an initial exploratory data analysis using Kohonen self-organizing maps; this allows us to make basic assumptions on the nature of tax compliant companies. Then we select a statistical model and propose an algorithm to estimate its parameters in unsupervised manner. Statistical approach appears to benefit the task of detecting tax evasion: our model outperforms the scoring model used by the State Revenue Committee of the Republic of Kazakhstan demonstrating significantly closer association between scores and audit results.

Keywords Self-organizing maps · Cluster analysis · Anomaly detection · Tax evasion detection

Z. Assylbekov (✉) · I. Melnykov · R. Bekishev
School of Science and Technology, Nazarbayev University, Astana, Kazakhstan
e-mail: zhassylbekov@nu.edu.kz

I. Melnykov
e-mail: igor.melnykov@nu.edu.kz

R. Bekishev
e-mail: rustam.bekishev@nu.edu.kz

A. Baltabayeva · D. Bissengaliyeva · E. Mamlin
State Revenue Committee, Ministry of Finance of Kazakhstan, Astana, Kazakhstan
e-mail: dbissengaliyeva@mgd.kz

D. Bissengaliyeva
e-mail: abaltabayeva@mgd.kz

E. Mamlin
e-mail: emamlin@mgd.kz

I. Czarnowski et al. (eds.), *Intelligent Decision Technologies 2016*,
Smart Innovation, Systems and Technologies 56,
DOI 10.1007/978-3-319-39630-9_4

1 Introduction

Tax revenue is considered to be one the most important financial resources of a government for achieving specific objectives. However, a large number of businesses frequently try to evade their payment of taxes in a proper way. In order to prevent these illegal actions, government tax authorities have to spend extra budgetary funds to detect and prevent illegal tax evasion activities. As a result, effective ways of detecting tax evasion cases are a challenging issue for tax authorities all over the world.

The State Revenue Committee (SRC) of the Ministry of Finance of the Republic of Kazakhstan has such a responsibility to prevent tax evasion and carry out regulatory, fiscal and control functions in the sphere of tax and customs affairs that involves ensuring the completeness and accuracy of tax receipts, tax computations, income retention, and many other functions. These events include R and D initiatives to achieve optimization in processes and gaining effectiveness. In order to maintain its functions on a high level, the Committee is always engaged in an ongoing process of operational improvements.

Currently SRC is working on establishing the analytic component of the Risk Management System. The committee recently established a new Risk Management and Analysis Department to assist with the implementation of improved risk-based measures to enhance their compliance strategy and base audits on objective measures. The usage of a risk management system bases audits on a number of indicators, which reduces arbitrary audit selection and increases focus on non-compliant taxpayers and many more. Due to the limitations on available resources, it is impossible to inspect all the taxpayers and identify all potential fraudsters. Thus, there appeared an urgent need to determine a more scientific approach to improve tax auditor's productivity in the tasks of tax evasion detection.

The goal of this paper is to describe the data analysis techniques that will help to detect value-added tax evasion by business entities of Kazakhstan. This project is a collaboration between the Nazarbayev University and the State Revenue Committee of the Ministry of Finance of the Republic of Kazakhstan aimed at the tax fraud detection. In this paper we first perform an initial exploratory data analysis using Kohonen maps in Sect. 3. Then we suggest a statistical distribution to describe the behavior of tax-compliant business entities and propose an algorithm to estimate the parameters of such distribution in an unsupervised manner in Sect. 4. The method is first tested on synthetic data designed to resemble the real tax data and then we perform experiments on the real-world data in Sect. 5. Section 6 is devoted to discussion and future work.

2 Related Work

Gupta and Nagadevara [5] used data mining techniques to find out the potential tax evaders from the selection of 2003–2004 business entities in the large amounts of tax returns in VAT system in India. Their paper developed prediction models which were based on various features such as gross profit, tax growth, turnover growth, etc. According to the available information, the whole selection of taxpayers was analyzed with the usage of discriminant function and classification tree models. Implemented techniques showed quite good strike rate and a reasonable prediction efficiency.

Williams et al. [17] applied self-organizing maps to visualize the structure of the real life data provided by the Australian Taxation office (ATO). The data characterized the taxpayers' behavior and consisted of about 6.5 millions entities with 89 features which were selected and designed by the ATO. The features represented income profile, market segments, lodgment profile, tax avoidance schemes involvement, etc. Z-score and min-max normalization techniques were also applied to achieve higher accuracy in clustering. This method was able to identify some abnormal clusters with unusual behavior which might carry out financial frauds.

González and Velásquez [4] described their study of an audit selection strategy for Chile, where taxes provide about 75 % of the resources for the expenses and investments of the whole country. This paper used the data on 2006 companies in Chile to figure out the relationship between their payment of taxes and usage of false invoices. For characterization of fraudulent behavior, they implemented data mining techniques, including self organizing maps, neural gas, and decision trees. The applied data mining techniques were able to detect suspicious activity on different stages of fraud detection.

Hsu et al. [6] described data mining approaches that were used for the improvement of the audit selection processes by the Minnesota Department of Revenue (DOR). In the study, a large variety of classification models were used in order to improve the process of audit selection.

3 Exploratory Data Analysis

3.1 Self-organizing Maps and Clustering

A self-organizing map (SOM) is a type of artificial neural network that is trained using unsupervised learning to produce a two-dimensional, discretized representation of the input space of the training samples, called a map [8]. Self-organizing maps are different from other artificial neural networks in the sense that they use a neighborhood function to preserve the topological properties of the input space. This makes SOMs useful for visualizing low-dimensional views of high-dimensional data, similar to multidimensional scaling.

Since SOMs perform projection to a two-dimensional map and have a topology retaining property, they are suitable for clustering data based on their spatial relationships. Existing SOM-based clustering methods can be categorized into visualization based clustering [7, 12], direct clustering [3, 10, 15], and two-level clustering [13, 14]. In our work we were using the two-level clustering, where the nodes of a trained SOM are treated as 'proto-clusters' serving as an abstraction of the data set. Their prototype vectors are clustered using a traditional clustering technique, such as hierarchical clustering [1], to form the final clusters. Each observation belongs to the same cluster as its BMU.

3.2 Selection of Features

In order to be able to compare our approach with some baseline, we decided to select those features which were already used by the SRC for scoring business entities in 2012 and 2013.[1] They are:

1. *coefficient of tax burden*, which is the ratio of the assessed amounts of tax revenue, without customs payments, to the gross annual income before adjustments;
2. *the industry average coefficient of tax burden*: this is the average coefficient of tax burden across the industry to which a business entity belongs;
3. *amount of transactions with false business entities, inactive taxpayers and taxpayers whose registration is declared invalid by the courts, as well as transactions which were recognized by the court to be concluded without intention to conduct entrepreneurial activities*,
4. *turnover as per the VAT declaration*,
5. *turnover as per the corporate income tax declaration*,
6. *wage fund*,
7. *total annual income*,
8. *assets*,
9. *whether the CEO and/or the founder is the CEO and/or the founder of (a) inactive taxpayers; (b) companies, registration of which is declared invalid by the courts; (c) taxpayers, whose transactions are recognized by the court to be concluded without intention to conduct entrepreneurial activities*,
10. *deregistration and registration with the tax authorities two or more times a year*.

[1]Joint Order of the Minister of Finance of the Republic of Kazakhstan dated September 16, 2011, # 468, and acting Minister of Economic Development and Trade of the Republic of Kazakhstan dated September 16, 2011, # 302, "On approval of the risk assessment criteria in the field of private enterprise on the execution of tax laws, and other laws of the Republic of Kazakhstan, control over the execution of which is entrusted to the tax authorities" (has been canceled on December 25, 2015).

It is important to notice here that initially we do not know the values of the response variable, i.e. *the amounts of additional VAT that would be payed by the business entities if they were tax audited.* Therefore, we are dealing with the problem of unsupervised learning—we need to find out patterns in the data associated with the companies which are likely or not likely to evade VAT.

3.3 Data Collection

The following sources were used to collect the data:

- VAT declarations (Form 300,00[2]);
- corporate income tax forms (100,00, 110,00);
- personal income tax and social tax forms (220,00, 200,00, 210,00);
- registration data;
- registers on purchased and sold goods, services, and works.

We collected the data on 57,874 and 58,620 legal entities which were active in 2011 and in 2012 correspondingly, and which had annual incomes not less than one million Kazakhstani tenge. The SRC audited some of these entities in 2012 and in 2013. The results of such audits are the amounts of additional VAT to be payed by the audited entities, i.e. the values of the response variable which was mentioned in the Sect. 3.2.

3.4 Visualization Using SOM

We trained SOMs separately for each year using `kohonen` package for R [16]. The results are provided in Figs. 1 and 2. The color of each node represents the number of observations from the training set which belong to that node. As we can see in all cases there are few dense nodes which comprise the big proportion of observations. Our main assumption is that *these nodes along with their neighborhoods include companies which are not likely to evade taxes*, and we will build our statistical model for detecting VAT evading entities based on this assumption. We believe that the majority of business entities do not have an intention to evade taxes on purpose, and therefore their standardized features should be more or less close to each other. On the other hand, an anomalous behavior of the features drives us towards questioning the integrity of a taxpayer.

[2]The RoK tax forms can be found at http://kgd.gov.kz/en/section/formy-nalogovoy-otchetnosti.

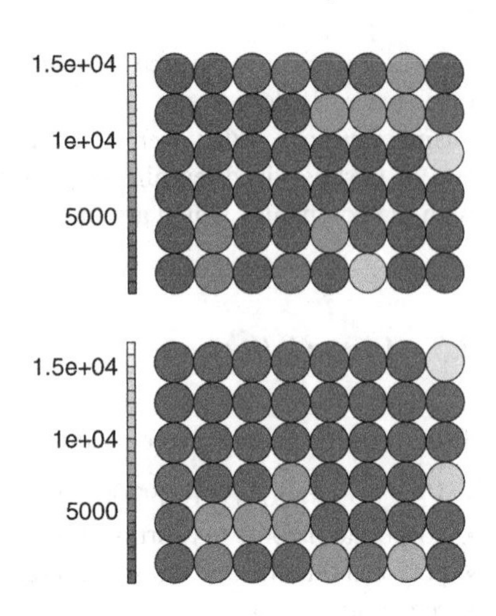

Fig. 1 Business entities of 2012

Fig. 2 Business entities of 2013

4 Statistical Model

4.1 Assumptions

As mentioned in Sect. 3.4, our basic assumption is that *business entities which are not likely to evade taxes are grouped compactly near each other*, whereas tax evading companies are likely to not conform to the behavior of the majority. These nonconforming patterns are often referred to as anomalies or outliers. Figure 3 illustrates anomalies in a simple two-dimensional data set. A comprehensive survey of the existing methods for anomaly detection task is given in [2].

We also assume that the features of the tax compliant business entities are distributed according to a multivariate Gaussian distribution $\mathcal{N}(\mu, \Sigma)$ with mean vector μ and covariance matrix Σ, i.e. the observations are distributed according to the pdf

$$\phi(x; \mu, \Sigma) = \frac{1}{\sqrt{(2\pi)^p|\Sigma|}} \exp\left(-\frac{1}{2}(x-\mu)^T\Sigma^{-1}(x-\mu)\right) \tag{1}$$

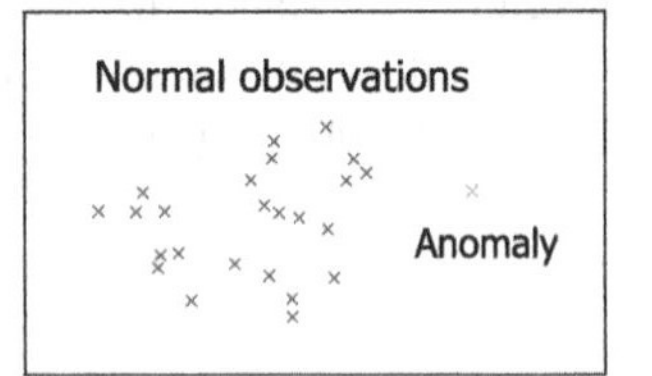

Fig. 3 An observed anomaly in a data set

4.2 Parameter Estimation

If every business entity in our data set was labeled as tax compliant or non-compliant, we could estimate the parameters μ and Σ from such annotated data using tax compliant entities. On the other hand, if we knew the parameters, we could automatically score business entities by applying $\phi(x; \mu, \Sigma)$ to each firm and then classify them into tax compliant or non-compliant using some threshold value. In other words: given the model, we could label our data; given the labeled data, we could estimate our models. Unfortunately, we have neither. This is a typical problem for machine learning: we want to estimate our models from incomplete data. Algorithm 1 addresses this situation of incomplete data.

Algorithm 1 Algorithm for estimating μ and $\boldsymbol{\Sigma}$

Require: matrix of observations $\mathbf{X}$
Ensure: estimates for μ and Σ
1: train SOM on centered and scaled version of $\mathbf{X}$
2: determine number of clusters using the 'elbow' method
3: cluster nodes of SOM using hierarchical clustering
4: $\mathbf{U} \leftarrow$ matrix of observations from the largest cluster
5: $\hat{\mu} \leftarrow \frac{1}{n} \sum_{i=1}^{n} \mathbf{u}_i$
6: $\hat{\Sigma} \leftarrow \frac{1}{n-1} \sum_{i=1}^{n} (\mathbf{u}_i - \mu)^T (\mathbf{u}_i - \mu)$

The algorithm first projects the observations onto two-dimensional space using Kohonen self-organizing maps (line 1), then clusters the prototype vectors (lines 2, 3), and finally estimates μ and Σ using the largest cluster (lines 4–6).

4.3 Scoring

Once we obtain the estimates $\hat{\mu}$, $\hat{\Sigma}$ of the parameters, we can use the pdf of the multivariate Gaussian distribution (1) to score each business entity by plugging in the values of its features x_i into this pdf. Our basic assumptions on the nature of tax compliant companies leads to the following principle: the lower the value of $\phi\left(x_i; \hat{\mu}, \hat{\Sigma}\right)$, the further the ith business entity is located from the center μ and therefore the higher the chance of this entity to be evading VAT (see Fig. 4).

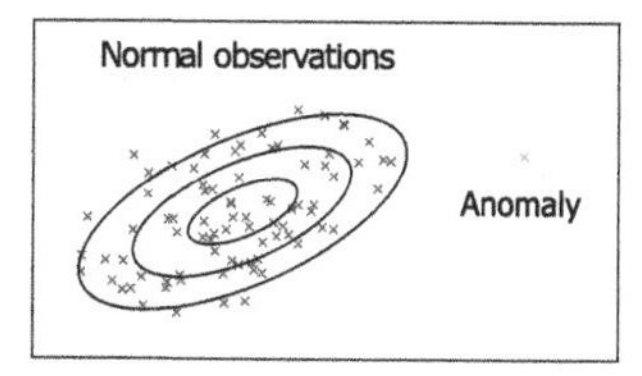

Fig. 4 Anomaly detection using multivariate Gaussian distribution

4.4 Testing on Synthetic Data

Before experimenting with the real-world data we decided to test our approach on synthetic data to make sure that it works when the basic assumptions from Sect. 4.1 are satisfied. For this purpose, a two-dimensional Gaussian mixture with ten components $f(x;\Omega) = \sum_{k=1}^{10} \tau_k \phi\left(x; \mu_k, \Sigma_k\right)$ was generated with the use of R package `MixSim` [11]. The maximum pairwise overlap was set equal to 0.05 and non-spherical covariance matrices of the general form were specified. The mixing proportions τ_k were defined to reflect the fact that in our application the cluster that represents law-abiding entities is normally the most populous, while clusters with various irregularities are much less prominent. Thus, we chose $\tau_1 = 0.80, \tau_2 = 0.03, \tau_3 = 0.04, \tau_4 = 0.01, \tau_5 = \tau_6 = \tau_7 = \tau_8 = \tau_9 = \tau_{10} = 0.02$.

The data set generated from this mixture included $n = 10000$ points. The structure of the data set can be seen in Fig. 5, where the cluster with the largest representation is located in the top left-hand corner and is shown in black.

The resulting SOM is shown in Fig. 6. The nodes representing different clusters in the solution are shown separated by the solid black lines. The 6×4 collection of

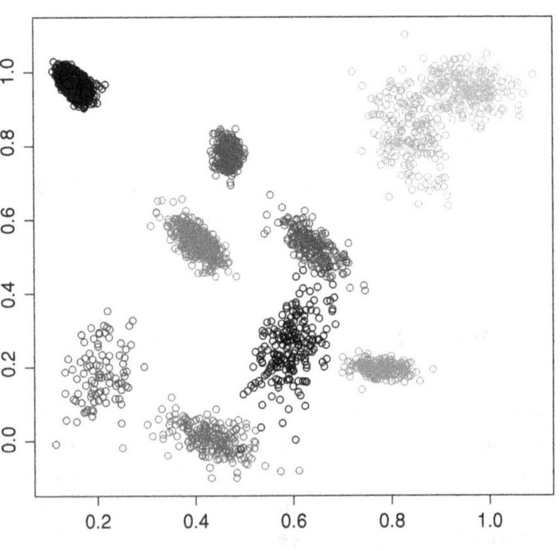
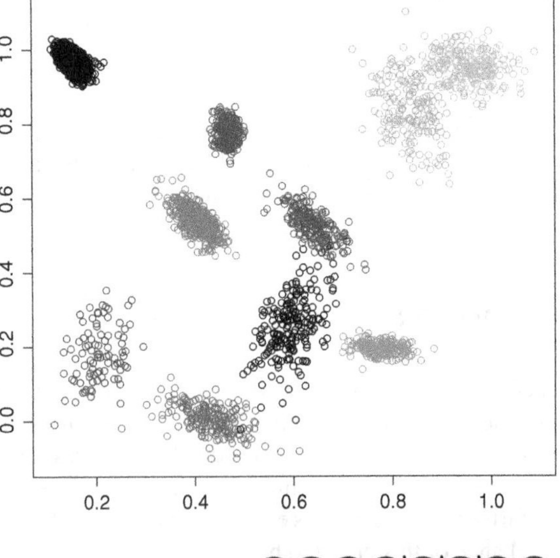

Fig. 5 Ten clusters modeled by a Gaussian mixture

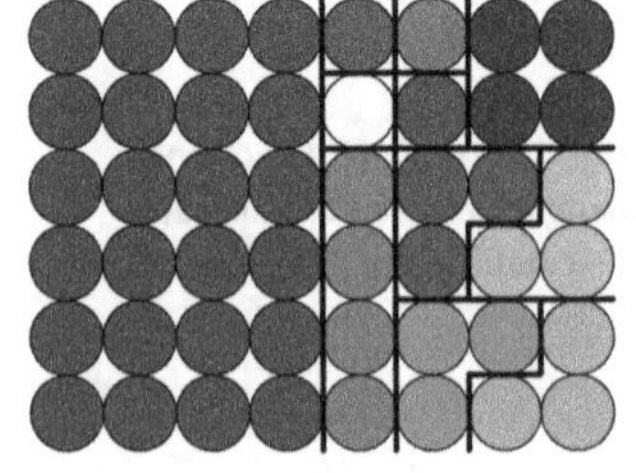

Fig. 6 SOM with clustering solutions shown

nodes on the left-hand side of the map corresponds to the dense cluster with the largest representation of points. It should be pointed out that this cluster consisted of 7970 points, all of which were correctly identified by the algorithm. The fact that there were no misclassifications recorded can be attributed to the small overlap with other clusters as well as the accuracy in the estimation of μ_k and Σ_k for this mixture component. The misclassifications that occurred among other clusters are not as important in our setting since we are more concerned with separating the group of compliant enterprises from the non-conforming entities than with identifying various groups among the entities that were non-compliant.

5 Experiments and Results

As described in Sect. 3.3, we collected the values of the 10 features, mentioned in Sect. 3.2, for 57,874 and 58,620 legal entities which were active in 2011 and in 2012 correspondingly, and which had annual incomes not less than one million Kazakhstani tenge. We applied Algorithm 1 to both 2011 and 2012 data sets. Self-organizing maps for both years were already mentioned in Sect. 3.4 (see Figs. 1 and 2). Within cluster sum of squares plots which were used to determine the optimal number of clusters in the "Elbow" method are given in Figs. 7 and 8.

Results of hierarchical clustering of SOM nodes are provided in Figs. 9 and 10. Clusters are separated by the solid black lines. The largest cluster in both cases was

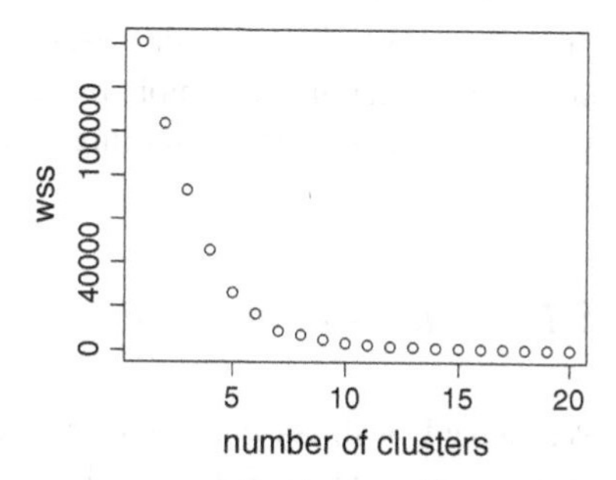

Fig. 7 Within cluster sum of squares for 2011 data

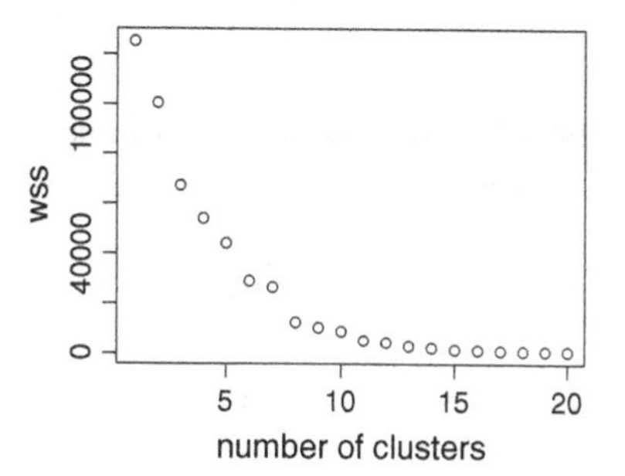

Fig. 8 Within cluster sum of squares for 2012 data

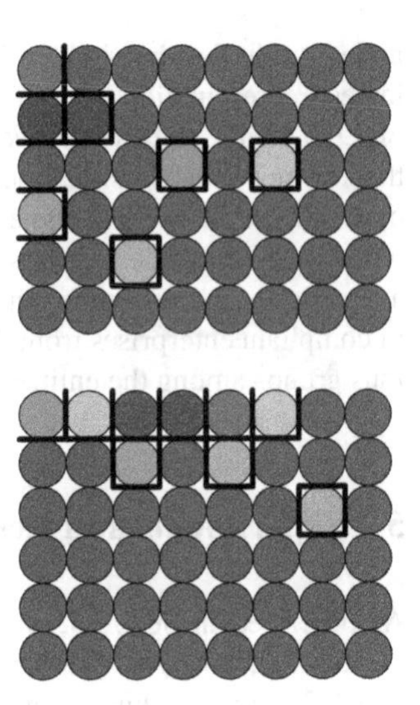

Fig. 9 Hierarchical clustering of SOM nodes for 2011 data

Fig. 10 Hierarchical clustering of SOM nodes for 2012 data

used to estimate the parameters of multivariate Gaussian pdf (1). This pdf is our scoring model which will be compared with the baseline scoring model currently used by the SRC.

From the set of entities which were active in 2011 (2012) SRC audited a selection of entities in 2012 (2013). The results of such audits are the amounts of additional VAT to be payed by audited entities. It is quite natural to consider the whole data set of all business entities as the training set, and the sample of audited entities as the test set. However, it is important to keep in mind that the audited companies do not represent a random sample from the population of all taxpayers, since they were selected using the current scoring model of the SRC.

5.1 Visual Comparison

We ranked business entities from the test sets using the baseline scoring model (which was used by SRC in 2012–13) and the new scoring model. Tables 1 and 2

Table 1 Scoring by the baseline model

Company ID	Score	Addit'l VAT	Cumul. VAT
1	120	600,000	600,000
2	120	900,000	1,500,000
3	120	400,000	1,900,000
⋮	⋮	⋮	⋮

Table 2 Scoring by the new model

Company ID	Score	Addit'l VAT	Cumul. VAT
2	0.001	900,000	900,000
1	0.002	600,000	1,500,000
4	0.005	500,000	2,000,000
⋮	⋮	⋮	⋮

give hypothetical examples of such rankings (due to confidentiality we cannot disclose the real data), and they also provide the cumulative amounts of VAT to be payed by the top business entities in each list.

The growth of the cumulative VAT for the 2012 and 2013 audit data is indicated in Figs. 11 and 12: the blue (lower) curve corresponds to the baseline scoring model used by the SRC, and the red (upper) curve corresponds to our new scoring model.

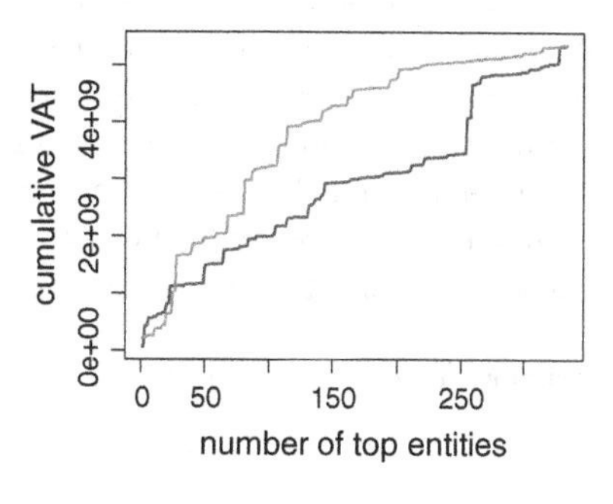

Fig. 11 Cumulative amounts of additional VAT to be payed by the entities audited in 2012

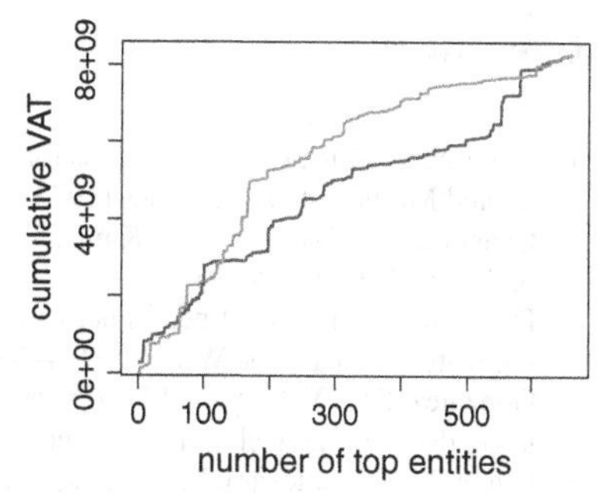

Fig. 12 Cumulative amounts of additional VAT to be payed by the entities audited in 2013

5.2 *Testing Correlations*

Using t-test we found out that the results of tax audit are more correlated with the scores given by the new model (−0.20, −0.15[3] for 2011/12 data) than with the scores given by the baseline model (0.07, 0.08 for 2011/2012 data) at 0.01 significance level.

6 Discussion and Future Work

The experiments have shown that our model outperforms the baseline model which was previously used by the SRC: given the list of companies to be audited our model ranks them better than the baseline model.

However, one needs to keep in mind that the suggested approach was tested on specially selected set of business entities: the baseline scoring model was applied to the whole population of business entities and then the companies with highest scores were selected for the audit. If our new model is applied to the whole population it will generate a different list of highest-score entities. In order to make a fair comparison between two models, one needs to audit companies from both lists and check whether the new model indeed provides scores which correspond better to the audit results (sums of additional VAT to be payed by the audited companies).

In the future, we are planning to apply our approach to other types of taxes (e.g. corporate income tax), as well as to other types of state revenues (e.g. customs dues).

Acknowledgments We would like to thank Inês Russinho Mouga for the thorough review of [9].

References

1. Anderberg, M.R.: Cluster Analysis for Applications. Monographs and Textbooks on Probability and Mathematical Statistics (1973)
2. Chandola, V., Banerjee, A., Kumar, V.: Anomaly detection: a survey. ACM Comput. Surv. (CSUR) **41**(3), 15 (2009)
3. Dolnicar, S.: The use of neural networks in marketing: market segmentation with self organising feature maps. Proc WSOM **97**, 4–6 (1997)
4. González, P.C., Velásquez, J.D.: Characterization and detection of taxpayers with false invoices using data mining techniques. Expert Syst. Appl. **40**(5), 1427–1436 (2013)
5. Gupta, M., Nagadevara, V.: Audit selection strategy for improving tax compliance: application of data mining techniques. In: Foundations of Risk-Based Audits. Proceedings of the eleventh International Conference on e-Governance, Hyderabad, India, December, pp. 28–30 (2007)
6. Hsu, K.W., Pathak, N., Srivastava, J., Tschida, G., Bjorklund, E.: Data mining based tax audit selection: a case study of a pilot project at the minnesota department of revenue. In: Real World Data Mining Applications, pp. 221–245. Springer (2015)

---

[3]Correlations are negative since the new model assigns *lower* scores to the companies which are *more likely* to evade VAT.

7. Iivarinen, J., Kohonen, T., Kangas, J., Kaski, S.: Visualizing the clusters on the self-organizing map (1994)
8. Kohonen, T.: The self-organizing map. Neurocomputing **21**(1), 1–6 (1998)
9. Lückeheide, S., Velásquez, J.D., Cerda, L.: Segmentación de los contribuyentes que declaran iva aplicando herramientas de clustering. Revista de Ingeniería de Sistemas **21**, 87–110 (2007)
10. Markey, M.K., Lo, J.Y., Tourassi, G.D., Floyd, C.E.: Self-organizing map for cluster analysis of a breast cancer database. Artif. Intell. Med. **27**(2), 113–127 (2003)
11. Melnykov, V., Chen, W.C., Maitra, R.: MixSim: an R package for simulating data to study performance of clustering algorithms. J. Stat. Softw. **51**, 1–25 (2012)
12. Pampalk, E., Rauber, A., Merkl, D.: Using Smoothed Data Histograms for Cluster Visualization in Self-organizing Maps. Springer (2002)
13. Squire, D.M., et al.: Visualization of Cluster Changes by Comparing Self-organizing Maps. Springer (2005)
14. Vesanto, J., Alhoniemi, E.: Clustering of the self-organizing map. IEEE Trans. Neural Netw. **11**(3), 586–600 (2000)
15. Viveros, M.S., Nearhos, J.P., Rothman, M.J.: Applying data mining techniques to a health insurance information system. VLDB 286–294 (1996)
16. Wehrens, R., Buydens, L.M., et al.: Self-and super-organizing maps in R: the kohonen package. J. Stat. Softw. **21**(5), 1–19 (2007)
17. Williams, G.J., Christen, P., et al.: Exploratory multilevel hot spot analysis: Australian taxation office case study. In: Proceedings of the Sixth Australasian conference on Data mining and analytics, vol. 70, pp. 77–84. Australian Computer Society, Inc. (2007)

On Classification of Linguistic Data—Case Study: Post-operative

Kalle Saastamoinen

Abstract This article presents simple yet efficient way to classify Post-operative patient data. The classification task of this database is to determine where patients in a postoperative recovery area should be sent to next. Because hypothermia is a significant concern after surgery, the attributes correspond roughly to the body temperature measurements. What makes classification task difficult here is that the most of the attributes are given by linguistic values. Method proposed in this article starts by representing linguistic variables by suitable numbers, which are later normalized into the values between 0 and 1. Next phase this data is classified using simple similarity classifier. Results are compared to the existing results and method presented in this paper provides mean accuracy of 65.23 % whereas second highest reported result is 62.67 % using similarity classifier with PCA and membership functions.

Keywords Linguistic data · Post-operative · Classification

1 Introduction

This article concerns two important issues in data mining that are comparison and handling of linguistic data. Comparison is very important task in all areas where decisions has to be done. The fields of problem solving, categorization, data mining, classification, memory retrieval, inductive reasoning, and generally cognitive

I feel gratitude to the National Defence University which have given me plenty of time to do my research.

K. Saastamoinen (✉)
Department of Military Technology, National Defence University,
P.O. Box 7, FI-00861 Helsinki, Finland
e-mail: kalle.saastamoinen@mil.fi
URL: http://www.puolustusvoimat.fi/en/

I. Czarnowski et al. (eds.), *Intelligent Decision Technologies 2016*,
Smart Innovation, Systems and Technologies 56,
DOI 10.1007/978-3-319-39630-9_5

processes require that we understand how to assess the sameness. Measures used for comparison can in general have many different forms depending the purpose of their utilization [1]. In this article it is used a simple similarity measure that originates from many valued logical structures. Similarity measures fulfill the metric axioms by their definition. Fuzzy logic has traditionally been used to transform linguistic data into the numerical values that are easy to handle with some computing system like for example Matlab. This transformation is usually carried out by using membership functions, which the most common ones are trapezoidal, triangular and Gaussian, where their names refers to their shapes. Idea of using membership functions is to represent data in a set where data points takes some membership degrees which again represents in what degree these points belongs in the sets which are characteristic for the problem in hand. Commonly if one uses membership functions one also need to use some defuzzification method in order to produce a quantifiable results. Interesting approach to classify Post-operative data using this kind of method was presented in [2]. Here it is presented simple approach, which does not need membership functions nor complicated defuzzification at to classify Post-operative data set and it proves to give better results than any method has been able to show before. The core of used method is similarity measure and likely the simplest meaningful transformation of Post-operative linguistic data into the numeric values that corresponds intuitively reality. Motivation for this article is to present simple and effective way to handle Post-operative type of linguistic data.

Article is organized as follows. In the first section Post operative data and of how this linguistics data has been handled is presented. The second section gives a little theory behind comparison measure used. The third section presents classification schemata. The fourth section presents results achieved and these results are compared versus to the previous results achieved. In the fifth section conclusions are done and some future directions are given.

2 Post-operative Data

Task of this data is to determine where patients in a postoperative recovery area should be sent to next. The attributes correspond roughly to the body temperature measurements. The number of Instances is 90. The number of attributes is 9 including the decision (class attribute). Attribute 8 has 3 missing values [3]. So each patient is defined by 8 attribute values, from which one should be able to make a decision which of 3 recovery areas (marked as 3 different classes) patient should be sent.

What makes this data challenging is that most of the attribute values are given by linguistic labels. Three attributes that gets values low, mid and high are patient's internal temperature (named as L-CORE), surface temperature (named as L-SURF) and last measurement of blood pressure (named as L-BP). Patient's oxygen saturation (named as L-O2) gets values excellent, good, fair and poor. Three attributes that gets values stable, mod-stable and unstable are stability of patient's surface temperature (named as SURF-STBL), stability of patient's core temperature (named as CORE-STBL) and stability of patient's blood pressure (named as BP-STBL). Attribute that describes patient's perceived comfort at discharge (named as COMFORT) is measured as an integer between 0 and 20. Last attribute describes discharge decision (named as decision ADM-DECS) it can get 3 values that are I (patient sent to Intensive Care Unit), S (patient prepared to go home) and A (patient sent to general hospital floor).

Comfort attribute has 3 missing values, which increases uncertainty in Post-operative. Furthermore there is no information what values experts used exactly to set up linguistic labels for the attribute values. It is only given some limits like for L-CORE high (>37), mid (≥36 and ≤37), low (<36). However no data is given what exactly patient's internal temperature was, when it is labeled as example to be Mid. For this reason these limits are useless. In fact it only creates more uncertainty if one starts to guess these values as it is done in case one starts to use some membership functions as is done in [2]. Below it is shown an example of Post-operative data concerning first patient:

mid, low, excellent, mid, stable, stable, stable, 15, A

Next 89 data cases are similar way stated. Now it is time to show how data was transformed to become computationally effective.

2.1 *Interpretation of Data*

Keeping in mind that one does not know how exactly experts have made decisions in 7 attributes out from 9, it seems reasonable to give all vague statements (low, mid, high, poor, fair, good, excellent, unstable, mod-stable, stable) only one value each, since there is no way one can know better without having original data in hand and preferable also possibility to consult some experts.

Linguistic data was first coded as follows: low = 1, mid = 3, high = 1, poor = 1, fair = 2, good = 3, excellent = 4, unstable = 1, mod-stable = 2, stable = 3. Then attribute called COMFORT was also re-scaled for the values from 1 to 4 so that 1 to 5 = 1, 6 to 10 = 2, 11 to 15 = 3 and 16 to 20 = 4. Finally classes were given numerical values. Used classification algorithm normalizes the data in between [0, 1]. This means that finally data is in the form where 0 means total absence of the membership in a set and 1 means full membership in a set. Now membership values symbolizes entities goodness to the patient e.g. mid gets relatively high value since it is good that temperature or blood pressure is not too low or too high.

3 Combined Lukasiewicz and Shweizer and Sklar Based Comparison Measure

The logical operation of a many-valued equivalence is commonly used when the comparison of two fuzzy propositions $a, b \in [0, 1]$ are required. Equivalence can then be interpreted to define the valuation of the two-way conditional proposition "a if and only if b". For this reason, it is naturally suitable for the comparison of different objects.

Jan Łukasiewicz used only implication and negation, when he studied many-valued logic. However Frink [4] used the term "Łukasiewicz arithmetical conjunction" from the $T(a, b) = \max\{0, a + b - 1\}$ for the first time this is also known as bounded difference. Nowadays, it is common to call bounded difference as Łukasiewicz conjunction or Łukasiewicz t-norm and bounded sum $S(a, b) = \min\{1, a + b\}$ as Łukasiewicz disjunction or Łukasiewicz t-conorm see, for example, Hájek [5], Klir [6], Kundu [7] etc.

Articles [8, 9] studies the use of the Łukasiewicz type of equivalence, with means and weights. In the article [10] this study was taken further by the use of a generalized mean and weights.

In this article it is used the comparison measure which rise from the functional definition for the implications given in [11]. It is noted that pseudo Łukasiewicz type 2 [6] and Shweizer and Sklar type 1 [6, 12] implications form the same equivalence measure when these equivalences are formed by using fuzzy conjunction to combine corresponding implications. Łukasiewicz equivalence is included to Shweizer and Sklar by taking the parameter values which go from negative side to the positive, so $p \in]-\infty, \infty[$.

The following demonstrates how to use similarity in order to find similar pairs. Here a chosen situation is examined where features of different objects can be expressed in values between [0,1]. Let X be the set of m objects. If the similarity value of the features are known $f_1, \ldots, f_n$ between objects, the object can be chosen that has the highest total similarity value. The problem is to find for object x_i a similar object x_j, where $1 \leq i, j \leq m$ and $i \neq j$. By choosing for example Łukasiewicz-structure for features of the objects n similarities are achieved for comparing the two objects (x_1, x_2)

$$S_{f_i}\langle x_1, x_2\rangle = E(x_1(f_i), x_2(f_i)), \tag{1}$$

where $x_1, x_2 \in X$ and $i \in \{1, \ldots, n\}$. Because Łukasiewicz-structure is chosen for the membership of objects, the similarity can be defined as follows

$$S\langle x_1, x_2\rangle = \frac{1}{n} \sum_{i=1}^{n} E(x_1(f_i), x_2(f_i)). \tag{2}$$

Different non-zero weights $(W_1, \ldots, W_n)$ can also be given to the different features in order to obtain the following formula, which again meets the definition of the similarity.

$$S\langle x_1, x_2\rangle = \frac{\sum_{i=1}^{n} W_i E(x_1(f_i), x_2(f_i))}{\sum_{i=1}^{n} W_i}. \tag{3}$$

In the ordinary Łukasiewicz-structure equivalence relation $E(x, y)$ as well as similarity $S(x, y)$ is defined as

$$E(x, y) = 1 - |x - y| = S(x, y). \tag{4}$$

In the case of the so called generalized Łukasiewicz-structure [6], the equivalence relation (or similarity in case of Łukasiewicz) is more complicated, i.e.

$$E(x, y) = (1 - |x^p - y^p|)^{\frac{1}{p}} = S(x, y). \tag{5}$$

This similarity has a clear connection with Minkowsky-metrics.

Lemma 1 *Consider n Łukasiewicz valued fuzzy similarities S_i, $i = 1, \ldots, n$ on a set X. Then …*

$$S\langle x, y\rangle = \frac{1}{n}\sum_{i=1}^{n} S_i\langle x, y\rangle$$

is a Łukasiewicz valued similarity on X.

Proof As all S_i, $i = 1, \ldots, n$ are reflexive and symmetric consequently S is also. The transitivity of S can be seen from the fact that arithmetic mean is monotonic. Look for example Carbonell et al. [13] or a bit different proof Turunen [14].

Since generalized mean is a monotonically increasing aggregation operator (so it preserves the order) the following result can be concluded:

Corollary 1 *Consider n Łukasiewicz valued fuzzy similarities S_i, $i = 1, \ldots, n$ on a set X. Then*

$$S\langle x, y\rangle = \left[\frac{1}{n}\sum_{i=1}^{n}\left(S_i\langle x, y\rangle\right)^m\right]^{\frac{1}{m}}$$

is a Łukasiewicz valued similarity on X.

So in this article used comparison measure takes the following form:

Definition 1 Comparison measure based on Shweizer and Sklar–Łukasiewicz:

$$E_{SSL}\left(f_1(i), f_2(i)\right) = \left(\sum_{i=1}^{n} w_i\left(1 - \left|f_1^p(i) - f_2^p(i)\right|\right)^{\frac{m}{p}}\right)^{\frac{1}{m}} \tag{6}$$

4 Classification

Many time there are given a set of data which is already grouped into classes and the problem is then to predict which class each new data belongs to. This is normally referred to as classification problem. First set of data is referred to as training set, while this new set of data is referred to as test set [15]. Classification is seen as comparison between training set and test set.

- **Post Operative**: Task of this database is to determine where patients in a post-operative recovery area should be sent to next. The attributes correspond roughly to body temperature measurements. The number of Instances is 90. The number of attributes is 9 including the decision (class attribute). Attribute 8 has 3 missing values. These 3 rows which included missing values has in this study been deleted.

4.1 Description of the ŁUkasiewicz Similarity Based Classifiers

Objects, each characterized by one feature vector in $[0, 1]^n$, is classified into different classes. The assumption that the vectors belong to $[0, 1]^n$ is not restrictive since the appropriate shift and normalization can be done for any space $[a, b]^n$. The comparison measures can be used to compare objects to classes. Below is the used classifier in the algorithmic form:

Similarity Based Classifier

Require: *data*
  scale *data* between [0, 1]
Require: *test*,*learn*[1...*n*],*weights*,*dim*
  for $i = 1$ to n **do**
    $idealvec[i] = IDEAL[learn[i]]$

$$maxcomp[i] = \left(\frac{1}{\dim}\right)^{1/m} \left(\sum_{j=1}^{\dim} weights\,[j]\,\left(SIM\left(idealvec\,[i,j]\,, test\,[j]\right)\right)^m\right)^{1/m}$$

  end for
  $class = \arg\max_i maxcomp[i]$

In the algorithm, the Łukasiewicz similarity (SIM) with a generalized mean is used. *IDEAL* is the vector that best characterizes the class i and here the generalized mean vector of the class as an *IDEAL*-operator has been used.

Evolutionary algorithm is used because of its diversity and robustness to find weights in classification process, information about evolutionary algorithms in general can be found for example from [16–19]. Obviously, other optimizers can be used as well. Evolutionary algorithm used here is based on differential evolution [20]. DE is a simple population based stochastic function minimizer. The objective of DE is to iterate each member of the population and compare its value to the trial member value, and the superior member stays for the next iteration. The evolution strategy defines the way in which a trial member is generated. DE tries to seek weights that

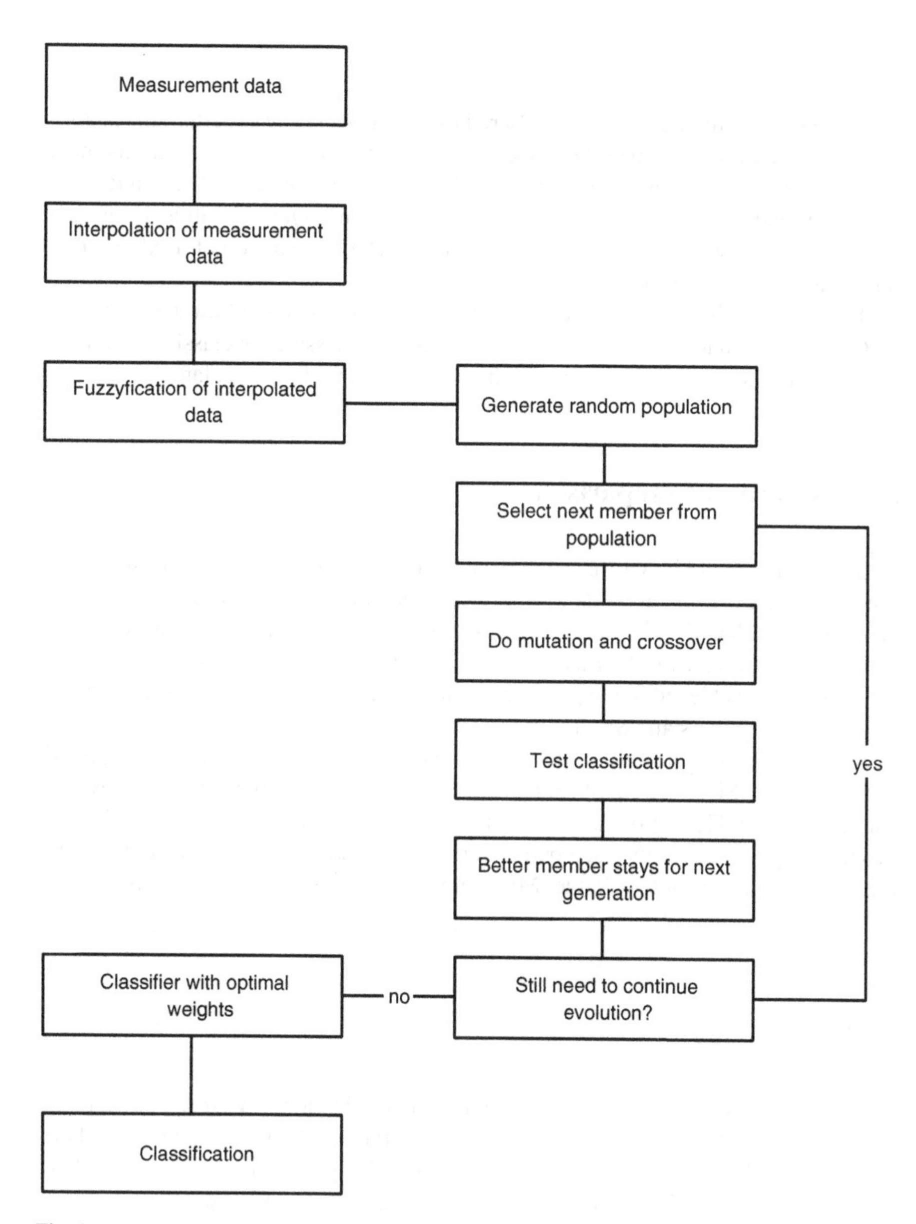

Fig. 1 Simplified computational model for DE

Table 1 Maximum and mean classification results (SSL) with the comparison of the previous best results (PFS)

Comparison measure	Max (%)	Mean (%)	p_{max}	p_{mean}	m
SSL	77.27	65.23	3.74	3.98	6.33
PFS	71.11	62.67	No	No	No

will give the maximal similarity compared to the values set by experts. This is done so that DE tries to minimize the value of the objective function with trial member values. The objective function is the total difference between classification defined by experts and the classification defined by similarity used here for all learning data sets. Finally, DE gives the optimal weight values. The basic action of used differential evolution is demonstrated in Fig. 1.

The classification task has been described more clearly in the flowchart (Fig. 2).

One can see that with respect to L, the number of classes, the classification time is $\mathcal{O}(L)$ after the parameters have been fixed and ideal vectors calculated.

5 Results and Comparison

In this section it is shown that classification results achieved using the schemata presented for Post-operative data set are considerable better than results achieved before in [2, 21] or [22]. In the Table 1 it is presented results that were achieved using similarity classifier with the Łukasiewicz equivalence marked as SSL (6) versus to the best results achieved in the previous studies in [2, 21] or [22] marked as PFS (as PCA, Fuzzy numbers and Similarity).

One can see from the Table 1 that the results achieved using the method presented in this article (SSL) were significantly better than the results achieved in the previous studies (PFS) in [2, 21] or [22]. Mean results in this study were reached doing 10 evolutionary rounds for the each parameter values p and m tested. In the Table 1 NO means that data was not available. Maximum variance using this method presented was 0.02.

6 Conclusions

In this paper it was proposed a method for the classification of Post-operative data. Method managed to give better results than any other method reported before. This method started by representing linguistic variables by suitable numbers, which then later were normalized into the values between 0 and 1. Next phase this data were classified using simple similarity classifier.

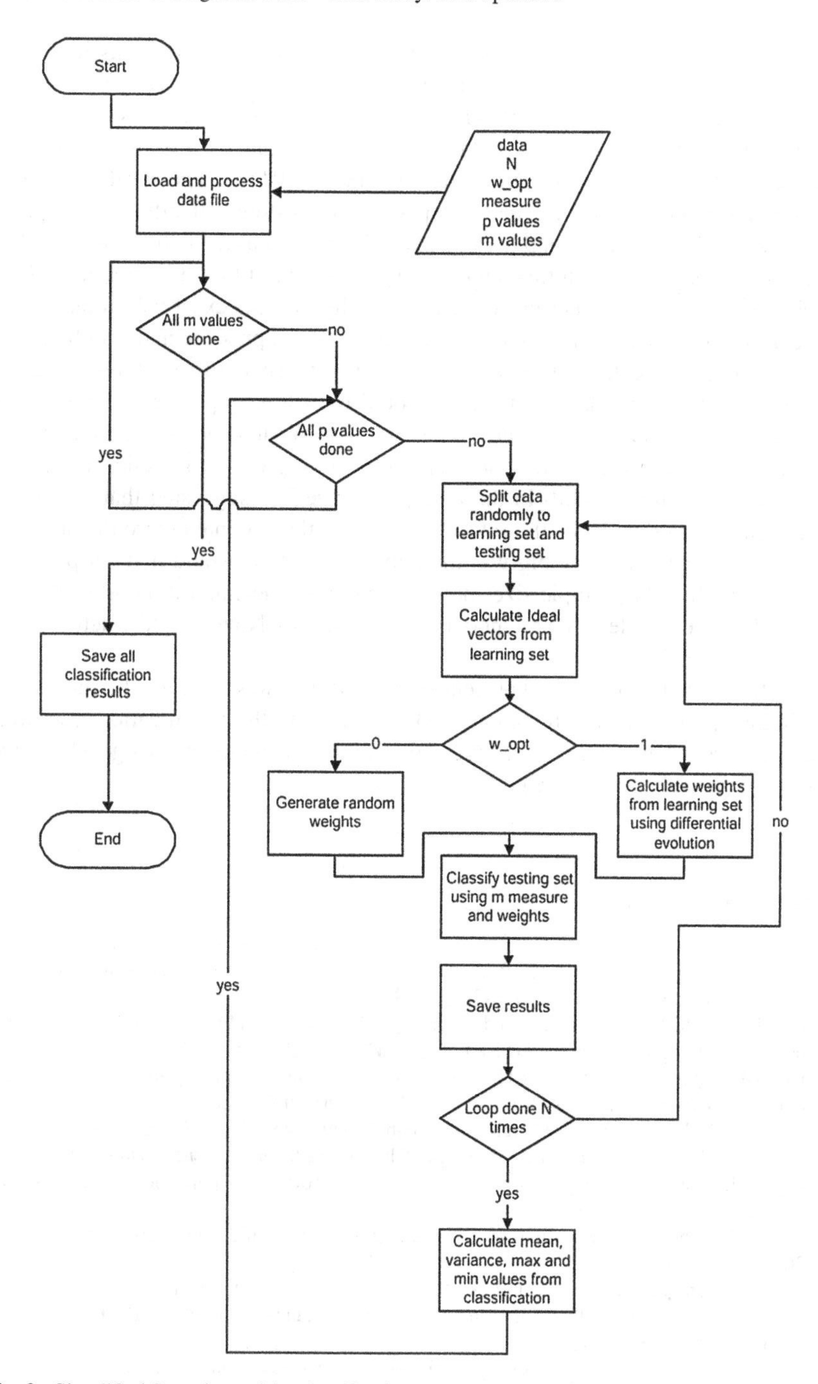

Fig. 2 Simplified flow chart of the classification procedure

Previous best average result reported was 62.67 % using similarity classifier with PCA and membership functions. In this paper it is reported best average result of 65.23 %. Likely the main reason that results in [2] are worse is that it used membership functions and Post-operative data set does not give any basis to use them.

Results were achieved using quite long step-size of 0.01 for p and m values. Likely results will get better using the smaller step-size. Also testing with different comparison measures than Shweizer and Sklar–Łukasiewicz (6) can lead to the better results.

There are many issues in this data which makes it hard to use in practice. Main problem is that it is not precisely written originally. It lacks exact values measured in the hospital for temperature, oxygen saturation, blood pressure and stability. It is necessary to get this data in order to make Post-operative classifier useful for medical practitioners. With this data in hand one could use for example fuzzy numbers in classification and reach more accurate classification. What comes to the data sample size of 90 this size is normally considered enough big that arithmetic means can be considered as meaningful. In the book [23] page 2 is suggested that 30 would be enough. Given data lacks the information about the original environment so no estimation can be done how big was the population which would make it possible to tell exactly how big sample size should be and how meaningful these results are statistically for example using method mentioned in [24]. Now sample is actually the whole data given in [3].

As a final conclusion method presented in this article was proven to be successful and similar approach can be recommended to be used for the classification cases like Post-operative, where one only has linguistic variables without knowledge what kind of measurements has been behind them.

References

1. Bouchon-Meunier, B., Rifqi, M.: Bothorel, Sylvie: Towards general measures of comparison of objects. Fuzzy Sets Syst. **84**(2), 143–153 (1996)
2. Luukka, P.: PCA for fuzzy data and similarity classifier in building recognition system for post-operative patient data. Expert Syst. Appl. **362**(1), 1222–1228 (2009)
3. UCI ML Rep., various authors.: UCI Repository of Machine Learning Databases network document. http://www.ics.uci.edu/126mlearn/MLRepository.html. Accessed 14 March 2004
4. Frink Jr., O.: New Algebras of Logic. Am. Math. Monthly **45**, 210–219 (1938)
5. Hájek, P.: Metamathematics of Fuzzy Logic. Kluwer Academic Publishers, Dordrecht (1998)
6. Klir, G.J., Yuan, B.: Fuzzy Sets and Fuzzy Logic: Theory and Applications, pp. 304–321. Prentice Hall, PTR (1995)
7. Kundu, S., Chen, J.: Fuzzy logic or Lukasiewicz logic: a clarification. Fuzzy Sets Syst. **95**, 369–379 (1998)
8. Saastamoinen K., Könönen V., Luukka P.: A classifier based on the fuzzy similarity in the Łukasiewicz-structure with different metrics. In: Proceedings of the FUZZ-IEEE 2002 Conference, Hawaii USA (2002)
9. Saastamoinen, K., Luukka, P.: Testing continuous t-norm called Łukasiewicz algebra with different means in classification. In: Proceedings of the FUZZ-IEEE 2003 Conference, St Louis, USA

10. Saastamoinen, K.: Semantic study of the use of parameterized s-implications and equivalences in comparison. In: Proceedings of the IEEE 2004 Conference on Cybernetic and Intelligent Systems, Singapore
11. Smets, P., Magrez, P.: Implication in fuzzy logic. Int. J. Approx. Reason. **1**(4), 327–347 (1987)
12. Whalen, T.: Parameterized R-implications. Fuzzy Sets Syst. **134**(2), 231–281 (2003)
13. Carbonell M., Mas M., Mayor G.: On a class of monotonic extended OWA operators. In: Proceedings of the 6th IEEE International Conference on Fuzzy Systems, vol. 3, pp. 1695–1700 (1997)
14. Turunen, E.: Mathematics Behind Fuzzy Logic. Advances in Soft Computing. Physica-Verlag, Heidelberg (1999)
15. Hastie, T., Tibshirani, R.: The Elements of Statistical Learning: Data Mining, Inference, and Prediction. Springer Series in Statistics. Springer, New York (2001)
16. Goldberg, D.E.: Real-coded genetic algorithms, virtual alphabets, and blocking. Technical Report 9001. University of Illinois at Urbana-Champain (1990)
17. Mantel, B., Periaux, J., Sefrioui M.: Gradient and genetic optimizers for aerodynamic desing. In: ICIAM 95 Conference, Hamburgh (1995)
18. Michalewics, Z.: Genetic Algorithms + Data Structures = Evolution Programs Artificial Intelligence. Springer, New York (1992)
19. Grefenstette, J.J.: Optimization of control parameters for genetic algorithms. IEEE Trans. Syst. Man Cybern. **16**(1), 122–128 (1986)
20. Price, K.V., Storn, R.M., Lampinen, J.A.: Differential Evolution—A Practical Approach to Global Optimization. Natural Computing Series. Springer, Berlin (2005)
21. Auephanwiriyakul, S., Theera-Umpon, N.: Comparison of linguistic and regular hard C-means in postoperative patient data. J. Adv. Comput. Intell. Intell. Inf. **8**(6), 599–605 (2004)
22. Woolery, L., Grzymala-Busse, J., Summers, S., Budihardjo, A.: The use of machine learning program LERS-LB 2.5 in knowledge acquisition for expert system development in nursing. Comput. Nursing **9**(6), 227–234 (1991)
23. Corder, G.W., Foreman, D.I.: Nonparametric statistics for non-statisticians: a step-by-step approach. Wiley (2009)
24. Krejcie, R.V., Morgan, D.W.: Determining sample size for research activities. Educ. Psychol. Meas. **30**(3), 607–610 (1970)

Solving Technician and Task Scheduling Problems with an Intelligent Decision Heuristic

Amy Khalfay, Alan Crispin and Keeley Crockett

Abstract This paper proposes a new approach, an intelligent decision (ID) heuristic, to solve a technician and task scheduling problem (TTSP) defined by the ROADEF 2007 challenge. The ID heuristic is unlike other approaches because at each stage the heuristic considers multiple scenarios of team configurations and job assignments. Within the ID heuristic, novel operators have been designed which focus on flexibility in team configurations. Furthermore, outsourcing is a sub-problem of the ROADEF 2007 challenge, so computational experiments have been performed to evaluate various strategies of outsourcing to utilize the ID heuristic. Results obtained using the ID heuristic have been compared against other researchers who have tackled this problem.

Keywords Technician and task scheduling problem (TTSP) · Intelligent decision (ID) heuristic and outsourcing

1 Introduction

Technician and task scheduling problems are present in many sectors such as telecommunications, health care and public utilities [1]. The difficulty in finding quality solutions to these constrained problems is recognized in [2] and highlights the benefits of optimized staff scheduling such as; customer and workforce

A. Khalfay (✉) · A. Crispin · K. Crockett
School of Computing, Mathematics and Digital Technology,
Manchester Metropolitan University, Manchester M1 5GD, UK
e-mail: a.khalfay@mmu.ac.uk; 11045871@stu.mmu.ac.uk

A. Crispin
e-mail: a.crispin@mmu.ac.uk

K. Crockett
e-mail: k.crockett@mmu.ac.uk

I. Czarnowski et al. (eds.), *Intelligent Decision Technologies 2016*,
Smart Innovation, Systems and Technologies 56,
DOI 10.1007/978-3-319-39630-9_6

satisfaction and economic savings. The ROADEF 2007 challenge, organized by the French Operational Research Society, encouraged researchers to find efficient ways of solving France Telecom's optimization problem [3]. France Telecom wished to protect their market share and maintain a high level of customer service whilst limiting the growth of their workforce [4]. Characteristics featured in the ROADEF 2007 challenge problem are still applicable to the problems faced today in many organisations [5].

The ROADEF 2007 challenge problem can be summarized as follows: each job has domain skill requirements that need a team to be built in order to satisfy the demand. A domain may be a particular area of expertise i.e. electrician and a skill level will represent the proficiency within that domain. Teams are made up of technicians who have intrinsic skill domain levels and days where they are unavailable. In addition, there are also dependency relationships between jobs, prohibiting some jobs being started until others have been completed. Jobs have a priority level, representing how important it is to serve the job as early as possible. Jobs have a completion time and must be started and finished on the same day. Furthermore, in some instances, there is an outsourcing budget available (outsourced jobs do not contribute to the objective function). The ROADEF 2007 challenge comprised of three sets of data; Set A, Set B and Set X, increasing in complexity. Set A instances range from 5–100 jobs, include precedence relationships but no outsourcing budget. Set B and Set X instances range from 110–800 and 100–800 jobs respectively, include precedence relationships and outsourcing budgets.

The ROADEF 2007 challenge attracted a lot of research attention and a number of solution approaches such as; adaptive large neighbourhood search [6], mixed integer programming [7, 8], local search heuristics [9, 10], greedy algorithms [11–13], and greedy randomized adaptive search algorithms [14] have been proposed. All of these approaches use procedures we would classify as single scenario (SS) heuristics. The approaches select a single seed job according to some criteria, and create a team able to service the seed job. The ID heuristic is a new approach, which at each stage evaluates many team and job configuration scenarios. The heuristic selects multiple seed jobs and creates a team able to service each job if possible. For each constructed team, the heuristic then checks which jobs could also be allocated to the team. A utility score is calculated for each scenario, which represents the quality of the job assignments to the team. The highest scoring scenario is then selected, the team is configured and job assignments are made.

In other work on the ROADEF 2007 challenge, team configurations have been rigid and difficult to alter. Three operators presented in this paper were developed to provide flexibility in team configurations using the ID heuristic. The first operator, move with team build, allows a single job to be moved onto a different day by creating a team to service the job. The second operator, decompose and rebuild, allows a single day to be rebuilt within the scheduling horizon. The last operator, decompose and rebuild N, allows N days in the scheduling horizon to be rebuilt. These operators allow distant solutions to be evaluated as they have the ability to change not only team configurations but also the allocation position of a job.

Experiments into strategies for outsourcing have previously not been carried out in the evaluation of other heuristics to determine whether the quality of the solution produced is dependent on the choice of outsourcing strategy.

The remainder of this paper is organized as follows. Section 2 presents the formulation of the problem. Section 3 describes the ID heuristic. Section 4 presents the outsourcing strategy testing and results. Section 5 outlines the local operators used within the ID heuristic and Sect. 6 shows the results obtained by the ID heuristic against other researchers work. Lastly, Sect. 7 draws conclusions about the performance of the ID heuristic and directions for further work are identified.

2 Problem Formulation

The aim of the ROADEF 2007 challenge problem is to construct teams over a scheduling horizon in order to service a set of jobs. A set of N jobs must be completed. Each job i has certain properties, a priority level p where $p = \{1 \ldots 4\}$, an execution time d_i, a domain skill requirement matrix $s^i_{\delta\alpha}$, an outsourcing cost c_i and a set of successor jobs σ_i. All jobs belonging to σ_i may not begin until job i has been completed. There are a set of technicians T, each technician t also has attributes; unavailable days and domain skill levels $v^t_{\delta\alpha}$. There are $D = \{1 \ldots \delta\}$ domains and $A = \{1 \ldots \alpha\}$ skill levels within each domain. The scheduling horizon $K = \{1 \ldots k\}$ represents an entire solution and each k represents a working day/schedule. Each day is limited to 120 time units with no overtime allowed. Each day k has a set of available technicians T_k, who make up teams $M = \{1 \ldots m\}$ and each team m will have job assignments and must stay together for the day. Let $x_{t,k,m} = 1$ if technician t belongs to team m on day k. Let $y_{i,k,m} = 1$ if job i is assigned to team m on day k. The start times of jobs are denoted as b_i. Let $u_{i,i'} = 1$ if jobs i and i' are assigned to the same team on the same day and i' begins after i is completed.

In the Set B and X instances there is an outsourcing budget available, C. Outsourced jobs must adhere to precedence constraints, so if a job is outsourced then so are all successor tasks. Let $z_i = 1$ if job i is outsourced. The objective function of a solution is calculated using Eq. (1), a weighted sum of the latest ending times of each priority type. The weightings in the objective function are given by $w_p = \{28, 14, 4, 1\}$ for $p = \{1 \ldots 4\}$ and e_p is the ending time of the latest job in the scheduling horizon of priority p. However, e_4 is the ending time of the latest job in the scheduling horizon over all priority types.

$$\sum_{p=1}^{4} w_p * e_p \tag{1}$$

Subject to the following constraints;

$$e_p \geq b_i + d_i \quad \forall p \in \{1, 2, 3\}, i \in N_p \tag{2}$$

$$e_4 \geq b_i + d_i \quad \forall\, i \in N \tag{3}$$

$$\sum_{m \in M} x_{t,k,m} \leq 1 \quad \forall k \in K,\ t \in T_k \tag{4}$$

$$\sum_{m \in M} x_{t,k,m} = 0 \quad \forall\, k \in K,\ t \in T \backslash T_k \tag{5}$$

$$z_i + \sum_{k \in K} \sum_{m \in M} y_{i,k,m} = 1 \quad \forall\, i \in N \tag{6}$$

$$y_{i,k,m} * s^i_{\delta\alpha} \leq \sum_{t \in T_k} v^t_{\delta\alpha} * x_{t,k,m} \quad \forall i \in N, k \in K, m \in M, \alpha \in A,\ \delta \in D \tag{7}$$

$$b_i + d_i \leq b_{i'} \quad \forall\, i \in N,\ i' \in \sigma_i \tag{8}$$

$$120(k-1) * \sum_{m \in M} y_{i,k,m} \leq b_i \quad \forall\, i \in N,\ k \in K \tag{9}$$

$$120(k) * \sum_{m \in M} y_{i,k,m} \geq b_i + d_i \quad \forall\, i \in N,\ k \in K \tag{10}$$

$$b_i + d_i - \left(1 - u_{i,i'}\right) M \leq b_{i'} \quad \forall\, i, i' \in N,\ i \neq i' \tag{11}$$

$$y_{i,k,m} + y_{i',k,m} - u_{i,i'} - \leq 1 \quad \forall\, i, i' \in N, i \neq i', k \in K, m \in M \tag{12}$$

$$\sum z_i * c_i \leq C \quad \forall\, i \in O \tag{13}$$

$$|\sigma_i| * z_i \leq \sum_{i' \in \sigma_i} z_{i'} \quad \forall i \in \mathrm{N} \tag{14}$$

with variables;

$$x_{t,k,m} = \{0, 1\} \quad \forall k \in K, m \in M,\ t \in T \tag{15}$$

$$y_{i,k,m} = \{0, 1\} \quad \forall i \in N, k \in K, m \in M \tag{16}$$

$$u_{i,i'} = \{0, 1\} \quad \forall i, i' \in N, i \neq i' \tag{17}$$

$$z_i = \{\ 0, 1\} \quad \forall i \in N \tag{18}$$

$$e_p \geq 0 \quad \forall p \in \{1, 2, 3, 4\} \tag{19}$$

$$b_i \geq 0 \quad \forall i \in N \tag{20}$$

Equation (2) states that the latest ending time for each priority group, N_p, must be greater than, or equal to, the start time of every job plus the duration of the job. Equation (3) ensures the latest ending time overall for all jobs is greater than, or equal to, the start time of every job belonging to the set of all jobs N, plus the duration of the job. Equation (4) guarantees that if a technician is available to work, then the technician may only be a member of one team that day. Conversely, Eq. (5) confirms that if a technician may not work, then the technician is not a member of any team. Equation (6) ensures that either a job is outsourced or a team completes it during the scheduling horizon. Equation (7) states that if a team completes a job, the team collectively has the skills necessary to service the job. Equation (8) shows that if a job is a successor it may not begin until the predecessor job has been completed. Equations (9) and (10) ensure that the start and end times of a job lie within the eligible working times of the day that the job is scheduled on. Equation (11) ensures time continuity. If a job is scheduled to be started after another job it does not begin until the other job has been completed. Equation (12) states that if two jobs are to happen sequentially then they must be both scheduled to be completed by the same team on the same day. Equation (13) ensures that the total sum of the outsourced jobs does not exceed the outsourcing budget available. Equation (14) guarantees that if a job i is outsourced, then all successor jobs belonging to σ_i are also outsourced. Equations (15)–(18) show that variables; $x_{t,k,r}$, $y_{i,k,r} u_{i,i'}$ and z_i are binary. Lastly, Eqs. (19) and (20) show that the starting and ending times of jobs are non-negative.

3 Intelligent Decision Heuristic

The variables associated with this heuristic are; the scheduling horizon K, which holds all schedules and is an entire solution, an array *AllJobs* containing all jobs that need to be scheduled, a schedule k that represents a day within the scheduling horizon K, a set of technicians T who are available for schedule k, an array *set* containing jobs of priority p that are under consideration for allocation, an array called *hypoteams* which contains a list of teams that could be made to service jobs belonging to *set*, an array called *OtherJobs* which contains further allocations that could be made teach hypothetical team, a team $T1$ which is the best team configuration to make according to a utilization score and lastly, a *PrecedenceArray*, which conins jobs which may not yet be allocated.

Figure 1 shows the pseudo codforhe ID construction heuristic. A scheduling horizon K is initialized, which Ids individual schedules and makes up an entire solution to the TTSP. The ID heuristic iterates through all jobs until they have been allocated to teams (i.e. the *AllJobs* array is empty).

```
Create Scheduling Horizon K
While (AllJobs>0)
  Create Schedule k, Add Techs T
  p=1
  While (p<= 4)
    set = AllJobs(p)
    hypoteams = MakeTeams (set)
    if (hypoteams != null)
        Otherjobs= findjobs (hypoteams)
        T1= HighestUtility(hypoteams)
        MakeTeam (t1)
        AddJobs(set)
        Update PrecedenceArray
    Else
        p=p+1
    End While
End While
Initial Solution Created K
```

Fig. 1 Pseudo code for the ID construction heuristic

A new day/schedule k is created by initializing all available technicians as single technician teams. The inner while loop is entered and the set of jobs with priority p. is found and stored in an array *set*. For each job belonging to *set*, a hypothetical team is made if possible, who if constructed has the time and skills to complete the job. In this heuristic, teams are made in a greedy fashion, at each step, the team member who covers the most skill and wastes the least skill is added as the next member of the team until the job requirements are fulfilled or no members can be added to the team.

A check is performed, if no teams can be created for jobs belonging to the set, then p is incremented and the loop is iterated through again. However, if a team can be created for any job belonging to the set of jobs with priority p, then the heuristic checks which other jobs from *set* could also be added onto each job list belonging to a hypothetical team. A utility score is calculated for each possible hypothetical team, Eq. (21). The utility function is made up of two components, the average over skill of the team to the jobs they would be allocated Eq. (22), and the wasted time Eq. (23). The highest scoring utility function is selected; the best hypothetical team is recorded as $T1$. Team $T1$ is constructed and added to the schedule k.

$$Utility = Skill\,(Scen) * \frac{1}{SlackTime\,(Scen)} \tag{21}$$

$$Skill\,(Scen) = \frac{1}{size} * \sum_{i=1}^{size} \frac{1}{overskill\,(hypotheticalteam,\ i)} \tag{22}$$

$$SlackTime\,(Scen) = 120 - \sum_{i=1}^{size} d_i \tag{23}$$

All job assignments from *set* are made to the team and the heuristic then checks whether any jobs are now eligible for allocation due to satisfied precedence constraints. Once no more jobs can be allocated to the current schedule, the heuristic checks whether all jobs have now been allocated, if not another schedule is created and the heuristic iterated through again. The construction heuristic terminates once all jobs have been allocated, and an initial solution has been created. The ID heuristic is used for the initial solution construction and it is used in the improvements phase. The ID heuristic is used by the local operators, which remove a job or selection of jobs and then reallocates them.

4 Outsourcing

In the ROADEF 2007 challenge, outsourcing is treated as a sub-problem and is solved before the scheduling process begins. The selection of outsourced jobs is final i.e. outsourced jobs do not enter the schedules. Experiments into outsourcing strategies have previously not been performed; therefore, the strategies featured in this work are "initial investigation" strategies. The strategies contain features such as the duration of a job, the skill requirements and outsourcing cost. Multiple outsourcing strategies were designed as shown in Table 1. Multiple instances were chosen for experiments from the Set B and Set X datasets, shown in Table 2, as one outsourcing may not be suitable for all problem instances.

Figure 2 shows that the strategy used for outsourcing can greatly affect the quality of the solution produced by the ID heuristic; it also shows that some datasets

Table 1 Outsourcing strategies

Strategy number	Description
1	Duration
2	Skill requirements
3	Outsourcing cost
4	Duration + Skill requirements
5	Duration + Skill requirements + Outsourcing cost
6	Duration + Outsourcing cost
7	Skill requirements + Outsourcing cost

Table 2 Datasets chosen for outsourcing experiments

Set	Data number	Jobs	Technicians
B	4	400	30
B	8	800	150
X	2	800	100
X	7	300	50
X	10	500	40

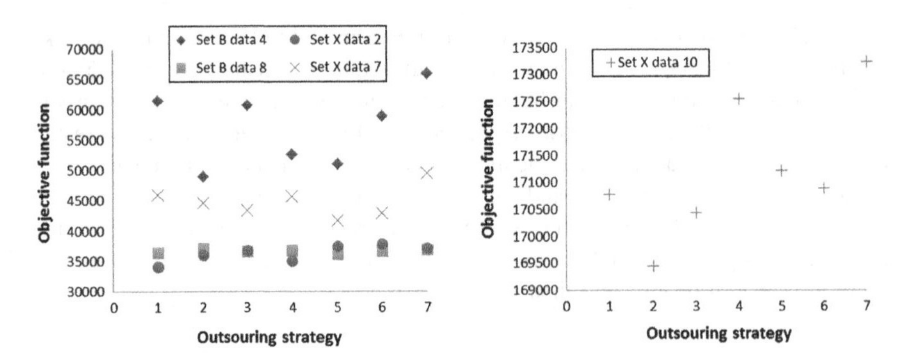

Fig. 2 Mean objective outsourcing strategy testing results

are more affected by the use of outsourcing strategy than other datasets. For example, the results produced for dataset B8 appear to be consistent; suggesting that the quality of solution produced is independent of the outsourcing strategy chosen. However, for dataset B4 the objective function appears to be heavily dependent on the outsourcing strategy used. It appears that the best strategies are 2, 4 and 5.

The results suggest the most important factors are the skill requirements and the duration of a job. In addition, for dataset X2 there seems to be a relationship between the strategy used and the objective value produced. Strategies 1 and 4 appear to produce the best results suggesting that the most important factors are duration of job followed by the skill requirements of a job.

Results produced by dataset X7 also appear to be dependent on the strategy used for outsourcing. The objective values found appear to be of better quality when using strategies 5 and 6, which both include duration time and outsourcing cost of the job in combination. Lastly, for dataset X10, strategy 2 appears to be the best strategy, which considers the skill requirements of the job.

As expected, these results agree with our original assumption that a single outsourcing strategy is not suitable for all datasets. Multiple outsourcing strategies are used on the ROADEF 2007 challenge datasets.

5 Local Operators

A variety of operators were used in this work, some from other combinatorial optimization work such as; move a job, swap two jobs, shuffle a job list, swap three jobs and swap job lists. The following operators where designed for this research and use the ID heuristic; move with team build, decompose and rebuild and decompose and rebuild N. The operator move with team build aims to reallocate a single job. The operator begins by selecting a job at random to reallocate. The ID heuristic then iterates through each day/schedule in the scheduling horizon and checks whether a team can be constructed in order to service the job.

The operator decompose and rebuild allows the structure of a day/schedule in the scheduling horizon to change. All jobs assigned to this day/schedule are removed and all team configurations are deconstructed. Each team member is assigned to his or her own individual team. The ID heuristic is then used to reallocate the removed jobs and to make team configurations able to satisfy the jobs. This operator is focused on exploring different team configurations.

The operator decompose and rebuild N allows the structure of a partial amount of the scheduling horizon to change. A value of N is selected which represents how many days/schedules in the scheduling horizon will be decomposed and rebuilt. The schedules are selected at random and ordered from the earliest to latest schedule. All jobs are removed from the schedules and teams are decomposed into single technician teams. The ID heuristic then iterates through the days/schedules, reallocating the jobs and constructs teams. This operator allows not only team configurations to change but also jobs to move across the scheduling horizon.

During the improvement phase of the ID heuristic, an operator is picked randomly (with equal probability) and applied to the current solution. The candidate solution is then evaluated using an Iterative Local Search (ILS) metaheuristic. ILS can be classified as a multi start technique and is conceptually simple [15]. ILS has similarities to hill climbing, however it contains a mechanism to escape locally optimal points within the search space. The ILS metaheuristic has two parameters, the step size N and the kick type. The step size N determines how many non-improving moves will be accepted before the heuristic is moved to another area of the search space. Once N non-improving moves have been reached, a kick is applied that transports us to another area of the search space. When implementing ILS many decisions are left to the developer in regards to the step size used and the type of kick to be performed [16].

6 Challenge Comparison

In the ROADEF 2007 challenge, a 20 min computational time limit was set. The heuristics were run on each dataset five times and the best score was recorded. The columns of Table 3 represent the datasets used, the best-known score, the results achieved by [8] (Hu), [6] (C), [9] (E), [7] (F), [10] (D), [13] (P), [11] (K), [14] (Ha) and the ID heuristic respectively.

For Sets B and X, in most instances outsourcing strategy 5 was used (time, skill and outsourcing cost) as testing showed this was generally the best strategy. However for some datasets B4, B5, B6, X3 and X10 strategy 2 was used (skill requirements) as this resulted in better quality solutions.

The ID heuristic appears to perform well on the Set A instances and matches the performance of the other researchers. In datasets A1–A4 and A6 the ID heuristic also finds the best reported values from the literature.

Table 3 ROADEF 2007 challenge comparison

Data	BKS	Hu	C	E	F	D	J	P	K	Ha	ID
A1	2340	2340	2340	2340	2340	2340	2490	2340	2340	2340	2340
A2	4755	5580	4755	4755	4755	4755	4755	4755	4755	4755	4755
A3	11880	12600	11880	11880	11880	13068	12600	11880	11880	11880	11880
A4	13452	13620	13452	14040	13452	13620	14040	14760	13452	13452	13452
A5	28845	30150	29335	29700	29335	31236	32400	33480	29335	28845	29040
A6	18795	20280	18795	18795	20005	21576	21120	22380	19935	18870	18795
A7	30540	32520	30540	30540	30960	40116	32520	33360	31050	30840	30660
A8	16920	18960	17700	20100	17355	23115	19380	21180	17587	17355	20100
A9	27348	29328	27692	28020	28280	34056	28280	30000	28028	27692	28020
A10	38296	40650	38636	38296	39300	52348	41580	42740	40350	40020	39000
B1	33900	34710	37200	34395	34575	58968	46995	44025	43620	43860	34410
B2	15870	17970	17070	15870	16775	28989	19890	21240	20010	20655	18600
B3	16005	18060	18015	16020	16275	34368	20340	20280	19575	20565	18210
B4	23775	26115	23775	25305	23925	56382	29460	31815	35385	26025	45855
B5	88680	94200	117540	89700	88920	N/A	100080	122760	119160	120840	119820
B6	26955	30450	27390	27615	28785	N/A	24230	37965	32760	34215	37755
B7	31620	33300	33900	38200	31620	N/A	36060	38820	41220	35460	37140
B8	33030	35490	33240	37440	35520	N/A	35550	34440	39240	33030	36000
B9	28080	28200	29760	32700	28080	N/A	29460	33360	30000	29550	33360
B10	34680	34680	35640	41280	35040	N/A	36960	44640	38040	34920	40680
X1	146220	151140	159300	188595	146220	N/A	N/A	N/A	N/A	181575	178560
X2	7260	9090	8280	8370	7740	N/A	N/A	N/A	N/A	7260	32925
X3	48720	50400	50400	50100	48720	N/A	N/A	N/A	N/A	52680	52920
X4	64600	65400	66780	68120	64600	N/A	N/A	N/A	N/A	72860	74880

(continued)

Table 3 (continued)

Data	BKS	Hu	C	E	F	D	J	P	K	Ha	ID
X5	144750	147000	157800	183700	144750	N/A	N/A	N/A	N/A	172500	182820
X6	9480	10320	9900	10440	9690	N/A	N/A	N/A	N/A	9480	13020
X7	32040	33240	47760	37200	32040	N/A	N/A	N/A	N/A	46680	40320
X8	23220	23460	24060	25480	23220	N/A	N/A	N/A	N/A	29070	27420
X9	122800	134760	152400	159660	122800	N/A	N/A	N/A	N/A	168240	159600
X10	120330	137040	140520	152040	120330	N/A	N/A	N/A	N/A	178560	160860

In the ROADEF 2007 challenge Set B instances there is variation in the results produced by the ID heuristic. In some datasets, the ID heuristic performs competitively B1–B3 and B7–B10 with regards to the solutions found by the other researchers. In other instances, particularly B4–B6, the ID is unable to perform competitively. Interestingly after examination it appears that these datasets have a high volume of precedence and successor relationships which form chains and result in producing elongated solutions. This suggests that a complex aspect of the ROADEF 2007 challenge problem is the interrelationships between jobs (successor and precedence).

The ROADEF 2007 challenge Set X instances were not tackled by [9–12]. For some datasets, the ID heuristic performs well, finding solutions that are competitive (X1, X7 and X9). For other datasets (X2, X3, X4 and X6), the ID heuristic found worse quality results than the other researchers, also suggesting there are further complexities to this problem that need to be studied.

7 Conclusion

The ID heuristic has matched some of the best-known solutions to the ROADEF 2007 challenge problem. In 23 out of 30 datasets, the ID heuristic has produced a competitive solution with regard to solutions found by other researchers.

This research has highlighted that there are many complexities in the ROADEF 2007 datasets which arise due to the real-world nature of the technician and task scheduling problem especially relating to the constraints and their relationships [2]. Future work will investigate the precedence relationships within the datasets to ascertain if results can be improved in instances which have a high number of precedence and successor relationships.

Our contributions in this paper to the field are; (i) the ID heuristic, which behaves in a different manner to the heuristics proposed by other researchers, (ii) outsourcing strategy testing which has shown dependency between the strategy chosen and the quality of result produced and (iii) novel operators designed to provide flexibility in team configurations.

Acknowledgments This research is sponsored by ServicePower Technologies PLC, a worldwide leader at providing innovative mobile workforce management solutions, in cooperation with MMU and KTP.

References

1. Pillac, V., Guéret, C., Medaglia, A.: On the technician routing and scheduling problem. In: The IX Metaheuristics International Conference, pp. S2–40, Italy (2011)
2. Ernst, A.T., Jiang, H., Krishnamoorthy, M., Sier, D.: Staff scheduling and rostering: a review of applications, methods and models. EJOR **153**(1), 3–27 (2004)

3. The ROADEF 2007 Challenge. http://challenge.roadef.org/2007/
4. Dutot, P.F., Laugier, A., Bustos, A.M.: Technicians and interventions scheduling for telecommunications (2007). http://challenge.roadef.org/2007
5. Montoya, C., Bellenguez-Morineau, O., Pinson, E., Rivreau, D.: Integrated column generation and lagrangian relaxation approach for the multi-skill project scheduling problem. In: Handbook on Project Management and Scheduling, vol. 1, pp. 565–586. Springer (2015)
6. Cordeau, J.F., Laporte, G., Pasin, F., Ropke, S.: Scheduling technicians and tasks in a telecommunications company. JOS **13**(4), 393–409 (2010)
7. Fırat, M., Hurkens, C.A.J.: An improved MIP-based approach for a multi-skill workforce scheduling problem. JOS **15**(3), 363–380 (2012)
8. Hurkens, C.A.: Incorporating the strength of MIP modeling in schedule construction. RAIRO Oper. Res. **43**(04), 409–420 (2009)
9. Estellon, B., Gardi, F., Nouioua, K.: High-performance local search for task scheduling with human resource allocation. In: Engineering Stochastic Local Search Algorithms. Designing, Implementing and Analyzing Effective Heuristics, pp. 1–15. Springer (2009)
10. Dongala, S.G.P.: The Problem of Scheduling Technicians and Interventions in a Telecommunications Company (2008)
11. Korteweg, P.: When to hire the A-team. ROADEF (2007). http://challenge.roadef.org/2007
12. Jaskowski, W., Wasik, S.: Efficient Greedy Algorithm with Hill Climbing for Technicians and Interventions Scheduling Problem (2007). http://challenge.roadef.org/2007
13. Pokutta, S., Stauffer, G.: France telecom workforce scheduling problem: a challenge. RAIRO Oper. Res. **43**(04), 375–386 (2009)
14. Hashimoto, H., Boussier, S., Vasquez, M., Wilbaut, C.: A GRASP-based approach for technicians and interventions scheduling for telecommunications. Ann. Oper. Res. **183**(1), 143–161 (2011)
15. Martí, R., Moreno-Vega, J.M., Duarte, A.: Advanced multi-start methods. In: Handbook of Metaheuristics, pp. 265–281. Springer, Boston (2010)
16. Lourenço, H.R., Martin, O.C., Stützle, T.: Iterated Local Search. Springer, Heidelberg (2003)

Scheduling System for Multiple Unmanned Aerial Vehicles in Indoor Environments Using the CSP Approach

Youngsoo Park, Yohanes Khosiawan, Ilkyeong Moon, Mukund Nilakantan Janardhanan and Izabela Nielsen

Abstract In recent years there has been an increased demand in use of multiple unmanned aerial vehicles (UAVs) for surveillance and material handling tasks in indoor environments. However, only a limited number of studies have been reported on UAV scheduling in an indoor 3D environment. This paper presents the indoor UAV scheduling problem and models it as a constraint satisfaction problem (CSP) to find a feasible solution in less computation time. A numerical example of the problem is presented to illustrate the proposed methodology.

Keywords Unmanned aerial vehicles · Indoor UAV scheduling · Constraint satisfaction problem

1 Introduction

In the recent technological development, Unmanned Aerial Vehicles (UAVs) take center stage for different applications in various domains, such as surveillance mission, material handling and quality control. Currently, UAVs are mainly used in

Y. Park (✉) · I. Moon
Department of Industrial Engineering, Seoul National University, Seoul 151-444, Republic of Korea
e-mail: simulacrum@snu.ac.kr

I. Moon
e-mail: ikmoon@snu.ac.kr

Y. Khosiawan · M.N. Janardhanan · I. Nielsen (✉)
Department of Mechanical and Manufacturing Engineering, Aalborg University, Aalborg, Denmark
e-mail: izabela@m-tech.aau.dk

Y. Khosiawan
e-mail: yok@m-tech.aau.dk

M.N. Janardhanan
e-mail: mnj@m-tech.aau.dk

I. Czarnowski et al. (eds.), *Intelligent Decision Technologies 2016*,
Smart Innovation, Systems and Technologies 56,
DOI 10.1007/978-3-319-39630-9_7

outdoor environment. However, due to the technological advancements (for instance, ultrasound usage for indoor positioning system), UAVs can be employed in indoor environment. Research on surveillance missions in indoor environments has been conducted by UAVs that use image recognition [1]. UAVs can facilitate material handling systems by moving materials/parts between different locations in a manufacturing plant. UAV applications in indoor environments are at an early stage and there is a requirement of developing an operating system which includes a scheduling system for indoor UAVs.

Utilization of the available resources should be maximized. Consequently, a scheduling system [2] that assigns tasks to UAVs is an essential requirement for the efficient management of the resources. The methodology for UAV scheduling covers routing, operation sequencing, assignment, and order splitting. Kim et al. [3] presented a UAV scheduling model, referred to as *split jobs*, in which the space-time mission trajectories are split into pieces. This method is useful because one UAV may not be able to cover the entire task due to fuel or battery constraints. Few researchers have developed scheduling methods for UAVs without including a travel time or distance restriction [3]. The major objective in UAV scheduling was the allocation and aggregation of the available UAVs based on the mission and likely scenarios. Some authors developed a mathematical model for UAV resource scheduling under mission synchronization based on single-objective non-linear integer programming [4]. The developed model was validated using a small problem. Other authors studied methods for scheduling UAVs without fuel limitations [5]. They developed a mathematical model to direct multiple UAVs for cooperative engagement with a ground-moving target. Assignment of multiple UAVs to perform tasks simultaneously is an NP-hard combinatorial optimization problem. A genetic algorithm was proposed to obtain feasible solutions [5].

Kim and Morrison [6] focused on scheduling capacitated UAVs with limited flight duration. This type of UAV should complete the tasks within its fuel (battery) capacity and should return to the base before the fuel runs out. They formulated a mixed integer linear program (MILP) to determine the types and numbers of UAVs, as well as locations and numbers of stations. A modified receding horizon task assignment algorithm and a branch and bound algorithm were developed and studied. Weinstein and Schumacher [7] presented UAV scheduling based on a vehicle routing problem with time windows. They developed an MILP to find the global optimal schedule. Kim et al. [8] considered a scheduling model with n tasks and m UAVs with a capacity limit of q each. Two situations were considered: no UAV return and UAV return. The model tries to minimize costs reflecting operating time and risk exposed. They first proposed a MILP formulation that exactly solves the problem and then proposed four alternative MILP formulations that are computationally less intensive than the original, and therefore, suitable for real-time purposes. Improvement to this model was proposed by Alidaee et al. [9] who minimized the number of variables and constraints. Kim et al. [3] proposed a UAV scheduling model that allows UAVs to recharge and return to service at multiple, shared service stations distributed throughout the field. Because they included a refuel/recharge process, the authors suggested that the problem can be included under the study of mobile robot logistics.

To deal with the complexity of the system using UAVs, methods such as an MILP and a genetic algorithm have been used. One alternative for these methods is a constraint satisfaction problem (CSP), which has been successfully implemented to solve specific types of scheduling problems (e.g., job shop, flow shop, and cyclic scheduling problems) [10–12]. A job shop scheduling problem which involves multiple resources (UAVs), is a typical NP-hard problem [11]. Due to the high complexity of the job shop scheduling problem, some authors have considered the CSP model which allows for modeling the problem in an intuitive way and helps to reduce the search space [11]. Therefore, the CSP is a promising approach for use in UAV scheduling in indoor environments.

A programming framework called *constraint programming* is used in the CSP paradigm. It provides an environment for the user to model the problem as a system of constraints, and the programming engine tries to solve the modeled problem by drawing inputs from artificial intelligence and operation research [13]. A selection of searching algorithm in the CSP is important in reducing the computation time. Some well-known search techniques reported in the literature are the depth-first search, breadth-first search, look-back algorithm, look-forward algorithm, forward-checking algorithm, and backtracking technique [14]. The backtracking technique solves the problem through an iterative selection operation. Some authors used depth-first backtrack search for the scheduling problem under time window constraint [15].

Although the demand for UAV applications in indoor environments is growing rapidly, relevant literature related to problem formulation of the respective UAV operation has not been reported to date. As an initial stage of the implementation, a deterministic system is proposed. Every order information is predefined and there is no uncertain event, neither from environment nor UAV. In the target indoor environment, UAVs are positioned by ultrasonic signal between ultrasonic transmitter mounted on the walls and sensor on UAVs. Tasks are scheduled by the scheduling system and transmitted to UAVs through radio frequency network. UAV system is centrally controlled by positioning system via radio frequency network.

Unlike outdoor environment, UAV scheduling problem in indoor environment is more complex due to more constraints and requirements: (1) to avoid a risk of damage on both UAVs and resources (e.g., human worker, machine on shop-floor), the areas should be preempted to a UAV at a time. (2) Compared to other automated transportation system such as automated guided vehicle (AGV) or overhead hoist transport (OHT) [16], the velocity of UAV is high, but the battery constraint is more critical and need to be considered as one of the main constraints. Unlike AGV, UAV cannot get a battery exchange and the recharging time is significantly longer than the flight time. (3) In addition, when UAVs are operated in indoor environment with other capital intensive resources, UAV schedule needs to fit well in the resource schedule to maintain resource utilization as high as possible. This incurs *time window* and *precedence relationship* between tasks. Consequently, these constraints complexify the scheduling problem and an agile algorithm is needed to respond the complicated scheduling problem. This paper proposes a CSP approach to solve a UAV scheduling problem.

The remainder of the paper is structured as follows: Sect. 2 explains the problem in detail, and Sect. 3 describes the CSP model for the presented problem. Section 4 discusses the proposed model with a numerical example, and Sect. 5 concludes the findings of this research.

2 Problem Definition

The manufacturing environment with multiple UAVs that handle multiple tasks is considered in this paper. Each UAV executes many rounds of flight where each round consists of several tasks. The proposed system features two types of tasks to be executed: surveillance inspection and material handling. The inspection tasks are performed in one position from which the UAV takes pictures of objects of interest for the purpose of quality inspection, while material handling tasks aim at transporting payloads from one position to another. Task and flight actions consist of some subactions: (1) flying down to reach an inspection or pick-up point, (2) flying up and (3) on a flight path including a straight and turn trajectory as well as hovering at a fixed position. Other task information related to UAV operations includes task execution time, release time, due time, and precedence relationships. Task execution time is the time needed to complete the task. Release time and due time form a duration called a *time window*, which is the available time period for the UAV to execute a particular task. As inferred, the release time is the opening time of a time window, and the due time is the deadline. There are precedence relationships between tasks because in manufacturing systems some tasks can only be executed upon the completion of predecessor tasks. UAVs implemented in indoor environments are mostly small multicopters. The multicopters used in this system are identical and can perform both types of tasks.

The following assumptions are considered in this work:

- The system is deterministic.
- The UAV has a constant flight speed and battery consumption rate, regardless of payload and current battery level.
- The payload of UAV is less than 10 % of the UAV's weight.
- There is no partial recharging.
- Every task time is shorter than the flight time limit.
- Tasks cannot be subdivided into subtasks.
- In every flight, a capacitated UAV has a fixed amount of
  - flight time (e.g., up to 20 min) and
  - recharge time (e.g., 45 min) at a predetermined recharging center.

The scheduling system develops a flight schedule and distributes the orders for the UAVs. A schedule contains times and tasks for each UAV. An example of the

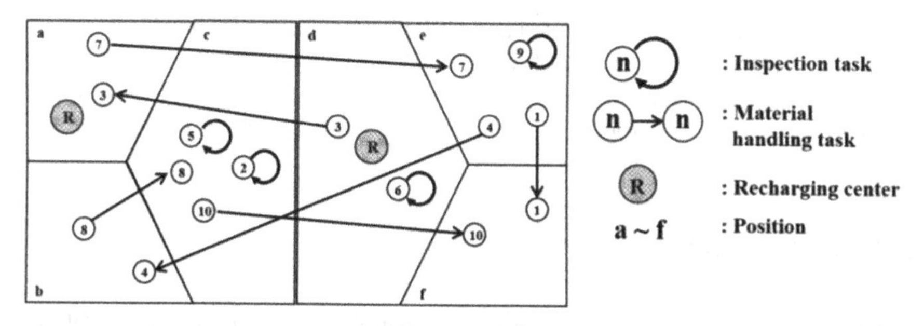

Fig. 1 Map of the system

<table>
<tr><th>Time</th><th>Origin</th><th>8:00:00</th><th>8:00:20</th><th>8:00:40</th><th>8:01:00</th><th>8:01:20</th><th>8:01:40</th><th>8:02:00</th><th>8:02:20</th><th>8:02:40</th><th>8:03:00</th><th>8:03:20</th><th>8:03:40</th><th>8:04:00</th><th>8:04:20</th><th>8:04:40</th><th>8:05:00</th><th>8:05:20</th><th>8:05:40</th><th>8:06:00</th><th>8:06:20</th><th>8:06:40</th><th>8:07:00</th><th>8:07:20</th><th>8:07:40</th><th>8:08:00</th><th>8:08:20</th><th>8:08:40</th><th>8:09:00</th><th>8:09:20</th><th>8:09:40</th><th>8:10:00</th><th>8:10:20</th><th>8:10:40</th><th>8:11:00</th><th>8:11:20</th><th>8:11:40</th><th>8:12:00</th><th>8:12:20</th><th>8:12:40</th><th>8:13:00</th><th>8:13:20</th><th>8:13:40</th><th>8:14:00</th><th>8:14:20</th><th>8:14:40</th><th>8:15:00</th><th>8:15:20</th><th>8:15:40</th><th>8:16:00</th><th>8:16:20</th><th>8:16:40</th><th>8:17:00</th></tr>
<tr><td>UAV 1</td><td>a</td><td colspan="6">a → c</td><td colspan="33">Task 2</td><td colspan="6">e → d</td><td colspan="7">Task 5</td></tr>
<tr><td>UAV 2</td><td>a</td><td colspan="2">H</td><td colspan="26">Task 1</td><td colspan="7">H</td><td colspan="4">a → e</td><td colspan="13">Task 4</td></tr>
<tr><td>UAV 3</td><td>d</td><td colspan="5">H</td><td colspan="5">d → b</td><td colspan="22">Task 3</td><td colspan="4">b → a</td><td colspan="5">Task 8</td><td colspan="11">Task 10</td></tr>
</table>

Fig. 2 Output of the scheduling system

map is depicted in Fig. 1. This map leaves out the details of physical paths so that the number of steps in the scheduling model is admissible for computation.

Abstractions of the map and the schedule decrease the domain size; detailed straight flights and turns are merged into tasks and flights, hence computational size and time are minimized. Figure 1 shows an example of the map containing 10 tasks performed at 6 positions. Numbers in the nodes represent the tasks at different positions and *R* represents the recharging center in the environment. Figure 2 shows the part of the flight schedule of 3 UAVs and 10 tasks which is studied in this paper.

3 CSP for UAV Scheduling for Assignment Phase

Concepts of the job shop scheduling problem can be applied as a part of the UAV scheduling system. Numerous studies on general job shop scheduling problems have dealt with different objective functions such as those for minimizing make-span, lateness, energy consumption, and maximizing utilization [11, 17, 18]. However, the MILPs or hybrid algorithms that have objective functions might not give a feasible solution, especially in a short computation time. However, the CSP focuses on providing a feasible solution within a short computation time.

The proposed problem is modeled as a CSP paradigm and expressed as a 3-tuple [10], $S=(V,D,C)$ where a set of variables $V=\{v_1,v_2,\ldots,v_m\}$ each having a corresponding domain $D=\{d_1,d_2,\ldots,d_m\}$ and a set of constraints $C=\{c_1,c_2,\ldots c_n\}$. The proposed CSP model is introduced in the following section.

3.1 CSP Model

Indices, decision variables, parameters and constraints of the proposed model are presented in this section.

Indices

- m, n index for a task, $m,n\in\{1,2,3,\ldots,N\}$, in which N is the number of total candidate tasks
- t a planning period, $t\in\{1,2,3,\ldots,T\}$, in which T is the last period of the planning horizon
- u index for a UAV, $u\in\{1,2,3,\ldots,U\}$, in which U is the number of UAVs
- q, r index for a position (node) $q,r\in\{1,2,3,\ldots,Q\}$, in which Q is the number of nodes
- N_q a set of tasks that have position q as starting or ending position

Decision Variables

$A_n = <T_n, u_n>$ task n is assigned to start at time T_n and UAV u_n

Parameters

- w_n task execution time of task n
- ds_n release time of task n
- dd_n due time of task n
- P_{mn} $=1$ if task m precedes task n
- qo_n starting position of task n
- qd_n ending position of task n
- d_{qr} distance between position q and r
- v_0 flight speed of UAV without payload

Constraints

$$[(ds_m \le T_m) \wedge ((T_m + w_n) \le dd_n)], \forall m \quad (1)$$

$$(P_{mn} = 1) \supset ((T_m + w_m < T_n)), \forall m \quad (2)$$

$$\left[(u_m = u_n) \supset \left(\left(w_m + \left(\frac{d_{qd_m qo_n}}{v_0} \right) < (T_n - T_m) \right) \wedge \left(w_n + \left(\frac{d_{qd_n qo_m}}{v_0} \right) < (T_m - T_n) \right) \right) \right], \forall m, n \tag{3}$$

$$\left[\left((m \in N_q) \wedge (n \in N_q) \right) \supset \left((w_m < (T_m - T_n)) \wedge (w_n < (T_n - T_m)) \right) \right], \forall m, n, q \tag{4}$$

Equation (1) describes the time window constraint. Equation (2) shows the precedence constraint. If such precedence relationship exists between task m and task n, then task m should be finished before task n is started. Equation (3) shows the UAV occupation constraint. If task m and task n are assigned to the same UAV, then the latter task can start after the former task is finished and the UAV moves from the end position of the former task to the start position of the latter task. Equation (4) describes the position occupation constraint. If task m and task n are held at the same position, then the latter task can start after the former task is finished.

3.2 Searching Algorithm

With the defined CSP model for the UAV scheduling problem, an efficient search technique prunes and explores the search space to get a better feasible solution quickly. Efficiency is especially important when one is trying to drive the search process to find not only a feasible solution but also near optimal solution. Several techniques to control searching by using heuristics have been discussed [11, 19, 20]. A major portion of the search space will be reduced according to due dates and precedence relationships to leave a focused set of start time intervals for the jobs. Although it is a significant factor in finding a solution, search space reduction needs to be researched more extensively because of the specific characteristics of the unique UAV scheduling problems.

In the aforementioned works on heuristics for the CSP search technique, contention was introduced to address the capacity constraint that involves a limited number of resources in the job scheduling problem. It measures the critical resources needed to complete many jobs and makes the scheduling system aware of it, so that the required prioritizations are accordingly assigned. In the presented work, a contention concept is involved by having a position occupation constraint to indicate that the machine is being used for a particular task at a certain time and thus avoids conflicts for task completion due to simultaneous occupation of machines. In the next section, an example is discussed that illustrates the backtracking technique [15]. The backtracking technique guarantees that a feasible solution is found quickly if it exists and increases system robustness; backtracking is useful while forward search techniques might require excessive computation time.

4 Numerical Example

The problem considered for the numerical example contains 10 tasks with precedence and time windows. Three UAVs travel through 6 positions in the system. Task information and flight time between positions are given in Tables 1 and 2. The position where the task is started and ended is respectively called as the *start* and *end* position. The time needed to execute a task, the release and due times, along with the precedence relationships among tasks, are presented in Table 1. For instance, task 4 can be executed only after task 1 is completed. The time bucket is 2 min and the starting positions of the UAVs are given. The backtracking technique is used as a searching algorithm for solving this example.

Figures 3, 4 and 5 display the basic steps of the searching algorithm. The earliest start time (*EST*), earliest finish time (*EFT*), latest start time (*LST*) and latest finish time (*LFT*) of each task are calculated based on release date, due date and

Table 1 Task information

Task	Start position	End position	Time needed	Release time	Due time	Precedence
1	e	f	2	0	5	–
2	c	c	7	3	10	–
3	d	a	6	0	45	–
4	e	b	5	9	20	1
5	c	c	2	5	20	2
6	d	d	2	10	44	2
7	a	e	4	5	40	4
8	b	c	3	0	15	4.5
9	e	e	3	30	40	7
10	c	f	3	20	50	6.8

Table 2 Flight time between positions

Flight time	a	b	c	d	e	f
a	0	1	1	2	3	3
b	1	0	1	2	3	3
c	1	1	0	1	2	2
d	2	2	1	0	1	1
e	3	3	2	1	0	1
f	3	3	2	1	1	0

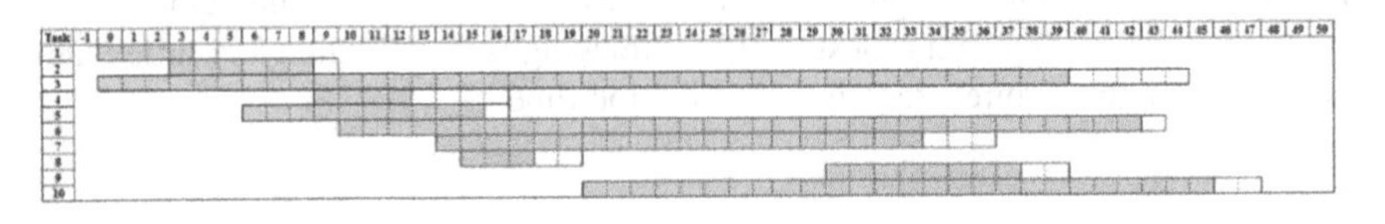

Fig. 3 Time windows for tasks

precedence relationship. *EST* and *EFT* are derived for each task according to a forward propagation, while *LST* and *LFT* are derived according to a backward propagation. Figure 3 shows the time windows including *EST* and *LST* for each task. The white area represents infeasible solutions, and grey area identifies the possible time period between *EST* to *LST*. The black area in Fig. 4 indicates the time period after *LST* to *LFT* for the latest possible task execution. To find a feasible solution, the backtracking technique is used as a searching algorithm. In the numerical example, starting from the latest possible time, tasks are assigned to the schedule following the *LSTs* and the cheapest available UAVs at the respective time. In other words, when there are multiple tasks to be assigned in the schedule, the cheapest flight time (cost) calculation of available UAVs for the task execution positions is used as a selector.

Figure 4a shows steps for assigning task 10 and Fig. 4b shows steps for assigning task 6 to UAV 1. In Fig. 4c, task 3 cannot be assigned to UAV 1 at period 39 because both tasks 3 and 6 occupy position *d*. In the searching algorithm, different UAVs are explored first and then time periods are explored thereafter. In Fig. 4d, periods 39, 38, and 37 are infeasible for all the UAVs and period 36 is infeasible for UAV 1 because it must fly from position *a* to position *d*; However, task 3 can be assigned to UAV 2 at period 36. Figure 5 shows one feasible solution (output) of the CSP model.

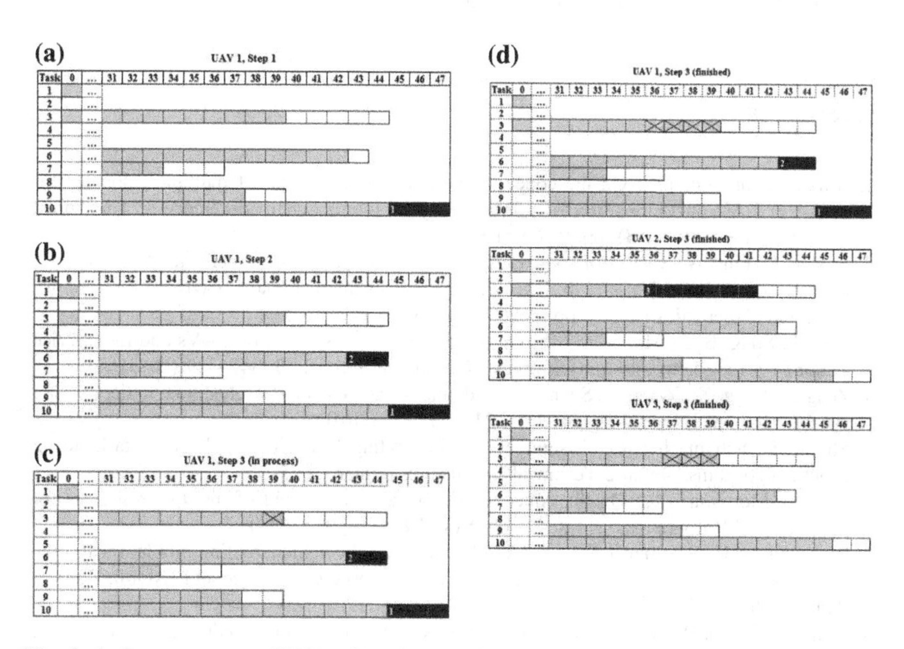

Fig. 4 Assignment steps. **a** UAV 1, Step 1, Assigning task 10 to UAV 1 in period 45. **b** UAV 1, Step 2, Assigning task 6 to UAV 1 in period 42. **c** UAV 1, Step 3 (in process), Task 3 cannot be assigned to UAV 1 in period 39. **d** UAV 1, 2, 3, Step 3 (finished) Assigning task 3 to UAV 2 in period 36

<table>
<tr><th>UAV</th><th>-1</th><th>0</th><th>1</th><th>2</th><th>3</th><th>4</th><th>5</th><th>6</th><th>7</th><th>8</th><th>9</th><th>10</th><th>11</th><th>12</th><th>13</th><th>14</th><th>15</th><th>16</th><th>17</th><th>18</th><th>19</th><th>20</th><th>...</th><th>30</th><th>31</th><th>32</th><th>33</th><th>34</th><th>35</th><th>36</th><th>37</th><th>38</th><th>39</th><th>40</th><th>41</th><th>42</th><th>43</th><th>44</th><th>45</th><th>46</th><th>47</th><th>48</th></tr>
<tr><td>1</td><td>a</td><td colspan="3">$a \to e$</td><td colspan="2">1</td><td></td><td colspan="2">$f \to c$</td><td colspan="2">2</td><td colspan="2">$c \to e$</td><td colspan="5">4</td><td colspan="3">8</td><td colspan="3">...</td><td>ca</td><td colspan="4">7</td><td></td><td colspan="3">9</td><td></td><td>ed</td><td colspan="2">6</td><td>dc</td><td colspan="3">10</td><td></td></tr>
<tr><td>2</td><td>d</td><td colspan="14"></td><td>dc</td><td colspan="2">5</td><td colspan="10">...</td><td>ed</td><td colspan="6">3</td><td colspan="7"></td></tr>
<tr><td>3</td><td>d</td><td colspan="41">...</td></tr>
</table>

Fig. 5 Feasible solution

5 Conclusion

In recent years, there has been a growing interest in the use of UAVs in indoor environments. Due to advances in the related technologies, indoor UAV applications are emerging but they are still in a very early stage. This paper presents a CSP model for scheduling system of UAVs in an indoor environment. The parameters and decision variables of the proposed model are discussed in this paper. A backtracking technique is incorporated to reduce the search space and minimize the computation time. The capability of finding a feasible solution is maintained by characteristics of the backtracking technique. A numerical example is used to illustrate the problem. For further research, heuristic and metaheuristic methods can be implemented and compared based on feasibility and optimality.

Acknowledgments This work has partly been supported by Innovation Fund Denmark under project UAWorld; grant agreement number 9-2014-3.

References

1. Kim, J., Morrison, J.R.: On the concerted design and scheduling of multiple resources for persistent UAV operations. In: IEEE Proceedings of International Conference on Unmanned Aircraft Systems (ICUAS), pp. 942–951 (2013)
2. Baxter, J., Findlay, S., Paxton, M., Berry, A.: Scheduling UAV surveillance tasks, lessons learned from trials with users. In: IEEE Proceedings of International Conference on Systems, Man, and Cybernetics (SMC), pp. 2606–2610 (2013)
3. Kim, J., Song, B.D., Morrison, J.R.: On the scheduling of systems of UAVs and fuel service stations for long-term mission fulfillment. J. Intell. Rob. Syst. **70**, 347–359 (2013)
4. Zeng, J., Yang, X., Yang, L., Shen, G.: Modeling for UAV resource scheduling under mission synchronization. J. Syst. Eng. Electron. **21**, 821–826 (2010)
5. Shima, T., Schumacher, C.: Assignment of cooperating UAVs to simultaneous tasks using genetic algorithms. Defense Technical Information Center (2005)
6. Kim, J., Morrison, J.R.: On the concerted design and scheduling of multiple resources for persistent UAV operations. J. Intell. Rob. Syst. **74**, 479–498 (2014)
7. Weinstein, A.L., Schumacher, C.: UAV scheduling via the vehicle routing problem with time windows. In: Proceedings of AIAA Infotech @ Aerospace Conference and Exhibit. Rohnert Park, California (2007)
8. Kim, Y., Gu, D.-W., Postlethwaite, I.: Real-time optimal mission scheduling and flight path selection. IEEE Trans. Autom. Control **52**, 1119–1123 (2007)
9. Alidaee, B., Wang, H., Landram, F.: A note on integer programming formulations of the real-time optimal scheduling and flight path selection of UAVs. IEEE Trans. Control Syst. Technol. **17**, 839–843 (2009)

10. Nielsen, I., Bocewicz, G., Do, N.A.D.: Production and resource scheduling in mass customization with dependent setup consideration. In: Brunoe, T.D., Nielsen, K., Joergensen, K.A., Taps, S.B. (eds.) Proceedings of the 7th World Conference on Mass Customization, Personalization, and Co-creation (MCPC 2014), pp. 461–472. Springer International Publishing, Aalborg, Denmark, Feb 4–7th 2014 (2014)
11. Fox, M.S., Sadeh, N.M.: Why is scheduling difficult? a CSP perspective. In: Proceedings of ECAI, pp. 754–767 (1990)
12. Bocewicz, G., Wójcik, R., Banaszak, Z.: Cyclic scheduling of multimodal concurrently flowing processes. Advances in Systems Science, pp. 587–598. Springer (2014)
13. Cohen, J.: Constraint logic programming languages. Commun. ACM **33**, 52–68 (1990)
14. Barták, R., Salido, M.A., Rossi, F.: Constraint satisfaction techniques in planning and scheduling. J. Intell. Manuf. **21**, 5–15 (2010)
15. Sadeh, N., Sycara, K., Xiong, Y.: Backtracking techniques for the job shop scheduling constraint satisfaction problem. Artif. Intell. **76**, 455–480 (1995)
16. Bocewicz, G., Wójcik, R., Banaszak, Z.A., Pawlewski, P.: Multimodal processes rescheduling: cyclic steady states space approach. Mathematical Problems in Engineering (2013)
17. Błażewicz, J., Domschke, W., Pesch, E.: The job shop scheduling problem: conventional and new solution techniques. Eur. J. Oper. Res. **93**, 1–33 (1996)
18. Le Pape, C., Baptiste, P.: Resource constraints for preemptive job-shop scheduling. Constraints **3**, 263–287 (1998)
19. Caseau, Y., Laburthe, F.: Improved CLP scheduling with task intervals. In: Proceedings of ICLP, pp. 369–383. Citeseer (1994)
20. Sadeh, N., Fox, M.: Preference propagation in temporal constraints graphs. Technical Report. Intelligent Systems Laboratory, The Robotics Institute, Carnegie-Mellon University Pittsburgh, PA (1988)

An Ontology Supporting Multiple-Criteria Decision Analysis Method Selection

Jarosław Wątróbski

Abstract In the last few years multiple-criteria decision analysis methods (MCDA) have received more and more attention from researchers. Consequently, the new methods have made it possible to solve a lot of real-world decision problems in the areas of private lives, management practice or science. However, proper selection of a MCDA method reflecting a decision problem and a decision-maker's preferences is a complex and important issue and inappropriate selection of a MCDA method may lead to generating improper recommendation. Therefore, this paper suggests using an ontology as a tool supporting multiple-criteria decision analysis method selection.

Keywords Multiple-criteria decision analysis · MCDA ontology · MCDA method selection

1 Introduction

The nature of a decision problem determines its multi-aspect approach. In almost every case the notion of a "good" decision requires considering decision variants in terms of many criteria. Criteria can also grade the quality of individual variants when all possibilities are acceptable and the problem is to subjectively choose the best one. Subjectivism, in this case, refers to the importance of individual criteria, since, usually, some factors are more important for the decision-maker than other ones. Furthermore, uncertainty and imprecision of data describing criteria values of variants [1] also influence subjectivity of evaluation. Therefore, under no circumstances can a multiple-criteria decision be recognized as fully objective. Here, the only objective element is assessment, generated with the use of a formalized computational procedure, of individual variants on the basis of criterion weights and criterion values of variants. The application of other criterion weights, even

J. Wątróbski (✉)
West Pomeranian University of Technology, Żołnierska 49, 71-210 Szczecin, Poland
e-mail: jwatrobski@wi.zut.edu.pl

I. Czarnowski et al. (eds.), *Intelligent Decision Technologies 2016*,
Smart Innovation, Systems and Technologies 56,
DOI 10.1007/978-3-319-39630-9_8

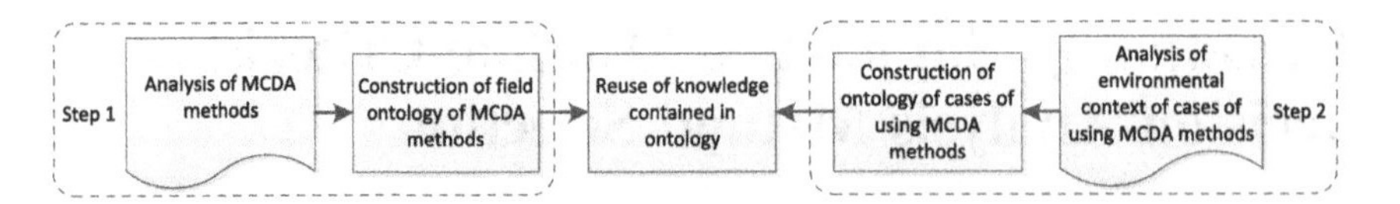

Fig. 1 Process of constructing an ontology of MCDA methods and their applications

when keeping variant values with relation to criteria, may obviously result in obtaining other solutions of a decision problem.

It is essential to note that even when applying the same criterion weights and without changing the evaluations of criterion variants, one can obtain a different decision. In such a case, the solution of a decision problem is influenced by a computational algorithm, which is called a multiple-criteria decision analysis (MCDA) method. The fact can be confirmed in scientific papers in which decision variant rankings, obtained with the use of different MCDA methods [2, 3], were examined. In such a situation the research issue is the selection of decision support methods suitable for a given problem, since only a properly selected method allows gaining a correct solution reflecting a decision-maker's preferences. Moreover, not all methods are able to encompass individual decision problems what results from the limitations of individual MCDA methods, for example, determining weights on an ordinal scale only.

The paper suggests using an ontology as a tool for selecting MCDA methods for a given decision problem. Figure 1 presents a diagram depicting the construction of the discussed ontology. The paper is development of research contained in [4, 5], where an ontology of available MCDA methods was presented and verified (see Fig. 1, step 1). On the basis of the ontology this article focuses on preparing a taxonomy and an ontology of environmental characteristics and abilities of individual MCDA methods (Fig. 1, step 2) as well as chosen literature reference cases of application of the MCDA methods with reference to defined characteristics. The possible construction of such a repository in the form of an ontology supports a decision-maker in selecting an appropriate MCDA method for a given decision problem and also it makes it possible to share and reuse that domain knowledge [6, 7].

2 Literature Review

Although there is a large number of MCDA methods, one needs to remember that no method is perfect and it cannot be considered appropriate for applying in every decision situation or for solving every decision problem [8]. A lot of effort is put into designing MCDA methods, but relatively little is done to evaluate the effectiveness of the methods or to define which methods should be used in certain circumstances [9]. However, if by means of different multiple-criteria contradictory results are obtained, then the correctness of the selection of each method can be questioned [10]. In this case an important research issue is the selection of a decision-making method suitable for a given problem, since only a properly selected method allows gaining an appropriate solution reflecting a decision-maker's preferences.

The complexity of the proper selection of a multiple-criteria method for solving a given decision problem results in many papers which deal with the issue. In Zanakis et al. [11] the benchmark results of 8 MCDA methods (Simple Additive Weighting, Multiplicative Exponential Weighting, TOPSIS, ELECTRE and 4 variants of AHP) were presented. Chang et al. [12] presented a selection procedure of a group fuzzy multiple-criteria method generating a preferred group ranking for a given problem. Hajkowicz and Higgins [9] compared rankings obtained by means of 5 methods (Simple Additive Weighting, Range of Value Method, Promethee II, Evamix, Compromise programming). Al-Shemmeri et al. [13] considered a selection issue of one out of seventeen MCDA methods on the basis of a multiple-criteria approach. Similarly, Celik et al. [14], who used the multiple-criteria approach, evaluated 24 methods in the field of operational research, management sciences and basic MCDA methods in the area of their applicability in the issue of operational maritime transport management. In Adil et al. [15] the selection of multiple-criteria methods for evaluating offers received in public tenders for the public sector was considered. On the other hand, De Montis et al. [16] considered the applicability of selected 7 MCDA methods in general multiple-criteria problems. Celik and Topcu [17] employed the IDEA method (Integrated Decision Aid) for the selection of one out of 24 MCDA methods for 8 management problems related to maritime transport, whereas Wątróbski and Jankowski suggested a rule—based selection of an MCDA method on the basis of abilities of the methods [18].

In the literature, one can observe attempts to develop models of knowledge representation of MCDA problems and methods areas. For example, the article [19] deals with an ontology designed to describe the structure of decision-making problems. It is an element of support for group decision-making processes. On the other hand, in [20] a set of inference rules as well as an ontological representation of the multi-criteria method of AHP were developed. Earlier studies of systematized knowledge about various aspects of decision-making are also presented [21, 22]. One of the approaches is founded on using an ontology knowledge model integrating knowledge about a decision-making process [21]. It includes elements such as a decision-making situation, a decision problem, a set of alternatives and evaluation criteria, rules, preferences and decision-makers. Then the developed model was confirmed within the problem of decision-making for the ERP system selection. Later, the proposed approach was extended by ontology components based on a generalized approach to formalization of methods of decision support [22]. The proposed ontologies focused on the knowledge that enables structuring knowledge about the decision problem. However, only to a small extent do they take the problem of systematization of knowledge about the various methods of multi-criteria decision support. Characterized ontologies do not contain knowledge about the characteristics of different MCDA methods and their environmental context and cases in which they were used [23]. Moreover, the ontologies do not support a decision-maker in the selection of an MCDA method for a given decision problem, for example by means of advanced reasoning. Nevertheless, in this area the ontologies offer a number of possibilities resulting from the fact that they are used to represent knowledge, but not to represent data.

3 Construction of an Ontology of MCDA Methods

In the constructed ontology, whose aim was to support a decision-maker in MCDA method selection suitable for a given decision problem, the features of the ontology as tools for representing knowledge were used. Nowadays, in the computer science literature the most often quoted definition of an ontology reads that it is a formal specification of conceptualization which can capture field knowledge suitable for multiple application [24]. Employing an ontology for creating databases results from a certain essential difference between knowledge bases and databases. Because of the difference when creating knowledge bases, one cannot use database schemes. Extracting information from databases is an algorithmic operation, while the same operation calculated on the basis of a knowledge base takes on the nature of logical reasoning [25, 26]. Furthermore, databases are usually non-monotonous, whereas knowledge bases are monotonous. Consequently, new data added to a database may change conclusions reached by a base (a base's response to a question asked) [27]. On the other hand, new knowledge added to a knowledge base cannot change conclusions which were drawn from it [7]. In other words, databases represent a closed-world assumption in which reasoning is based on the principle that the base contains a complete description of the world. Therefore, if it is impossible to confirm a fact, it is considered false (it is so-called default negation, negation as failure). On the other hand, knowledge bases most often represent an open-world assumption, in which reasoning is based on the principle that the base contains an incomplete description of the world. Therefore, if a fact is openly denied, it is not considered false. What is more, the impossibility of proving some fact does not make it false (it is so-called classical negation) [7]. The presented features of an ontology are coherent with the rules of propagation for MCDA methods described in [1, 8]. The rules state that a functionally-richer method may be used for solving problems suitable in terms of their complexity as well as less complex problems. Instantly, by means of the Electre III method one can solve a problem in which criterion weights are different as well as a problem where there are no weights (all criteria are equally important). By means of a less complex method one cannot handle a more complex problem without information losses. For example, with the use of the Electre IV method we cannot solve, without information losses, a decision problem in which individual criteria have different weights, since the method does not take criterion weights into account. As for an ontology, given cohesion with rules of propagation is manifested by the fact that when defining requirements of a decision problems, the ontology points to all MCDA methods fulfilling these requirements including methods which have more functionalities.

Table 1 Abilities of individual MCDA methods

No.	Method name	Ability 1	Ability 2	Ability 3	Ability 4	Ability 5
1	Additive fuzzy weight method	T	T	T	F	T
2	Additive weight method	T	T	T	F	T
3	AHP	T	T	T	T	T
4	ANP	T	T	T	T	T
5	ARGUS	T	T	F	F	F
6	COMET	T	F	F	F	T
7	Electre I	T	T	T	F	F
8	Electre II	T	T	T	F	F
9	Electre III	T	T	T	F	T
10	Electre IS	T	T	T	F	T
11	Electre IV	T	F	F	F	F
12	Electre TRI	T	T	T	F	T
13	EVAMIX	T	T	T	F	T
14	Fuzzy AHP	T	T	T	T	T
15	Fuzzy ANP	T	T	T	T	T
16	Fuzzy methods of extracting the minimum and maximum values of the attribute	T	F	F	F	F
17	Fuzzy PROMETHEE I	T	T	T	F	T
18	Fuzzy PROMETHEE II	T	T	T	F	T
19	Fuzzy TOPSIS	T	T	T	F	T
20	IDRA	T	T	T	F	T
21	Lexicographic method	T	T	F	F	F
22	MACBETH	T	T	T	T	T
23	MAPPAC	T	T	T	F	T
24	MAUT	T	T	T	F	T
25	MAVT	T	T	T	F	T
26	Maximin	T	F	F	F	F
27	Maximin fuzzy method	T	T	T	F	T
28	MELCHIOR	T	T	F	F	T
29	Methods of extracting the minimum and maximum values of the attribute	T	F	F	F	F
30	NAIADE I	T	T	T	F	T
31	NAIADE II	T	T	T	F	T
32	ORESTE	T	T	F	F	T
33	PACMAN	T	T	T	F	T
34	PAMSSEM I	T	T	T	F	T
35	PAMSSEM II	T	T	T	F	T
36	PRAGMA	T	T	T	F	T
37	PROMETHEE I	T	T	T	F	T

(continued)

Table 1 (continued)

No.	Method name	Ability 1	Ability 2	Ability 3	Ability 4	Ability 5
38	PROMETHEE II	T	T	T	F	T
39	QUALIFLEX	T	T	F	F	F
40	REGIME	T	T	F	F	F
41	SMART	T	T	T	F	T
42	TACTIC	T	T	T	F	T
43	TOPSIS	T	T	T	F	T
44	UTA	T	T	T	F	T

Table 2 Selected cases of using MCDA methods

No	Abilities/ requirements					Remarks	Recommended MCDA method	MCDA requirements meeting	References
	1	2	3	4	5				
1	T	T	T	T	T	Factory/manufacturer selection for producing a given product	Fuzzy AHP (14)	3, 4, 14, 15, 22	[29]
2	T	T	T	F	F	Selecting a material for production	Electre II (8)	1–4, 7–10, 12–15, 17–20, 22–25, 27, 30, 31, 33–38, 41–44	[30]
3	T	F	F	F	T	Selecting glass for assembly	Electre III (9)	1–4, 6, 9–10, 12–15, 17–20, 22–25, 27, 28, 30–38, 41–44	[31]
4	T	T	T	F	T	Evaluation of outsourcing companies	Promethee I (37)	1–4, 9–10, 12–15, 17–20, 22–25, 27, 30, 31, 33–38, 41–44	[32]
5	T	T	T	T	T	Evaluation of the performance of ports	MACBETH (22)	3, 4, 14, 15, 22	[33]
6	T	F	F	F	F	Selecting a material for production of sensitive components	Electre IV (11)	1–44	[34]
7	T	T	T	F	T	Choice of storage location	Topsis (43)	1–4, 9–10, 12–15, 17–20, 22–25, 27, 30, 31, 33–38, 41–44	[35]
8	T	F	F	F	T	Choice of airport for airlines	AHP (3)	1–4, 6, 9–10, 12–15, 17–20, 22–25, 27, 28, 30–38, 41–44	[36]

The basis for enlarging the authors' ontology, which is presented in detail in [4, 5], is to define a complete set of abilities of MCDA methods. The set (abilities) was determined when analyzing guidelines presented in [1, 8, 12, 13] and using research results in [18] and it includes:

- Ability 1—Possibility of comparing all alternatives according to all criteria
- Ability 2—Applying weights for criteria
- Ability 3—Ability to define quantitative importance of each criterion
- Ability 4—Ability to use relative quantitative importance of each criterion
- Ability 5—Ability to possibility of comparing all alternatives according to all criteria on quantitative scale

The functionalities, with respect to individual MCDA methods, are shown in Table 1.

In the next step, with the use of taxonomy presented in Table 1, an analysis of selected reference examples of using MCDA methods in solving specific decision problems was conducted. A partial list as well as requirements, referring to individual methods' abilities, of a decision problem are included in Table 2.

In the constructed ontology individual methods and their abilities are resented as concepts, while cases of using individual methods are presented as individuals. The abilities of individual MCDA methods and their coherent requirements of decision problems contained in the cases are presented as attributes. In order to do that a binary relation has Ability was employed. Moreover, the ontology contains characteristics of individual MCDA methods, as shown in [4, 5].

4 Research Results

The constructed ontology is available online [28]. Its tests were carried out with the use of competence questions [24]. Simple questions were used in which MCDA methods fulfilling individual abilities as well as cases taking account individual requirements of decision problems were indicated. Figure 2a, b display answers of the ontology to competence questions regarding methods and cases they were used in, for which fulfilling abilities 1, 2, 3, 5 were taken into consideration.

```
hasAbility some CompAllAlternativesAccToAllCriteria
and (hasAbility some ApplyingCriteriaWeights)
and (hasAbility some QuantitativeImportanceOfEach-
Criterion)
and (hasAbility some CompAllAlternativesAccToAll-
CriteriaOnQuantitativeScale)
```

The analysis of the results of the competence questions indicates that in accordance with the assumptions, by means of open-world assumption and classical negation, the ontology chooses all methods meeting special requirements. It is a broader solution compared to that presented in [18], where, because of the use of

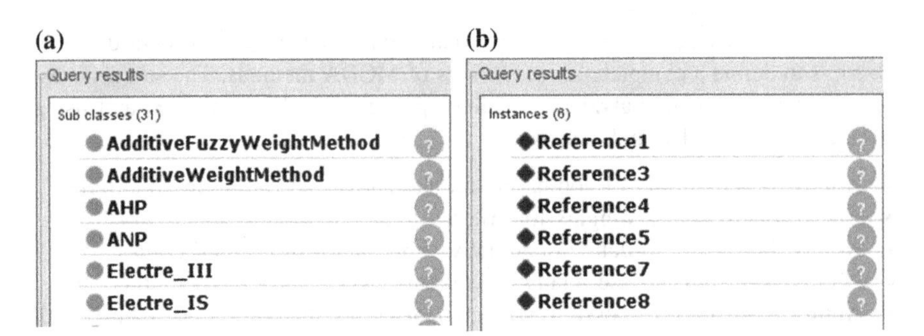

Fig. 2 Results of competence questions regarding methods and cases in which they were used fulfilling determined abilities

default negation and closed-world assumption, only methods closely related to the solution of a given decision problem were chosen.

A proper construction of the ontology allowed verifying if the applied method met expected requirements of a decision problem. For instance, a decision-maker faced a decision problem requiring abilities 1, 2 and 5 by a method. By defining a new case of the application of the method, the decision-maker wants to test if the COMET method meets the requirements. The situation written in the ontology has the form of:

$\top \equiv \text{Ability} \cup \text{MCDA\_Method}$	(1)
$\text{ApplyingCriteriaWeights} \subseteq \text{Ability}$	(2)
$\text{MCDA\_ApplyingCriteriaWeights} \subseteq \text{MCDA\_Method}$	(3)
$\text{COMET} \subseteq \text{MethodName} \subseteq \text{MCDA\_Method}$	(4)
$\text{MCDA\_ApplyingCriteriaWeights} \equiv \text{MCDA\_Method} \cap \exists \text{hasAbility.Applying CriteriaWeights}$	(5)
COMET(test)	(6)
$\exists \text{hasAbility.ApplyingCriteriaWeights(test)}$	(7)

Description: test
Types
COMET
hasAbility some ApplyingCriteriaWeights
hasAbility some CompAllAlternativesAccToAllCriteria
hasAbility some CompAllAlternativesAccToAllCriteriaOnQuantitativeScale
MCDA_ApplyingCriteriaWeights

Fig. 3 Results of reasoner's usage

Starting a reasoner in the ontology has a result presented in Fig. 3. The analysis of Fig. 3 makes it possible to notice that the ontology points to requirements of a decision problem, which are not met by a chosen MCDA method. In the presented situation, the method does not allow determining the weights of criteria.

5 Summary

The selection of a proper MCDA method is a key issue in a decision-making process. The fact is confirmed by the literature presented in the paper. It is worth mentioning that in spite of many tests, the problem has not been explicitly solved.

The article presents an approach to MCDA method selection with the use of an ontology. The construction process of a domain ontology was carried out in three stages: the identification of detailed characteristics of individual MCDA methods, the construction of taxonomy on their basis and, in consequence, the ontology of MCDA methods. Broadening the area of knowledge of the prepared ontology with example practical reference cases expands the area of practical application of the proposed solution. The correctness of the prepared solution was verified by means of competence questions and reasoning. The ontology itself is publicly available online [28].

The future research should include broadening the ontology with complete reference sets of using MCDA methods for specific domains such as economy, management, logistics, environment, medicine, etc. What seems to be interesting is expanding the ontology with a set of characteristics of MCDA methods describing the environmental context of their use as it was suggested in [8, 23].

References

1. Roy, B.: Paradigms and challenges. In: Figueira, J., Greco, S., Ehrgott, M. (eds.) Multiple Criteria Decision Analysis. State of the Art Surveys, pp. 3–24. Springer, Boston (2005)
2. Wang, X., Triantaphyllou, E.: Ranking irregularities when evaluating alternatives by using some ELECTRE methods. Omega **36**, 45–63 (2008)
3. Peng, Y., Wang, G., Wang, H.: User preferences based software defect detection algorithms selection using MCDM. Inf. Sci. **191**, 3–13 (2012)
4. Wątróbski, J., Jankowski, J.: Knowledge management in MCDA domain. In: IEEE Proceedings of the Federated Conference on Computer Science and Information Systems. Annals of Computer Science and Information Systems, vol. 5, pp. 1445–1450 (2015)
5. Wątróbski, J., Jankowski, J.: An ontology-based knowledge representation of MCDA methods. In: ACIIDS 2016, Part I, LNAI 9621, pp. 245–255. Springer, Heidelberg (2016)
6. Ziemba, P., Wątróbski, J., Jankowski, J., Wolski, W.: Construction and restructuring of the knowledge repository of website evaluation methods. In: Ziemba, E. (ed.) Information Technology for Management, LNiBIP 243, pp. 29–52. Springer, Heidelberg (2016)

7. Hepp, M.: Ontologies: state of the art, business potential, and grand challenges. In: Hepp, M., de Leenheer, P., de Moor, A., Sure, Y. (eds.) Ontology Management. Semantic Web, Semantic Web Services, and Business Applications, pp. 2–23. Springer, Heidelberg (2008)
8. Guitouni, A., Martel, J.M.: Tentative guidelines to help choosing an appropriate MCDA method. Eur. J. Oper. Res. **109**, 501–521 (1998)
9. Hajkowicz, S., Higgins, A.: A comparison of multiple criteria analysis techniques for water resource management. Eur. J. Oper. Res. **184**, 255–265 (2008)
10. Jankowski, J.: Integration of collective knowledge in fuzzy models supporting web design process. In: Proceedings of the Third International Conference Computational Collective Intelligence Technologies and Applications, LNCS, vol. 6923, pp. 395–404 (2011)
11. Zanakis, S.H., Solomon, A., Wishart, N., Dublish, S.: Multi-attribute decision making: a simulation comparison of select methods. Eur. J. Oper. Res. **107**, 501–521 (1998)
12. Chang, Y.H., Yeh, C.H., Chang, Y.W.: A new method selection approach for fuzzy group multicriteria decision making. Appl. Soft Comput. **13**, 2179–2187 (2013)
13. Al-Shemmeri, T., Al-Kloub, B., Pearman, A.: Model choice in multicriteria decision aid. Eur. J. Oper. Res. **97**, 550–560 (1997)
14. Celik, M., Cicek, K., Cebi, S.: Establishing an international MBA program for shipping executives: managing OR/MS foundation towards a unique curriculum design. In: International Conference on Computers and Industrial Design (CIE), pp. 459–463 (2009)
15. Adil, M., Baptista Nunes, M., Peng, G.C.: Identifying operational requirements to select suitable decision models for a public sector e-procurement decision support system. JISTEM—J. Inf. Syst. Technol. Manag. **11,** 211–228 (2014)
16. De Montis, A., De Toro, P., Droste-Franke, B., Omann, I., Stagl, S.: Assessing the quality of different MCDA methods. In: Getzner, M., Spash, C.L., Stagl, S. (eds.) Alternatives for Environmental Valuation, pp. 99–133. Routledge, London (2004)
17. Celik, M., Topcu, I.: Analytical modelling of shipping business processes based on MCDM methods. Marit. Policy Manag. **36**, 469–479 (2009)
18. Wątróbski, J., Jankowski, J.: Guideline for MCDA method selection in production management area. In: Różewski, P., Novikov, D., Bakhtadze, N., Zaikin, O. (eds.) New Frontiers in Information and Production Systems Modelling and Analysis. Intelligent Systems Reference Library, vol. 98, pp. 119–138. Springer, Heidelberg (2016)
19. Chai, J., Liu, J.N.K.: An Ontology-driven framework for supporting complex decision process. In: World Automation Congress (WAC) (2010)
20. Liao, X.Y., Rocha Loures, E., Canciglieri, O., Panetto, H.: A novel approach for ontological representation of analytic hierarchy process. Adv. Mater. Res. **988**, 675–682 (2014)
21. Kornyshova, E., Deneckere, R.: Using an ontology for modeling decision-making knowledge. Frontiers Artif. Intell. Appl. **243**, 1553–1562 (2012)
22. Kornyshova, E., Deneckere, R.: Decision-making ontology for information system engineering. LNCS, vol. 6412, pp. 104–117. Springer, Heidelberg (2010)
23. Wątróbski, J., Jankowski, J., Piotrowski, Z.: The selection of multicriteria method based on unstructured decision problem description. In: Computational Collective Intelligence, ICCCI 2014. LNAI, vol. 8733, pp. 454–465. Springer, Heidelberg (2014)
24. Gruber, T.R.: A translation approach to portable ontology specifications. Knowl. Acquis. **5**, 199–220 (1993)
25. Ziemba, P., Jankowski, J., Wątróbski, J., Piwowarski, M.: Web projects evaluation using the method of significant website assessment criteria detection. Trans. Comput. Collect. Intell. **22**, 167–188 (2016)
26. Ziemba, P., Jankowski, J., Wątróbski, J., Becker, J.: Knowledge management in website quality evaluation domain. In: Nunez, M. et al. (eds.) Computational Collective Intelligence. ICCCI 2015, Part II. LNAI, vol. 9330, pp. 75–85. Springer, Heidelberg (2015)
27. Jankowski, J., Michalski, R., Bródka, P., Kazienko, P., Utz, S.: Knowledge acquisition from social platforms based on network distributions fitting. Comput. Hum. Behav. **51**, 685–693 (2015)
28. http://tinyurl.com/ontoMCDAabil

29. Mikhailov, L.: Fuzzy analytical approach to partnership selection in formation of virtual enterprises. Omega **30**, 393–401 (2002)
30. Shanian, A., Savadogo, O.: A material selection model based on the concept of multiple attribute decision making. Mater. Des. **27**, 329–337 (2006)
31. Marzouk, M.M.: ELECTRE III model for value engineering applications. Autom. Constr. **20**, 596–600 (2011)
32. Araz, C., Ozfirat, P.M., Ozkarahan, I.: An integrated multicriteria decision-making methodology for outsourcing management. Comput. Oper. Res. **34**, 3738–3756 (2007)
33. Madeira Junior, A.G., Cardoso Junior, M.M., Neyra Belderrain, M.C., Correia, A.R., Schwanz, S.H.: Multicriteria and multivariate analysis for port performance evaluation. Int. J. Prod. Econ. **140**, 450–456 (2012)
34. Shanian, A., Savadogo, O.: A methodological concept for material selection of highly sensitive components based on multiple criteria decision analysis. Expert Syst. Appl. **36**, 1362–1370 (2009)
35. Ozcan, T., Celebi, N., Esnaf, S.: Comparative analysis of multi-criteria decision making methodologies and implementation of a warehouse location selection problem. Expert Syst. Appl. **38**, 9773–9779 (2011)
36. Janic, M., Reggiani, A.: An application of the multiple criteria decision making (MCDM) analysis to the selection of a new hub airport. Eur. J. Transp. Infrastruct. Res. **2**, 113–142 (2002)

A Hybrid Approach to Decision Support for Resource-Constrained Scheduling Problems

Paweł Sitek, Izabela Nielsen, Jarosław Wikarek and Peter Nielsen

Abstract Resource-constrained scheduling problems are commonly found in various areas, such as project management, manufacturing, transportation, software engineering, computer networks, and supply chain management. Its problem models involve a large number of constraints and discrete decision variables, including binary and integer. In effect, the representation of resource allocation, for instance, is often expressed using binary or integer decision variables to form several constraints according to the respective scheduling problem. It significantly increases the number of decision variables and constraints as the problem scales; such kind of traditional approaches based on operations research is insufficient. Therefore, a hybrid approach to decision support for resource-constrained scheduling problems which combines operation research (OR) and constraint logic programming (CLP) is proposed. Unlike OR-based approaches, declarative CLP provides a natural representation of different types of constraints. This approach provides: (a) decision support through the answers to the general and specific questions, (b) specification of the problem based on a set of facts and constraints, (c) reduction to the combinatorial solution space. To evaluate efficiency and applicability of the proposed hybrid approach and implementation platform,

P. Sitek (✉) · J. Wikarek
Department of Information Systems, Kielce University of Technology,
Kielce, Poland
e-mail: sitek@tu.kielce.pl

J. Wikarek
e-mail: j.wikarek@tu.kielce.pl

I. Nielsen · P. Nielsen
Department of Mechanical and Manufacturing Engineering, Aalborg University,
Aalborg, Denmark
e-mail: izabela@m-tech.aau.dk

P. Nielsen
e-mail: peter@m-tech.aau.dk

I. Czarnowski et al. (eds.), *Intelligent Decision Technologies 2016*,
Smart Innovation, Systems and Technologies 56,
DOI 10.1007/978-3-319-39630-9_9

implementation examples of job-shop scheduling problem are presented separately for the three environments, i.e., Mathematical Programming (MP), CLP, and hybrid implementation platform.

Keywords Decision support · Constraint logic programming · Resource-constrained scheduling problem · Mathematical programming · Hybridization

1 Introduction

In this highly competitive economy era, the information technology system is demanded to make optimal decisions in the shortest time possible on behalf of managers and owners (humans). It is to make and execute decisions that will take effect on the business operations in this global market. It means, there is no room for mistake. Successful decisions depend on optimal and fast allocation of financial, technological, and human resources at the optimal cost and guaranteed feasible execution time to meet the actual requirements at hand. Very often, the decision-making problems are the allocation of different types of resources, which must satisfy a set of different constraints, where some constraints may conflict with each other, over time. In this case, we can talk about a general form of resource-constrained scheduling problem (RCSP), which covers flow-shop, job-shop and open-shop problems, the project scheduling problem, and production scheduling problems. These problems appear in decisions problems in supply chain, manufacturing, transportation, distribution, logistic, computer networks, and construction engineering area. The RCSPs involve various integer, logic, linear and other types of constraint. Hence, to create an efficient, fast, and easy ways of modeling and solving different types of constraint is a key issue. The simplest methods for modeling and solving linear constraints are operation research (OR) methods and mathematical programming (MP) [1]. For the remaining types of constraints (e.g., binary, integer, logical, non-linear constraints), MP methods shows weak performance and efficiency. For instance, one way of modeling the allocation was to introduce 0–1 decision variables and constraints.

Introduction of 0–1 decision variables and constraints to the model muddle the structure of the model and increases the search space. Such difficulties do not appear in declarative environments. Unlike mathematical programming approach, declarative constraint logic programming (CLP) provides an easy and intuitive specification of different types of constraints. CLP is a form of constraint programming (CP), in which logic programming (LP) is extended to include concepts from constraint satisfaction problem (CSP) [2]. Formally speaking, a CSP is defined as a triple $(X_i, D^i_{om}, C^i_{st})$, where X_i is a set of variables, D^i_{om} is a domain of values, and C^i_{st} is a set of constraints. Constraint satisfaction problems (CSPs) are the subject of intense restudy in artificial intelligence, operations research and soft computing [2–5]. CSPs are typically solved using a different form of search. The most commonly used methods include variants of backtracking algorithms, constraint propagation, and local search. CLP-based approach is very effective for solving binary constraints (binding at most two decision variables). If there are more than two decision variables in the constraints, CLP efficiency decreases dramatically. Moreover, discrete optimization is not a strong point of CLP. On the other hand, declarative approach undoubtedly suitable to build a decision support system. This is due to the easy modeling of any constraints, questions and facts.

Based on [3–6] and our previous study on integration CLP and MP [7–9] some strengths and weaknesses of the aforementioned environments have been observed. The integration of CLP/MP can help modeling and solving decision and optimization problems for resource-constrained scheduling which are difficult to solve with one of two methods alone. This integration, enriched with problem transformation, has been called a hybrid approach [7, 8].

The main contribution and also the motivation of this study is to apply a hybrid approach to decision support for RCSPs. The proposed approach enables managers and users to ask different kinds of questions (specific, general, search). Moreover, this approach allows easy and effective modeling and solving of decision and optimization models for scheduling problems.

2 Resource-Constrained Scheduling Problems

Scheduling processes are common to many different management and engineering areas. Scheduling process answers "How" and "When" it will be done. Additionally, these problems include many types and forms of constraint (e.g., temporal, precedence, availability). One of the most important types of scheduling problems is a resource-constrained scheduling problem. In its most general form, the resource-constrained scheduling problem [10–12] is defined as follows.

Given a set of activities (e.g., orders, machine operations, services) that must be done, a set of resources (e.g., processors, machines, tools, employees, row materials, electricity) to execute activities, a set of different types of constraints (e.g., capacity, allocation, precedence) must be satisfied, and a set of objectives (e.g., costs, makespan, tardiness) to evaluate a schedule's execution with. There may also exist several logical conditions. For instance, a condition where two resources cannot be used simultaneously. Two main concerns about such constraint: to decide the best way to assign available resources to the activities at specific times such that all of the constraints are satisfied and to decide the best objective measures (in regards to optimization problems).

Objectives and constraints are defined during the construction of the model. Objectives define the "optimality" of the schedule while constraints define the "feasibility" of the schedule.

Constraints appear in many areas and forms. Precedence constraints define the order in which activities can be executed. Temporal constraints limit the times at which resources may be used and/or activities may be done. There are linear, integer, and logic constraints.

3 A Hybrid Approach to Decision Support for RCSPs-Hybrid Implementation Platform

The hybrid approach to decision support for RCSPs is able to fill the gaps and eliminate the shortcomings that occur in both MP-based and CLP-based approaches when they are used separately. A hybrid approach is not just a simple integration of the two environments (CLP/MP) but also contains the process model transformation [13]. In short, the transformation changes the representation of the problem using the data about the problem (facts) and the characteristics and possibilities of CLP. The result of the transformation is to reduce the number of decision variables and constraints of the problem, which is particularly important and useful for optimization because it reduces the search space. To support and illustrate the concept of hybrid approach to resource constrained scheduling problems, a hybrid implementation platform is proposed (Fig. 1).

The important assumptions of the hybrid implementation platform (HIP) include the following:

- specification of the problem can be stated as integrity, binary, linear, logic and symbolic constraints;
- data instances of the problem are stored as sets of facts (Appendix A);
- structure of the problem reflects the relationship between the sets of facts (Fig. 2);
- two environments, i.e. CLP and MP are integrated;
- platform supports the transformation of the problem [13] as an integral part of the hybrid approach;
- decision support is based on the answers the questions being asked;
- different types of questions can be asked (general, specific, wh-questions, etc.);
- general questions (T1a): *Whether it is possible …? Is it feasible …? Whether for such parameters it is possible..?*
- specific questions (T2a): *What is the maximum/minimum makespan …? What is the minimum/maximum number …? What is the configuration… for..?*
- the logic questions can be asked (T1b,T2b): *What is the minimum/maximum.. If..cannot be used simultaneously?, Is it possible.. If..cannot be used simultaneously?* (the list of the sample questions for illustrative example is shown in Table 1);
- general questions are implemented as CLP predicates (CLP predicate is a collection of clauses. The idea of clause is to define that something is true);
- specific questions and optimization are performed by the MP environment;

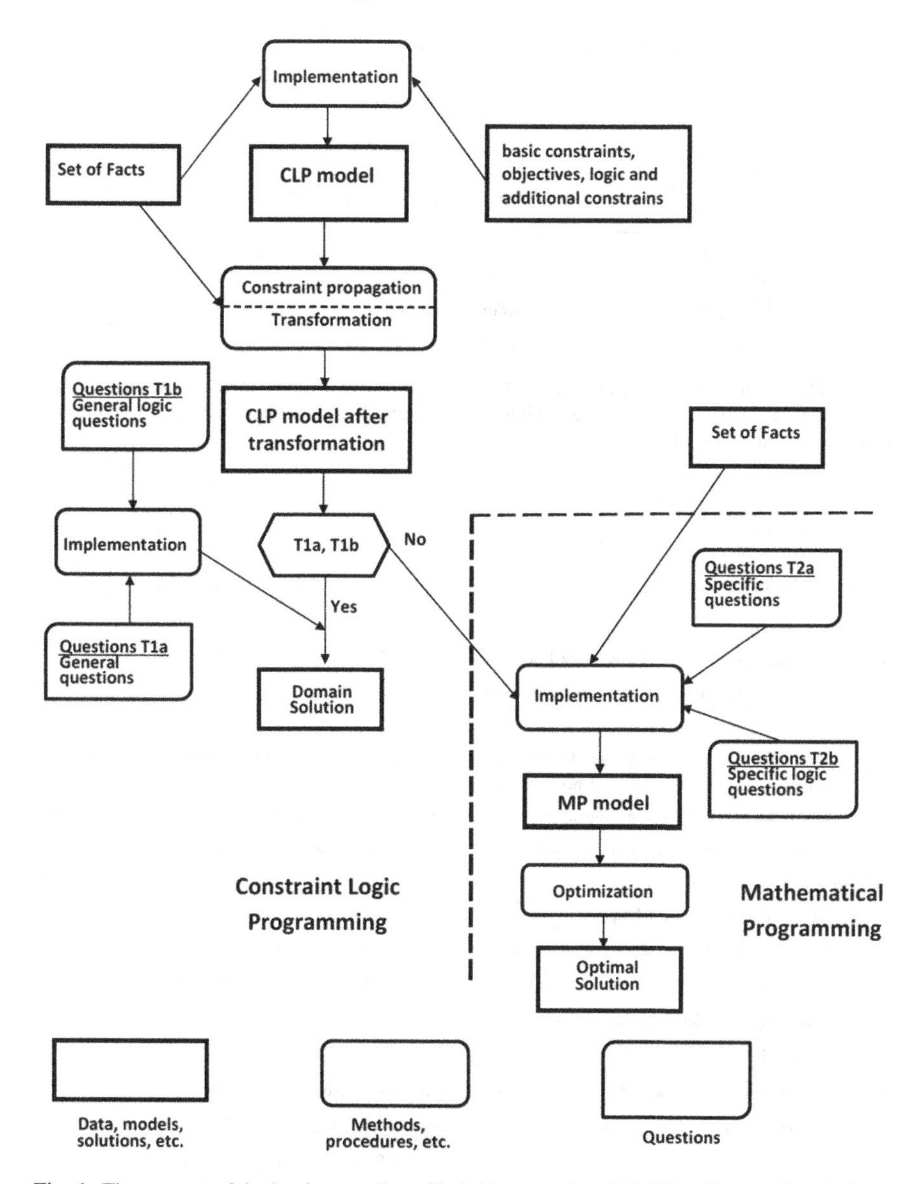

Fig. 1 The concept of the implementation of hybrid approach as hybrid implementation platform (HIP)

- domain solution and model after transformation are transferred from CLP to the MP environment;
- problem is generally modeled and pre-solved using the CLP and finally solved by the MP environment.

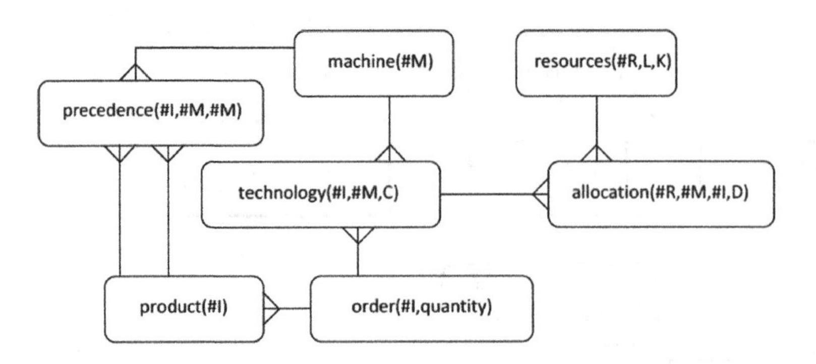

Fig. 2 The relationship between the sets of facts for illustrative example (#-ID of the fact, L-limit of additional resources, K-cost of the additional resources, D-the number of allocated additional resources, C-execution time)

Table 1 Model of illustrative example (decision variables, constraints and questions)

Decision variables	
V_1	The start time and the end time of each job for activity
V_2	The set of additional resources assigned to each job
Constraints	
C_1	Specify the possible values for the start and end times of jobs on machine and the relations between the start and end time of two jobs (precedence and temporal constraints).
C_2	Specify the possible set of additional resources for each job (resource constraint).
C_3	Limit the available capacity of each additional resource over time (capacity constraint)
Questions	
Q1	What is the min C_{max} (makespan)?
$Q2_A$	What is the min C'_{max} if the set of additional resources is R = R1 = R2 = R3 = R4 = R5 = 2?
$Q2_B$	What is the min C'_{max} if the set of additional resources is R = R1 = R2 = R3 = R4 = R5 = 3?
$Q3_A$	What is the minimum set of resources R1 at C'_{max}?
$Q3_B$	What is the minimum set of resources R3 at C'_{max}?
Q4	Is it possible to schedule orders in C'_{max} and what are the sets of resources R1..R5?
Q5	Is it possible to schedule orders in C'_{max} if the set of resources R1, R2, R3, R4, R5?
Q6	Is it possible to schedule orders in C'_{max} if resources R1 and R4 cannot be used simultaneously
Q7	Is it possible to schedule orders in C'_{max} if machines M1 and M2 cannot be used simultaneously
Q8	What is the min C'_{max} if resources R1 and R2 cannot be used simultaneously
Q9	What is the min C'_{max} if machines M1 and M3 cannot be used simultaneously

The schema of implementation platform for the hybrid approach has been shown at Fig. 1. From a variety of tools for the implementation of the CLP in the platform (HIP), the ECL^{i}PS^e software [14] is used. The area of mathematical programming (MP) in the HIP was implemented using the LINGO by LINDO Systems [15].

4 Illustrative Example and Computational Experiments

Job-shop scheduling problem with additional resources. (employees, tools, etc.) is used as an illustrative example. It is a well-known example from the literature about the appropriate degree of difficulty and complexity [10–12]. To evaluate the presented hybrid implementation platform for decision support, a number of numerical experiments were performed for the illustrative example.

The main assumptions of the illustrative example in the form of decision variables, main constraints, and questions are shown in Table 1. The structure of the facts and their relationships for illustrative example are presented at Fig. 2. The numerical experiments are performed for the data instance consisting of seven products (A, B, C, D, E, F, G), eight machines (M1..M8), and five different sets of the additional resources (R1, R2, R3, R4, R5).

The set of facts for illustrative example is shown in Appendix A. The obtained results for each of the questions Q1..Q9 are presented in Table 2 and the corresponding schedules for questions Q1, Q2, Q9 in Figs. 3, 4 and 5.

These studies and numerical experiments have shown great possibilities of the implementation platform for supporting various types of decision by asking questions Q1..Q9. Questions are implemented easily and quickly as a result of declarativeness of the CLP. There were general (Q4, Q5, Q6) specific (Q1, Q2A, Q2B, Q3A, Q3B) and logic (Q7, Q8, Q9) questions. The results for question Q1..Q9 shown in Table 2. In addition, schedules obtained as a result of answers to the questions Q1 and Q9 shown in Figs. 3, 4 and 5. All numerical experiments were performed using an ordinary PC with the following parameters -Intel core (TM2), 2.4 GHz, 2 GB of RAM.

In the second phase of the study, efficiency of the presented hybrid implementation platform (HIP) is evaluated relative to the MP-based and CLP-based implementations. For this purpose, question Q1, which has the largest computing requirements, is implemented and asked in three environments, CLP, MP, and in the HIP. Numerous experiments were performed with varied parameter R. The obtained results is presented in Table 3.

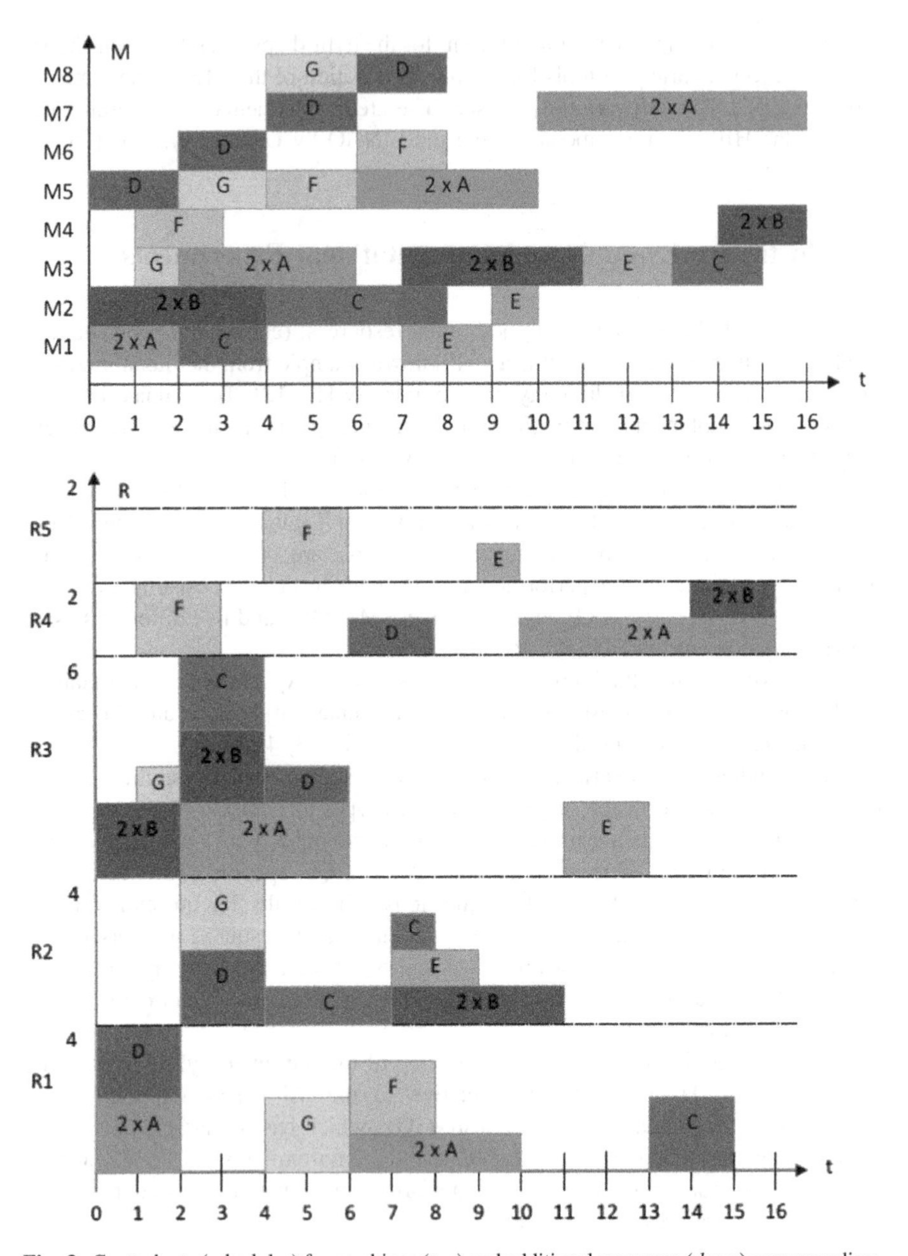

Fig. 3 Gantt charts (schedules) for machines (*top*) and additional resources (*down*) corresponding to the question Q1 for R $\geq$ 10

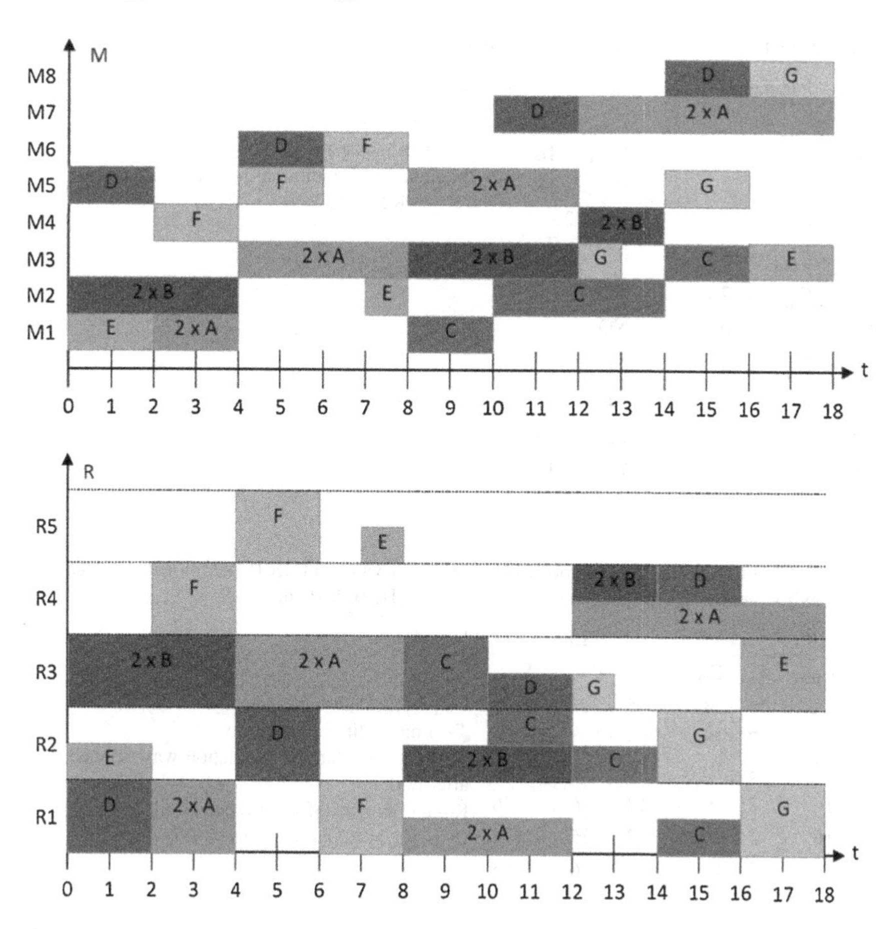

Fig. 4 Gantt charts (schedules) for machines (*top*) and additional resources (*down*) corresponding to the question Q2 for R = 2

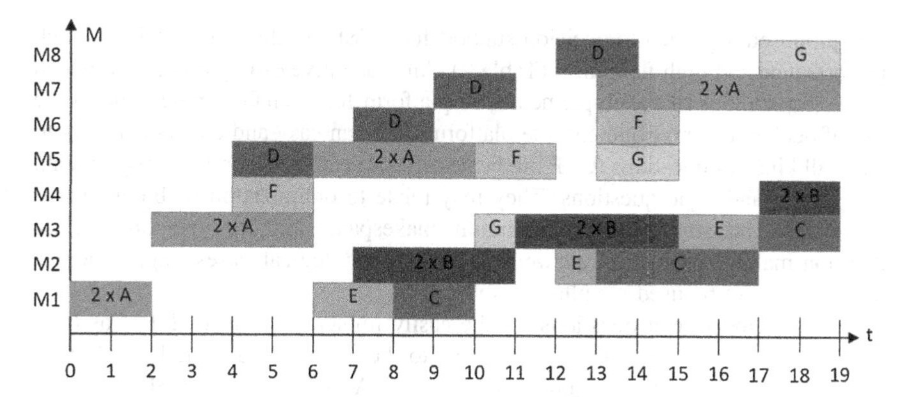

Fig. 5 Gant chart (schedule) for machines corresponding to the question Q9

Table 2 Obtained results of the numerical experiments for asked questions Q1..Q9 to HIP

Q	Parameters	Result	Description
Q1	–	$C_{max} = 16$	C_{max}—optimal makspan C'_{max}—given makspan R—the number of additional resources (R = R1 = R2 = R3 = R4 = R5)
$Q2_A$	R = 2	$C_{max} = 18$	
$Q2_B$	R = 3	$C_{max} = 18$	
$Q3_A$	$C'_{max} = 16$	R1 = 2	
$Q3_B$	$C'_{max} = 16$	R3 = 4	
Q4	$C'_{max} = 17$ R = 3	NO	
Q5	$C'_{max} = 20$, R = 2	NO	
Q6	$C'_{max} = 25$	NO	
Q7	$C'_{max} = 19$	YES	
Q8	–	$C_{max} = 28$	
Q9	–	$C_{max} = 19$	

Table 3 Obtained results for questions Q1 at different values of set R (additional resources) in three environments: Hybrid Implementation Platform (HIP), MP, and CLP

R	HIP		MP		CLP		Description
	C_{max}	T	C_{max}	T	C_{max}	T	
2	18	26	18	285	20*	600*	C_{max}—optimal makespan T—time of finding solution *—feasible solution, calculation was stopped after 600 s R—the number of additional resources (R = R1 = R2 = R3 = R4 = R5)
3	18	17	18	254	20*	600*	
4	16	2	16	166	16	356	
5	16	2	16	13	16	236	
6	16	2	16	12	16	134	
7	16	2	16	8	16	63	

5 Conclusion

The proposed approach to decision support for RCSPs is characterized by its high performance and high flexibility (Table 2). An illustrative example is given, which shows capabilities of the implementation platform for a small sample. Due to the use of declarative environment, the platform offers an easy and convenient way to carry all kinds of questions (CLP predicates). These can be general, wh-type, what.. if, specific, and logic questions. They may relate to optimization problems (minimum cost, maximum profit, minimum makespan, minimum resources etc.), decision-making problems (answers: Yes/No), and logical cases (e.g., where a resource cannot be used simultaneously.).

The aforementioned questions can be easily implemented in ECL^iPS^e as CLP predicates. Unfortunately, in CLP, answers to the specific and some logical questions (in the illustrative example for Q1, Q2A, Q2B, Q3A, Q3B and Q8, Q9–Table 3) are inefficient. Hence, the concept of replacing the classical declarative distribution of variables [7, 8] by the OR-based methods and transformation of the problem is the basis for the hybrid implementation platform. Evaluation of performance and effectiveness have been studied using job-shop scheduling optimization problem with additional resources (Q1) with regard to the CLP-based and MP-based approaches (Table 3).

It is found that even in the small illustrative example presented, it is 4–80 times faster relative to the MP, and many times faster relative to the CLP (Table 3). It shows that the proposed implementation platform gives significantly better results in the search time than the use of each of its components alone (Table 3).

Further studies will include implementation of various types of problems in the area of supply chain management, capacity vehicle routing problems [16], resource capacity project scheduling, multi-project scheduling problems [17], and multi-assortment repetitive production. The hybrid implementation platform will also be supplemented by the paradigm of Fuzzy Logic and Answer Set Programming [18, 19].

Appendix A Data Instances for Illustrative Example (Sets of Facts)

```
%machine (#M).
machine (M1). machine (M2). machine (M3). machine (M4).
machine (M5). machine (M6). machine (M7). machine (M8).
%product (#I).
product (A). product (B). product (C). product (D).
product (E). product (F). product (G).
%technology (#I,#M,C).
technology (A,M1,1). technology (A,M3,2). technology
(A,M5,2).
technology (A,M7,3). technology (B,M2,2). technology
(B,M3,2).
technology (B,M4,1). technology (C,M1,2) technology
(C,M2,4).
technology (C,M3,2). technology (D,M5,2). technology
(D,M6,2).
technology (D,M7,5). technology (D,M8,2). technology
(E,M1,2).
technology (E,M2,1). technology (E,M3,2). technology
(F,M4,2).
technology (F,M5,2). technology (F,M6,2). technology
(G,M3,1).
technology (G,M5,2). technology (G,M8,2).
```

```
%resources (#R,L,K).
resources   (R1,12,40).   resources   (R2,12,30).resources
(R3,12,30).
resources (R4,12,20). resources (R5,12,20).
%allocation (#R,#M,#I,D)
allocation (O1,M1,A,2). allocation (O3,M3,A,2).
allocation (O1,M5,A,1). allocation (O4,M7,A,1).
allocation (O3,M2,B,2). allocation (O2,M3,B,1).
allocation (O4,M4,B,1). allocation (O3,M1,C,2).
allocation (O2,M2,C,1). allocation (O1,M3,C,2).
allocation (O1,M5,D,2). allocation (O2,M6,D,2).
allocation (O3,M7,D,1). allocation (O4,M8,D,1).
allocation (O2,M1,E,1). allocation (O5,M2,E,1).
allocation (O3,M3,E,2). allocation (O4,M4,F,2).
allocation (O5,M5,F,2). allocation (O1,M6,F,2).
allocation (O3,M3,G,1). allocation (O2,M5,G,2).
allocation (O1,M8,G,2).
%precedence (#P,#M,#M).
precedence  (A,M1,M2).  precedence  (A,M2,M3).precedence
(A,M3,M7).
precedence  (B,M2,M3).  precedence  (B,M3,M4).precedence
(C,M1,M2).
precedence  (C,M2,M3).  precedence  (D,M5,M6).precedence
(D,M6,M7).
precedence  (D,M7,M8).  precedence  (E,M1,M2).precedence
(E,M2,M3).
precedence  (F,M4,M5).  precedence  (F,M5,M6).precedence
(G,M3,M5).
precedence (G,M5,M8).
Order (#P,quantity).
order(A,2). order(B,2). order(C,1). order(D,1).
order(E,1). order(F,1). order(G,1).
```

References

1. Schrijver, A.: Theory of Linear and Integer Programming. Wiley, New York (1998)
2. Rossi, F., Van Beek, P., Walsh, T.: Handbook of Constraint Programming (Foundations of Artificial Intelligence). Elsevier Science Inc., New York (2006)
3. Apt, K., Wallace, M.: Constraint Logic Programming using Eclipse. Cambridge University Press, Cambridge (2006)
4. Milano, M., Wallace, M.: Integrating operations research. Constraint Program. Ann. Oper. Res. **175**(1), 37–76 (2010)
5. Achterberg, T., Berthold, T., Koch, T.: Wolter K: Constraint integer programming, a new approach to integrate CP and MIP. Lect. Notes Comput. Sci. **5015**, 6–20 (2008)

6. Bocewicz, G., Banaszak, Z.: Declarative approach to cyclic steady states space refinement: periodic processes scheduling. Int. J. Adv. Manuf. Technol. **67**(1–4), 137–155 (2013)
7. Sitek, P., Wikarek, J.: A Hybrid Approach to the Optimization of Multiechelon Systems. Mathematical Problems in Engineering, Article ID 925675, Hindawi Publishing Corporation, (2014). doi:10.1155/2014/925675
8. Sitek, P., Nielsen I.E., Wikarek, J.: A Hybrid Multi-agent Approach to the Solving Supply Chain Problems. Procedia Computer Science KES, pp. 1557–1566 (2014)
9. Sitek, P., Wikarek J.: A hybrid framework for the modelling and optimisation of decision problems in sustainable supply chain management. Int. J. Prod. Res. 1–18 (2015). doi:10.1080/00207543.2015.1005762
10. Guyon, O., Lemaire, P., Pinson, Ă., Rivreau, D.: Solving an integrated job-shop problem with human resource constraints. Ann. Oper. Res. **213**(1), 147–171 (2014)
11. Blazewicz, J., Lenstra, J.K., Rinnooy Kan, A.H.G.: Scheduling subject to resource constraints: classification and complexity. Discret. Appl. Math. **5**, 11–24 (1983)
12. Lawrence, S.R., Morton, T.E.: Resource-constrained multi-project scheduling with tardy costs: comparing myopic, bottleneck, and resource pricing heuristics. Eur. J. Oper. Res. **64**(2), 168–187 (1993)
13. Sitek, P.: A hybrid CP/MP approach to supply chain modelling, optimization and analysis. In: Proceedings of the 2014 Federated Conference on Computer Science and Information Systems, Annals of Computer Science and Information Systems, vol. 2, pp. 1345–1352 (2014). doi:10.15439/2014F89
14. Lindo Systems INC, LINDO™, www.lindo.com, Accessed Dec 4 (2015)
15. Eclipse–The Eclipse Foundation open source community website, www.eclipse.org, Accessed Dec 4 (2015)
16. Toth, P., Vigo, D.: Models, relaxations and exact approaches for the capacitated vehicle routing problem. Discret. Appl. Math. **123**(1–3), 487–512 (2002)
17. Coelho, J., Vanhoucke, M.: Multi-mode resource-constrained project scheduling using RCPSP and SAT solvers. Eur. J. Oper. Res. **213**, 73–82 (2011)
18. Relich, M.: A computational intelligence approach to predicting new product success. In: Proceedings of the 11th International Conference on Strategic Management and its Support by Information Systems, pp. 142–150 (2015)
19. Wang, J., Liu, C.: Fuzzy Constraint Logic Programming with Answer Set Semantics. Lecture Notes in Computer Science, pp. 52–60 (2007). doi:10.1007/978-3-540-76719-0_9

Prediction of Length of Hospital Stay in Preterm Infants a Case-Based Reasoning View

Ana Coimbra, Henrique Vicente, António Abelha, M. Filipe Santos, José Machado, João Neves and José Neves

Abstract The length of stay of preterm infants in a neonatology service has become an issue of a growing concern, namely considering, on the one hand, the mothers and infants health conditions and, on the other hand, the scarce healthcare facilities own resources. Thus, a pro-active strategy for problem solving has to be put in place, either to improve the quality-of-service provided or to reduce the inherent financial costs. Therefore, this work will focus on the development of a diagnosis decision support system in terms of a formal agenda built on a Logic Programming approach to knowledge representation and reasoning, complemented with a case-based problem solving methodology to computing, that caters for the handling of incomplete, unknown, or even contradictory information. The proposed model has been quite accurate in predicting the length of stay (overall accuracy of 84.9 %) and by reducing the computational time with values around 21.3 %.

A. Coimbra
Departamento de Informática, Universidade do Minho, Braga, Portugal
e-mail: cecilia.coimbra@hotmail.com

H. Vicente
Departamento de Química, Escola de Ciências e Tecnologia
Universidade de Évora, Évora, Portugal
e-mail: hvicente@uevora.pt

H. Vicente · A. Abelha · M. Filipe Santos · J. Machado · J. Neves (✉)
Centro Algoritmi, Universidade do Minho, Braga, Portugal
e-mail: jneves@di.uminho.pt

A. Abelha
e-mail: abelha@di.uminho.pt

M. Filipe Santos
e-mail: mfs@dsi.uminho.pt

J. Machado
e-mail: jmac@di.uminho.pt

J. Neves
Drs. Nicolas & Asp, Dubai, United Arab Emirates
e-mail: joaocpneves@gmail.com

I. Czarnowski et al. (eds.), *Intelligent Decision Technologies 2016*,
Smart Innovation, Systems and Technologies 56,
DOI 10.1007/978-3-319-39630-9_10

Keywords Preterm infants • Length of stay • Neonatology • Knowledge representation and reasoning • Logic programming • Case-based reasoning

1 Introduction

In the current century hospital deliveries increased significantly and, consequently, issues related with the Length of Stay (LoS) of subsequent postpartum hospitalization became of the utmost importance. As a result, indicators that may assess the effectiveness of organizations and promote the improvement in hospital quality-of-care, long-term resource planning and family counseling were enforced [1–4].

There are different opinions about the time considered appropriate to hospitalization, ranging from drive-through deliveries (with only a few hours of LoS) to 14 (fourteen) day lying-in periods [1]. With the early discharge the family ties are promoted, the patient satisfaction is improved, the risk of iatrogenic infection is diminished and the hospitalization care and patients' costs are reduced [5, 6]. However, some aspects should not be ruled out, namely the fact that until two or more days after delivery some adverse events are not noticeable, which may lead to an increased readmission rate. On the other hand these readmission episodes may cause anxiety in parents and induce the premature cessation of breastfeeding [5, 6]. In an ideal scenario the LoS should be reduced to the strictly necessary time, without endangering the mother's and infant's health, allowing identification of early problems and ensuring that the family is able and prepared to care for the infant at home. Thus, the LoS should be customized for each mother/infant dyad, taking into consideration the health of the mother, the health of the infant and the conditions that they will have outside the hospital [7, 8].

One of the main causes of 4 million neonatal deaths that annually occur all over the world stands for premature birth [9]. Premature or preterm labor may be defined as one in which pregnancy ends between the 20th and the 37th week. The LoS are more than six times greater for extremely preterm infants than for late ones and, therefore, the hospitalization costs are significantly higher under the former case [2]. In fact, the prediction of the LoS of preterm infants in a neonatology service can contribute to the improvement of the long-term planning, leading to a cost reduction, and it also can identify modifiable risk factors that can support quality improvement initiatives. Thus, this paper addresses the length of hospital stay in the preterm infants theme and describes an attempt to predict such period, using a Case-Based Reasoning (CBR) approach to problem solving [10, 11]. CBR allows handling new problems by reusing knowledge acquired from past experiences, namely when similar cases have similar terms and solutions [10 11]. Actually, its use may be found in different arenas, namely in online dispute resolution [12, 13], medicine [14, 15] or education [16], among others. An approach that allows one to deal with incomplete, contradictory or even unknown information.

2 Knowledge Representation and Reasoning

The Logic Programming (LP) paradigm has been used for knowledge representation and reasoning in different areas, like Model Theory [17, 18], and Proof Theory [19, 20]. In the present work the proof theoretical approach is followed in terms of an extension to LP. An Extended Logic Program is a finite set of clauses in the form:

$$\{\ p \leftarrow p_1, \cdots, p_n, not\ q_1, \cdots, not\ q_m$$
$$?\,(p_1, \cdots, p_n, not\ q_1, \cdots, not\ q_m)\ \ (n, m \geq 0)$$
$$exception_{p_1}\ \ \ldots\ \ exception_{p_j}\ \ (0 \leq j \leq k),\ being\ k\ an\ integer\ number$$
$$\}::\ scoring_{value}$$

where "?" is a domain atom denoting falsity, the pi, qj, and p are classical ground literals, i.e., either positive atoms or atoms preceded by the classical negation sign ¬ [19]. Indeed, under this formalism, every program is associated with a set of abducibles [17, 18], given here in the form of exceptions to the extensions of the predicates that make the program, i.e., clauses of the form:

$$exception_{p_1}\ \ \cdots\ \ exception_{p_j}\ (0 \leq j \leq k),\ \ being\ k\ an\ integer\ number$$

that stand for information or knowledge that cannot be ruled out. On the other hand, clauses of the type:

$$?\,(p_1,\ \ldots, p_n, not\ q_1,\ \ldots, not\ q_m)\ \ \ (n, m \geq 0)$$

also named invariants or restrictions to complain with the universe of discourse, set the context under which it may be understood. The term $scoring_{value}$ stands for the relative weight of the extension of a specific *predicate* with respect to the extensions of the peers ones that make the overall program.

In order to evaluate the knowledge that can be associated to a logic program, an assessment of the *Quality-of-Information* (*QoI*), given by a truth-value in the interval [0, 1], that stems from the extensions of the predicates that make a program, inclusive in dynamic environments, is set [21, 22]. Thus, $QoI_i = 1$ when the information is *known* (*positive*) or *false* (*negative*) and $QoI_i = 0$ if the information is unknown. Finally for situations where the extension of $predicate_i$ is unknown but can be taken from a set of terms, $QoI_i \in]0, 1[$. Thus, for those situations, the *QoI* is given by:

$$QoI_i = 1/Card \tag{1}$$

where *Card* denotes the cardinality of the *abducibles* set for *i*, if the *abducibles* set is disjoint. If the *abducibles* set is not disjoint, the clause's set is given by $C_1^{Card} + \cdots + C_{Card}^{Card}$, under which the *QoI* evaluation takes the form:

$$QoI_{i_{1 \leq i \leq Card}} = 1/C_1^{Card}, \ldots, 1/C_{Card}^{Card} \quad (2)$$

where C_{Card}^{Card} is a card-combination subset, with *Card* elements. The objective is to build a quantification process of *QoI* and *DoC* (Degree of Confidence), being the later a measure of one's confidence that the argument values or attributes of the terms that make the extension of a given predicate, with relation to their domains, fit into a given interval [23]. The *DoC* is evaluated as depicted in [23] and computed using $DoC = \sqrt{1 - \Delta l^2}$, where Δl stands for the argument interval length, which was set in the interval [0, 1]. Thus, the universe of discourse is engendered according to the information presented in the extensions of such predicates, according to productions of the type:

$$predicate_i - \bigcup_{1 \leq i \leq m} clause_j((QoI_{x_1}, DoC_{x_1}), \ldots, (QoI_{x_m}, DoC_{x_m})) :: QoI_i :: DoC_i \quad (3)$$

where $\cup$ and m stand, respectively, for *set union* and the *cardinality* of the extension of $predicate_i$. QoI_i and DoC_i stand for themselves [23].

As an example, let us consider the logic program given by:

$$\{$$

$$\neg f_1 \left((QoI_{x_1}, DoC_{x_1}),\ (QoI_{y_1}, DoC_{y_1}),\ (QoI_{z_1}, DoC_{z_1}) \right)$$

$$\leftarrow not \left((QoI_{x_1}, DoC_{x_1}),\ (QoI_{y_1}, DoC_{y_1}),\ (QoI_{z_1}, DoC_{z_1}) \right)$$

$$f_1 \underbrace{\left((QoI_{[7,10]}, DoC_{[7,10]}),\ (QoI_{\perp}, DoC_{\perp}),\ (QoI_{1.5}, DoC_{1.5}), \right)}_{attribute\text{'}s\ values} :: QoI :: DoC$$

$$\underbrace{[5, 35] \qquad [0, 10] \qquad [0.5, 2.5]}_{attribute\text{'}s\ domains}$$

$$exception_{f_{1,1}} \left((QoI_{18}, Doc_{18}), (QoI_{[1,2]}, DoC_{[1,2]}), (QoI_{\perp}, DoC_{\perp}) \right) :: QoI :: DoC$$

$$\ldots$$

$$exception_{f_{1,k}} \left((QoI_{\perp}, DoC_{\perp}), (QoI_5, DoC_5), (QoI_{[0.6,1]}, DoC_{[0.6,1]}) \right) :: QoI :: DoC$$

$\}$ :: 1 (*once the universe of discourse is set in terms of the extension of only one predicate*)

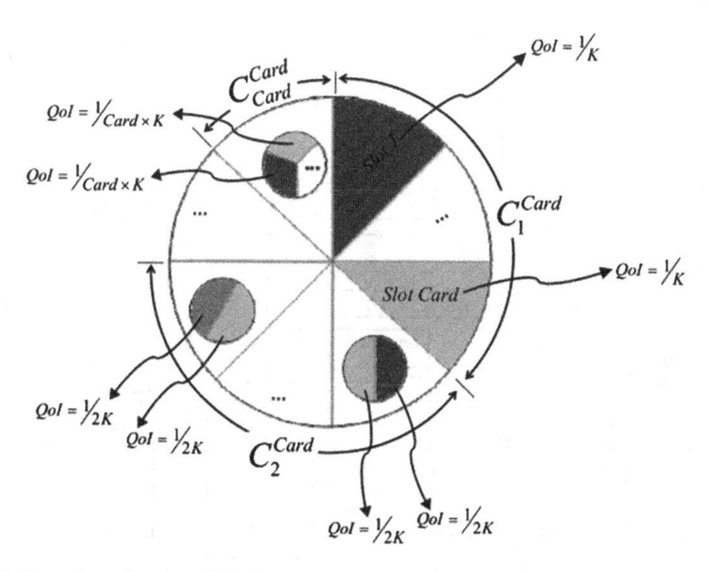

Fig. 1 *QoI's* values for the abducible set of clauses referred to above, where the clauses cardinality set, *K*, is given by the expression $C_1^{Card} + C_2^{Card} + \cdots + C_{Card}^{Card}$

where ⊥ denotes a null value of the type unknown. It is now possible to split the abducible or exception set into the admissible clauses or terms and evaluate their QoI_i. A pictorial view of this process is given in Fig. 1, as a pie chart.

3 Methods

Aiming to develop a predictive model to estimate the length of hospital stay in preterm infants, a database was set with a total of 284 cases. The data was taken from the health records of patients at a major health care institution in the north of Portugal. This section demonstrates briefly the process of extraction, transformation and loading. Moreover, shows how the information comes together and how it is processed.

3.1 Case Study

Nowadays information systems are often confused with computer systems, however, they are two distinct things, because information systems are as old as the institutions themselves, which has varied is the technology which supports them

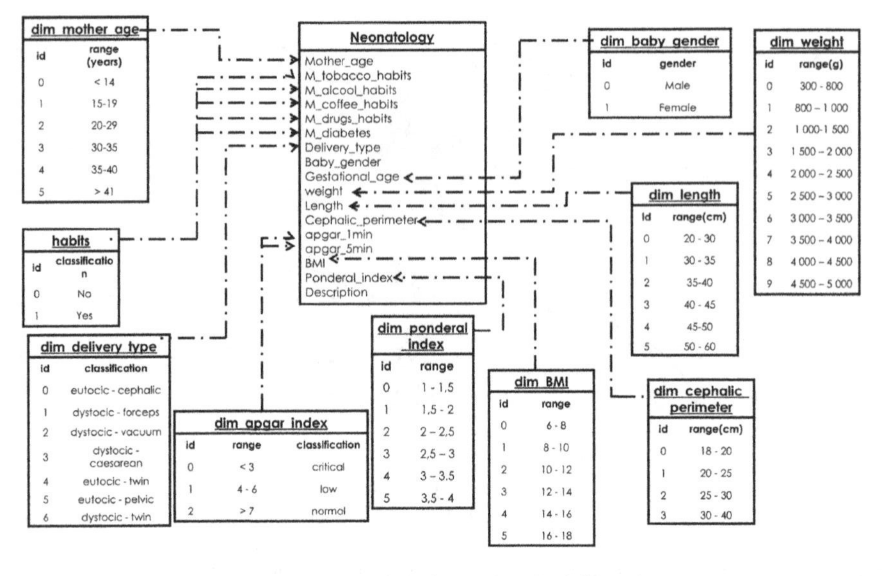

Fig. 2 An overview of the rational model used to gather the information

[24]. The main purpose of Hospital Information Systems is to store the patients' treatment information, and other associated medical needs [25]. Nonetheless, there was a need to organize the information, get only the necessary in order to facilitate its access. Hence, the information was organized as a *star schema* (Fig. 2), which consists of a collection of tables that are logically related to each other. To obtain a star schema it was essential to follow a few steps. In the former one it was necessary to understand the problem under study and gather the parameters that have influence in the final outcome. The following stage was related with the dimensions that would be needed to set these parameters on the facts table. Finally, information from several sources was collected, transformed according the fact and dimension table and loaded into the fact table [26].

The star schema conceived for this study (Fig. 2) takes into account the variables that allow one to estimate the LoS in preterm infants (Neonatology Table), where Dim Tables show how the data regarding those variables were pre-processed (e.g., to babies with apgar index <3 (critical), ranging between [4, 6] (low), and >7 (normal) were assigned the values 0 (zero), 1 (one) or 2 (two), respectively).

3.2 Data Processing

Based on the star schema presented in Fig. 2, it is possible to build up a knowledge database given in terms of the tables depicted in Fig. 3, which stand for a situation where one has to manage information aiming to estimate the LoS. Under this

Infant Related Factors									
#	Gestational Age	Gender	Weight	Length	Cephalic Perimeter	Apgar 1 min	Apgar 5 min	BMI	Ponderal Index
1	35	0	5	4	[2, 3]	⊥	2	2	2
2	32	0	2	3	2	2	2	1	2
…	…	…	…	…	…	…	…	…	…
n	34	1	4	4	3	1	2	2	2

Mother Related Factors						
#	Age	Diabetes	Coffee Consumption	Tobacco Consumption	Alcohol Consumption	Drugs Consumption
1	3	0	⊥	⊥	⊥	⊥
2	2	0	1	1	0	0
…	…	…	…	…	…	…
n	1	1	1	0	0	0

General Information			
#	Date	Delivery Type	Description
1	2015/03/14	1	*Description 1*
2	2015/02/02	3	*Description 2*
…	…	…	…
n	2015/01/03	0	*Description n*

Length of Stay														
#	Delivery Type	Gestational Age	Gender	Weight	Length	Cephalic Perimeter	Apgar 1 min	Apgar 5 min	BMI	Ponderal Index	Mother Age	Mother Diabetes	Mother Habits	Description
1	1	35	0	5	4	[2, 3]	⊥	2	2	2	3	0	⊥	*Description 1*
2	3	32	0	2	3	2	2	2	1	2	2	0	2	*Description 2*
…	…	…	…	…	…	…	…	…	…	…	…	…	…	…
n	0	34	1	4	4	3	1	2	2	2	1	1	1	*Description n*

Fig. 3 A fragment of the knowledge base to predict the length of stay in preterm infants

scenario some incomplete and/or unknown data is also available. For instance, in case 1, the *Apgar Index 1 min* is unknown, which is depicted by the symbol , while the *Cephalic Perimeter* ranges in the interval [2, 3].

The *General Information* and *Mother/Infant Related Factors* tables are filled according to Fig. 2. The values presented in the *Mother Habits* column of *Length of Stay* table are the sum of the columns *Coffee/Tobacco/Alcohol/Drugs Consumption*, ranging between [0, 4]. The *Descriptions* column stands for free text fields that allow for the registration of relevant case's features.

Applying the reduction algorithm presented in [23] to the fields that make the knowledge base for LoS estimation (Fig. 3), excluding of such a process the *Description* ones, and looking to the DoC_s values obtained as described in [23], it is possible to set the arguments of the predicate ***length_of_stay*** (*los*) referred to below, that also denotes the objective function with respect to the problem under analyze:

$$\begin{aligned} los\colon & D_{elivery}T_{ype}, G_{estational}A_{ge}, Gen_{der}, W_{eight}, L_{ength}, \\ & C_{ephalic}P_{erimeter}, A_{pgar}1_{min}, A_{pgar}5_{min}, BMI, P_{onderal}I_{ndex}, M_{other}A_{ge}, \\ & M_{other}D_{iabetes}, M_{other}H_{abits} \rightarrow \{0, 1\} \end{aligned}$$

where 0 (zero) and 1 (one) denote, respectively, the truth values *false* and *true*.

Exemplifying the application of the reduction algorithm presented in [23], to a term (case) that presents feature vector ($D_{elivery}T_{ype} = 1$, $G_{estational}A_{ge} = 33$, $Gen_{der} = 0$, $W_{eight} = 3$, $L_{ength} = 3$, $C_{ephalic}P_{erimeter} = [2, 3]$, $A_{pgar}1_{\ min} = 1$, $A_{pgar}5_{\ min} = 2$, $BMI = 2$, $P_{onderal}I_{ndex} = 2$, $M_{other}A_{ge} = 5$, $M_{other}D_{iabetes} = 1$, $M_{other}H_{abits} = \perp$), one may have:

Begin (DoCs evaluation),

The predicate's extension that maps the Universe-of-Discourse for the term under observation is set ←

$$\{$$

$$\neg\, los\left((QoI_{DT}, DoC_{DT}),\ \cdots, (QoI_{CP}, DoC_{CP}),\ \cdots, (QoI_{MH}, DoC_{MH})\right)$$

$$\leftarrow not\ los\left((QoI_{DT}, DoC_{DT}),\ \cdots, (QoI_{CP}, DoC_{CP}),\ \cdots, (QoI_{MH}, DoC_{MH})\right)$$

$$los\ \underbrace{\left((1_1, DoC_1),\ \cdots, \left(1_{[2,\,3]}, DoC_{[2,\,3]}\right),\ \cdots, (1_\perp, DoC_\perp)\right)}_{attribute`s\ values} :: 1 :: DoC$$

$$\underbrace{[0,6] \qquad \cdots \qquad [0,3] \qquad \cdots \qquad [0,4]}_{attribute`s\ domains}$$

$$\} :: 1$$

The attribute's values ranges are rewritten ←

$$\{$$

$$\neg\, los\left((QoI_{DT}, DoC_{DT}),\ \cdots, (QoI_{CP}, DoC_{CP}),\ \cdots, (QoI_{MH}, DoC_{MH})\right)$$

$$\leftarrow not\ los\left((QoI_{DT}, DoC_{DT}),\ \cdots, (QoI_{CP}, DoC_{CP}),\ \cdots, (QoI_{MH}, DoC_{MH})\right)$$

$$los\ \underbrace{\left(\left(1_{[1,1]}, DoC_{[1,1]}\right),\ \cdots, \left(1_{[2,\,3]}, DoC_{[2,\,3]}\right),\ \cdots, \left(1_{[0,4]}, DoC_{[0,4]}\right)\right)}_{attribute`s\ values\ ranges} :: 1 :: DoC$$

$$\underbrace{[0,6] \qquad \cdots \qquad [0,3] \qquad \cdots \qquad [0,4]}_{attribute`s\ domains}$$

$$\} :: 1$$

The attribute's boundaries are set to the interval [0, 1] $\leftarrow$

{

$$\neg\, los\left((QoI_{DT}, DoC_{DT}),\ \cdots,(QoI_{CP}, DoC_{CP}),\ \cdots,(QoI_{MH}, DoC_{MH})\right)$$

$$\leftarrow not\ los\left((QoI_{DT}, DoC_{DT}),\ \cdots,(QoI_{CP}, DoC_{CP}),\ \cdots,(QoI_{MH}, DoC_{MH})\right)$$

$$los\underbrace{\left(\left(1_{[0.2,0.2]}, DoC_{0.2,0.2]}\right),\cdots,\left(1_{[0.7,1]}, DoC_{[0.7,1]}\right),\cdots,\left(1_{[0,1]}, DoC_{[0,1]}\right)\right)}_{attribute`s\ values\ ranges\ once\ normalized} :: 1 :: DoC$$

$$\underbrace{[0,1]\qquad\cdots\qquad[0,1]\qquad\cdots\qquad[0,1]}_{attribute`s\ domains\ once\ normalized}$$

}:: 1

The DoC's values are evaluated $\leftarrow$

{

$$\neg\, los\left((QoI_{DT}, DoC_{DT}),\ \cdots,(QoI_{CP}, DoC_{CP}),\ \cdots,(QoI_{MH}, DoC_{MH})\right)$$

$$\leftarrow not\ los\left((QoI_{DT}, DoC_{DT}),\ \cdots,(QoI_{CP}, DoC_{CP}),\ \cdots,(QoI_{MH}, DoC_{MH})\right)$$

$$los\underbrace{\left((1,\ 1),\ \cdots,\ (1,\ 0.94),\ \cdots,\ (1,\ 0)\right)}_{\substack{attribute`s\ quality-of-information\\ and\ respective\ confidence\ values}} :: 1 :: 0.92$$

$$\underbrace{[0.2,0.2]\ \cdots\ [0.7,1]\ \cdots\ [0,1]}_{attribute`s\ values\ ranges\ once\ normalized}$$

$$\underbrace{[0,1]\ \cdots\ [0,1]\ \cdots\ [0,1]}_{attribute`s\ domains\ once\ normalized}$$

}:: 1

End.

This approach allows the representation of the case repository in a graphic form, showing each case in the Cartesian plane in terms of its *QoI* and *DoC*. Thus, the data can be presented in two different forms, i.e., one that is comprehensible to the user and the normalized one that speeds up the retrieval process.

4 Case-Based Reasoning

CBR methodology for problem solving stands for an act of finding and justifying the solution to a given problem based on the consideration of similar past ones, by reprocessing and/or adapting their data or knowledge [11, 27]. In *CBR—the cases —*are stored in a *Case Base*, and those cases that are similar (or close) to a new one are used in the problem solving process. The typical CBR cycle presents the mechanism that should be followed to have a consistent model. In fact, it is an iterative process since the solution must be tested and adapted while the result of applying that solution is inconclusive. In the final stage the case is learned and the knowledge base is updated with the new case [10, 11]. Despite promising results, the current CBR systems are neither complete nor adaptable enough for all domains. In some cases, the user is required to follow the similarity method defined by the system, even if it does not fit into their needs [27]. Moreover, other problems may be highlighted. On the one hand, the existent CBR systems have limitations related to the capability of dealing with unknown, incomplete and contradictory information. On the other hand, an important feature that often is discarded is the ability to compare strings. In some domains strings are important to describe a situation, a problem or even an event [11, 27].

Contrasting with other problem solving methodologies (e.g., those that use *Decision Trees* or *Artificial Neural Networks*), relatively little work is done offline. Undeniably, in almost all the situations, the work is performed at query time. The main difference between this new approach and the typical CBR one relies on the fact that not only all the cases have their arguments set in the interval [0, 1] but it also allows for the handling of incomplete, unknown, or even contradictory data or knowledge [27]. The classic CBR cycle was changed in order to include a normalization phase aiming to enhance the retrieve process (Fig. 4). The *Case Base* will be given in terms of triples that follow the pattern:

$$Case = \{\langle Raw_{case}, Normalized_{case}, Description_{case} \rangle\}$$

where Raw_{case} and $Normalized_{case}$ stand for themselves, and $Description_{case}$ is made on a set of strings or even in free text, which may be analyzed with string similarity algorithms.

When confronted with a new case, the system is able to retrieve all cases that meet such a structure and optimize such a population, i.e., it considers the attributes *DoC*'s value of each case or of their optimized counterparts when analysing similarities among them. Thus, under the occurrence of a new case, the goal is to find similar cases in the *Case Base*. Having this in mind, the reductive algorithm given in [23] is applied to the new case that presents feature vector ($D_{elivery}T_{ype} = 0$, $G_{estational}A_{ge} = [34, 35]$, $Gen_{der} = 1$, $W_{eight} = [2, 3]$ $L_{ength} = 3$, $C_{ephalic}P_{erimeter} = \perp$, $A_{pgar}1_{min} = \perp$, $A_{pgar}5_{min} = 2$, $BMI = 2$, $P_{onderal}I_{ndex} = 2$, $M_{other}A_{ge} = 3$, $M_{other}D_{iabetes} = 0$, $M_{other}H_{abits} = 2$, $Description = Description\ new$), with the results:

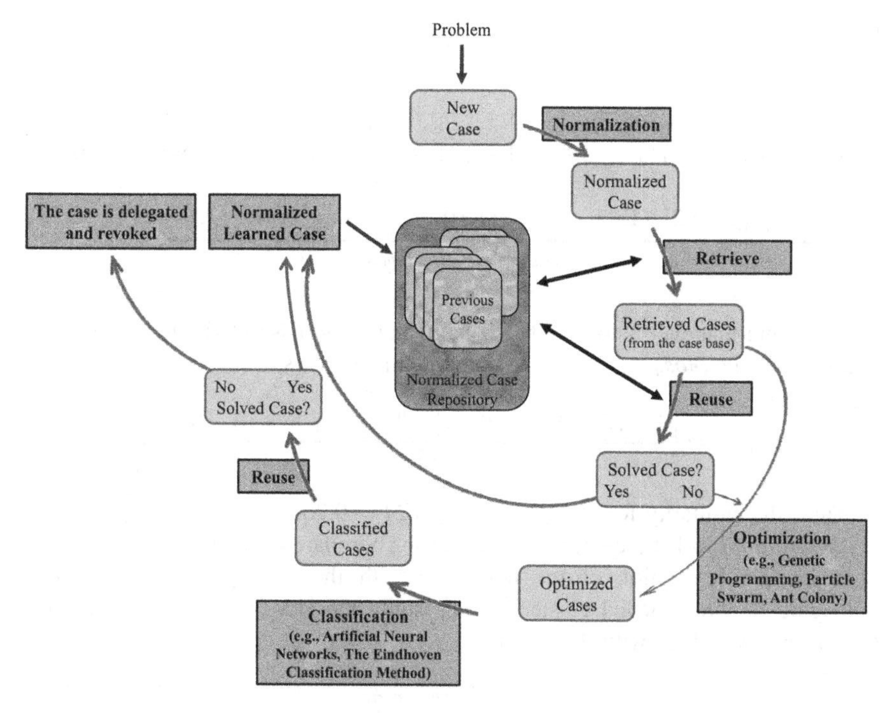

Fig. 4 An extended view of the CBR cycle [27]

$$\underbrace{los_{new}((1,1),\ (1,0.99),\ (1,1),\ \cdots,\ (1,0),(1,0),\cdots,\ (1,1),\ (1,1))::1::0.84}_{new\ case}$$

Thus, the *new case* can be depicted on the Cartesian plane in terms of its *QoI* and *DoC*, and through clustering techniques, it is feasible to identify the clusters that intermingle with the new one (symbolized as a star in Fig. 5). In present work the technique used to induce clusters was the k-means clustering method [28].

The *new case* is compared with every retrieved case from the cluster using a similarity function *sim*, given in terms of the average of the modulus of the arithmetic difference between the arguments of each case of the selected cluster and

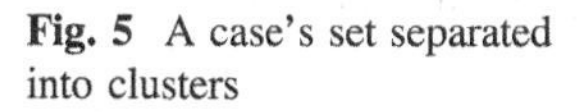

Fig. 5 A case's set separated into clusters

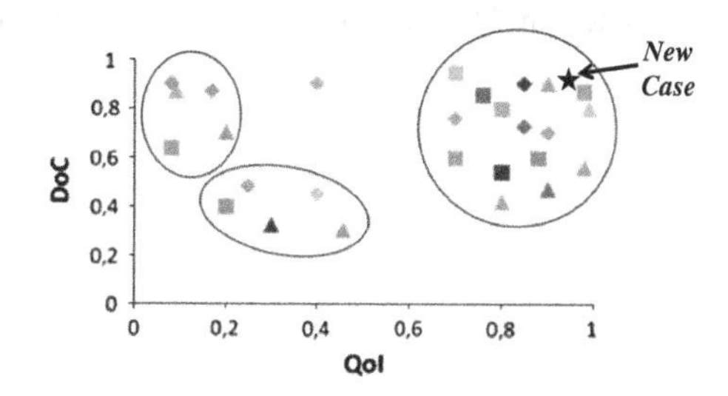

those of the *new case* (once *Description* stands for free text, its analysis is excluded at this stage). Thus, one may have:

$$\begin{array}{l} los_1((1,1),(1,0.98),(1,0), \quad \cdots, \quad (1,1),(1,1),\cdots, \quad (1,1),(1,0)) :: 1 :: 0.88 \\ los_2((1,1),(1,1),(1,1), \quad \cdots, \quad (1,0),(1,1),\cdots, \quad (1,1),(1,0.95)) :: 1 :: 0.92 \\ \quad \vdots \\ los_j \underbrace{((1,1),(1,0.92),(1,1), \quad \cdots, \quad (1,0),(1,0),\cdots, \quad (1,1),(1,0))}_{normalized\ cases\ from\ retrieved\ cluster} :: 1 :: 0.78 \end{array}$$

Assuming that every attribute has equal weight, the dissimilarity between los_{new} and the los_1, i.e., $los_{new \to 1}$, may be computed as follows:

$$los_{new \to 1} = \frac{\|1-1\| + \|0.99-0.98\| + \|1-0\| + \cdots + \|1-0\|}{13} = 0.24$$

Thus, the similarity for $los_{new \to 1}$ is $1 - 0.24 = 0.76$.

Descriptions will be compared using String Similarity Algorithms, in order to compare the description of the new case with the descriptions of the cases belonging to the retrieved cluster (in this study the strategy used was the Dice Coefficient one [29]), with the results:

$$los_{new \to 1}^{Description} = 0.79$$

With these similarity values it is possible to get a global similarity measure, i.e.,

$$los_{new \to 1} = \frac{0.76 + 0.79}{2} = 0.775$$

These procedures should be applied to the remaining cases of the retrieved cluster in order to obtain the most similar ones, which may stand for the possible solutions to the problem.

In order to evaluate the performance of the proposed model the dataset was divided in exclusive subsets through the ten-folds cross validation. In the implementation of the respective dividing procedures, ten executions were performed for each one of them. To ensure statistical significance of the attained results, 30 (thirty) experiments were applied in all tests. The model accuracy was 84.9 % (i.e., 241 instances correctly classified in 284). Moreover, the computational time was shortened in 21.3 %, when compared with classic CBR implementations.

5 Conclusions

The decision support system to estimate the length of hospital stay in preterm infants, presented in this work, is centred on a formal framework based on Logic Programming for Knowledge Representation and Reasoning, complemented with a *CBR* approach to problem solving that caters for the handling of incomplete, unknown, or even contradictory information. Under this approach the cases' retrieval and optimization phases were heightened and the time spent on those tasks shortened in 21.3 %, when compared with existing systems. On the other hand the overall accuracy was around 84.9 %. The proposed method allows also for the analysis of free text attributes using *String Similarities Algorithms*, which fulfils a gap that is present in almost all *CBR* software tools. Additionally, under this approach the users may define the weights of the cases' attributes on the fly, letting them to choose the most appropriate strategy to address the problem (i.e., it gives the user the possibility to narrow the search space for similar cases at runtime).

Acknowledgments This work has been supported by COMPETE: POCI-01-0145-FEDER-007043 and FCT—Fundação para a Ciência e Tecnologia within the Project Scope: UID/CEC/00319/2013.

References

1. Watt, S., Sword, W., Krueger, P.: Longer postpartum hospitalization options—who stays, who leaves, what changes? BMC Pregnancy Childbirth **5**, 1–10 (2005)
2. Hintz, S., Bann, C., Ambalavanan, N., Cotten, C., Das, A., Higgins, R.: Predicting time to hospital discharge for extremely preterm infants. Pediatrics **125**, e146–e154 (2010)
3. Adebanji, A., Adeyemi, S., Gyamfi, M.: Empirical analysis of factors associated with neonatal length of stay in Sunyani, Ghana. J. Public Health Epidemiol. **7**, 59–64 (2015)
4. Goyal, N., Zubizarreta, J., Small, D., Lorch, S.: Length of stay and readmission among late pre term infants: an instrumental variable approach. Hosp. Pediatr. **3**, 7–15 (2013)
5. Gupta, P., Malhotra, S., Singh, D., Dua, T.: Length of postnatal stay in healthy newborns and re-hospitalization following their early discharge. Indian J. Pediatr. **73**, 897–900 (2006)
6. Farhat, R., Rajab, M.: Length of postnatal hospital stay in healthy newborns and re-hospitalization following early discharge. North Am. J. Med. Sci. **3**, 146–151 (2011)
7. American Academy of Pediatrics: Committee on Fetus and Newborn: hospital stay for healthy term newborns. Pediatrics **113**, 1434–1436 (2004)
8. American Academy of Pediatrics: Committee on Fetus and Newborn: hospital stay for healthy term newborns. Pediatrics **125**, 405–409 (2010)
9. Niknajad, A., Ghojazadeh, M., Sattarzadeh, N., Hashemi, F., Shahgholi, F.: Factors affecting the neonatal intensive care unit stay duration in very low birth weight premature infants. J. Caring Sci. **1**, 85–92 (2012)
10. Aamodt, A., Plaza, E.: Case-based reasoning: Foundational issues, methodological variations, and system approaches. AI Commun. **7**, 39–59 (1994)
11. Richter, M.M., Weber, R.O.: Case-Based Reasoning: A Textbook. Springer, Berlin (2013)

12. Carneiro, D., Novais, P., Andrade, F., Zeleznikow, J., Neves, J.: Using case based reasoning to support alternative dispute resolution. In: Carvalho, A.F., Rodríguez-González, S., Paz-Santana, J.F., Corchado-Rodríguez, J.M. (eds.) Distributed Computing and Artificial Intelligence, Advances in Intelligent and Soft Computing, vol. 79, pp. 123–130. Springer, Berlin (2010)
13. Carneiro, D., Novais, P., Andrade, F., Zeleznikow, J., Neves, J.: Using case-based reasoning and principled negotiation to provide decision support for dispute resolution. Knowl. Inf. Syst. **36**, 789–826 (2013)
14. Guessoum, S., Laskri, M.T., Lieber, J.: Respidiag: a case-based reasoning system for the diagnosis of chronic obstructive pulmonary disease. Expert Syst. Appl. **41**, 267–273 (2014)
15. Ping, X.-O., Tseng, Y.-J., Lin, Y.-P., Chiu H.-J., Feipei Lai, F., Liang J.-D., Huang, G.-T., Yang, P.-M.: A multiple measurements case-based reasoning method for predicting recurrent status of liver cancer patients. Comput. Ind. **69**, 12–21 (2015)
16. Tsinakos, A.: Asynchronous distance education: teaching using case based reasoning. Turk. Online J Distance Educ. **4**, 1–8 (2003)
17. Kakas, A., Kowalski, R., Toni, F.: The role of abduction in logic programming. In: Gabbay, D., Hogger, C., Robinson, I. (eds.) Handbook of Logic in Artificial Intelligence and Logic Programming, vol. 5, pp. 235–324. Oxford University Press, Oxford (1998)
18. Pereira, L., Anh, H.: Evolution prospection. In: Nakamatsu, K. (ed.) New Advances in Intelligent Decision Technologies—Results of the First KES International Symposium IDT 2009, Studies in Computational Intelligence, vol. 199, pp. 51–64. Springer, Berlin (2009)
19. Neves, J.: A logic interpreter to handle time and negation in logic databases. In: Muller, R., Pottmyer, J. (eds.) Proceedings of the 1984 Annual Conference of the ACM on the 5th Generation Challenge, pp. 50–54. Association for Computing Machinery, New York (1984)
20. Neves, J., Machado, J., Analide, C., Abelha, A., Brito, L.: The halt condition in genetic programming. In: Neves, J., Santos, M.F., Machado, J. (eds.) Progress in Artificial Intelligence. LNAI, vol. 4874, pp. 160–169. Springer, Berlin (2007)
21. Machado, J., Abelha, A., Novais, P., Neves, J.: Quality of service in healthcare units. In Bertelle, C., Ayesh, A. (eds.) Proceedings of the ESM 2008, pp. 291–298. Eurosis—ETI Publication, Ghent (2008)
22. Lucas, P.: Quality checking of medical guidelines through logical abduction. In: Coenen, F., Preece, A., Mackintosh, A. (eds.) Proceedings of AI-2003 (Research and Developments in Intelligent Systems XX), pp. 309–321. Springer, London (2003)
23. Fernandes, F., Vicente, H., Abelha, A., Machado, J., Novais, P., Neves J.: Artificial Neural Networks in Diabetes Control. In: Proceedings of the 2015 Science and Information Conference (SAI 2015), pp. 362–370, IEEE Edition (2015)
24. Almeida, A.: The management systems of archival information on Portuguese public hospitals. Master's thesis, University of Lisbon (2011)
25. Cardoso, M.: Auditing an hospital information system—SAM. Master's thesis, Polytechnic Institute of Bragança (2010)
26. O'Neil, P., O'Neil, B., Chen, X.: Star schema Benchmark. Revision 3, June 5, 2009. http://www.cs.umb.edu/~poneil/StarSchemaB.pdf
27. Neves, J., Vicente, H.: Quantum approach to Case-Based Reasoning (in preparation)
28. MacQueen J.B.: Some methods for classification and analysis of multivariate observations. In: Proceedings of the 5th Berkeley Symposium on Mathematical Statistics and Probability, vol. 1, pp. 281–297. University of California Press (1967)
29. Dice, L.R.: Measures of the amount of ecologic association between species. Ecology **26**, 297–302 (1945)

The Shapley Value on a Class of Cooperative Games Under Incomplete Information

Satoshi Masuya

Abstract In this paper, we study cooperative TU games in which the worths of some coalitions are not known. We investigate superadditive games and the Shapley values on a class of cooperative games under incomplete information. We show that the set of the superadditive complete games and the set of the Shapley values which can be obtained from a given incomplete game are polytopes and propose selection methods of the one-point solution from the set of the Shapley values.

Keywords Cooperative game · Partially defined cooperative game · Superadditive game · Shapley value

1 Introduction

The cooperative game theory provides useful tools to analyze various cost and/or surplus allocation problems, the distribution of voting power in a parliaments or a country, and so on. The problems to be analyzed by the cooperative game theory include n entities called players and are usually expressed by characteristic functions which map each subset of players to a real number. The solutions to the problems are given by a set of n-dimensional real numbers or value functions which assign a real number to each player. Such a real number can show the cost borne by the player, power of influence, an allocation of the shared profits, and so on. Several solution concepts for cooperative games have been proposed. As representative examples of solution concepts, the core, the Shapley value [5] and the nucleolus [4] are well-known. The core can be represented by a set of solutions while the Shapley value and the nucleolus are one-point solutions.

A classical approach of von Neumann and Morgenstern [6] to cooperative games assumes that the worths of all coalitions are given. However, in the real world

S. Masuya (✉)
Faculty of Business Administration, Daito Bunka University, 1-9-1, Takashimadaira Itabashi-ku, Tokyo 175-8571, Japan
e-mail: masuya@ic.daito.ac.jp

I. Czarnowski et al. (eds.), *Intelligent Decision Technologies 2016*,
Smart Innovation, Systems and Technologies 56,
DOI 10.1007/978-3-319-39630-9_11

problems, there may exist situations in which the worths of some coalitions are unknown. Such cooperative games under incomplete information have not yet investigated considerably.

Cooperative games under incomplete information are first considered by Willson [7] which are called partially defined cooperative games. In Willson [7] he proposed the generalized Shapley value which is obtained by using only known coalitional worths of a game and he axiomatized the proposed Shapley value. After that, Housman [1] continued the study of Willson [7]. Housman [1] characterized the generalized Shapley value by Willson [7]. However, the simplest interpretation of the generalized Shapley value is that it coincides with the ordinary Shapley value of a game whose coalitional worths take zero if they are unknown and given values otherwise. Such a game usually dissatisfies natural properties such as superadditivity. As another approach for such a game, Polkowski and Araszkiewicz [3] considered the estimation problem of coalitional worths as well as the Shapley value of such a game from the partial data about coalition structures.

In our previous paper [2], we propose an approach to partially defined cooperative games from a different angle from Willson [7] and Housman [1]. Moreover, we have treated a different problem setting from Polkowski and Araszkiewicz [3]. We have assumed that some coalitional worths are known but the others are unknown. For the sake of simplicity, in that paper, a partially defined cooperative game, i.e., a cooperative game under incomplete information is called "an incomplete game" while a cooperative game with complete information is called "a complete game". Although we have defined the lower and upper games associated with the given general incomplete game which is assumed the superadditivity, we have considered the simplest case when only worths of singleton coalitions and the grand coalition are known for the investigation of the solution concepts to incomplete games. In other words, the case with minimal information is considered. Therefore, it remains many unsolved problems such as other or general games under incomplete information.

Then, in this paper, we investigate superadditive games and the Shapley values for more general incomplete games. In this case, we show that the set of the superadditive complete games which can be obtained from a given incomplete game is a polytope. Further, we show the set of the Shapley values which can be obtained from a given incomplete game assumed the superadditivity. Furthermore, we propose selection methods of the one-point solution from the set of the Shapley value and investigate the properties of the selected solutions.

This paper is organized as follows. In Sect. 2, we introduce the classical cooperative game and well-known solution concepts. Further, we present partially defined cooperative games, the lower and upper games and their properties by [2]. In Sect. 3, we investigate the set of superadditive games which can be obtained from a given incomplete game. In Sect. 4, we show the set of the Shapley values which can be obtained from a given incomplete game and we propose two selection methods of the one-point solution from the set of the Shapley values. Further, we investigate the properties of the selected solutions. In Sect. 5, concluding remarks and future directions are given.

2 Classical Cooperative Games and Partially Defined Cooperative Games

Let $N = \{1, 2, \ldots, n\}$ be the set of players and $v: 2^N \to \mathbb{R}$ such that $v(\emptyset) = 0$. A classical cooperative game, i.e., a coalitional game with transferable utility (a TU game) is characterized by a pair (N, v). A set $S \subseteq N$ is regarded as a coalition of players and the number $v(S)$ represents a collective payoff that players in S can gain by forming coalition S. For arbitrary coalition S, the number $v(S)$ is called the worth of coalition S.

A game (N, v) is said to be superadditive if and only if

$$v(S \cup T) \geq v(S) + v(T),\ \forall S,\ T \subseteq N \quad \text{such that} \quad S \cap T = \emptyset. \tag{1}$$

Superadditivity is a natural property that gives each player an incentive to form a larger coalition.

Now let us introduce basic solution concepts in cooperative games. In cooperative games, it is assumed that the grand coalition N forms. The problem is how to allocate the collective payoff $v(N)$ to all players. A solution is a vector $x = (x_1, x_2, \ldots, x_n) \in \mathbb{R}^n$ where each component $x_i \in \mathbb{R}$ represents the payoff to player i. Many solution concepts have been proposed. We describe the Shapley value.

A solution x is efficient in a game (N, v) iff $\sum_{i \in N} x_i = v(N)$. A set of requirements $x_i \geq v(\{i\}), \forall i \in N$ are called the individual rationality. Let $I(N, v)$ denote the set of payoff vectors which satisfy efficiency and individual rationality in (N, v), or "imputations"

The Shapley value is a well-known one-point solution concept. It selects an imputation when the game is superadditive.

Let $G(N)$ be the set of all cooperative games with the player set N. For convenience, because the set of players is fixed as N, cooperative game (N, v) is denoted simply by v.

The Shapley value is defined by

$$\phi_i(v) = \sum_{\substack{S \subseteq N \\ S \ni i}} \frac{(|S| - 1)!(n - |S|)!}{n!}(v(S) - v(S \setminus i)),\ \forall i \in N. \tag{2}$$

In the rest of this section, we present the results on partially defined cooperative games by Masuya and Inuiguchi [2]. In classical cooperative games, we assume that worths of all coalitions are known. However, in the real world problems, there may exist situations in which worths of some coalitions are unknown. To avoid the confusion, we call such games "incomplete games" and the conventional cooperative games "complete games".

The incomplete games can be characterized by a set of players $N = \{1, 2, \ldots, n\}$, a set of coalitions whose worths are known, say $\mathcal{K} \subseteq 2^N$, and a function $v : \mathcal{K} \to \mathbb{R}$, with $\emptyset \in \mathcal{K}$ and $v(\emptyset) = 0$. We assume that worths of singleton coalitions and the grand coalition are known and the worths of singleton coalitions are non-negative, i.e., $\{i\} \in \mathcal{K}$ and $v(\{i\}) \geq 0, i \in N$ and $N \in \mathcal{K}$. Moreover, we assume that v is superadditive in the following sense,

$$v(S) \geq \sum_{i=1}^{s} v(T_i),\ \forall S, T_i \in \mathcal{K},\ i = 1, 2, \ldots, s \text{ such that } \bigcup_{i=1,2,\ldots,s} T_i = S \text{ and } T_i,\ i = 1, 2, \ldots, s \text{ are disjoint.} \tag{3}$$

A triple $(N, \mathcal{K}, v)$ identifies an incomplete game. When we consider only games under fixed N and $\mathcal{K}$, incomplete game $(N, \mathcal{K}, v)$ is simply written as v.

Given an incomplete game $(N, \mathcal{K}, v)$, we define two associated complete games $(N, \underline{v})$ and $(N, \overline{v})$:

$$\underline{v}(S) = \max_{\substack{T_i \in \mathcal{K},\ i=1,2,\ldots,s \\ \cup_i T_i = S,\ T_i \text{ are disjoint}}} \sum_{i=1}^{s} v(T_i), \tag{4}$$

$$\overline{v}(S) = \min_{\hat{S} \in \mathcal{K}, \hat{S} \supseteq S} \left(v(\hat{S}) - \underline{v}(\hat{S} \setminus S) \right) \tag{5}$$

From the superadditivity of v, we have $\underline{v}(S) = v(S)$ and $\overline{v}(S) = v(S)\ \forall S \in \mathcal{K}$.

A complete game (N, v) such that $v(T) = v(T), \forall T \in \mathcal{K}$ is called a complete extension of $(N, \mathcal{K}, v)$, or simply a complete extension of v. As shown in the following theorem, $\underline{v}(S)$ is the minimal payoff of coalition S among superadditive complete extensions of $(N, \mathcal{K}, v)$. On the other hand, $\overline{v}(S)$ is the maximal payoff of coalition S among superadditive complete extensions of $(N, \mathcal{K}, v)$.

Theorem 1 ([2]) *Let $(N, \mathcal{K}, v)$ be an incomplete game, and $(N, \underline{v})$ and $(N, \overline{v})$ the complete extensions defined by (4) and (5). For an arbitrary superadditive complete extension (N, v) of $(N, \mathcal{K}, v)$, we obtain*

$$\underline{v}(S) \leq v(S),\ \forall S \subseteq N, \tag{6}$$

$$\overline{v}(S) \geq v(S),\ \forall S \subseteq N. \tag{7}$$

Therefore, we call complete games $(N, \underline{v})$ and $(N, \overline{v})$ "lower game" and "upper game" associated with $(N, \mathcal{K}, v)$, respectively. When there is no confusion of the underlying incomplete game, those games are simply called the lower game and the upper game.

Using the incomplete information expressed by v, we may consider the set $V(v)$ of possible complete games. More explicitly let the set of superadditive completions of v be

$$V(v) = \{v : 2^N \to \mathbb{R} \mid v \text{ is superadditive}, v(S) = v(S),\ \forall S \in \mathcal{K}\}. \tag{8}$$

3 The Set of Superadditive Complete Extensions of Partially Defined Cooperative Games

In the previous section, we present the definitions and properties of the cooperative game under incomplete information. The result which is introduced in the previous section holds for general superadditive games with arbitrary $\mathcal{K}$. In this section, we investigate the set of superadditive complete games $V(v)$ which can be obtained from a given incomplete game $(N, \mathcal{K}, v)$. Because it is difficult to investigate $V(v)$ for general $(N, \mathcal{K}, v)$, we assume two properties for $\mathcal{K}$ below which includes $\{\{1\}, \ldots, \{n\}, N\}$.

Assumption 1 (*Symmetry among players*) For two players $i \in N, j \in N$, we consider a coalition $S \ni i$ such that $S \not\ni j$ and $S \subseteq N$. Further, let S' be a coalition interchanging i and j in S. That is, we consider $S' = S \cup \{j\} \setminus \{i\}$. Then, if $S \in \mathcal{K}$, $S' \in \mathcal{K}$ holds.

Assumption 1 claims that if a coalition $S \in \mathcal{K}$ holds, all coalitions which have the same cardinality as S are included in $\mathcal{K}$. This means that the known set of coalitions $\mathcal{K}$ is symmetric with respect to players and the solution of the incomplete game is not disadvantageous (or advantageous) for some players from the missing coalitional worths.

Assumption 2 (*The belonging possibility of small coalitions*) For an arbitrary coalition $S \subset N (S \neq N)$, $T \in \mathcal{K} \forall T \subset S$ if $S \in \mathcal{K}$.

If we consider production planning games or spanning tree games, the computation of values of small coalitions is easier than that of large coalitions. From this point of view, Assumption 2 claims that if a coalitional worth is known, then its subcoalitional worths are also known.

When $\mathcal{K}$ satisfies the Assumptions 1 and 2, it can be represented by an integer k $(1 \leq k \leq n-1)$ as follows:

$$\mathcal{K} = \{S \subseteq N \mid |S| \leq k\} \cup \{N\}. \tag{9}$$

Indeed, we can establish $k = \max\{|S| \mid S \in \mathcal{K}, S \neq N\}$. Therefore, we call an incomplete game $(N, \mathcal{K}, v)$ satisfying Assumptions 1 and 2 an (N, k)-incomplete game. $1 \leq k \leq n-1$ holds. When $k = n-1$, $(N, \mathcal{K}, v)$ is a complete game and when $k = 1$, the game is an incomplete game where only the worths of the grand coalition and singleton coalitions are known.

Now, we investigate the set of superadditive complete extensions $V(v)$ of an (N, k)-incomplete game.

Consider a finite set of coalitions $\mathcal{T} = \{T_1, \ldots, T_m\}$ whose cardinalities are no less than $k+1$ and each coalition in the set is not included each other. That is, we consider $\mathcal{T}$, for $T_p \in \mathcal{T}$, $|T_p| \geq k+1$ holds and for arbitrary $T_p, T_s \in \mathcal{T}$ $(p \neq s)$, $T_p \nsubseteq T_s$ and $T_s \nsubseteq T_p$ holds. Note that the number of T_p which is denoted by m is not constant but varies corresponding to $\mathcal{T}$. Although a number of the set $\mathcal{T}$ exists, the number of $\mathcal{T}$ is finite. The set of all $\mathcal{T}$ is denoted by $\Gamma(N, k+1)$. We define the complete game dependent on $\mathcal{T} \in \Gamma(N, k+1)$ as follows:

$$v^{\mathcal{T}}(S) = \begin{cases} \overline{v}(S), & \text{if } n > |S| > k \text{ and } \exists T \in \mathcal{T},\ S \supseteq T, \\ \underline{v}(S), & \text{if } n > |S| > k \text{ and } \forall T \in \mathcal{T},\ S \nsupseteq T, \\ v(S), & \text{otherwise.} \end{cases} \tag{10}$$

The complete game $v^{\mathcal{T}}$ which is given by (10) has the values if its coalitional worths are known, otherwise, has the worths of the upper game when there are some coalitions in $\mathcal{T}$ which are included in the coalition, and those of the lower game when there are no coalitions in $\mathcal{T}$ which are included in the coalition. In other words, the complete game $v^{\mathcal{T}}$ is the game which has the highest coalitional worths when the all members of some coalitions in $\mathcal{T}$ are joined in the coalition. That is, $\mathcal{T}$ is the list of coalitions which make a significant contribution.

Then, the next lemma is obtained.

Lemma 1 *Let v be an (N, k)-incomplete game satisfying $k \geq \lceil \frac{n}{2} \rceil$. Then, the complete game $(N, v^{\mathcal{T}})$, $\forall \mathcal{T} \in \Gamma(N, k+1)$ is superadditive.*

From the above lemma, the next theorem is obtained.

Theorem 2 *Let v be an (N, k)-incomplete game satisfying $k \geq \lceil \frac{n}{2} \rceil$. Then, the set of superadditive complete games $V(v)$ is a polytope whose extreme points are $v^{\mathcal{T}}$, $\forall \mathcal{T} \in \Gamma(N, k+1)$ which is defined in (10). That is, the next expression holds.*

$$V(v) = \left\{ v : 2^N \to \mathbb{R} \;\middle|\; v = \sum_{\mathcal{T} \in \Gamma(N,k+1)} c_{\mathcal{T}} v^{\mathcal{T}}, \sum_{\mathcal{T} \in \Gamma(N,k+1)} c_{\mathcal{T}} = 1, c_{\mathcal{T}} \geq 0, \right.$$
$$\left. \forall \mathcal{T} \in \Gamma(N, k+1) \right\}.$$

Roughly speaking, Theorem 2 means that when the worths of coalitions which are formed by more than half of all players are unknown and the others are known, $v^{\mathcal{T}}$, $\forall \mathcal{T} \in \Gamma(N, k+1)$ is the extreme points of $V(v)$ and all superadditive complete games can be obtained by the convex combinations of those extreme points.

Example 1 Let $N = \{1, 2, 3, 4\}$, and let $\mathcal{K} = \{S \subseteq N \,||S| \leq 2\} \cup \{\{1, 2, 3, 4\}\}$. Then $\Gamma(N, k+1) = \{\mathcal{T} \subseteq 2^n \,||T| = 3, \forall T \in \mathcal{T}\} \cup \{N\}$. We consider an incomplete game $v : \mathcal{K} \to \mathbb{R}$ with

$$v(\{1\}) = 8, v(\{2\}) = 7, v(\{3\}) = 3,$$
$$v(\{4\}) = 1,$$
$$v\{1, 2\} = 18, v\{1, 3\} = 16, v\{1, 4\} = 10,$$

$$v(\{2,3\}) = 14, v(\{2,4\}) = 12, v(\{3,4\}) = 9,$$
$$v(\{1,2,3,4\}) = 30.$$

This game is an (N, k)-incomplete game satisfying $k \geq \lceil \frac{n}{2} \rceil$. Then, using (4) and (5), the lower game $\underline{v}$ and the upper game $\overline{v}$ are obtained as follows:

$$\underline{v}(\{1\}) = 8, \underline{v}(\{2\}) = 7, \underline{v}(\{3\}) = 3, \underline{v}(\{4\}) = 1,$$
$$\underline{v}(\{1,2,3,4\}) = 30,$$
$$\underline{v}(\{1,2\}) = 18, \underline{v}(\{1,3\}) = 16, \underline{v}(\{1,4\}) = 10,$$
$$\underline{v}(\{2,3\}) = 14, \underline{v}(\{2,4\}) = 12, \underline{v}(\{3,4\}) = 9,$$
$$\underline{v}(\{1,2,3\}) = 23, \underline{v}(\{1,2,4\}) = 20,$$
$$\underline{v}(\{1,3,4\}) = 17, \underline{v}(\{2,3,4\}) = 16,$$
$$\overline{v}(\{1\}) = 8, \overline{v}(\{2\}) = 7, \overline{v}(\{3\}) = 3, \overline{v}(\{4\}) = 1,$$
$$\overline{v}(\{1,2,3,4\}) = 30,$$
$$\overline{v}(\{1,2\}) = 18, \overline{v}(\{1,3\}) = 16, \overline{v}(\{1,4\}) = 10,$$
$$\overline{v}(\{2,3\}) = 14, \overline{v}(\{2,4\}) = 12, \overline{v}(\{3,4\}) = 9,$$
$$\overline{v}(\{1,2,3\}) = 29, \overline{v}(\{1,2,4\}) = 27,$$
$$\overline{v}(\{1,3,4\}) = 23, \overline{v}(\{2,3,4\}) = 22,$$

4 Selection Methods of the One-Point Solution from the Set of the Shapley Values and Its Properties

In this section, we deal with (N, k)-incomplete games such as $k \geq \lceil \frac{n}{2} \rceil$. First, we derive the set of all the Shapley values which can be obtained from an (N, k)-incomplete game. After that, we propose two selection methods of the one-point solution from the set of all the Shapley values which is denoted by $\Phi(v)$ and investigate its properties.

From the linearity of the Shapley value and Theorem 2, the set of all the Shapley values which can be obtained from an (N, k)-incomplete game v is the polytope whose extreme points are $\phi(v^{\mathcal{T}})$, $\forall \mathcal{T} \in \Gamma(N, k+1)$.

Considering the applications of incomplete games, it is necessary to select the rational one-point solution in some sense from the set of all the Shapley values. In the rest of this section, we propose two selection methods of the rational one-point solution from all the Shapley values $\Phi(v)$ which is obtained from an (N, k)-incomplete game v.

First selection method is to select the center of gravity. Since the set of the Shapley values which can be obtained from v is a polytope, we propose the center of gravity of $\Phi(v)$ as the rational Shapley value $\tilde{\phi}(v)$. That is, $\tilde{\phi}(v)$ is defined as follows.

$$\tilde{\phi}(v) = \frac{\sum_{\mathcal{T} \in \Gamma(N,k+1)} \phi(v^{\mathcal{T}})}{|\Gamma(N, k+1)|} \tag{11}$$

Second selection method is the way of the minimization of the maximal excess. This is the application of the idea from the definition of the nucleolus. Particularly, we define the excess by the difference between an interior point of the maximal and the minimal Shapley value and the payoff of each player. Further, we propose the payoff vector of the minimization of the maximal excess as the rational Shapley value.

Given an incomplete game v, let $x = (x_1, \dots, x_n)$ be an n-dimensional payoff vector where $x \in \Phi(v)$. Let $\overline{\phi}_i(v)$ be the maximal Shapley value of player i, and $\underline{\phi}_i(v)$ be the minimal Shapley value of player i. Let α be a real number which satisfies $0 \leq \alpha \leq 1$. Then, the excess of player i to payoff vector x is defined by

$$e_i^{\alpha}(x) = \alpha\overline{\phi}_i(v) + (1-\alpha)\underline{\phi}_i(v) - x_i. \tag{12}$$

Further, we arrange $e_i^{\alpha}(x)$ for all players $i \in N$ in non-increasing order. Then the arranged vector are denoted by $\theta(x)$.

Moreover, when two payoff vectors x, y satisfies the next condition, we call x is lexicographically smaller than y.

$$\begin{gathered} \theta_1(x) < \theta_1(y), \text{ or} \\ \exists h \in \{1, 2, \dots, n\} \text{ such that} \\ \theta_i(x) = \theta_i(y), \forall i < h \text{ and } \theta_h(x) < \theta_h(y) \end{gathered} \tag{13}$$

Finally, the Shapley value which is lexicographically minimized is called the minimized maximal excess Shapley value $\hat{\phi}^{\alpha}(v)$ or the min-max excess Shapley value in short.

We obtain the following theorem.

Theorem 3 *Let v be an (N, k)-incomplete game satisfying $k \geq \lceil \frac{n}{2} \rceil$. Let $\overline{\phi}_i(v)$ be the maximal Shapley value which player i can obtain and $\underline{\phi}_i(v)$ be the minimal Shapley value. Further, let α be a real number satisfying $0 \leq \alpha \leq 1$. Then, $\hat{\phi}^{\alpha}(v)$ is given by*

$$\begin{gathered} \hat{\phi}_i^{\alpha}(v) = \alpha\overline{\phi}_i(v) + (1-\alpha)\underline{\phi}_i(v) \forall i \in N, \\ \textit{where} \sum_{j \in N} \left(\alpha\overline{\phi}_j(v) + (1-\alpha)\underline{\phi}_j(v) \right) = v(N). \end{gathered} \tag{14}$$

Note that the number α satisfying the equation $\sum_{j \in N} \left(\alpha\overline{\phi}_j(v) + (1-\alpha)\underline{\phi}_j(v) \right) = v(N)$ is given uniquely for arbitrary v. Therefore, from Theorem 3, it can be said that the min-max excess Shapley value can be obtained easily.

Next, we investigate the properties of the one-point solutions $\tilde{\phi}(v)$ and $\hat{\phi}^{\alpha}(v)$ respectively. In particular, we axiomatize $\tilde{\phi}(v)$.

Definition 1 Let v be an (N, k)-incomplete game satisfying $k \geq \lceil \frac{n}{2} \rceil$. Then we consider an arbitrary $S \subseteq N \backslash \{\mathrm{i}, \mathrm{j}\}$ satisfying $|S| \leq k-1$ and two players $i, j \in N$. Then C$v(S \cup i)$ and $v(S \cup j)$ are known, and if $v(S \cup i) = v(S \cup j) \forall |S| \leq k-1$ holds, we call players i, j are symmetric in v.

Axiom 1 (*v-consistent symmetry*) Let v be an (N, k)-incomplete game satisfying $k \geq \lceil \frac{n}{2} \rceil$. We assume that two players $i, j \in N$ are symmetric in v. Then, we call that the Shapley value $\pi(v)$ satisfies v-consistent symmetry if the following holds.

$$\pi_i(v) = \pi_j(v) \tag{15}$$

Axiom 1 means that if two players i and j are symmetric in an incomplete game, the two players are symmetric in the superadditive complete extension of the game and the Shapley value of player i coincides with that of player j.

Axiom 2 (*the principle of insufficient reason*) Let v be an (N, k)-incomplete game and the Shapley value $\pi \in \Phi(v)$ be $\pi = \sum_{\mathcal{T} \in \Gamma(N,k+1)} \phi(c_{\mathcal{T}} v^{\mathcal{T}})$ where $\sum_{\mathcal{T} \in \Gamma(N,k+1)} c_{\mathcal{T}} = 1$ and $c_{\mathcal{T}} \geq 0 \forall \mathcal{T} \in \Gamma(N, k+1)$. Then, with respect to two families of coalitions $\mathcal{T}_1, \mathcal{T}_2 \in \Gamma(N, k+1)$, if $\mathcal{T}_2$ is not an exchange of symmetric players belonged to $\mathcal{T}_1$, the following holds:

$$c_{\mathcal{T}_1} = c_{\mathcal{T}_2}.$$

$c_{\mathcal{T}}$ can be interpreted as the reward by forming the family of coalitions $\mathcal{T}$. Therefore, if there is no additional information on the relation of two families of coalitions such that one family can be obtained by an exchange of symmetric players belonged to the other family, there is no sufficient reason why one family can have more reward than the other family.

Then the following theorem is obtained.

Theorem 4 *$\tilde{\phi}(v)$ is the unique Shapley value in $\Phi(v)$ satisfying Axioms 1 and 2.*

From Theorem 4, the gravity center of the Shapley values $\tilde{\phi}(v)$ is axiomatized. Finally, we obtain the following theorem.

Theorem 5 *Let v be an (N, k)-incomplete game satisfying $k \geq \lceil \frac{n}{2} \rceil$, and let α be real number satisfying $0 \leq \alpha \leq 1$ and $\sum_{j \in N} \left(\alpha \overline{\phi}_j(v) + (1-\alpha) \underline{\phi}_j(v) \right) = v(N)$. Then the min-max excess Shapley value $\hat{\phi}^{\alpha}(v)$ satisfies Axiom 1.*

Theorem 5 shows that both $\tilde{\phi}(v)$ and $\hat{\phi}^{\alpha}(v)$ satisfy v-consistent symmetry.

Example 2 Using the result of Example 1, we obtain the Shapley values of each $v^{\mathcal{T}}$ and their gravity center $\tilde{\phi}(v)$ as Table 1.

Table 1 The set of coalitions $\mathcal{T}$ C the Shapley values of each player $\phi(v^{\mathcal{T}})$ and $\tilde{\phi}(v)$

$\mathcal{T}$	$\phi_1(v^{\mathcal{T}})$	$\phi_2(v^{\mathcal{T}})$	$\phi_3(v^{\mathcal{T}})$	$\phi_4(v^{\mathcal{T}})$
{1,2,3,4}	10.33	9.67	6.5	3.5
{1,2,3}	10.83	10.17	7	2
{1,2,4}	10.92	10.25	4.75	4.08
{1,3,4}	10.83	8.17	7	4
{2,3,4}	8.83	10.17	7	4
{1,2,3}, {1,2,4}	11.42	10.75	5.25	2.58
{1,2,3}, {1,3,4}	11.33	8.67	7.5	2.5
{1,2,3}, {2,3,4}	9.33	10.67	7.5	2.5
{1,2,4}, {1,3,4}	11.42	8.75	5.25	4.58
{1,2,4}, {2,3,4}	9.42	10.75	5.25	4.58
{1,3,4}, {2,3,4}	9.33	8.67	7.5	4.5
{1,2,3}, {1,2,4}, {1,3,4}	11.92	9.25	5.75	3.08
{1,2,3}, {1,2,4}, {2,3,4}	9.92	11.25	5.75	3.08
{1,2,3}, {1,3,4}, {2,3,4}	9.83	9.17	8	3
{1,2,4}, {1,3,4}, {2,3,4}	9.92	9.25	5.75	5.08
{1,2,3}, {1,2,4}, {1,3,4}, {2,3,4}	10.42	9.75	6.25	3.58
$\tilde{\phi}(v)$	10.38	9.71	6.38	3.54

Moreover, using Theorem 3, the min-max excess Shapley value $\hat{\phi}^{\alpha}(v)$ can be obtained as follows:

$$\hat{\phi}^{\alpha}(v) = (10.38, 9.71, 6.38, 3.54), \text{ where } \alpha = 0.5.$$

In this example, we can see that $\tilde{\phi}(v) = \hat{\phi}^{\alpha}(v)$.

5 Concluding Remarks

In this paper, we investigated the superadditive complete extensions and the Shapley values of the cooperative game under incomplete information which is called an (N, k)-incomplete game. We have shown that the set of superadditive complete extensions and the set of the Shapley values of an incomplete game are polytopes respectively. Further, we proposed two selection methods of the one-point solution from the set of the Shapley values and investigated its properties.

References

1. Housman, D.: Linear and symmetric allocation methods for partially defined cooperative games. Int. J. Game Theor. **30**, 377–404 (2001)
2. Masuya, S., Inuiguchi, M.: A fundamental study for partially defined cooperative games. Fuzzy Optim. Decis. Making (2015). doi:10.1007/s10700-015-9229-1
3. Polkowski, L., Araszkiewicz, B.: A rough set approach to estimating the game value and the Shapley value from data. Fundamenta Informaticae **53**, 335–343 (2002)
4. Schmeidler, D.: The nucleolus of a characteristic function game. SIAM J. Appl. Math. **17**, 1163–1170 (1969)
5. Shapley L.S.: A value for *n*-person games. In: Kuhn, H. Tucker, A. (eds.) Contributions to the Theory of Games II, pp. 307–317. Princeton (1953)
6. Von Neumann, J., Morgenstern, O.: Theory of Games and Economic Behavior. Princeton University Press (1944)
7. Willson, S.J.: A value for partially defined cooperative games. Int. J. Game Theor. **21**, 371–384 (1993)

Modeling and Property Analysis of E-Commerce Logistics Supernetwork

Chuanmin Mi, Yinchuan Wang and Yetian Chen

Abstract More and more social participants are involved in logistics network, making it growingly complicated. This paper puts forward supernetwork research approach, builds a three-layer logistics supernetwork model of supplier, distributor and consumer, sets maximum profits as objective, expresses market behaviors of decision makers of supernetwork model with income and fee functions, builds objective function, discusses conditions of balance in logistics network enterprises, and finally has case study of Amazon Xi'an operation center E-commerce logistics.

Keywords Electronic commerce logistics · Supernetwork · Variational inequality

1 Introduction

As the fast development of E-commerce in China, the logistics services such as storage and delivery are becoming more and more important. Competition among E-commerce enterprises will eventually turn to back-end logistics [1]. Enterprises with better logistics service will win more clients and consumption. Development of logistics is far less than satisfactory for needs of E-commerce [2]. According to data and statistics, China's E-commerce is developing at a speed of 200–300 %, compared to only 40 % of logistics, enlarging the gap between them [3]. More and More research focus on e-commerce logistics problems [4–6]. In our daily life, it is difficult to describe network characteristics of the real world with simple

C. Mi (✉) · Y. Wang · Y. Chen
College of Economics and Management, Nanjing University of Aeronautics and Astronautics, Nanjing, Jiangsu, China
e-mail: michuanmin@163.com

Y. Wang
e-mail: 18262620070@163.com

Y. Chen
e-mail: bigsillybear@163.com

I. Czarnowski et al. (eds.), *Intelligent Decision Technologies 2016*, Smart Innovation, Systems and Technologies 56,
DOI 10.1007/978-3-319-39630-9_12

networking relationships. Nested networks or interweaved multi-layer networks often appear in study of certain super-scale network problems. American scholar Sheffi proposed "Hypernetworks" [7], and Nagurney put forward the concept of "Supernetwork" as a tool of studying inter-influence and reaction within the networks, which is able to have quantitative analysis and calculation of variables in complex networks with such mathematical tools as variational inequality equation and visualized graphics [8]. Nagurney and her group studied supply chain related problems based on supernetwork theory [9, 10]. In China, Wang first introduced supernetwork, and studied the supply chain which two channels which considering E-commerce [11]. But few studies focus on using supernetwork to solve reality e-commerce logistics problems.

This paper treats E-commerce logistics network as supernetwork, builds a three-layer supernetwork model of supplier, distributor and consumer market, sets objective function, has optimal analysis of the objective function with variational inequality equation, discusses conditions of a balanced logistics network, and takes Xi'an Operation Center of Amazon E-commerce logistics as an example to have quantitative analysis of relations between product price and quantity in different layers of decision makers.

2 Modeling of E-Commerce Logistics Supernetwork

The traditional supply-demand chain consisting of producers and supply-demand market is difficult to completely and clearly depict current supply chain network relations. In this paper, we assume that there are three layers of decision makers in supply chain, namely, supplier, distributor and consumer market. Supplier supplies product and under certain circumstances, delivers product. While distributor stores, delivers and sells product. This supernetwork consists of three layers: l represents supplier participants of logistics network, $l \in \{1, \ldots L\}$; m represents distributor participants of logistics network, $m \in \{1, \ldots M\}$; n represents consumers, $n \in \{1, \ldots N\}$. See Fig. 1. Full line represents trade network flow among supplier, distributor and consumer; dotted line represents correlation. In logistics supernetwork, flow medium is distribution and storage of product; in supply chain supernetwork, flow medium is sales of product. There are 2 kinds of flow according to flow nodes:

(1) **Supplier-Distributor-Consumer market**

Distributor provides logistics service. There are 2 payment approaches according to fee payer: first, distributor buys product from supplier; product is stored in supplier's warehouse; logistics fee is undertaken by distributor; product is then delivered to consumer by distributor. Second, distributor builds E-commerce platform as third-party logistics provider; supplier will store the product on distributor's platform, via which it is delivered to consumer. Logistics fee is paid by supplier to distributor.

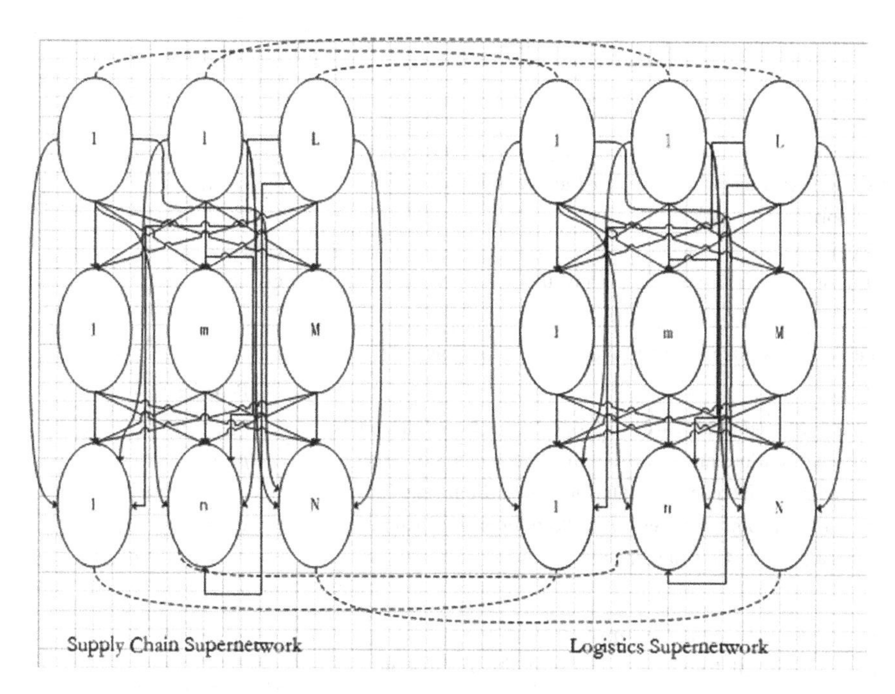

Fig. 1 E-commerce Logistics Supernetwork Model

(2) **Supplier-Consumer market**

Product is stored and supplied to consumer by supplier; logistics fee is undertaken by supplier.

3 Analysis of E-Commerce Logistics Supernetwork Model

3.1 *Model Symbols*

q_{1lm} Quantity of product sold by Supplier l to Distributor m

q_{1lmn} Quantity of product sold by Supplier l to Consumer n via Distributor m, and logistics service is provided by distributor

q_{2lmn} Quantity of product sold by Supplier l to Consumer n via Distributor m, and logistics service is provided by supplier

q_{3lmn} Quantity of product traded between Distributor m and Consumer n

p_{1lm} Unit price of product traded between Supplier l and Distributor m

p_{2ln} Unit price of product traded between Supplier l and Consumer n

p_{3mn} Unit price of product traded between Distributor m and Consumer n

f_{lm} Commission paid between Supplier l and Distributor m for platform trade

g_l Logistics cost of Supplier

g_{lm} Logistics fee of product trade between Supplier l and Distributor m platform

Table 1 Supplier fee and income function table

Items	Function
Income from product sold from supplier l to distributor m	$p_{1lm} * q_{1lm}$
Income from product sold from supplier l to consumer n	$p_{2ln} * (q_{1lmn} + q_{2lmn})$
Commission paid from supplier l to distributor m	$f_{lm} * (q_{1lmn} + q_{2lmn})$
Supply cost of supplier l	$h_l * (q_{1lm} + q_{1lmn} + q_{2lmn})$
Logistics cost of supplier l	$g_l * q_{2lmn}$
Logistics fee from supplier l via distributor m	$g_{lm} * q_{2mn}$

Table 2 Distributor fee and income function table

Items	Function
Cost fee of product distributor m purchases from Supplier l	$p_{1lm} * q_{1lm}$
Income from product sold from distributor m to consumer n	$p_{3mn} * q_{3mn}$
Commission income distributor m collects from supplier l	$f_{lm} * (q_{1lmn} + q_{2lmn})$
Logistics income distributor m collects from distributor l	$g_l * q_{1lmn}$
Logistics cost of distributor m	$r_m * (q_{1lmn} + q_{3mn})$

h_l Supply cost of Supplier l
r_m Cost of logistics service of Distributor m

3.2 *Supplier Trade Behavior Function Expression*

See Table 1.

3.3 *Distributor Trade Behavior Function Expression*

See Table 2.

4 Optimal Condition Analysis of E-Commerce Logistics Supernetwork

4.1 *Optimal Condition Analysis of Layers of Decision Makers*

(1) **Optimal Condition Analysis of Supplier**

According to income and cost functions in Table 1, we can get max profit function of supplier.

$$\max\Big(\sum_{m=1}^{M} q_{1lm}p_{1lm} + \sum_{m=1}^{M}\sum_{n=1}^{N} p_{2ln}(q_{1lmn}+q_{2lmn}) - \sum_{m=1}^{M}\sum_{n=1}^{N} h_l(q_{1lm}+q_{1lmn}+q_{2lmn})$$
$$- \sum_{m=1}^{M}\sum_{n=1}^{N} f_{lm}(q_{1lmn}+q_{2lmn}) - \sum_{m=1}^{M}\sum_{n=1}^{N} g_l(q_{2lmn}) - \sum_{m=1}^{M}\sum_{n=1}^{N} g_{lm}(q_{1lmn})\Big)$$
$$s.t.\ q_{1lm}, q_{1lmn}, q_{2lmn}, p_{1lm}, p_{2ln} \geq 0 \tag{1}$$

For all suppliers, transfer Eq. (1) to Variational Inequality.

$$\sum_{l=1}^{L}\sum_{m=1}^{M}\sum_{n=1}^{N}\left\{\frac{h_l(q_{1lm}+q_{1lmn}+q_{2lmn})}{\partial_{q_{1lm}}}\right\}\times(q_{1lm}-q_{1lm}^{*})$$
$$+\sum_{l=1}^{L}\sum_{m=1}^{M}\sum_{n=1}^{N}\left\{\frac{h_l(q_{1lm}+q_{1lmn}+q_{2lmn})}{\partial_{q_{1lm}}}+\frac{f_{lm}(q_{1lmn}+q_{2lmn})}{\partial_{q_{1lmn}}}+\frac{g_{lm}(q_{1lmn})}{\partial_{q_{1lmn}}}-p_{2ln}\right\}\times(q_{1lmn}-q_{1lmn}^{*})$$
$$+\sum_{l=1}^{L}\sum_{m=1}^{M}\sum_{n=1}^{N}\left\{\frac{h_l(q_{1lm}+q_{1lmn}+q_{2lmn})}{\partial_{q_{2lmn}}}+\frac{f_{lm}(q_{1lmn}+q_{2lmn})}{\partial_{q_{2lmn}}}+\frac{g_l(q_{2lmn})}{\partial q_{2lmn}}-p_{2ln}\right\}\times(q_{2lmn}-q_{2lmn}^{*})$$
$$-\sum_{l=1}^{L}\sum_{m=1}^{M}(p_{1lm}-p_{1lm}^{*})\times q_{1lm}^{*}-\sum_{l=1}^{L}\sum_{m=1}^{M}\sum_{n=1}^{N}\{p_{2ln}-p_{2ln}^{*}\}\times(q_{1lmn}^{*}-q_{2lmn}^{*})\geq 0$$
$$s.t.\ q_{1lm}, q_{1lmn}, q_{2lmn}, p_{1lm}, p_{2ln} \geq 0 \tag{2}$$

Quantity with * mark is the optimal solution, the same below.

(2) **Optimal Condition Analysis of Distributor**

According to income items and fee functions of distributor in Table 2, we can get max profit function of distributor.

$$\max\Big(\sum_{n=1}^{N} q_{3mn}p_{3mn} - \sum_{l=1}^{L} q_{1lm}p_{1lm} + \sum_{l=1}^{L}\sum_{n=1}^{N} f_{lm}(q_{1lmn}+q_{2lmn})$$
$$- \sum_{l=1}^{L}\sum_{n=1}^{N} g_{lm}(q_{1lmn}) - \sum_{l=1}^{L}\sum_{n=1}^{N} r_m(q_{1lmn}+q_{3mn})\Big) \tag{3}$$
$$s.t.\ q_{3mn}, q_{1lm}, q_{1lmn}, q_{2lmn}, p_{3mn}, p_{1lm} \geq 0$$

For all distributors, transfer Eq. (3) to Variational Inequality.

$$\begin{aligned}
&\sum_{l=1}^{L}\sum_{m=1}^{M}\sum_{n=1}^{N}\left\{\frac{r_m(q_{1lmn}+q_{3lmn})}{\partial_{q_{1lmn}}}-\frac{f_{lm}(q_{1lmn}+q_{2lmn})}{\partial_{q_{1lmn}}}-\frac{g_{lm}(q_{1lmn})}{\partial_{q_{1lmn}}}\right\}\times\left(q_{1lmn}-q_{1lmn}^{*}\right)\\
&-\sum_{l=1}^{L}\sum_{m=1}^{M}\sum_{n=1}^{N}\frac{f_{lm}(q_{1lmn}+q_{2lmn})}{\partial_{q_{2lmn}}}\times\left(q_{2lmn}-q_{2lmn}^{*}\right)\\
&+\sum_{l=1}^{L}\sum_{m=1}^{M}\sum_{n=1}^{N}\left\{\frac{r_m(q_{1lmn}+q_{3lmn})}{\partial_{q_{1lmn}}}-p_{3mn}\right\}\times\left(q_{3mn}-q_{3mn}^{*}\right)\\
&-\sum_{m=1}^{M}\sum_{n=1}^{N}\left(p_{3mn}-q_{3mn}^{*}\right)\times q_{3mn}^{*}+\sum_{m=1}^{M}\sum_{n=1}^{N}\left(p_{1lm}-p_{1lm}^{*}\right)\times q_{1lm}^{*}\geq 0
\end{aligned}\tag{4}$$

$s.t.\ q_{3mn}, q_{1lm}, q_{1lmn}, q_{2lmn}, p_{3mn}, p_{1lm}\geq 0$

4.2 Analysis of Balance Conditions

Conditions for balance status of E-commerce logistics supernetwork are: trade product quantity and unit price of supplier; distributor and consumer are sum of Eqs. (2) and (4). Equation (5) is the sum of variational inequalities.

$$\begin{aligned}
&\sum_{l=1}^{L}\sum_{m=1}^{M}\sum_{n=1}^{N}\left\{\frac{h_l(q_{1lm}+q_{1lmn}+q_{2lmn})}{\partial_{q_{1lm}}}\right\}\times\left(q_{1lm}-q_{1lm}^{*}\right)\\
&+\sum_{l=1}^{L}\sum_{m=1}^{M}\sum_{n=1}^{N}\left\{\frac{h_l(q_{1lm}+q_{1lmn}+q_{2lmn})}{\partial_{q_{1lmn}}}-\frac{g_{lm}(q_{1lmn})}{\partial_{q_{1lmn}}}-p_{2ln}\right\}\times\left(q_{1lmn}-q_{1lmn}^{*}\right)\\
&+\sum_{l=1}^{L}\sum_{m=1}^{M}\sum_{n=1}^{N}\left\{\frac{h_l(q_{1lm}+q_{1lmn}+q_{2lmn})}{\partial_{q_{2lm}}}+\frac{g_l(q_{2lmn})}{\partial_{q_{2lmn}}}-p_{2ln}\right\}\times\left(q_{2lmn}-q_{2lmn}^{*}\right)\\
&+\sum_{l=1}^{L}\sum_{m=1}^{M}\sum_{n=1}^{N}\left\{\frac{r_m(q_{1lmn}+q_{3mn})}{\partial_{q_{3mn}}}-p_{3mn}\right\}\times\left(q_{3mn}-q_{3mn}^{*}\right)\\
&-\sum_{l=1}^{L}\sum_{m=1}^{M}\sum_{n=1}^{N}\left(p_{2ln}-p_{2ln}^{*}\right)\times\left(q_{1lmn}^{*}-q_{2lmn}^{*}\right)\\
&-\sum_{m=1}^{M}\sum_{n=1}^{N}\left(p_{3mn}-p_{3mn}^{*}\right)\times p_{3mn}^{*}\geq 0
\end{aligned}\tag{5}$$

$s.t.\ q_{3mn}, q_{1lm}, q_{1lmn}, q_{2lmn}, p_{3mn}, p_{1lm}, p_{2ln}\geq 0$

5 Case Analysis

5.1 Case Background

There are 11 operational centers of Amazon in China, two in Beijing, other nine in Suzhou, Guangzhou, Chengdu, Wuhan, Shenyang, Xi'an, Xiamen, Kunshan and Tianjin separately. These 11 centers constitute China's biggest and most advanced operation network of E-commerce. Xi'an operation center is the 8th of Amazon China and was put to use in August 2010. It shortens order delivery time of local consumers averagely by 1–2 days and guarantees delivery quality.

There are over 30 suppliers of body wash in Xi'an operation center, such as Unilever, Johnson & Johnson and P&G. In this case study, we choose these four suppliers, which represented by $l = 1, 2, 3, 4$.

Similar to national market, second and third-tier cities of Central China are battle fields of Integrated mall B2C E-commerce enterprises, among which Amazon China, JD, TMall and Dangdang take the majority of local market. They are in competition relationships. These four competitors represented by $m = 1, 2, 3, 4$. The commission coefficient of these four distributors are 0.06, 0.04, 0.04, and 0.05.

We chooses 100 consumers of body wash within the last quarter operation period of 2015 by random sampling, represented by $n = 1, 2, \ldots 100$.

5.2 *Result Analysis*

According to fee functions of suppliers and distributors provided by the operation center, the following conclusions can be got based on variational inequality model and solutions.

(1) **Quantity of product supplier sells to consumer drops with the rise of price**

p_{2ln} and $\sum_M q_{1lmn}$, p_{2ln} and $\sum_M q_{2lmn}$ have opposite change trends; and q_{1lmn} is more influenced than q_{2lmn}. It indicates that quantity of product supplier sells to consumer drops with the rise of price; if q_{1lmn} is equally influenced to q_{2lmn}, quantity of product of supplier's self-built logistics is more influenced than quantity of product sold with distributor providing logistics service.

(2) **Supplier is more inclined to sell product via distributor**

p_{1lm} and $\sum_M q_{1lm}$, p_{2ln} and $\sum_M (q_{1lmm} + q_{2lmn})$ have opposite change trends, and q_{1lm} is more influenced than $(q_{1lmm} + q_{2lmn})$. It indicates that compared to selling on its own platform, supplier is more inclined to provide product to distributor for sales. Distributor has better sales advantage over supplier regarding consumer.

(3) **Consumer is more sensitive to self-built logistics**

p_{2ln} and $\sum_M q_{1lmn}$, p_{3mn} and $\sum_M (q_{2lmm} + q_{3lmn})$ have opposite change trends, and $\sum_M q_{1lmn}$ is more influenced than $\sum_M (q_{2lmm} + q_{3lmn})$. It indicates that quantity of product sold via supplier's self-built logistics is more fluctuated than quantity of product sold with distributor providing logistics service. Consumer is more sensitive to supplier's self-built logistics.

(4) **Fluctuation of supplier's price will influence quantity of product distributor buys from supplier**

p_{1lm} and $\sum_M q_{1lm}$, p_{3mn} and $\sum_M q_{3lmn}$ have opposite change trends, and $\sum_M q_{1lm}$ is more influenced than $\sum_M q_{3lmn}$. It indicates that when product price of supplier changes, distributor's buying from supplier will be greatly influenced. It will decrease sales and lower profit instead of transferring growth of price to consumer; or else, distributor will increase sales and profit instead of cutting the price.

6 Conclusion

This paper focuses on market behaviors of E-commerce logistics network participants, takes the research approach of supernetwork to study complicated E-commerce logistics network, builds a three-layer logistics supernetwork model consisting of supplier, distributor and different consumers, sets income and fee functions for different participants, analyzes optimization conditions of participants and transfers these conditions to variational inequalities to get condition of balance for the entire complicated logistics network, and analyzes operation of E-commerce logistics based on case study.

However, the following aspects need to be further studies regarding the building of model: (1) Model of this paper only takes sales and supply of single product into consideration; sales and supply of multiple products can be considered in further researches. Inter-influence and gaming exist in sales and supply of multiple products. (2) Function of this paper only takes maximum profit into consideration to build objective function; other factors like value of relations among layers of decision-makers, and risk relations in gaming process can be considered in further researches.

Acknowledgements The relevant researches are supported by the Fundamental Research Funds for the Central Universities of China (NS2016078).

References

1. Sink, H.L., Langley, C.J.: A managerial framework for the acquisition of third-party logistics service. J. Bus. Logist. **18**(2), 183–168 (1997)
2. Charles, C.P., Michael, J.B.: E-supply chain: using the internet to revolutionize your business: how market leaders focus their entire organization on driving value to customers. J. Bus. Logist. **24**(2), 259–260 (2003)
3. China Internet Network Information Center (2016): The 37th Statistical Report for China Internet Development (2016) (in Chinese)
4. Wang Z. P., Wang Z. T. Supernetworks: Theory and Application. Science Press, Beijing (2008) (in Chinese)

5. Jiang, Z.Z., Wang, D.W.: Model and algorithm of logistics delivery route optimization in B2C E-commerce. Inf. Control **34**, 481–484 (2005)
6. Wang, Q., Yan, L.J.: Study of logistics delivery route optimization under E-commerce environment. J. Tianjin Univ. Commerce. **30**(3), 27–30 (2010) (in Chinese)
7. Sheffi, Y., Daganzo, C.: Hypernetworks and supply-demand equilibrium with disaggregate demand models. Transp. Res. Rec. **673**, 1213–1333 (1979)
8. Nagurney, A., Dong, J.: Supernetworks: Decision-Making for the Information Age. Edward Elgar Publishing, Cheltenham (2002)
9. Dong, J., Nagurney, A.: A general multitiered supply chain network model of quality competition with suppliers. Int. J. Prod. Econ. **170**, 336–356 (2015)
10. Nagurney, A., Dong, J.: Equilibria and dynamics of supply chain network competition with information asymmetry in quality and minimum quality standards. CMS **11**(3), 285–315 (2014)
11. Wang, X.B., Li, Y.J.: Model and assessment approach of choosing location of logistics delivery center under E-commerce environment. Syst. Eng. Theories Appl. **15**(3), 199– 204 (2006) (in Chinese)

Maximum Lifetime Problem in Sensor Networks with Limited Channel Capacity

Zbigniew Lipiński

Abstract We analyze the maximum lifetime problem in sensor networks with limited channel capacity for multipoint-to-multipoint and broadcast data transmission services. For the transmission model in which the transmitter adjust the power of its radio signal to the distance to the receiver we propose a new Signal to Interference plus Noise Ratio function and use it to modify the Shannon-Hartley channel capacity formula. We show, that in order to achieve an optimal data transmission regarding considered the maximum lifetime problem we cannot allow for any interference of signals. For considered transmission model and the modified capacity formula we solve the maximum lifetime problem in one dimensional regular sensor network for both discussed data transmission services.

Keywords Wireless communication · Sensor network lifetime · Energy management · Channel capacity

1 Introduction

Characteristic feature of sensor networks is that these consist of small electronic devices with limited power and computational resources. Most of the energy the nodes of the sensor network possess is being utilized in the process of data transmission. The energy consumption depends on the size of the network, amount of data transmitted over the network and also on the presence of noise and interference of signals within the transmission channels. On the other hand, from the sensors a long term operating time is expected, that the functional lifetime and services delivered by the network can be available for a long time. We define a sensor network lifetime as the time until the first node of the network runs out of energy, [1–3]. Equivalently, the network lifetime can be defined as the number of cycles the network can perform its functions. Namely, if each node of the network has a battery with an initial

Z. Lipiński (✉)
Institute of Mathematics and Informatics, Opole University, Opole, Poland
e-mail: zlipinski@math.uni.opole.pl

I. Czarnowski et al. (eds.), *Intelligent Decision Technologies 2016*,
Smart Innovation, Systems and Technologies 56,
DOI 10.1007/978-3-319-39630-9_13

energy E_0, then by finding the optimal energy utilization of each node in one cycle of network lifetime E_i^{opt} we can determine the number of cycles $N_{\mathrm{cycles}} = [\frac{E_0}{E_{i'}^{\mathrm{opt}}}]$ the network can perform its functions until the most overloaded node i' runs out of its energy. From the above definitions follows that, to extend the sensor network lifetime it is necessary to minimize the maximum energy utilized by the network nodes.

The energy consumption model in wireless sensor networks can be expressed in terms of power P_i and the data transmission time t_i of each sensor. For the optimization problems in which we minimize the energy consumed by each node

$$E_i = P_i\, t_i, \tag{1}$$

we should assume that capacities of the transmission channels are limited, otherwise the minimum energy of each node is reached for $t_i = 0$. The capacity of a noisy channel can be described by the well know Shannon-Hartley formula, [4],

$$C = B \log(1 + \frac{P}{N_o}), \quad [b/s], \tag{2}$$

where B is the channel bandwidth, P the signal and N_o the noise power. An impact on the channel capacity has not only the level of a noise but also the interference of signals simultaneously generated by nodes in a given network. To take into account the interference of signals in the network the quotient $\frac{P}{N_o}$ in the Shannon-Hartley formula (2) should be replaced by a Signal to Interference plus Noise Ratio function. Let us denote by S_N the wireless sensor network built of N nodes. We assume, that the nodes are located at the points x_i of d-dimensional space R^d. By $P(x_i)$ we denote the transmission power of the i-th node and by $\gamma(x_i, x_j)$ the signal gain function between two nodes located at the points x_i and x_j. Because of the loss factor $\gamma(x_i, x_j)$ the strength of signal of the i-th node detected by the j-th node is equal to $P(x_i)\gamma(x_i, x_j)$. The Signal to Interference plus Noise Ratio (SINR) function defined in [5], see also [4, 6–8] is given by the following formula

$$s(x_i, x_j, U_i) = \frac{P(x_i)\gamma(x_i, x_j)}{N_o + \sum_{k \in U_i} P(x_k)\gamma(x_k, x_j)}, \tag{3}$$

where $U_i \subset S_N$ is the set of nodes which signals interfere with the signal of given i-th node. By definition $i \notin U_i$ for any $i \in S_N$. Note that, when the nodes use the omnidirectional antennas and their transmission signal is detected by any other node of the network, then U_i is the set of nodes which transmit data simultaneously with the i-th node. The formula (3) allows to estimate the conditions under which the transmission from the i-th node is successfully received by the j-th node. Instead of restrictions on the channel capacity functions in literature there is commonly utilized a threshold for the SINR function $\beta \leq s(x_i, x_j, U_i)$ below which the data cannot be transmitted. The value of β depends on the sensors transmission model in the data link layer, [6]. From the formula (3) it follows that the parameter β limits the number of nodes in the set U_i. The Shannon-Hartley formula modified by the SINR function (3) can be written in the form

$$C(x_i, x_j, U_i) = B \log(1 + s(x_i, x_j, U_i)). \tag{4}$$

The formula (4) allows to determine the maximum achievable transmission rate between the i-th and j-th node in the presence of other transmitters. If the sensors utilize the point-to-point data transmission in the physical layer then the amount of data $q_{i,j}$ transmitted between i-th and j-th node is a product of the transmission rate $c_{i,j}$ and transmission time $t_{i,j}$ of the i-th node

$$q_{i,j} = c_{i,j}\, t_{i,j}. \tag{5}$$

In optimization problems discussed in this paper the transmission rate matrix elements $c_{i,j}$, used as parameters, satisfy the inequality

$$0 \leq c_{i,j} \leq C(x_i, x_j, U_i), \tag{6}$$

where $C(x_i, x_j, U_i)$ is given by (4).

In the paper we consider two types of the maximum lifetime problems in sensor networks with limited channel capacity caused by the presence of noise and interference. The first problem is related to the optimization of the data transmission from a given set of sensors to the set of data collectors in S_N. This type of problem we call later on a maximum lifetime problem for the multipoint-to-multipoint data transmission service. The second problem describes the maximum lifetime problem for the broadcast transmission, [9, 10]. The objective function of the maximum lifetime problem for both types of services has the form

$$E_{i'} = \max_{i \in S_N} \{E_i\}, \tag{7}$$

where E_i is the energy consumed by the i-th node of the network S_N. According to [1, 11], we define the energy consumed by each node in terms of data transmission cost energy matrix $E_{i,j}$ and the data flow matrix $q_{i,j}$. Elements of the matrix $E_{i,j}$ define the energy required to transmit one unit of data between two nodes. The energy necessary to transmit the amount of data $q_{i,j}$ from the i-th node to the j-th node is a product $E_{i,j} q_{i,j}$ and the total the energy E_i consumed by the i-th node to transmit all of its data is given by the formula

$$E_i = \sum_j q_{i,j} E_{i,j}. \tag{8}$$

For the multipoint-to-multipoint data transmission service the minimum of the objective function (7) is determined under the data flow constraint

$$\sum_i q_{i,j} = Q_i + \sum_j q_{j,i}. \tag{9}$$

This equation states that the amount of data generated by the i-th node Q_i and the amount of data received from other nodes $\sum_j q_{j,i}$ must be equal to the amount of data which the node can send $\sum_i q_{i,j}$. By definition the data collectors does not generate, transmit or retransmit any data, i.e., for a data collector $k \in S_N$, $\forall_{j \in S_N}\ q_{k,j} = 0$. For the broadcast data transmission service, when the k-th node sends the amount Q_k of data to all other nodes of the network, the requirement that each node of S_N receives this data can be written in the form

$$\forall_{j \in S_N, j \neq k} \sum_{i, i \neq j} q_{i,j} = Q_k. \tag{10}$$

The above formulas are valid provided that in the physical layer the nodes use the point-to-point data transmission. For such transmission the sender transmits data to the unique receiver and the transmitted data are not copied.

In this paper we show that the optimal transmission for the maximum lifetime problem with limited channel capacity is achieved when there is no interference of signals in the transmission channels. For models in which the nodes use the omni-directional antennas and the threshold β is equal to zero, the requirement that there is no interference means that the data is transmitted by the nodes sequentially. We propose a new point-to-point transmission model with SINR function in which the signal power of the transmitting node $P(x_i)$ depends on the distance to the receiver and has the form $P(x_i) = P_0 \gamma(x_i, x_j)$, where $\gamma(x_i, x_j)$ is the signal gain function. We give the solution of the maximum lifetime problem with the new SINR function in the one dimensional regular sensor network L_N for both the multipoint-to-multipoint and broadcast data transmission services and show that these solutions coincide with the analytical solution of the problems discussed in [1, 12, 13]. We also show, that when the distance between sensor nodes is sufficiently large, the solution of the maximum network lifetime problem with Gupta-Kumar SINR function (3) coincides with the solution of the problem written in terms of $E_{i,j}$ and $q_{i,j}$ matrices discussed in [1, 12, 13].

2 Data Transmission Model with the SINR Function

We assume, that the power of the transmitting signal at the receiver must have some minimal level P_0. In other words, P_0 is the minimal signal power the receiver node can hear. This requirement means, that the transmitting node must generate the signal with the strength

$$P_{i,j} = P_0\, \gamma_{i,j}^{-1}, \tag{11}$$

where $\gamma_{i,j} = \gamma(x_i, x_j)$ is a signal gain function between two nodes located at the points x_i and x_j of R^d. The energy required to send data between the i-th and j-th node in the time $t_{i,j}$ equals

$$E_{i,j} = P_0 \gamma_{i,j}^{-1} t_{i,j}. \tag{12}$$

When sensors use in the physical layer the point-to-point data transmission the total energy consumed by the i-th node is given by the formula

$$E_i(t) = P_0 \sum_{j \in S_N} \gamma_{i,j}^{-1} t_{i,j}. \tag{13}$$

We denote by $U \subset S_N \times S_N$ the set of transmitter-receiver pairs $(i,k),\ i,k \in S_N$ such that for two elements (i,k) and (i',k') of U the nodes i and i' transmit data to the k and k' nodes respectively and their signals interfere. We assume, that $i \neq i', k \neq k'$. For wireless networks in which the nodes use the omnidirectional antennas and the signal is received in the whole network U can be defined as a set of node pairs which transmitters transmit data simultaneously, i.e.,

$$U = \{(i,k), (i',k') \in S_N \times S_N | t_{i,k}^{(s)} = t_{i',k'}^{(s)},\ t_{i,k}^{(e)} = t_{i',k'}^{(e)}\}, \tag{14}$$

where $t_{i,k}^{(s)}$ and $t_{i,k}^{(e)}$ is the start and the end of transmission time between i-th and k-th node. If $(i,k) \in U$ then by $U_{i,k}$ we denote the set $U_{i,k} = U \backslash \{(i,k)\}$. We assume, that each node cannot receive data simultaneously from many transmitters and cannot send data to several receivers, i.e. there is no wireless multicast advantage [16]. For the transmission model (11) the maximum achievable transmission rate between the i-th and j-th nodes can be given by the formula

$$C(x_i, x_j, U_{i,j}^n) = B \log(1 + \frac{P_0}{N_o + P_0 \sum_{(k,m) \in U_{i,j}^n} \gamma(x_k, x_m)^{-1} \gamma(x_k, x_j)}), \tag{15}$$

where $U_{i,j}^n$ is a given set of transmitter-receiver pairs which signal of the transmitters interfere with the signal of the i-th node.

In the next sections we show, that the solutions of the maximum network lifetime problem with the channel capacity (15) for the multipoint-to-multipoint and the broadcast data transmission services for both energy consumption models (1) and (8) are equivalent. We show that, the solution of the maximum network lifetime problem with maximum transmission rate (15) for both types of services in the one dimensional regular sensor network L_N can be obtained from the analytical solutions of the problems given in [1, 12, 13].

3 Multipoint-to-multipoint Data Transmission Service

The maximum network lifetime problem for the multipoint-to-multipoint data transmission service with the channel capacity given by (15) is defined by the objective function (7) in which the energy E_i of each node is given by the formula

$$E_i(t) = P_0 \sum_{j,n} \gamma_{i,j}^{-1} \, t_{i,j}^n, \tag{16}$$

where $t_{i,j}^n$ is the data transmission time between the i-th and j-th nodes in the presence of other transmitters from the set $U_{i,j}^n$. We minimize the objective function (7), i.e.,

$$E_{i'}(t_0) = \min_t \max_{i \in S_N} \{E_i(t)\},$$

with the data flow transmission constraints

$$\begin{cases} \sum_{i,n} c_{i,j}^n t_{i,j}^n = Q_i + \sum_{j,n} c_{j,i}^n t_{j,i}^n, \\ 0 \le c_{i,j}^n \le C_{i,j}^n, \end{cases} \tag{17}$$

where $C_{i,j}^n$ is the maximum transmission rate (the channel capacity) between i-th and j-th nodes calculated from (15).

For the maximum network lifetime problem of the multipoint-to-multipoint data transmission service with limited channel capacity we deduct the following

Theorem 1 *The optimal data transmission for the maximum network lifetime problem (7), (16), (17), (15) is the transmission without interference.*

Proof The energy consumed by the i-th node $E_i(t) = P_0 \sum_{j,n} t_{i,j}^n \gamma_{i,j}^{-1}$ is minimal when the data transmission time $t_{i,j}^n$ is minimal. Because of the constrain (17), the time parameter $t_{i,j}^n$ reaches the minimal value when $c_{i,j}^n = C_{i,j}^n$ and the capacity $C_{i,j}^n$ of the channel (i,j) is maximal. The $C_{i,j}^n$ is maximal when the interference factor in the SINR function $s(x_i, x_j, U_i^n)$ or $s(x_i, x_j, U_{i,j}^n)$ is equal to zero. It means, that the set U_i^n in (3) and $U_{i,j}^n$ in (15) must be empty and there is no interference of signals in the network. ◇

The result of Theorem 1 we use to redefine the maximum network lifetime problem of the multipoint-to-multipoint data transmission service to the form

$$\begin{cases} \min_t \max_i \{E_i(t)\}, \\ E_i(t) = P_0 \sum_j \gamma_{i,j}^{-1} \, t_{i,j}, \\ c_0 \sum_i t_{i,j} = Q_i + c_0 \sum_j t_{j,i}, \\ 0 \le c_0 \le C_0, \\ t_{i,j} \ge 0, \quad i,j \in [1,N], \end{cases} \tag{18}$$

where by C_0 we denoted the maximal transmission rate in (15), achieved when $U_{i,j}^n = \emptyset$, i.e.,

$$\forall_{i,j}\ C(x_i, x_j) = C_0 = B\log(1+\frac{P_0}{N_o}). \tag{19}$$

In the next lemma we give the solution of the maximum network lifetime problem (18) for a one dimensional, regular sensor network L_N with one data collector for arbitrary signal gain function of the form

$$\gamma^{-1}(\bar{a}, \bar{\lambda}, x_i, x_j) = \sum_{n=0}^{\infty} \lambda_n d(x_i, x_j)^{a_n},$$

where $d(x_i, x_j)$ is Euclidean distance in R and $\sum_{n=0}^{\infty} \lambda_n = 1,\ \lambda_n \geq 0,\ a_n \geq 1$. We assume, that sensors of the L_N network are located at the points $x_i = i$ of a line and the data collector at the point $x_0 = 0$. For $d_{i,j} = |i-j|$ any two sensors in L_N, which distance between them is equal to $|i-j| = r$ the signal gain $\gamma_{i,j}$ can be written in the form

$$\gamma_r^{-1}(\bar{a}, \bar{\lambda}) = \sum_{n=0}^{\infty} \lambda_n r^{a_n},\quad r \in [1, N], \tag{20}$$

where $\sum_{n=0}^{\infty} \lambda_n = 1$, $\lambda_n \geq 0$ and $a_n \geq 1$. In the following lemma we give the solution of (18) for any gain function (20).

Lemma 1 *The solution of the maximum lifetime problem for the multipoint-to-multipoint data transmission service in the one dimensional regular sensor network L_N, with the channel gain function (20) and the channel capacity (15) is given by the following set of data transmission trees*

$$T^1 = T_{1,0},\quad T^i = \{T^i_{i,0}, T^i_{i,i-1}\},\quad i \in [2, N],$$

and weights

$$\begin{cases} t_{1,0} = \frac{1}{C_0}Q_N + \frac{1}{C_0}\sum_{j=2}^{N} Q_{j-1}\prod_{r=j}^{N}(1-\frac{\gamma_1^{-1}}{\gamma_r^{-1}}), \\ t_{2,0} = \frac{1}{C_0}\frac{\gamma_1^{-1}}{\gamma_2^{-1}}Q_1, \\ t_{i,0} = \frac{1}{C_0}\frac{\gamma_1^{-1}}{\gamma_i^{-1}}(Q_{i-1} + \sum_{j=1}^{i-2} Q_{i-j-1}\prod_{r=1}^{j}(1-\frac{\gamma_1^{-1}}{\gamma_{i-r}^{-1}})), \quad i \in [3, N], \\ t_{2,1} = \frac{1}{C_0}Q_2 - \frac{1}{C_0}\frac{\gamma_1^{-1}}{\gamma_2^{-1}}Q_1, \quad N = 2, \\ t_{2,1} = \frac{1}{C_0}\sum_{k=2}^{N} Q_k - \frac{1}{C_0}\sum_{k=2}^{N}\frac{\gamma_1^{-1}}{\gamma_k^{-1}}Q_{k-1} \\ -\frac{1}{C_0}\sum_{k=3}^{N}\frac{\gamma_1^{-1}}{\gamma_k^{-1}}\sum_{j=1}^{k-2} Q_{k-j-1}\prod_{r=1}^{j}(1-\frac{\gamma_1^{-1}}{\gamma_{k-r}^{-1}}), \quad N \geq 3, \\ t_{i,i-1} = \frac{1}{C_0}\sum_{k=i}^{N} Q_k - \frac{1}{C_0}\sum_{k=i}^{N}\frac{\gamma_1^{-1}}{\gamma_k^{-1}}(Q_{k-1} + \sum_{j=1}^{k-2} Q_{k-j-1}\prod_{r=1}^{j}(1-\frac{\gamma_1^{-1}}{\gamma_{k-r}^{-1}})), \\ i \in [3, N]. \end{cases} \tag{21}$$

Proof It is easy to show that the formulas given in (18) by a simple transformation $q_{i,j} \to P_0 t_{i,j}$ and $E_{i,j} \to \gamma_{i,j}^{-1}$ and $Q_i \to \frac{P_0}{C_0} Q_i$ can be obtained from the formulas (8) from [12], and the solution (21) can be obtained by these transformation from the solution (16) given in [12]. Detailed proof that the set of trees and weight polynomials (21) in the Lemma 1 is optimal can be carried out analogously to the proof given in [12]. ◇

As an simple example, let us assume that the gain is quadratic function $\gamma_r = r^{-2}$ of the distance and each sensor sends the same amount Q_0 of data. According to the Theorem 1 and Lemma 1 the i-th sensor, $i \in [1, N]$, must transmit sequentially data to the data collector for the time

$$t_{i,0} = \frac{Q_0}{C_0} \frac{i - H_i}{i(i-1)}, \; i > 1,$$

where H_i is the i-th harmonic number. The rest of the data with the C_0 transmission rate it is transmitted to the $(i-1)$-th sensor. Similar result for the maximum lifetime problem was obtained in [14].

4 Broadcast Data Transmission Service

To define the maximum network lifetime problem for the broadcast data transmission we represent the sensor network S_N as a directed, weighted graph $G_N = \{S_N, V, E\}$, in which S_N is the set of graph nodes, V is the set edges and E the set of weights. Each directed edge $T_{i,j} \in V$ defines a communication link between i-th and j-th node of the network. To each edge $T_{i,j}$ we assign a weight $E_{i,j}$, which is the cost of transmission of one unit of data between i-th and j-th node. By $U_i^{(out)} \subseteq S_N$ we denote a set of the network nodes to which the i-th node can sent the data $U_i^{(out)} = \{x_j \in S_N | \; \exists \; T_{i,j} \in V\}$. The set $U_i^{(out)}$ defines the maximal transmission range of the i-th node. We assume, that each node of S_N can send data to any other nodes of the network, i.e.,

$$\forall_{i \in [1,N]}, \quad U_i^{(out)} = S_N. \tag{22}$$

We describe the data flow in the network S_N in terms of spanning trees of the graph G_N. The set of all spanning trees $T^{k,r}$, which begin in the k-th node we denote by V_k. If the assumption (22) is satisfied, then G_N is a complete graph and the number of spanning trees rooted at the k-th node $|V_k|$ is equal to N^{N-2}, [15]. By $q_{i,j}^{k,r}$ we denote the amount of data which is transmitted along the edge $T_{i,j}^{k,r}$. The amount of data $q_{i,j}^k$ send by the i-th node to the j-th node along all trees is given by the formula

$$q_{i,j}^k = \sum_r q_{i,j}^{k,r} T_{i,j}^{k,r}. \tag{23}$$

We require, that along each tree the transmitted data is the same. This means that, for fixed k and r, the weights $q_{i,j}^{k,r}$ of the edges $T_{i,j}^{k,r}$ are equal. We denote these weights by q_r^k, i.e.,

$$\forall_{i,j}\; q_{i,j}^{k,r} = q_r^k. \tag{24}$$

In the energy consumption model (1), when the data is transmitted without interference, the parameter q_r^k will be replaced by the parameter t_r^k which defines the transmission time for the amount q_r^k of data along the r-th tree. Because we assumed, that all data for the broadcast data transmission service is transmitted along some set of trees, we must modify in the channel capacity formula (15) the definition of the set $U_{i,j}^n$. We replace it by $U_{i,j}^{k,n,r}$, the set of sensors which signal interferes with the signal of the i-th node when the broadcasted data is transmitted along the tree $T^{k,r}$. We denote the new channel capacity function by $C(x_i, x_j, U_{i,j}^{k,n,r})$.

The maximum network lifetime problem of the broadcast data transmission service with the channel capacity (15) with the set $U_{i,j}^{k,n,r}$ is defined by the objective function (7) in which the energy of each node is given by the formula

$$E_i(t^k) = P_0 \sum_{n,r,j} \gamma_{i,j}^{-1}\, t_{i,j}^{k,n,r}\, T_{i,j}^{k,r}, \tag{25}$$

where $\gamma_{i,j}^{-1}$ is given by (20). The data flow transmission constraint (10), which is the requirement that each node of the network must receive data broadcasted by the k-th node, can be written in the form

$$\begin{cases} \sum_{n,r,i} c_{i,j}^{k,n,r}\, t_{i,j}^{k,n,r}\, T_{i,j}^{k,r} = Q_k, & j \in [1, N], \\ 0 \le c_{i,j}^{k,n,r} \le C_{i,j}^{k,n,r}, & \\ t_{i,j}^{k,n,r} \ge 0, \quad T^{k,r} \in V_k, & r \in [1, N^{N-2}], \end{cases} \tag{26}$$

where $C(x_i, x_j, U_{i,j}^{k,n,r}) \equiv C_{i,j}^{k,n,r}$ is given by (15). For the maximum network lifetime problem of the broadcast data transmission service defined by (7), (25), (26) we propose a theorem, similar to the Theorem 1 for the multipoint-to-multipoint data transmission service.

Theorem 2 *The optimal data transmission for the maximum network lifetime broadcasting problem (7), (25), (26) is the transmission without interference.*

Proof The proof is analogous to the proof of Theorem 1. ◇

From the Theorem 2 it follows that $\forall_{n,r,i,j} C_{i,j}^{k,n,r} = C_0$ and C_0 is given by (19). Because the transmission time between i-th and j-th node, when the data is transmitted along the edge $T_{i,j}^{k,n,r}$ must be the same then we have the following relation

$$\forall_{n,i,j}\; t_{i,j}^{k,n,r} = t_r^k.$$

The formula (5) written in terms of trees for each set $U_{i,j}^{k,n,r}$ has the form

$$\forall_{(i,j)\in T^{k,r}} \; q_r^k = c_{i,j}^{k,n,r} t_{i,j}^{k,n,r}.$$

The maximum network lifetime problem of the broadcast data transmission service without interference can be written in the form

$$\begin{cases} \min_{t^k} \max_{i\in S_N} \{E_i(t^k)\}_{i=1}^N, \; k \in [1,N], \\ E_i(t^k) = P_0 \sum_{j,r} \gamma_{i,j}^{-1} \, t_r^k \, T_{i,j}^{k,r}, \\ c_0 \sum_{i,r} t_r^k \, T_{i,j}^{k,r} = Q_k, \; j \in [1,N], \\ 0 \le c_0 \le C_0, \\ t_r^k \ge 0, \; T^{k,r} \in V_k, \; r \in [1, N^{N-2}]. \end{cases} \tag{27}$$

The following lemma gives the solution of the problem (27) for the internal nodes of the one dimensional regular sensor network L_N.

Lemma 2 *The solution of the maximum network lifetime problem for the broadcast data transmission service (27) in the one dimensional regular sensor network L_N with the channel gain function (20) and channel capacity (19) is given by the following set of data transmissions trees T^r*

$$\begin{array}{ll} (T_{k,k-1}^r, \ldots, T_{2,1}^r, T_{r,k+1}^r, T_{k+1,k+2}^r, \ldots, T_{N-1,N}^r), & r \in [1, k-1], \\ (T_{k,k-1}^k, \ldots, T_{2,1}^k, T_{k,k+1}^k, \ldots, T_{N-1,N}^k\}, & r = k, \\ (T_{k,k+1}^r, \ldots, T_{N-1,N}^r, T_{r,k-1}^r, T_{k-1,k-2}^r, \ldots, T_{2,1}^r), & r \in [k+1, N], \end{array} \tag{28}$$

and weights t_r^k

$$t_r^k = \begin{cases} \frac{\gamma_1^{-1}}{\gamma_k^{-1}} t_k^k + \frac{\gamma_1^{-1}}{\gamma_k^{-1}} \frac{1}{C_0} Q_k, & r = 1, \\ \frac{\gamma_1^{-1}}{\gamma_{k+1-r}^{-1}} t_k^k, & r \in [2, k-1], k \ge 3, \\ \frac{\gamma_1^{-1}}{\gamma_{r-k+1}^{-1}} t_k^k, & r \in [k+1, N-1], k \le N-2, \\ \frac{\gamma_1^{-1}}{\gamma_{N-k+1}^{-1}} t_k^k + \frac{\gamma_1^{-1}}{\gamma_{N-k+1}^{-1}} \frac{1}{C_0} Q_k, & r = N, \end{cases} \tag{29}$$

where

$$t_k^k = \frac{1 - \frac{\gamma_1^{-1}}{\gamma_k^{-1}} - \frac{\gamma_1^{-1}}{\gamma_{N-k+1}^{-1}}}{-1 + \sum_{i=1}^k \frac{\gamma_1^{-1}}{\gamma_i^{-1}} + \sum_{i=1}^{N-k+1} \frac{\gamma_1^{-1}}{\gamma_i^{-1}}} \frac{1}{C_0} Q_k. \tag{30}$$

and γ_r^{-1}, $r \in [1,N]$ is given by (20).

Proof of this lemma can be found in [13]. For the boundary nodes $i = 1, N$ of L_N the solution of (27) is trivial. Proof of this fact can be found in [13].

Let us assume that, the k-th node broadcasts the amount Q_0 of data and the gain function has the form $\gamma_r = r^{-2}$, then according to Theorem 2 and Lemma 2, the nodes must transmit the data along the r-th tree sequentially, $r \in [2, k-1]$, for the time

$$t_r^k = \frac{1}{(k+1-r)^2} \frac{1 - \frac{1}{k^2} - \frac{1}{(N-k+1)^2}}{H_{k,2} + H_{N-k+1,2} - 1} \frac{Q_0}{C_0},$$

where $H_{k,2}$ is the harmonic number of order 2.

5 Gupta-Kumar Model Without Interference

We apply the results of Theorems 1 and 2 to the maximum network lifetime problem with channel capacity function given by (3). If nodes of the network transmit data with the some constant power $P(x_i) = P_0$ and there is no interference, then the maximum transmission rate in (3) is given by the expression

$$C(x_i, x_j) = B \log(1 + \frac{P_0}{N_o} \gamma(x_i, x_j)). \tag{31}$$

From (31) it follows that even when there is no interference the function $C(x_i, x_j)$ still depends on the distance between transmitter and receiver. In models, in which the SINR function (3) is utilized, the nodes of the sensor network do not adjust the transmission power to the distance of the receiver. Thus, we must modify the formulas for the energy of the node (16) and (25). The maximum network lifetime problem of the multipoint-to-multipoint transmission service with the channel capacity (31) has the form

$$\begin{cases} \min_t \max_i \{E_i(t)\}, \\ E_i(t) = P_0 \sum_j t_{i,j}, \\ \sum_i c_{i,j} t_{i,j} = Q_i + \sum_j c_{j,i} t_{j,i}, \\ 0 \le c_{i,j} \le C_{i,j}, \\ t_{i,j} \ge 0, \quad i, j \in [1, N], \end{cases} \tag{32}$$

where $C_{i,j} \equiv C(x_i, x_j)$ is given by (31). In (32) we put $c_{i,j} = C_{i,j}$ and resize the time variables as follows $t_{i,j} \to \frac{C_0}{C_{i,j}} t_{i,j}$, where C_0 is an arbitrary positive number of dimension $[b/s]$. Because $\log(1+x) \approx x$ for $|x| < 1$, then for $B \frac{P_0}{N_o} \gamma(x_i, x_j) < 1$ we can write

$$C_{i,j} = B \log(1 + \frac{P_0}{N_o} \gamma(x_i, x_j)) \approx B \frac{P_0}{N_o} \gamma(x_i, x_j).$$

This means, that the maximum lifetime problem for the multipoint-to-multipoint transmission service can be transformed to the form (18), where $E_i(t) = N_o C_0 \sum_j \gamma_{i,j}^{-1}$

$t_{i,j}$ and $c_0 = C_0$. The requirement $B\frac{P_0}{N_o}\gamma(x_i, x_j) < 1$, for $\gamma_{i,j} = (d_{i,j})^{-a}$ can be written as $B\frac{P_0}{N_o} < (d_{i,j})^a$, which means that for a sufficiently large network the two models (3), (4) and (15) coincide.

Similar results can be obtained for the broadcast data transmission service with the channel capacity function (3). For the broadcast data transmission service the third formula in (32) should be replaced by the third formula in (27) and the energy of each node by $E_i(t^k) = P_0 \sum_{j,r} t_r^k \, T_{i,j}^{k,r}$.

6 Conclusions

In the paper we analyzed the maximum network lifetime problem in sensor networks with limited channel capacity for multipoint-to-multipoint and broadcast data transmission services. We showed, that for the optimal data transmission of the maximum lifetime problem there is no interference of signals in the network. We introduced new Signal to Interference plus Noise Ratio function which was defined under the assumption that the signal power $P(x_i)$ of the transmitting node depends on the signal gain $\gamma(x_i, x_j)$ and location of the receiver $P(x_i) = P_0\gamma(x_i, x_j)$. For such transmission model with the channel capacity defined by the new SINR function we solved the maximum lifetime problem in one dimensional regular sensor network L_N for both data transmission services. We showed, that these solutions coincide with the solutions of the maximum lifetime problem analyzed in [1, 11–13], where the problem was defined in terms of data transmission cost energy matrix $E_{i,j}$ and data flow matrix $q_{i,j}$. This way we find the relation between the energy consumption model written in terms of $E_{i,j}$ and $q_{i,j}$ and the model defined in terms of signal power, transmission time and channels capacities.

There are several differences between the SINR functions $s(x_i, x_j, U_{i,j}^n)$ defined in (15) and proposed by Gupta and Kumar in (3). It seems that, the most interesting future of the proposed SINR functions is that for a gain function of the form $\gamma(x_k, x_m) = d(x_k, x_m)^{-a}$ the formula (15) is invariant under the rescaling of the distance between network nodes $d'(x_k, x_m) = \lambda d(x_k, x_m)$. From which follows, that the solutions of the maximum lifetime problem for two networks with proportional distance between nodes are related by a constant factor. For example, when we extend the distance between all nodes by the same factor $d(x_k, x_m) \rightarrow \lambda d(x_k, x_m)$, and the node energy for the distance $d(x_k, x_m)$ is equal $E_i(t) = P_0 \sum_{j,n} d_{i,j}^a \, t_{i,j}^n$, then the nodes must increase their power signal and the energy consumption increases to $E_i(t) = \lambda^a P_0 \sum_{j,n} d_{i,j}^a \, t_{i,j}^n$.

References

1. Giridhar, A., Kumar, P.R.: Maximizing the functional lifetime of sensor networks. In: Proceedings of the 4-th International Symposium on Information Processing in Sensor Networks. IEEE Press, Piscataway, NJ, USA (2005)
2. Acharya, T., Paul, G.: Maximum lifetime broadcast communications in cooperative multihop wireless ad hoc networks: centralized and distributed approaches. Ad Hoc Netw. **11**, 1667–1682 (2013)
3. Dietrich, I., Dressler, F.: On the lifetime of wireless sensor networks. ACM Trans. Sens. Netw. **5**(1), Article 5 (2009)
4. Franceschetti, M., Meester, R.: Random Networks for Communication. Cambridge University Press (2007)
5. Gupta, P., Kumar, P.R.: The capacity of wireless networks. IEEE Trans. Inf. Theor. **46**(2), 388–404 (2000)
6. Baccelli, F., Blaszczyszyn, B.: Stochastic Geometry and Wireless Networks, vol. 1, 2. Now Publishers Inc (2009)
7. Grossglauser, M., Tse, D.: Mobility increases the capacity of ad-hoc wireless networks. IEEE/ACM Tran. Netw. **10**(4), 477–486 (2002)
8. Toumpis, S., Goldsmith, A.: Capacity regions for wireless ad hoc networks. IEEE Trans. Wirel. Commun. **2**, 4 (2003)
9. Kang, I., Poovendran, R.: Maximizing network lifetime of broadcasting over wireless stationary ad hoc networks. Mob. Netw. Appl. **10**, 879896 (2005)
10. Deng, G., Gupta, S.K.S.: Maximizing broadcast tree lifetime in wireless ad hoc networks. In: Global Telecommunications Conference, GLOBECOM, IEEE (2006)
11. Chang, J.H., Tassiulas, L.: Energy conserving routing in wireless ad-hoc networks. In: Proceedings INFOCOM 2000, pp. 22–31 (2000)
12. Lipiński, Z.: Stability of routing strategies for the maximum lifetime problem in ad-hoc wireless networks (2014). http://arxiv.org/abs/1407.3646
13. Lipiński, Z.: Maximum lifetime broadcasting problem in sensor networks (2015). http://arxiv.org/abs/1511.05587
14. Cichoń, J., Gębala,M., Kutyłowski,M.: On optimal one-dimensional routing strategies in sensor networks. In: 4th International Conference on Broadband Communication, BroadBandCom'09 Wrocław, Poland (2009)
15. Berge, C.: Graphs, North Holland (1989)
16. Wieselthier, J.E., Nguyen, G.D., Ephremides, A.: Algorithms for energy-efficient multicasting in static ad hoc wireless networks. Mob. Netw. Appl. **6**, 251–263 (2001)

Statistical Method for the Problem of Bronchopulmonary Dysplasia Classification in Pre-mature Infants

Wiesław Wajs, Hubert Wojtowicz, Piotr Wais and Marcin Ochab

Abstract The problem of data classification with a statistical method is presented in the paper. Described classification method enables calculation of probability of disease incidence. A case of disease incidence is described with two parameters expressed in real numbers. The case can belong to a known set of cases where the disease occurred or to the set where the disease did not occur. A method for calculating probability with which a given case belongs to the set labeled as "1" or "0" is proposed. Source data used in the paper come from medical databases and are original. The algorithm of the method was checked on clinical cases. Correlation method was used for generating respective statistics. The calculated correlation at a level of 0.8 is indicative of disease occurrence, whereas the correlation coefficient at a level of 0.0 is indicative of the lack of disease. This property is used in the classification algorithm. It is frequent in the clinical practice that we have one test case and we try to determine whether or not that case describes symptoms of liability to the disease. Classification is related with the occurrence of Bronchopulmonary dysplasia, which is analyzed in a 3 to 4 week period preceding the disease incidence.

Keywords Bronchopulmonary dysplasia · Classification · Statistical methods

W. Wajs · H. Wojtowicz (✉)
The University of Rzeszów, 16c Al. Rejtana 35-959, Rzeszów, Poland
e-mail: hubert.wojtowicz@gmail.com; hwmo@horrify.org

W. Wajs
e-mail: wwa@agh.edu.pl

P. Wais
State Higher Vocational School in Krosno, Rynek 1 38-400, Krosno, Poland
e-mail: waisp@poczta.onet.pl

M. Ochab
AGH University of Science and Technology, 30 Mickiewicza 30-059, Kraków, Poland
e-mail: mj.ochab@labor.rzeszow.pl

I. Czarnowski et al. (eds.), *Intelligent Decision Technologies 2016*,
Smart Innovation, Systems and Technologies 56,
DOI 10.1007/978-3-319-39630-9_14

1 Introduction

Bronchopulmonary dysplasia (*BPD*) is a chronic lung disease that develops in neonates treated with oxygen and positive pressure ventilation, [3]. The majority of *BPD* cases occur in pre-mature infants, usually those who have gestational age under $ge \leq 34$ weeks and birth weight less than $bw \leq 1500$ g. These babies are more likely to be affected by a condition known as infant Respiratory Distress Syndrome, which occurs as a result of tissue damage to the lungs from mechanical ventilation for a significant amount of time [2, 4]. Although mechanical ventilation is essential to their survival over time, the pressure from the ventilation and excess oxygen intake can injure a newborn's delicate lungs. If symptoms of respiratory distress syndrome persist then the condition is considered Bronchopulmonary Dysplasia. Important factors in diagnosing *BPD* are: premature birth, infection, mechanical ventilator dependence, and oxygen exposure. The primary focus of all treatment associated with premature infants is on prevention of *BPD*. Surfactant replacement, invasive and noninvasive ventilation techniques, management of the patient ductus arteriosus, cautious management of oxygen therapy, caffeine, inhaled nitric oxide, and changes in delivery room practice have been studied to assess their effects on the development of the disease. The definition of *BPD* has many forms. Several factors found to have led to *BPD* in diverse populations of patients have made this difficult to accomplish. In 2001 the National Institute of Child Health and Human Development defined and classified *BPD* by gestational age and supplemental oxygen requirement [3]. Infants <32 weeks postmenstrual age with clinical manifestations of the disease, requiring supplemental oxygen at 28 days of life, and who were weaned to room air by 36 weeks or at discharge were considered to have mild *BPD*. Infants requiring <30 % continuous oxygen at 36 weeks postmenstrual age or at discharge were considered to have moderate disease [2]. Infants remaining on >30 % oxygen and on continuous positive airway pressure (CPAP) were considered to have a severe form of the disease. For infants >32 weeks gestation, the identical oxygen requirement was implemented at the 56th day of life [1]. The clinical definition of *BPD* determines the extent of the disease based on the level of oxygen administration. Walsh employed a psychologic test and response to room air for infants 36 weeks ±1 week [6]. The test consisted of challenging patients by conducting a 30 min room—air trial if the patient was on <30 % oxygen or was >96 % saturated on >30 % [3]. Extreme prematurity in very-low birth-weigh infants and implementation of general patient care strategies such as ventilator management, oxygen administration, and their associated response have played a role in the prevalence of *BPD* [5].

2 Classification Method

The problem of data classification is often encountered in medical practice. This issue is analyzed when a given case of disease is described with two parameters: birth weight (*bw*) and gestational age (*ge*). Classification lies in determining whether a given case of disease belongs to the set labeled "1" or "0". The proposed method enables calculation of probability that a given case belongs to the set labeled "1", and to the set labeled "0". Qualification of a case of disease as belonging to the set of recognized cases has a practical significance. There are numerous known methods on the basis of which a given case can be qualified to a known set of data, e.g. Artificial Neural Network and also Artificial Immune System. These methods do not determine probability with which a given case belongs to the set of data. The proposed method employs a method correlating two parameters (bw, ge) of the same case of disease.

$$\rho_{bw,ge} = \frac{\text{cov}(bw, ge)}{std_{bw} std_{ge}}, -1 \leq \rho_{bw,ge} \leq 1 \tag{1}$$

$$\text{cov}(bw, ge) = \frac{1}{n} \sum_{i=1}^{i=n} (bw_i - mean_{bw})(ge_i - mean_{ge}).$$

Calculated correlation coefficients form a probability density distribution function. We propose two algorithms: training and testing. The influence of each pair of parameters (bw, ge) on probability density distribution is analyzed by removing one pair of parameters from the set of data, each time a different pair, to finally obtain a distribution with the aid of a training algorithm. Training algorithm makes use of thus calculated distribution of probability. Testing algorithm calculates the influence of a case determined by two parameters bw and ge on the probability density distribution. Testing lies in substituting each pair of parameters by a testing pair of parameters in the training set. The probability density distribution calculated with the training algorithm can be compared with the one calculated employing the testing algorithm. Two plots of probability density distribution of correlation coefficients are presented in Fig. 1a. The plot, in Fig. 1a, on the left shows probability density distribution calculated with the use of the training algorithm for data describing cases

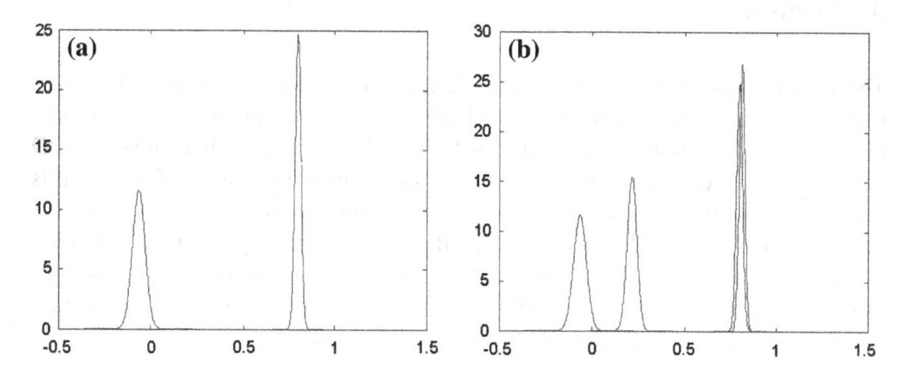

Fig. 1 **a** Probability distributions of correlation coefficients calculated by the training algorithm for parameters: $bw = 600$ and $ge = 24$, calculated by the set labeled "0" *left*, and by the set labeled "1" *right*. **b** Probability distributions of correlation coefficients calculated by training and testing algorithms for parameters: $bw = 600$ and $ge = 24$

without incidence of the disease. The plot on the right illustrates probability density calculated with the training algorithm and represents cases with *BPD* incidence. These cases are labeled "1". Four plots are presented in Fig. 1b, two of which have been already given in Fig. 1a. Plots on the left of Fig. 1b have a negligible common area, unlike the remaining two. New plots in Fig. 1b were obtained with the use of the testing algorithm. The analysis of probability density plots reveals that the test case determined by parameters $bw = 600$ and $ge = 24$ does not belong to the set labeled "0", but to set labeled "1". Mean value for the plot $mean_{tr1} = 0.8007$ (mean calculated by the training algorithm for the set labeled 1). The plot of probability density calculated with testing algorithm is to the right. The mean value for the plot is $mean_{te1} = 0.8141$ (mean calculated by the testing algorithm for the set labeled 1). The common area for these plots is considerable, i.e. ca. 0.7. The probability that a given case belongs to the set labeled "1" can be calculated through integration. Integration limits determine two points of intersection of respective probability density plots. The proposed classification method assumes normal probability density distribution. When the probability density plot calculated with the testing algorithm is to the right of the probability density plot calculated with the training algorithm, the classification condition may assume the form of inequality. For the set labeled "1" there are two probability density distributions: one calculated with the training algorithm and the other with the testing algorithm. Mean values and standard deviations are denoted as: $mean_{tr1}$, $mean_{te1}$, std_{tr1} and std_{te1}, respectively. Points of intersection of two plots are denoted as $x1_1$ and $x2_1$. For the set labeled "0" there are two probability density distributions: one calculated with the training algorithm and the other with the testing algorithm. Similarly, for the set labeled "0" there are two probability density distributions. Mean values and standard deviations are denoted as: $mean_{tr0}$, $mean_{te0}$, std_{tr0} and std_{te0}, respectively. Points of intersection of two plots are denoted as $x1_0$, $x2_0$.

3 Training Algorithm

The training algorithm is used for calculating correlation coefficients. The set of correlation coefficients forms the probability density distribution. The first probability density distribution is calculated for data, which are surely indicative of *BPD* incidence, these data are labeled "1". The other probability density distribution is calculated for data labeled "0". We are positive about these data that they do not describe any *BPD* case. It is assumed that the mean probability density distribution obtained for the set labeled "1" differs from the mean probability density distribution obtained for the set labeled "0". Following equations are used for calculating probability density distribution of correlation coefficients:

$$f_{tr1}(x) = \frac{1}{std_{tr1}\sqrt{(2\pi)}}\exp(\frac{-(x - mean_{tr1})^2}{2std_{tr1}^2}) \quad (2)$$

$$f_{tr0}(x) = \frac{1}{std_{tr0}\sqrt{(2\pi)}}\exp(\frac{-(x - mean_{tr0})^2}{2std_{tr0}^2}) \tag{3}$$

It has been arbitrarily assumed that the bigger is the difference between the mean values for sets, and the smallest are the standards deviation values, the more effective is the classification method. The correlation equation used for calculating respective probability density distributions is given with formula (1). The training algorithm is used for the set labeled 1 and for the set labeled 0. Each time the training algorithm is used, a different pair of parameters is removed from the set labeled 0, and different pair is removed from the set labeled 1. When the pair of parameters is removed a new correlation coefficient is calculated. Finally, we can calculate *mean* and *std* values. We run the training algorithm twice. One time for the set labeled 1, and one time for the set labeled 0.

4 Testing Algorithm

The testing algorithm enables calculation of probability with which a given test case belongs to the set labeled "1" and to the set labeled "0". The testing algorithm is used for calculating two probability density distributions. Following equations are used for calculating probability density distribution of correlation coefficients:

$$f_{te1}(x) = \frac{1}{std_{te1}\sqrt{(2\pi)}}\exp(\frac{-(x - mean_{te1})^2}{2std_{te1}^2}) \tag{4}$$

$$f_{te0}(x) = \frac{1}{std_{te0}\sqrt{(2\pi)}}\exp(\frac{-(x - mean_{te0})^2}{2std_{te0}^2}) \tag{5}$$

The testing algorithm requires substitution of each pair of parameters into the source data set by the pair of test parameters bw, ge. When we replace the pair of parameters into source data by the test one we calculate a new value of correlation coefficient. Like the training algorithm, we run the testing algorithm twice. First run is for the set labeled 1, and second for the set labeled 0. At the end of the training algorithm we calculate: $mean_{te1}, std_{te1}, mean_{te0}, std_{te0}$.

5 Integral Limits Calculation

The presented problem lies in calculating probability with which a given test case described with parameters *bw* and *ge* belongs to the set labeled "1". Calculation of points of intersection of probability density plots obtained with testing and training algorithms allows determination of the integral limits. Two probability density plots have two points of intersection. We have two functions:

$$\frac{1}{std_{te1}\sqrt{(2\pi)}}\exp(\frac{-(x-mean_{te1})^2}{2std_{te1}^2}) = \frac{1}{std_{tr1}\sqrt{(2\pi)}}\exp(\frac{-(x-mean_{tr1})^2}{2std_{tr1}^2})$$

After ln operations

$$\ln\frac{1}{std_{te1}\sqrt{(2\pi)}} + (\frac{-(x-mean_{te1})^2}{2std_{te1}^2}) = \ln\frac{1}{std_{tr1}\sqrt{(2\pi)}} + (\frac{-(x-mean_{tr1})^2}{2std_{tr1}^2})$$

we receive:

$$-(x^2 - 2mean_{tr1}x + mean_{tr1}^2)std_{te1}^2 + (x^2 - 2mean_{te1}x + mean_{te1}^2)std_{tr1}^2$$

$$-2std_{tr1}^2 std_{te1}^2(\ln\frac{1}{std_{te1}\sqrt{(2\pi)}} - \ln\frac{1}{std_{tr1}\sqrt{(2\pi)}}) = 0.$$

Coefficients a, b, and c of the quadratic equation are given by

$$a = std_{tr}^2 - std_{te}^2, b = 2(mean_{tr}std_{te}^2 - mean_{te}std_{tr}^2) \quad (6)$$

$$c = 2std_{tr}^2 std_{te}^2(\ln\frac{1}{std_{tr}\sqrt{2\pi}} - \ln\frac{1}{std_{te}\sqrt{2\pi}}) - mean_{te}^2 std_{tr}^2 + mean_{tr}^2 std_{te}^2 \quad (7)$$

Without losing the general perspective, let's assume that the plots obtained with the use of testing algorithms are right to the plots obtained with the use of training algorithms. The classification criterion can be formulated: if dependence (8) is fulfilled for an arbitrary positive value Δ, then the test case meets the condition of belonging to the set labeled "1".

$$(\frac{1}{std_{te1}}\int_{-\infty}^{x1_1}\exp(\frac{-(x-mean_{te1})^2}{2std_{te1}^2})dx + \frac{1}{std_{tr1}}\int_{x1_1}^{x2_1}\exp(\frac{-(x-mean_{tr1})^2}{2std_{tr1}^2})dx$$

$$+\frac{1}{std_{te1}}\int_{x2_1}^{\infty}\exp(\frac{-(x-mean_{te1})^2}{2std_{te1}^2})dx - (\frac{1}{std_{te0}}\int_{-\infty}^{x1_0}\exp(\frac{-(x-mean_{te0})^2}{2std_{te0}^2})dx$$

$$+\frac{1}{std_{tr0}}\int_{x1_0}^{x2_0}\exp(\frac{-(x-mean_{tr0})^2}{2std_{tr0}^2})dx$$

$$+\frac{1}{std_{te0}}\int_{x2_0}^{\infty}\exp(\frac{-(x-mean_{te0})^2}{2std_{te0}^2})dx))\frac{1}{\sqrt{(2\pi)}}\geq \Delta. \quad (8)$$

Condition (8) can be written depending on the position with respect to mutual position of density plots of correlation coefficients. Mutual position of probability density plots determines calculated mean values. Four systems of probability density plots depend on calculated means for training and testing:

$$mean_{te0} > mean_{tr0} \text{ and } mean_{te1} > mean_{tr1}, \quad (9)$$

$$mean_{te0} < mean_{tr0} \text{ and } mean_{te1} < mean_{tr1}, \quad (10)$$

$$mean_{te0} > mean_{tr0} \text{ and } mean_{te1} < mean_{tr1}, \quad (11)$$

$$mean_{te0} < mean_{tr0} \text{ and } mean_{te1} > mean_{tr1}. \quad (12)$$

The position of probability density plots for correlation coefficients has influence on the sequence of integration operations for the classification condition form (8) and for respective probability density plots. For instance, for the relation (10) classification condition can be written in the form:

$$(\frac{1}{std_{tr1}}\int_{-\infty}^{x1_1}\exp(\frac{-(x-mean_{tr1})^2}{2std_{tr1}^2})dx+\frac{1}{std_{te1}}\int_{x1_1}^{x2_1}\exp(\frac{-(x-mean_{te1})^2}{2std_{te1}^2})dx$$

$$+\frac{1}{std_{tr1}}\int_{x2_1}^{\infty}\exp(\frac{-(x-mean_{tr1})^2}{2std_{tr1}^2})dx-(\frac{1}{std_{tr0}}\int_{-\infty}^{x1_0}\exp(\frac{-(x-mean_{tr0})^2}{2std_{tr0}^2})dx$$

$$+\frac{1}{std_{te0}}\int_{x1_0}^{x2_0}\exp(\frac{-(x-mean_{te0})^2}{2std_{te0}^2})dx$$

$$+\frac{1}{std_{tr0}}\int_{x2_0}^{\infty}\exp(\frac{-(x-mean_{tr0})^2}{2std_{tr0}^2})dx)\frac{1}{\sqrt{(2\pi)}}\geq \Delta \quad (13)$$

6 Case Analysis $bw = 600, ge = 24$

Data describing *BPD* disease were collected as a set of pairs of parameters *bw* and *ge*. The set of source data covers 52 pairs of data for the training algorithm and a few pairs of data for the testing algorithm. Source data are listed in Table 1. There are 26 pairs of parameters describing cases with confirmed *BPD* labeled "1", and 26 pairs of parameters for the set labeled "0". Training and testing algorithms are then used to bring about appropriate probability distributions. The training algorithm is used 26 times for the set labeled "1" and 26 times for the set labeled "0". Each time the

Table 1 Source data describing cases of BPD disease

No.	*bw*	*ge*	*bw*	*ge*
	Label 1	Label 1	Label 0	Label 0
	BPD	*BPD*	No *BPD*	No *BPD*
1	890	25	1360	29
2	700	24	1400	31
3	1100	28	880	28
4	760	28	985	32
5	1200	29	1100	28
6	700	25	1100	28
7	960	26	1300	29
8	760	25	900	28
9	860	27	1100	30
10	600	24	1400	30
11	860	28	1100	28
12	1300	29	1000	30
13	1400	31	880	30
14	940	28	985	32
15	800	28	1100	32
16	1000	27	1100	27
17	600	25	1300	27
18	950	28	900	29
19	1095	28	1100	32
20	800	25	1400	28
21	760	25	1100	26
22	770	24	1200	28
23	1500	30	900	28
24	1200	30	1250	31
25	650	28	1250	30
26	720	25	930	32

Table 2 Training phase data

No.	$\rho_{bw,ge}$ Label 1	$\rho_{bw,ge}$ Label 0
1	0.8115	−0.0551
2	0.7950	−0.1375
3	0.7985	−0.1126
4	0.8270	−0.0183
5	0.7912	−0.0678
6	0.7938	−0.0678
7	0.8080	−0.0569
8	0.7972	−0.1080
9	0.8023	−0.0629
10	0.7846	−0.0948
11	0.8117	−0.0678
12	0.7928	−0.0545
13	0.7623	−0.0452
14	0.8037	−0.0183
15	0.8202	−0.0612
16	0.8023	−0.0718
17	0.7928	−0.0064
18	0.8030	−0.0767
19	0.7984	−0.0612
20	0.8006	−0.0136
21	0.7972	−0.0772
22	0.8055	−0.0497
23	0.7842	−0.1080
24	0.7878	−0.0968
25	0.8512	−0.0770
26	0.7947	0.0026

training algorithm is run, a different pair of data is removed from the set labeled "0", 26 correlation coefficients $\rho_{bw,ge}$ are calculated from Eq. (1). Analogously for the set labeled "1", 26 values of correlation coefficient are calculated. The results of these calculations are presented in Table 2. Values in row 1 in Table 2 were calculated from Eq. (1) when row 1 was removed from Table 1. Parameters $bw = 890$ and $ge = 25$ were removed from the set labeled "1", and parameters $bw = 1360$ and $ge = 29$ from the set labeled "0". The testing algorithm is used 26 times for the set labeled "1" and 26 times for the set labeled "0". Table 3 comprises results of operation of the testing algorithm. The testing algorithm requires substitution of each pair of parameters in Table 1 by test case parameters $bw = 600$ and $ge = 24$. Data listed in Table 3 give probability density distribution describing a change in the set of source data caused

Table 3 Testing phase data

No.	$\rho_{bw,ge}$	$\rho_{bw,ge}$
	Label 1	Label 0
1	0.8242	0.2201
2	0.8099	0.1625
3	0.8114	0.1891
4	0.8386	0.2517
5	0.8047	0.2098
6	0.8085	0.2098
7	0.8206	0.2158
8	0.8114	0.1906
9	0.8154	0.2086
10	0.8009	0.1923
11	0.8241	0.2098
12	0.8064	0.2182
13	0.7787	0.2332
14	0.8165	0.2517
15	0.8321	0.2159
16	0.8151	0.2136
17	0.8081	0.2632
18	0.8158	0.2095
19	0.8114	0.2159
20	0.8144	0.2570
21	0.8114	0.2199
22	0.8193	0.2226
23	0.7994	0.1906
24	0.8016	0.1851
25	0.8616	0.1982
26	0.8092	0.2704

by substitution of each of the 26 pairs of test parameters: $bw = 600$, $ge = 24$. Values given in row 1 of Table 3 were calculated from Eq. (1) when parameters $bw = 890$ and $ge = 25$ in Table 1 in row 1 were substituted with respective parameters $bw = 600$ and $ge = 24$ for the set labeled "1", and parameters $bw = 1360$ and $ge = 29$ with respective parameters $bw = 600$ and $ge = 24$ for the set labeled "0". Similarly, parameters $bw = 700$ and $ge = 24$ in row 2 were substituted with parameters $bw = 600$ and $ge = 24$ for the set labeled "1", and parameters $bw = 1400$ and ge = 31 with parameters $bw = 600$ and $ge = 24$ for the set labeled "0". Probability distributions of correlation coefficients for the case $bw = 600$ and $ge = 24$ are presented in Fig. 2a, that concerns the data of the cases for which Bronchopulmonary Dysplasia didn't occur. Points of intersection marked with "o" for the data labeled "0" are $x1_0 = 0.0951$ and $x2_0 = 1.0694$.

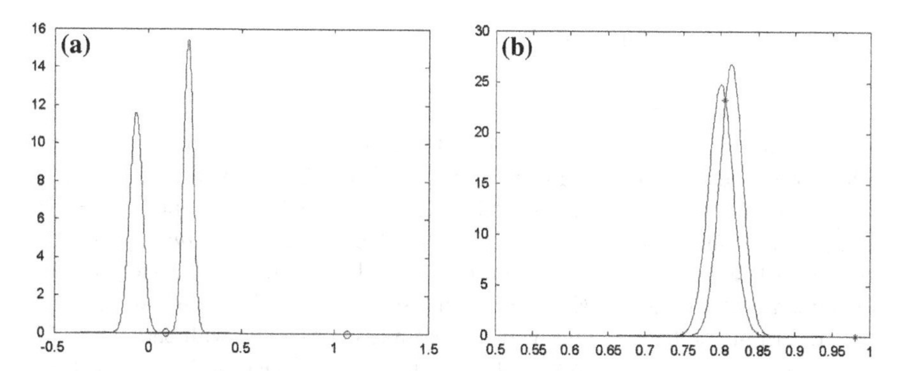

Fig. 2 Probability distributions calculated by training and testing algorithms for parameters: $bw = 600$, $ge = 24$, by the set labeled: (**a**) "0", (**b**) "1"

Two probability distributions of correlation coefficients for the case $bw = 600$ and $ge = 24$ are presented in Fig. 2b, that concerns the data of the cases for which Bronchopulmonary Dysplasia occurred. Points of intersection marked with "*" for the data labeled "1" are $x1_1 = 0.8062$ and $x2_1 = 0.9802$, respectively. These points enable calculation of probability with which a given case described with parameters $bw = 600$ and $ge = 24$ belongs to the set of the data labeled "1" and probability with which a given case belongs to the set of data labeled "0". The common area of both plots on the right of Fig. 2b can be calculated by using three integrals, integrating the probability density function. By adding values: $0.29798 + 0.361655 + 0.119188 \times 10^{-28}$ we obtain probability with which a given case $bw = 600$ and $ge = 24$ belongs to the set labeled "1":

$$\frac{1}{0.0149\sqrt{(2\pi)}}\int_{-\infty}^{0.8062}\exp(\frac{-(x-0.8143)^2}{2\times 0.0149^2})dx = 0.29798 \tag{14}$$

$$\frac{1}{0.0161\sqrt{(2\pi)}}\int_{-0.8062}^{0.9802}\exp(\frac{-(x-0.8007)^2}{2\times 0.0161^2})dx = 0.361655 \tag{15}$$

$$\frac{1}{0.0149\sqrt{(2\pi)}}\int_{9802}^{\infty}\exp(\frac{-(x-0.8143)^2}{2\times 0.0149^2})dx = 0.119188\times 10^{-28} \tag{16}$$

Two points of intersection for sets labeled "1" are $x1_1 = 0.8062$ and $x2_1 = 0.9802$. The mean probability density calculated with the training algorithm for the set labeled "1" is $mean_{tr1} = 0.8007$. The value calculated with the testing algorithm is $mean_{te1} = 0.8141$. The value *std* calculated with the testing algorithm for the set labeled "1" is $std_{te1} = 0.0149$. The value of std calculated with the training algorithm for the set labeled "1" is $std_{tr1} = 0.0161$.

7 Case Analysis $bw = 900, ge = 28$

The analyzed case is described with parameters $bw = 900$, $ge = 28$. The mean value of correlation coefficient $mean_{tr0} = -0.0640$ was calculated with the training algorithm for the set labeled "0", and mean value of correlation coefficient $mean_{tr1} = 0.8007$ for the set labeled "1". The testing algorithm enables calculation of the correlation mean value $mean_{te1} = 0.7947$ for the set labeled "1" and $mean_{te0} = -0.0252$ for the set labeled "0". The probability density plots for the set labeled "0" have a common area with plots calculated for the sets labeled "1". These two common parts determine probability with which a given case described with parameters $bw = 900$ and $ge = 28$ belongs to the set labeled "0" and set labeled "1". Comparison of the case $bw = 600$ and $ge = 24$ with the described case $bw = 900$ and $ge = 28$ from Fig. 3 shows a clear difference. In Fig. 3 probability distributions of correlation coefficients have similar values, means and values of standard deviations. The described case $bw = 900$ and $ge = 28$ is a hard case for the classification problem. In

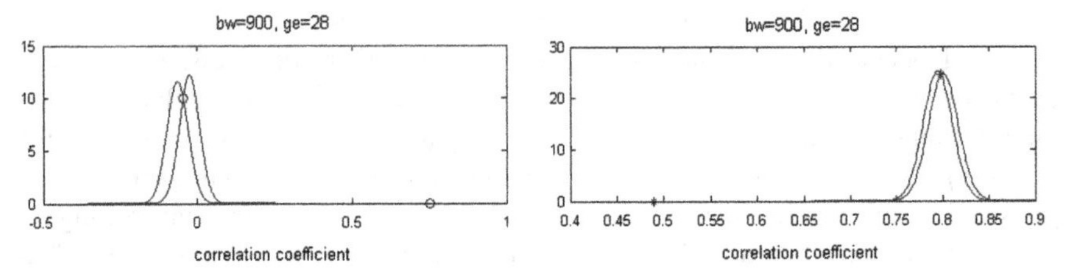

Fig. 3 Probability distributions calculated by training and testing algorithm for parameters: $bw = 900, ge = 28$

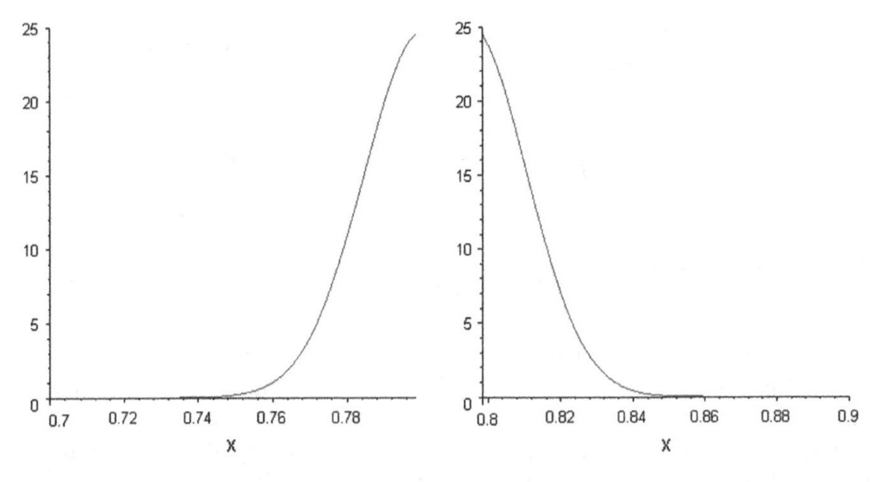

Fig. 4 Probability distributions calculated by training and testing algorithm for parameters: $bw = 900, ge = 28$, labeled "1". Partial integral value on the *left* is equal to 0.4456. Partial integral value on the *right* is equal to 0.4049

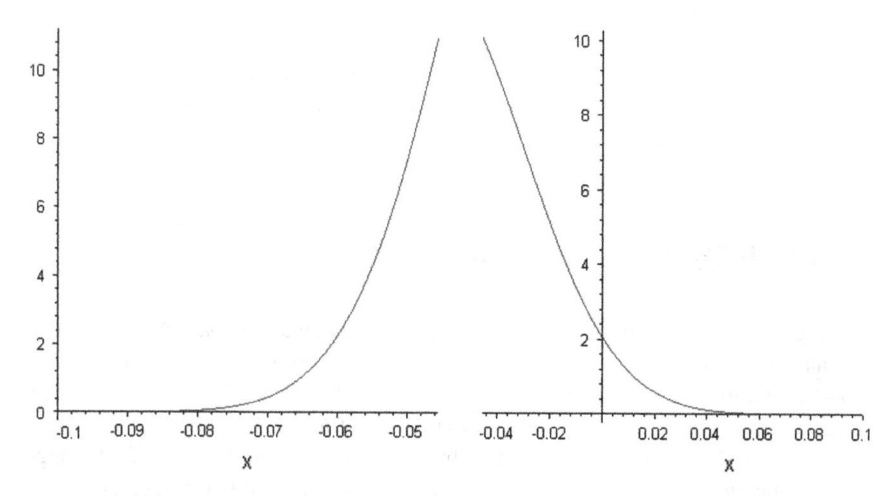

Fig. 5 Probability distributions calculated by training and testing algorithm for parameters: $bw = 900$, $ge = 28$, labeled "0". Partial integral value on the *left* is equal to 0.2664. Partial integral value on the *right* is equal to 0.2964

case when two of the integrals from Eq. (8) have negligibly low values it is possible to use Eqs. (17), (18), (19) and (20) instead and achieve approximate classification criterion with lower computational overhead. This situation is shown in Figs. 4 and 5. In this case only one intersection point of probability density plots is used.

8 Probability for Sets Labeled "1"

By integrating probability density functions we calculate probability with which an analyzed case belongs to the set labeled "1". Calculating the point of intersection of probability density plots at points $x1_1 = 0.4891$ and $x2_1 = 0.7985$ one can calculate the common part for plots calculated with training and testing algorithms. The probability with which a given case, described by parameters $bw = 900$, $ge = 28$, belongs to the set of cases labeled "1" is approximated to 0.85 as a sum of two integrals, as the integral calculated for limits $-\infty$ to 0.4891 is negligible

$$\frac{1}{0.0161\sqrt{(2\pi)}}\int_{0.4891}^{0.7985} \exp(\frac{-(x-0.8007)^2}{2\times 0.0161^2})dx = 0.4456 \tag{17}$$

$$\frac{1}{0.0158\sqrt{(2\pi)}}\int_{0.7985}^{\infty} \exp(\frac{-(x-0.7947)^2}{2\times 0.0158^2})dx = 0.4049 \tag{18}$$

Simplifying, we assumed that only radical $x2_1 = 0.7985$ is analyzed due to the negligible influence of integral calculated accounting for point $x1_1 = 0.4891$. If we take into account only one radical, only two Eqs. (17) and (18) can be written in the simplification.

9 Probability for the Sets Labeled "0"

By integrating probability density functions we can calculate probability with which the analyzed case belongs to the set labeled "0". By calculating place of intersection of probability density plots at points $x1_0 = -0.0456$ and $x2_0 = 0.7489$ one can calculate the common part for plots calculated with training and testing algorithms. Probability with which a case described by parameters $bw = 900$, $ge = 28$ belongs to the set labeled "0" and is approximated to 0.56 as a sum of two integrals, owing to the negligibly low value of integral calculated within limits 0.7489 to ∞. For $bw = 900$ $ge = 28$ the test reveals a considerable degree of belonging to the set labeled "1" and to the set labeled "0".

$$\frac{1}{0.0327\sqrt{(2\pi)}}\int_{-\infty}^{-0.0456} \exp(\frac{-(x-(-0.0252))^2}{2\times 0.0327^2})dx = 0.2664 \tag{19}$$

$$\frac{1}{0.0344\sqrt{(2\pi)}}\int_{-0.0456}^{0.7489} \exp(\frac{-(x-(-0.0640))^2}{2\times 0.0344^2})dx = 0.2964 \tag{20}$$

10 Summary

With the presented classification method four general cases can be analyzed. In the first case the probability density plots for cases labeled "1", obtained with the training and testing algorithm, have a significant common part, and the respective plots for sets labeled "0" have a small common area. It can be certified with some probability that the case belongs to the set labeled "1". In the second general case the probability density has a considerable common part for the set labeled "0" and a small common area for the set labeled "1". In this case it is highly probable that the analyzed case belongs to the set labeled "0". It is generally assumed in the third case that the probability density plots have a considerable common part of sets labeled "1", and a considerable common part for sets labeled "0". In this case probability can be calculated from the difference of integrals calculated with the use of probability density functions. In the fourth case the probability density plots have a small common area for sets labeled "0" and a small common part for the sets labeled "1". For the collected original data only the first case evidently belongs to the sets labeled "1". In the three remaining cases the probability can be estimated.

References

1. Cerny, I., Torda, J.S., Kehan, V.K.: Prevention and treatment of bronchopulmonary dysplasia: contemporary status and future outlook. Lung **186**(2), 75–89 (2008)
2. Cunha, G.S., Mezzacappa-Fihlo, F., Ribeiro, J.D.: Risk factors for bronchopulmonary dysplasia in very low birth weight newborns treated with mechanical ventilation in the first week of life. J. Trop. Pediatr. **51**(6), 334–340 (2005)
3. Deakins, K.M.: Bronchopulmonary displasia. Respir. Care **54**(9), 1252–1262 (2009)
4. Tapia J. L., Agost D., Alegria A., Standen J., Escobar E., Grandi C., Musante G., Zegarra J., Estay A., Ramirez R.: Bronchopulmonary displasia: incidence, risk factors and resource utilization in a population of South-American very low birth weight infants. J. Pediatric. **82**(1), (2006)
5. Wajs W., Stoch P., Kruczek P.: Bronchopulmonary dysplasia prediction using logistic regresion. In: Proceedings of the 6th International IEEE Conference on Intelligent Systems Design and Applications (ISDA'06), vol. 3, pp. 98–102 (2006)
6. Walsh, M.C., Yao, Q., Gettner, P., Hale, E., Collins, M., Hansman, A., et al.: Impact of a physiologic definition on bronchopulmonary dysplasia rates. Pediatrics **114**(5), 1305–1311 (2004)

The Rank Reversals Paradox in Management Decisions: The Comparison of the AHP and COMET Methods

Wojciech Sałabun, Paweł Ziemba and Jarosław Wątróbski

Abstract Making decisions include many areas of human activity, as well as the management of the organization. Actually, decision-making requires a consideration of rapidly changing options from multiple sources. For this reason, the decision-making process requires modification, which allows processing of a set of alternatives on the fly. The most popular method in this field is the AHP method. However, a serious shortcoming is known, which does not allow to reliably carry out this process. This problem is known as the *RankReversals* phenomenon. The paper identifies the problem and highlights the importance in the context of numerical examples. These examples are also solved by using the COMET method, which uses a pairwise comparison also. The COMET method is completely free of rank reversal paradox and can be used in exchange for the AHP method.

Keywords Rank reversal · AHP · MCDA · COMET · Fuzzy logic

1 Introduction

Multi Criteria Decision Analysis (MCDA) methods become more common with the increasing complexity of the decision. In practice, they have used a variety of computational methods to obtain a rank of available alternatives or to indicate the best

W. Sałabun (✉) · J. Wątróbski
West Pomeranian University of Technology, Szczecin al. Piastów 17,
70-310 Szczecin, Poland
e-mail: wsalabun@wi.zut.edu.pl

J. Wątróbski
e-mail: jwatrobski@wi.zut.edu.pl

P. Ziemba
The Jacob of Paradyż University of Applied Sciences in Gorzów Wielkopolski,
ul. Chopina 52, 66-400 Gorzów Wielkopolski, Poland
e-mail: pziemba@pwsz.pl

I. Czarnowski et al. (eds.), *Intelligent Decision Technologies 2016*,
Smart Innovation, Systems and Technologies 56,
DOI 10.1007/978-3-319-39630-9_15

solution [1, 2]. Facilitated access to information and rapid changes in input data require increased automation of decision-making process.

The Analytic Hierarchy Process (AHP) is one of the most popular decision making method. The main areas of application are supplier evaluation and selection, planning and scheduling, project management, managing the supply chain [3, 4]. The AHP is so popular for its intuitive and easy to use. It requires only pairwise comparisons for all alternatives, which simplifies the decision-making process. Unfortunately, the AHP method has a serious shortcoming as the Rank Reversal paradox [5–7].

The *RankReversals* paradox is a significant problem. In the paper, we consider one type of this phenomenon. We assume that one alternative is removed from the set of alternatives. This action does not influence on the rest of the values in the comparison matrix because this means only plotting the one row and column in the initial matrix. However, this change may cause the final revision of the order. As the result, we may obtain two matrices and two different rankings. There is an important issue, which of these two contradictory rankings is correct?

The rest of the paper is organized as follows: First, we give an outline of a literature review for the Rank Reversals paradox. Then, we describe the COMET method as a tool for identification of the decision model in the space of the problem. Then, the experimental study is presented, where the AHP and COMET methods are compared in respect to rank reversal. Finally, we discuss results and their importance.

2 Literature Review

The analytic hierarchy process has been many times criticized for its possible rank reversal phenomenon. This paradox means changes of the relative rankings of the other alternatives after a decision variant is added or removed. This issue was first noticed and pointed out by Belton and Gear [8]. In this way, Belton and Gear have started long-lasting debate about the validity of the AHP method. It was continued by many researchers, e.g., Dyer [9, 10], Harker and Vargas [11, 12], Leung and Cao [13], and Saaty [14, 15]. Extremely interesting thread of the discussion was a part about the legitimacy of rank reversal [5, 16–20].

Belton and Gear [8] suggested how to avoid rank reversal in the AHP. However, Saaty and Vargas [19] provided a counterexample to show that Belton and Gear had been wrong. Schoner and Wedley [21] presented a referenced AHP to avoid rank reversal paradox. This approach requires the modification of criteria weights always when an alternative is added or removed. Schoner et al. [22] also suggested a method of normalization to avoid rank reversal. Barzilai and Golany [23] showed that no normalization could prevent rank reversal. They suggested that a multiplicative aggregation rule helps to avoid rank reversal. Lootsma [24] and Barzilai and Lootsma [25] suggested a multiplicative AHP for rank preservation. However, Vargas [26] provided a practical counterexample to show the invalidity of the multiplicative AHP. Afterwards, Triantaphyllou [7] offered two new cases to demonstrate that the rank reversals do not occur with the multiplicative AHP, but do occur with the AHP and some of its additive variants. Leung and Cao [13] showed that Sinarchy, a particular form of analytic network process (ANP), could prevent rank reversal. But in fact, the AHP still suffers from rank reversal paradox [27].

This literature review shows that the rank reversal phenomenon has not been perfectly resolved since the eighties. There still exist debates about the ways of avoiding this paradox. So, this paper shows a short discussion on comparison AHP and COMET methods with respect to rank reversal problem.

3 The Characteristic Objects Method

The COMET method is completely free of the Rank Reversal phenomenon. The basic concept of the COMET method was proposed by prof. Piegat [28, 29]. In previous works, the accuracy of the COMET method was verified [30]. The formal notation of the COMET method should be shortly recalled [31–34].

Step 1. Define the space of the problem—the expert determines the dimensionality of the problem by selecting the number r of criteria, $C_1, C_2, \ldots, C_r$. Then, the set of fuzzy numbers for each criterion C_i is selected (1):

$$C_r = \{\tilde{C}_{r1}, \tilde{C}_{r2}, \ldots, \tilde{C}_{rc_r}\} \tag{1}$$

where $c_1, c_2, \ldots, c_r$ are numbers of the fuzzy numbers for all criteria.

Step 2. Generate characteristic objects—The characteristic objects (CO) are obtained by using the Cartesian Product of fuzzy numbers cores for each criteria as follows (2):

$$CO = C(C_1) \times C(C_2) \times \cdots \times C(C_r) \tag{2}$$

Step 3. Rank the characteristic objects—the expert determines the Matrix of Expert Judgment (MEJ). It is a result of pairwise comparison of the COs by the problem expert. The MEJ matrix contains results of comparing characteristic objects by the expert, where α_{ij} is the result of comparing CO_i and CO_j by the expert. The function f_{exp} denotes the mental function of the expert. It depends solely on the knowledge of the expert and can be presented as (3). Afterwards, the vertical vector of the Summed Judgments (SJ) is obtained as follows (4).

$$\alpha_{ij} = \begin{cases} 0.0, f_{exp}(CO_i) < f_{exp}(CO_j) \\ 0.5, f_{exp}(CO_i) = f_{exp}(CO_j) \\ 1.0, f_{exp}(CO_i) > f_{exp}(CO_j) \end{cases} \tag{3}$$

$$SJ_i = \sum_{j=1}^{t} \alpha_{ij} \tag{4}$$

Finally, values of preference are approximated for each characteristic object. As a result, the vertical vector P is obtained, where i-th row contains the approximate value of preference for CO_i.

Step 4. The rule base—each characteristic object and value of preference is converted to a fuzzy rule as follows (5):

$$IF\ C(\tilde{C}_{1i})\ AND\ C(\tilde{C}_{2i})\ AND\ \ldots\ THEN\ P_i \tag{5}$$

In this way, the complete fuzzy rule base is obtained.

Step 5. Inference and final ranking—each alternative is presented as a set of crisp numbers (e.g., $A_i = \{a_{1i}, a_{2i}, \ldots, a_{ri}\}$). This set corresponds to criteria $C_1, C_2, \ldots, C_r$. Mamdani's fuzzy inference method is used to compute preference of $i - th$ alternative. The rule base guarantees that the obtained results are unequivocal. The bijection makes the COMET a completely rank reversal free.

4 Experimental Study

As it was said in introduction section, the AHP method is commonly used in many problems from management field. The AHP frequently plays a critical role in shaping overall business performance. Therefore, this section presents two numerical examples, where each one is solved by using AHP and COMET method. It shows difference between AHP and COMET methods in respect to rank reversal paradox.

4.1 The Problem of Selecting Providers

Let's consider a example of the decision-making problem comprising of two criteria (Table 1). The problem of selecting providers is presented with respect to the cost (C_1) and the quality (C_2) of the service. These two criteria are mostly used in many MCDA problems of management. The set of four alternatives is presented specifically in Table 2. We assume that each criterion has equal weight. M_1 and M_2 matrices (6) are obtained on the basis of expert knowledge. Afterwards, we considered four subsets of the alternatives set. M_3 and M_4 matrices are obtained by plotting the first column and row from matrices (6). This means that the alternative A_1 is removed. M_5 and M_6 matrices are obtained by plotting the second column and row from matrices (6), which means that the alternative A_2 is deleted. The rest of matrices are obtained in the same way (accordingly plotting for alternatives A_3 and A_4). The consistency

Table 1 The set of four alternatives in the problem of selecting providers

A_i	C_1: cost of service	C_2: quality of service
A_1	60	70
A_2	75	50
A_3	100	75
A_4	140	80

Table 2 Consistency ratios for matrices M_1–M_{10}

M_i	M_1	M_2	M_3	M_4	M_5	M_6	M_7	M_8	M_9	M_{10}
c.r.	0.0292	0.0390	0.0332	0.0251	0.0462	0.0079	0.0251	0.0810	0.0158	0.0462

Table 3 The priority vectors for matrices (9–13)

A_i	(9)	(10)	(11)	(12)	(13)
A_1	0.3406	–	0.4272	0.4321	0.4553
A_2	0.1830	0.3590	–	0.2100	0.2042
A_3	0.2097	0.3001	0.2573	–	0.3405
A_4	0.2667	0.3408	0.3155	0.3580	–

ratio (*c.r.*) is calculated for each matrix (M_1–M_{10}). The summary of this calculation is presented in Table 2. All matrices M_i are enough consistent for further calculations.

$$M_1 = \begin{pmatrix} 1 & 2 & 4 & 6 \\ \frac{1}{2} & 1 & 3 & 5 \\ \frac{1}{4} & \frac{1}{3} & 1 & 3 \\ \frac{1}{6} & \frac{1}{5} & \frac{1}{3} & 1 \end{pmatrix}_{C_1} \quad M_2 = \begin{pmatrix} 1 & 5 & \frac{1}{2} & \frac{1}{3} \\ \frac{1}{5} & 1 & \frac{1}{5} & \frac{1}{6} \\ 2 & 5 & 1 & \frac{1}{2} \\ 3 & 6 & 2 & 1 \end{pmatrix}_{C_2} \tag{6}$$

Priority vectors are computed as normalized Eigen vectors of each comparison matrix (Table 3). Detailed results are presented in the Table 4. As we can see, there is one Rank Reversal situation. The ranking for full set is as (7). However, if we consider the full set but without the alternative A_1, we obtain the changed order of the priority, where the worst alternative A_2 is becoming the best one (8). We can try increase consistency of matrices by using the least logarithmic square method (LLSM) to checked how will this affect to the solution. First, we transform the matrices M_1 and M_2 to obtain fully consistent matrices (9–10). Then, all steps are repeated as previously, and c.r. factors for all matrices are equal zero. The new priority vectors are presented in Table 5. On the basis of these data, we can see that level of consistency has not got influence on an occurrence of this paradox. As the result, the difference between the maximum and minimum values of each priority vector is increased.

Table 4 The priority vectors for matrices (9–13) after LLSM

A_i	(9)	(10)	(11)	(12)	(13)
A_1	0.3867	–	0.5097	0.4197	0.4820
A_2	0.1377	0.4144	–	0.1400	0.1473
A_3	0.1455	0.2092	0.1554	–	0.3707
A_4	0.3301	0.3764	0.3349	0.4343	–

Table 5 The results of the COMET method for the decisional model

CO_i	C_1	C_2	SJ	Rank	P
CO_1	50	0	2.5	2	$\frac{2}{3}$
CO_2	50	100	3.5	1	1
CO_3	150	0	0.5	4	0
CO_4	150	100	1.5	3	$\frac{1}{3}$

$$A_1 > A_4 > A_3 > A_2 \tag{7}$$

$$A_2 > A_4 > A_3 \tag{8}$$

$$M_1^{LLSM} = \begin{pmatrix} 1 & 2.5298 & 13.8564 & 65.7267 \\ 0.39528 & 1 & 5.4772 & 25.9808 \\ 0.072169 & 0.18257 & 1 & 4.7434 \\ 0.015215 & 0.03849 & 0.21082 & 1 \end{pmatrix}_{C_1} \tag{9}$$

$$M_2^{LLSM} = \begin{pmatrix} 1 & 11.1803 & 0.40825 & 0.15215 \\ 0.089443 & 1 & 0.036515 & 0.013608 \\ 2.4495 & 27.3861 & 1 & 0.37268 \\ 6.5727 & 73.4847 & 2.6833 & 1 \end{pmatrix}_{C_2} \tag{10}$$

This same problem can be solved by using COMET method. We assume that all fuzzy numbers, which are presented as (11), are the triangular. The domain for C_1 is a interval from 50 to 150, and for C_2 from 0 to 100. In this situation, we will have four characteristic objects (COs). The MEJ matrix is presented as (12), and the rest results are presented in Table 6. Finally, we obtain the decisional model as a simple rule base (13) with four fuzzy rules. We calculate values of priority for each alternative separately by using the obtained model. We obtain the assessment for alternatives as following: A_1—0.8333, A_2—0.6667, A_3—0.5833, and A_4—0.3333. These values indicate only one order: $A_1 > A_2 > A_3 > A_4$. Therefore, the COMET method is a completely rank reversal free (Table 7).

Table 6 The set of four alternatives in the problem of quality assessment

A_i	Profit criterion C_1	Profit criterion C_2	Profit criterion C_3
A_1	90	50	45
A_2	80	60	60
A_3	70	80	75
A_4	60	70	90

Table 7 The priority vectors for five sets, i.e., all alternatives, without A_1, without A_2, without A_3, and without A_4

A_i	All	Without A_1	Without A_2	Without A_3	Without A_4
A_1	0.2088	–	0.2766	0.2753	0.2555
A_2	0.1822	0.2693	–	0.2648	0.2616
A_3	0.3007	0.3650	0.3634	–	0.4829
A_4	0.3084	0.3658	0.3600	0.4599	–

$$\begin{matrix} C_1 = \{\tilde{C}_{11}, \tilde{C}_{12}\} \\ C_2 = \{\tilde{C}_{21}, \tilde{C}_{22}\} \end{matrix} \tag{11}$$

$$MEJ = \begin{pmatrix} 0.5 & 0 & 1 & 1 \\ 1 & 0.5 & 1 & 1 \\ 0 & 0 & 0.5 & 0 \\ 0 & 0 & 1 & 0.5 \end{pmatrix}_{C_2} \tag{12}$$

$$\begin{matrix} R_1 : IF\ C_1 \sim 50\ \ AND\ C_2 \sim 0\ \ \ THEN\ \frac{2}{3} \\ R_2 : IF\ C_1 \sim 50\ \ AND\ C_2 \sim 100\ THEN\ 1 \\ R_3 : IF\ C_1 \sim 150\ AND\ C_2 \sim 0\ \ \ THEN\ 0 \\ R_4 : IF\ C_1 \sim 150\ AND\ C_2 \sim 100\ THEN\ \frac{1}{3} \end{matrix} \tag{13}$$

4.2 *The Problem of Quality Assessment*

Let's consider an example of the decision-making problem comprising of three criteria. The task of quality assessment is presented with respect to three criteria (C_1—C_3). The set of alternatives is presented in Table 6. We assume that each criterion has equal weight. The W_1, W_2 and W_3 matrices (14) are filled by using expert answers. Afterwards, we considered four subsets of the alternatives set as in subsection 4.1. Matrices W_1—W_{15} are obtained by plotting sequentially each alternative. The consistency ratio ($c.r.$) is calculated for each matrix (W_1—W_{15}). The summary of this

Table 8 Consistency ratios for matrices W_1–W_{15}

W_i	W_1	W_2	W_3	W_4	W_5	W_6	W_7	W_8	W_9	W_{10}	W_{11}	W_{12}	W_{13}	W_{14}	W_{15}
$c.r.$	0.01	0.01	0.04	0.01	0.01	0.03	0.00	0.00	0.06	0.00	0.01	0.06	0.01	0.00	0.03

Table 9 The result of the COMET method for the decisional model

CO_i	C_1	C_2	C_3	SJ	Rank	P
CO_1	0	0	0	0.5	8	0
CO_2	0	0	100	3.5	5	$\frac{3}{7}$
CO_3	0	100	0	2.5	6	$\frac{2}{7}$
CO_4	0	100	100	6.5	2	$\frac{6}{7}$
CO_5	100	0	0	1.5	7	$\frac{1}{7}$
CO_6	100	0	100	5.5	3	$\frac{5}{7}$
CO_7	100	100	0	4.5	4	$\frac{4}{7}$
CO_8	100	100	100	7.5	1	1

calculation is presented in Table 9. On the basis of these data we can said that all matrices are enough consistent for further calculations. Priority vectors are computed by using normalized Eigen vectors. Detailed results are presented in the Table 8. As we can see, there is two Rank Reversal situation, for removed A_2 and A_4.

$$W_1 = \begin{pmatrix} 1 & 2 & 3 & 5 \\ \frac{1}{2} & 1 & 2 & 3 \\ \frac{1}{3} & \frac{1}{2} & 1 & 2 \\ \frac{1}{5} & \frac{1}{3} & \frac{1}{2} & 1 \end{pmatrix}_{C_1} \quad W_2 = \begin{pmatrix} 1 & \frac{1}{2} & \frac{1}{5} & \frac{1}{3} \\ 2 & 1 & \frac{1}{3} & \frac{1}{2} \\ 5 & 3 & 1 & 2 \\ 3 & 2 & \frac{1}{2} & 1 \end{pmatrix}_{C_2} \quad W_3 = \begin{pmatrix} 1 & \frac{1}{3} & \frac{1}{5} & \frac{1}{7} \\ 3 & 1 & \frac{1}{3} & \frac{1}{5} \\ 5 & 3 & 1 & \frac{1}{3} \\ 7 & 5 & 3 & 1 \end{pmatrix}_{C_3} \tag{14}$$

This problem will be solved also by using the COMET method. We assume that all fuzzy numbers, which are presented as (15), are the triangular. The domain for each criterion is a interval from 0 to 100. In this case, we will have eight characteristic objects (COs). The MEJ matrix is presented as (16), and the rest results are presented in Table 9. Finally, we obtain the decisional model as a rule base (17) with eight fuzzy rules. We calculate values of priority for each alternative by using identified model. We obtain the assessment for alternatives as following: A_1—0.5607, A_2—0.6491, A_3—0.7707, and A_4—0.7906. These values indicate only one order: $A_4 > A_3 > A_2 > A_1$.

$$\begin{array}{l} C_1 = \{\tilde{C}_{11}, \tilde{C}_{12}\} \\ C_2 = \{\tilde{C}_{21}, \tilde{C}_{22}\} \\ C_3 = \{\tilde{C}_{31}, \tilde{C}_{32}\} \end{array} \tag{15}$$

$$MEJ = \begin{pmatrix} 0.5 & 0 & 0 & 0 & 0 & 0 & 0 & 0 \\ 1 & 0.5 & 1 & 0 & 1 & 0 & 0 & 0 \\ 1 & 0 & 0.5 & 0 & 1 & 0 & 0 & 0 \\ 1 & 1 & 1 & 0.5 & 1 & 1 & 1 & 0 \\ 1 & 0 & 0 & 0 & 0.5 & 0 & 0 & 0 \\ 1 & 1 & 1 & 0 & 1 & 0.5 & 1 & 0 \\ 1 & 1 & 1 & 0 & 1 & 0 & 0.5 & 0 \\ 1 & 1 & 1 & 1 & 1 & 1 & 1 & 0.5 \end{pmatrix} \quad (16)$$

$$\begin{array}{l} R_1 : IF\ C_1 \sim 0 \quad AND\ C_2 \sim 0 \quad AND\ C_3 \sim 0 \quad THEN\ 0 \\ R_2 : IF\ C_1 \sim 0 \quad AND\ C_2 \sim 0 \quad AND\ C_3 \sim 100\ THEN\ \frac{3}{7} \\ R_3 : IF\ C_1 \sim 0 \quad AND\ C_2 \sim 100\ AND\ C_3 \sim 0 \quad THEN\ \frac{2}{7} \\ R_4 : IF\ C_1 \sim 0 \quad AND\ C_2 \sim 100\ AND\ C_3 \sim 100\ THEN\ \frac{6}{7} \\ R_5 : IF\ C_1 \sim 100\ AND\ C_2 \sim 0 \quad AND\ C_3 \sim 0 \quad THEN\ \frac{1}{7} \\ R_6 : IF\ C_1 \sim 100\ AND\ C_2 \sim 0 \quad AND\ C_3 \sim 100\ THEN\ \frac{5}{7} \\ R_7 : IF\ C_1 \sim 100\ AND\ C_2 \sim 100\ AND\ C_3 \sim 0 \quad THEN\ \frac{4}{7} \\ R_8 : IF\ C_1 \sim 100\ AND\ C_2 \sim 100\ AND\ C_3 \sim 100\ THEN\ 1 \end{array} \quad (17)$$

5 Conclusions

The rank reversals paradox is a serious problem and obstacle to automatize the decision-making process for management by using the AHP method. The presented simple examples show that for a few alternatives we may have already problems with the order of the obtained ranking. The order for the main set and its subsets can be different. These simple experiments show that the solution obtained by the AHP method may be not stable.

We present the new method, which allows automatized decision-making process. The COMET method is very easy to use, and the simplest models given reliable results. First of all, this approach is rank reversals free. This is caused by the fact that we obtain not only discrete value of priority but the mathematical function, which can be used to calculate the priority for all alternatives from the space of the problem.

References

1. Wątróbski, J., Jankowski, J.: Knowledge management in MCDA domain. In: 2015 Federated Conference on Computer Science and Information Systems (FedCSIS), pp. 1445–1450. IEEE (2015)
2. Wątróbski, J., Ziemba, P., Wolski, W.: Methodological aspects of decision support system for the location of renewable energy sources. In: 2015 Federated Conference on Computer Science and Information Systems (FedCSIS), pp. 1451–1459. IEEE (2015)

3. Ho, W., Xu, X., Dey, P.K.: Multi-criteria decision making approaches for supplier evaluation and selection: a literature review. Eur. J. Oper. Res. **202**(1), 16–24 (2010)
4. Subramanian, N., Ramanathan, R.: A review of applications of analytic hierarchy process in operations management. Int. J. Prod. Econ. **138**(2), 215–241 (2012)
5. Millet, I., Saaty, T.L.: On the relativity of relative measures accommodating both rank preservation and rank reversals in the AHP. Eur. J. Oper. Res. **121**(1), 205–212 (2000)
6. Triantaphyllou, E., Mann, S.H.: Using the analytic hierarchy process for decision making in engineering applications: some challenges. Int. J. Ind. Eng.: Appl. Pract. **2**(1), 35–44 (1995)
7. Triantaphyllou, E.: Two new cases of rank reversals when the AHP and some of its additive variants are used that do not occur with the multiplicative AHP. J. Multi Criteria Decis. Anal. **10**(1), 11–25 (2001)
8. Belton, V., Gear, T.: On a short-coming of Saaty's method of analytic hierarchies. Omega **11**(3), 228–230 (1983)
9. Dyer, J.S.: Remarks on the analytic hierarchy process. Manag. Sci. **36**(3), 249–258 (1990)
10. Dyer, J.S.: A clarification of remarks on the analytic hierarchy process. Manag. Sci. **36**(3), 274–275 (1990)
11. Harker, P.T., Vagas, L.G.: The theory of ratio scale estimation: saaty's analytic hierarchy proces. Manag. Sci. **33**, 1383–1403 (1987)
12. Harker, P.T., Vargas, L.G.: Reply to remarks on the analytic hierarchy process by J.S. Dyer. Manag. Sci. **36**, 269–273 (1990)
13. Leung, L.C., Cao., D.: On the efficacy of modeling multi-attribute decision problems using AHP and Sinarchy. Eur. J. Oper. Res. **132**, 39–49 (2001)
14. Saaty, T.L.: Axiomatic foundation of the analytic hierarchy process. Manag. Sci. **32**, 841–855 (1986)
15. Saaty, T.L., Vargas, L.G., Wendell, R.E.: Assessing attribute weights by ratios. Omega **11**, 9–13 (1983)
16. Forman, E.H.: AHP is intended for more than expected value calculations. Decis. Sci. **36**, 671–673 (1990)
17. Saaty, T.L.: Rank generation, preservation, and reversal in the analytic hierarchy decision process. Decis. Sci. **18**, 157–177 (1987)
18. Saaty, T.L.: Decision making, new information, ranking and structure. Math. Model. **8**, 125–132 (1987)
19. Saaty, T.L., Vargas, L.G.: The legitimacy of rank reversal. Omega **12**(5), 513–516 (1984)
20. Schoner, B., Wedley, W.C., Choo, E.U.: A rejoinder to Forman on AHP, with emphasis on the requirements of composite ratio scales. Decis. Sci. **23**, 509–517 (1992)
21. Schoner, B., Wedley, W.C.: Ambiguous criteria weights in AHP: consequences and solutions. Decis. Sci. **20**, 462–475 (1989)
22. Schoner, B., Wedley, W.C., Choo, E.U.: A unified approach to AHP with linking pins. Eur. J. Oper. Res. **64**, 384–392 (1993)
23. Barzilai, J., Golany, B.: AHP rank reversal, normalization and aggregation rules. INFOR **32**(2), 57–63 (1994)
24. Lootsma, F.A.: Scale sensitivity in the multiplicative AHP and SMART. J. Multi-Criteria Decis. Anal. **2**, 87–110 (1993)
25. Barzilai, J., Lootsma, F.A.: Power relations and group aggregation in the multiplicative AHP and SMART. J. Multi-Criteria Decis. Anal. **6**, 155–165 (1997)
26. Vargas, L.G.: Why the multiplicative AHP is invalid: a practical example. J. Multi-Criteria Decis. Anal. **6**(3), 169–170 (1997)
27. Wang, Y.M., Elhag, T.M.: An approach to avoiding rank reversal in AHP. Decis. Support Syst. **42**(3), 1474–1480 (2006)
28. Piegat, A., Sałabun, W.: Identification of a multicriteria decision-making model using the characteristic objects method. Appl. Comput. Intell. Soft Comput. (2014)
29. Piegat, A., Sałabun, W.: Nonlinearity of human multi-criteria in decision-making. J. Theor. Appl. Comput. Sci. **6**(3), 36–49 (2012)

30. Piegat, A., Sałabun, W.: Comparative analysis of MCDM methods for assessing the severity of chronic liver disease. In: Artificial Intelligence and Soft Computing, pp. 228-238. Springer International Publishing (2015)
31. Sałabun, W.: Application of the fuzzy multi-criteria decision-making method to identify nonlinear decision models. Int. J. Comput. Appl. **89**(15), 1–6 (2014)
32. Sałabun, W.: Reduction in the number of comparisons required to create matrix of expert judgment in the comet method. Manag. Prod. Eng. Rev. **5**(3), 62–69 (2014)
33. Sałabun, W.: The characteristic objects method: a new distance based approach to multicriteria decision making problems. J. Multi Criteria Decis. Anal. **22**(1–2), 37–50 (2015)
34. Sałabun, W.: The use of fuzzy logic to evaluate the nonlinearity of human multi-criteria used in decision making. Przeglad Elektrotechniczny (Electrical Review) **88**(10b), 235–238 (2012)

A New Approach to a Derivation of a Priority Vector from an Interval Comparison Matrix in a Group AHP Framework

Jiri Mazurek

Abstract The aim of the article is to present a new approach to a derivation of a priority vector form an interval comparison matrix in a group AHP framework. It is supposed that preferences of individual decision makers are aggregated into a group interval comparison matrix, and the priority weights of all alternatives are estimated via the geometric mean method generalized to interval numbers with the use of interval arithmetic. This approach differs from usual solutions of the problem based on linear programming methods or a decomposition of the interval comparison matrix into crisp matrices, followed by the eigenvalue method. This new approach is demonstrated on an example, and a comparison with a standard group AHP is provided as well.

Keywords AHP • Group AHP • Group decision making • Interval AHP

1 Introduction

The analytic hierarchy process (AHP) proposed by T.L. Saaty in 1980s is one of the most successful tools for multiple criteria decision making. Its essential part consists of pair-wise comparisons of alternatives and criteria regarding a superior element in a given hierarchy with the use of Saaty's fundamental scale, see [7–10]. However, in real-world decision making problems uncertainty is often present, which is usually modeled by fuzzy or interval numbers. Also, a group decision making is becoming more important as the complex nature of many present problems (in environmental protection, economics, engineering law, medicine, etc.) requires a cooperation of experts in various fields.

J. Mazurek (✉)
Silesian University in Opava, School of Business Administration in Karvina, University Square 1934/3, Karvina, Czech Republic
e-mail: mazurek@opf.slu.cz

I. Czarnowski et al. (eds.), *Intelligent Decision Technologies 2016*, Smart Innovation, Systems and Technologies 56,
DOI 10.1007/978-3-319-39630-9_16

An extension of the AHP to a group AHP (GAHP) was proposed in [10] and followed by many others; see e.g. [1, 3] or [6].

The interval AHP was proposed by Saaty and Vargas in [9]. In their pioneer work Monte Carlo simulations were applied to obtain weights of alternatives or criteria. Later other methods based mainly on linear programming (LP) or matrix decomposition were developed both for the interval AHP and the interval group AHP (IGAHP), see e.g. [2, 5] or [12, 13].

The aim of this paper is to present a modification of the interval group analytic hierarchy process. Crisp judgments of decision makers (DM) are converted into a group interval comparison matrix, and alternatives' weights with regard to all criteria are estimated via the geometric mean method with the use of proper interval arithmetic. In the final step all alternatives (their interval weights) are compared pairwise with the use of a possibility measure, and a total or partial ordering of all alternatives is established at the end.

While GAHP is computationally simple and easy comprehensible, it also has some drawbacks that might be avoided by the use of IGAHP. Firstly, GAHP is not robust, as adding one more decision maker into a group results in a change of weights of alternatives and their ranking in general. Furthermore, by GAHP a costumer acquires crisp weights of alternatives and their ranking, but information on variability or diversity among decision makers' judgments gets lost in the process, creating a false image of a precise output arising from a complex input (DMs' judgments might be very dissimilar).

IGAHP might be more convenient than GAHP, because it is more robust, as adding a decision maker into a group might not change a final result (see later), and also the priority vector in a form of an interval reflects more realistically decision makers' diversity. On the other hand, a general disadvantage of IGAHP methods is that one may not obtain a final total ordering of all alternatives, as circular problems of the type A is preferred to B, B is preferred to C and C is preferred to A may occur.

The paper is organized as follows: Sect. 2 provides a brief description of AHP and GAHP, in Sect. 3 requisite interval arithmetic is introduced, the proposed method is described in Sect. 4 and in Sect. 5 an illustrative example is provided. Conclusions close the article.

2 Analytic Hierarchy Process and Group Analytic Hierarchy Process

In an AHP, the elements such as goal, criteria or alternatives form a strictly hierarchical structure, where elements from the same level are regarded independent. The relative importance of elements from a given level of hierarchy with respect to an element on a higher level is expressed by a number s_{ij} on Saaty's fundamental scale. The number $s_{ij} = v_i / v_j$ is the ratio of importance of element i compared to element j, $s_{ij} \in \{1/9, 1/8, \ldots 1, \ldots, 8, 9\}$, where v_i are (unknown) weights of compared elements. It is assumed that $s_{ij} = 1/s_{ji}$ for all i and j, hence the matrix S (s_{ij}) is reciprocal. The weights of elements ν_i can be determined by the right principal eigenvector related to the largest (positive) eigenvalue of S; hence the vector of weights ν satisfies the equation:

$$S\nu = \lambda_{\max}\nu. \tag{1}$$

In the next step the vector of weights ν from (1) is normalized so that $\|\nu\| = 1$. The existence of the largest positive eigenvalue is assured by Perron-Frobenius theorem.

Also, the geometric mean of the row elements of S (the logarithmic least squares method) can be used for derivation of weights of n elements with the use of the following formula:

$$\nu_i = \left(\prod_{j=1}^{n} s_{ij}\right)^{1/n} \Big/ \sum_{k=1}^{n} \left(\prod_{j=1}^{n} s_{kj}\right)^{1/n}. \tag{2}$$

In the case of $n = 3$ both methods provide the same results [7].

In the group AHP preferences of decision makers can be aggregated into one group preference matrix A^{group} at the beginning of the process by the weighted geometric mean. Then, a group priority vector is estimated usually by the eigenvalue method. Another often followed approach is to estimate priority vectors of individual DMs, and then aggregating them into the group priority vector by the weighted arithmetic or geometric mean, see e.g. [3] or [13].

In the interval AHP preferences of decision makers are converted into an interval comparison matrix which expresses the natural diversity of DMs' judgments. For deriving priority weights from interval comparison matrices several methods were proposed. One approach is based on linear programming methods, other approaches, such as MEDINT or ADEXTREME, aim at a decomposition of an interval matrix into two crisp matrices, and the final priority vector is obtained by a combination of their priority vectors, see [5] or [13].

3 Interval Arithmetic

Detailed principles and rules of interval arithmetic are described in [4]. Though interval arithmetic is highly non-trivial due to cases including non-bounded intervals, negative numbers or 0, for our purposes it suffices to define elementary operations (addition, subtraction, multiplication and division) for closed and bounded intervals $I \subseteq [1/9, 9]$, which are bounded by the least and largest values of Saaty's fundamental scale. Let $I_1 = [a, b] \subseteq [1/9, 9]$ and $I_2 = [c, d] \subseteq [1/9, 9]$. Then:

$$I_1 + I_2 = [a, b] + [c, d] = [a + c, b + d], \tag{3a}$$

$$I_1 - I_2 = [a, b] - [c, d] = [a - d, b - c], \tag{3b}$$

$$I_1 \cdot I_2 = [a, b] \cdot [c, d] = [a \cdot c, b \cdot d], \tag{3c}$$

$$I^n = [a, b]^n = [a^n, b^n], n \in (0, \infty), \tag{3d}$$

$$\frac{I_1}{I_2} = \frac{[a, b]}{[c, d]} = \left[\frac{a}{d}, \frac{b}{c}\right]. \tag{3e}$$

In the above relations, (3d) follows immediately from (3c). Also, relation (3d), which is a power law, enables one to introduce a root of an interval as well.

4 Interval Group Analytic Hierarchy Process

Let's consider a three-level AHP problem (goal-criteria-alternatives) with K decision makers, n feasible alternatives and M criteria. It is assumed that all alternatives are pairwise comparable with respect to all criteria by all decision makers. Each decision maker provides his/her crisp pairwise comparison matrix for a given criterion:

$$PCM^i = \begin{bmatrix} 1 & s^i_{12} & s^i_{13} & \cdots \\ s^i_{21} & 1 & s^i_{23} & \cdots \\ s^i_{31} & \cdots & 1 & \cdots \\ \cdots & \cdots & \cdots & \cdots \end{bmatrix}, i \in \{1, \ldots, K\}.$$

Then, a pairwise comparison matrix (PCMG) of the whole group (for a given criterion) is constructed as an interval matrix with lower and upper bounds of each interval corresponding to minimal and maximal values s_{ij} provided by all DMs:

$$PCMG = \begin{bmatrix} 1 & [l_{12}, u_{12}] & [l_{13}, u_{13}] & \cdots \\ [l_{21}, u_{21}] & 1 & [l_{23}, u_{23}] & \cdots \\ [l_{31}, u_{31}] & \cdots & 1 & \cdots \\ \cdots & \cdots & \cdots & \cdots \end{bmatrix},$$

where $u_{ij} = \max_k \left\{ s^k_{ij} \right\}$, $l_{ij} = \min_k \left\{ s^k_{ij} \right\}$, $k \in \{1, \ldots, K\}$, $\forall i, j \in \{1, \ldots, n\}$.

The matrix PCMG expresses an aggregated pairwise comparison of all alternatives by the group of DMs. It should be noted that some decision makers may diametrically differ in their judgments from other group members, so a check for potential 'outliers' would be appropriate before continuing into the next step.

Also, PCMG is 'reciprocal': $l_{ij} = \frac{1}{u_{ji}}, \forall i, j \in \{1, \ldots, n\}$. To estimate weights and ranking of all alternatives, interval arithmetic and interval generalization of the geometric mean is applied.

With the use of relation (3d) the geometric mean for intervals can be defined. Let I_i be real intervals, then interval geometric mean is defined as follows:

$$\bar{I}_G = \left(\prod_{i=1}^{n} I_i \right)^{1/n}. \tag{4}$$

Formula (4) enables us to obtain the overall weights (or scores, because the weights do not satisfy the so called interval probabilities constraints, see [2]) of alternatives, which, in turn, can be compared and ranked regarding to a goal.

Let $I_{ij} = [l_{ij}, u_{ij}]$, $i, j \in \{1, 2, \ldots, n\}$, be elements of a PCMG matrix. Then the weight (score) of each alternative i under criterion k is given as follows:

$$w_i^k = \frac{\left(\prod_{j=1}^{n} I_{ij}^k \right)^{1/n}}{\sum_{m=1}^{n} \left(\prod_{j=1}^{n} I_{mj}^k \right)^{1/n}}. \tag{5}$$

Let v_k be the weights of criteria and let w_i^k be the weight of an alternative i with respect to a criterion k. Then the weight of an alternative i with regard to a goal is given as:

$$w_i = \sum_{k=1}^{K} w_i^k \cdot v_k. \tag{6}$$

This value enters the dominance relation (7), so alternatives can be ranked from the best to the worst (but generally only quasi-ordering of alternatives is provided as some alternatives might be tied).

Since 1980s many procedures for ordering of interval numbers or fuzzy numbers were proposed. In this paper a formula based on the possibility theory proposed by Xu [11] is applied. Let $I_1 = [a, b]$ and $I_2 = [c, d]$ be intervals of real numbers. Then the degree of possibility that $I_1 \geq I_2$ is defined as:

$$p(I_1 \geq I_2) = \max\left\{ 1 - \max\left(\frac{d-a}{b+d-a-c}, 0 \right), 0 \right\}. \tag{7}$$

Relation (7) has some desirable properties [11]:

(i) $0 \leq p(I_k \geq I_l) \leq 1, \forall k, l$,
(ii) $p(I_k \geq I_l) + p(I_l \geq I_k) = 1, \forall k, l$,
(iii) If $p(I_k \geq I_l) \geq 0.5$ and $p(I_l \geq I_m) \geq 0.5$, then $p(I_k \geq I_m) \geq 0.5$.

Therefore, relation (7) is nothing else than a dominance relation. Trivially, if $I_1 = [a, b]$, $I_2 = [c, d]$ and $c > b$ (intervals are not overlapping), then $I_2 > I_1$.

5 A Numerical Example

In this section a numerical example is provided. Consider a three level (goal-criteria-alternatives) hierarchy problem with two criteria C_1 and C_2, where the weights of criteria are $\nu_1 = 0.6, \nu_2 = 0.4$; four alternatives A, B, C and D; and three decision makers DM1, DM2 and DM3 (with the same weight). Pairwise comparison matrices of all alternatives for both criteria by all DMs are provided in Figs. 1, 2 and 3. The solution of the problem is provided by a GAHP with the weighted arithmetic mean aggregation of individual priorities (GAHP-WAM) and by the proposed geometric mean IGAHP (GM-IGAHP) method.

5.1 Solution by GAHP-WAM Method

From DMs' pairwise comparison matrices in Figs. 1, 2 and 3 weights w_i^k, where k denotes a decision maker and i denotes a criterion, of all alternatives were estimated by the eigenvalue method:

Fig. 1 A comparison matrix provided by DM1 for criteria C_1 (*left*) and C_2 (*right*)

$$\begin{bmatrix} 1 & 2 & 1/3 & 3 \\ 1/2 & 1 & 1/5 & 1 \\ 3 & 5 & 1 & 7 \\ 1/3 & 1 & 1/7 & 1 \end{bmatrix}, \begin{bmatrix} 1 & 3 & 1 & 2 \\ 1/3 & 1 & 1/3 & 1 \\ 1 & 3 & 1 & 2 \\ 1/2 & 1 & 1/2 & 1 \end{bmatrix}$$

Fig. 2 A comparison matrix provided by DM2 for criteria C_1 (*left*) and C_2 (*right*)

$$\begin{bmatrix} 1 & 3 & 1/4 & 4 \\ 1/3 & 1 & 1/5 & 2 \\ 4 & 5 & 1 & 8 \\ 1/4 & 1/2 & 1/8 & 1 \end{bmatrix}, \begin{bmatrix} 1 & 4 & 1/2 & 3 \\ 1/4 & 1 & 1/8 & 1 \\ 2 & 8 & 1 & 6 \\ 1/3 & 1 & 1/6 & 1 \end{bmatrix}$$

Fig. 3 A comparison matrix provided by DM3 for criteria C_1 (*left*) and C_2 (*right*)

$$\begin{bmatrix} 1 & 2 & 1/3 & 4 \\ 1/2 & 1 & 1/6 & 2 \\ 3 & 6 & 1 & 8 \\ 1/4 & 1/2 & 1/8 & 1 \end{bmatrix}, \begin{bmatrix} 1 & 1 & 1/2 & 1/3 \\ 1 & 1 & 1/2 & 1/3 \\ 2 & 2 & 1 & 1 \\ 3 & 3 & 1 & 1 \end{bmatrix}$$

$$w_1^1 = (0.220, 0.104, 0.590, 0.086),\ w_2^1 = (0.354, 0.131, 0.354, 0.161),$$
$$w_1^2 = (0.226, 0.103, 0.612, 0.060),\ w_2^2 = (0.279, 0.075, 0.559, 0.087),$$
$$w_1^3 = (0.222, 0.111, 0.605, 0.062),\ w_2^3 = (0.144, 0.144, 0.320, 0.392).$$

Priority vectors of individual decision makers $w^i = \sum_{j=1}^{2} \nu_j w_j^i$:

$$w^1 = (0.273, 0.115, 0.496, 0.116),\ w^2 = (0.247, 0.092, 0.590, 0.071),$$
$$w^3 = (0.191, 0.124, 0.491, 0.194).$$

By the aggregation of individual priority vectors into the group priority vector by the weighted arithmetic mean the following result is obtained: $w = (0.237, 0.111.0.526, 0.127)$.

Therefore, the final ranking of all four alternatives with regard to the goal is: $C \succ A \succ D \succ B$.

5.2 Solution by GM-IGAHP Method

Interval weights of each alternative with regard to the criterion C_1 (C_2) are estimated by relation (5), results are provided in the right hand side of Fig. 4 (Fig. 5).

In the next step final weights of alternatives are obtained via relation (6):

$$w_A = \sum_{k=1}^{2} \nu_k \cdot w_A^k = [0.176, 0.275] \cdot 0.6 + [0.084, 0.707] \cdot 0.4 = [0.139, 0.448],$$

$$w_B = \sum_{k=1}^{2} \nu_k \cdot w_B^k = [0.077, 0.130] \cdot 0.6 + [0.042, 0.320] \cdot 0.4 = [0.063, 0.206],$$

$$w_C = \sum_{k=1}^{2} w_C^k \cdot \upsilon_k = [0.510, 0.723] \cdot 0.6 + [0.157, 1.189] \cdot 0.4 = [0.369, 0.909],$$

$$w_D = \sum_{k=1}^{2} w_A^k \cdot \upsilon_k = [0.056, 0.091] \cdot 0.6 + [0.064, 0.658] \cdot 0.4 = [0.059, 0.318].$$

Fig. 4 The interval comparison matrix of all DMs with regard to the criterion C_1

$$\begin{bmatrix} 1 & [2,3] & [1/4,1/3] & [3,4] \\ [1/3,1/2] & 1 & [1/6,1/5] & [1,2] \\ [3,4] & [5,6] & 1 & [7,8] \\ [1/4,1/3] & [1/2,1] & [1/8,1/7] & 1 \end{bmatrix} \begin{bmatrix} 0.176, 0.275 \\ 0.077, 0.130 \\ 0.510, 0.723 \\ 0.056, 0.091 \end{bmatrix}$$

Fig. 5 The interval comparison matrix of all DMs with regard to the criterion C_2

$$\begin{bmatrix} 1 & [1,4] & [1/2,1] & [1/3,3] \\ [1/4,1] & 1 & [1/8,1/2] & [1/3,1] \\ [1,2] & [2,8] & 1 & [1,6] \\ [1/3,3] & [1,3] & [1/6,1] & 1 \end{bmatrix} \begin{bmatrix} 0.084, 0.707 \\ 0.042, 0.320 \\ 0.157, 1.189 \\ 0.064, 0.658 \end{bmatrix}$$

In the final step all alternatives are compared pair-wise. We trivially obtain that C dominates B ($C \succ B$) and C dominates D ($C \succ D$), as alternatives' weights are non-overlapping intervals. For the remaining four pairwise comparisons of alternatives, relation (7) is used.

(i) The possibility that A dominates B is given as:

$$p(A \geq B) = \max\left\{1 - \max\left(\frac{0.206 - 0.139}{0.448 + 0.206 - 0.139 - 0.063}, 0\right), 0\right\} = 0.852,$$

$p(B \geq A) = 0.148$; therefore A dominates B.

Applying the same procedure we obtain the following results:

(ii) The possibility that A dominates C: $p(A \geq C) = 0.093$,

(ii) The possibility that A dominates D: $p(A \geq D) = 0.685$,

(iv) The possibility that B dominates D: $p(B \geq D) = 0.366$.

From points (i)–(iv) the final ranking of all four alternatives can be inferred: $C \succ A \succ D \succ B$. The alternative C is the best.

This final ranking is the same as the ranking obtained by GAHP-WAM method in the previous section. In this method alternatives D and B are almost tied for the third spot (their weights are 0.111 and 0.127 respectively).

GM-IGAHP method provides more information: alternative D dominates B only with a possibility of 0.685. There is more than 30 % possibility of B dominating D, which should be taken into account when executing a final decision.

6 Conclusions

The aim of the paper was to propose a modification to IGAHP methods which uses the geometric mean for the derivation of alternatives' weights. The advantage of the proposed approach is its natural formulation and simplicity: a solution can be achieved without computer software; therefore it might be more convenient for a practical use. The disadvantage of the proposed method resides in the fact that only possibilities that some alternatives are better than others are obtained in the end. While LP methods and methods based on the weighted arithmetic (geometric) mean aggregation of individual priority vectors are considered standard tools for the IGAHP solution, the presented method (along with other recently proposed approaches such as MEDINT or ADEXTREME) can be reckoned as an alternative or a complementary approach. Future research may focus on (dis)similarities among the aforementioned methods under different conditions.

Acknowledgements This research was supported by the grant project of GACR No. 14-02424S.

References

1. Dyer, R.F., Forman, E.H.: Group decision support with the analytic hierarchy process. Decis. Support Syst. **8**(2), 99–124 (1992)
2. Entani, T.: Interval AHP for a group of decision makers. In: IFSA-EUSFLAT, pp. 155–160 (2009)
3. Grošelj, P., Zadnik Stirn, L., Ayrilmis, N., Kuzman, M.K.: Comparison of some aggregation techniques using group analytic hierarchy process. Expert Syst. Appl. **42**(4), 2098–2204 (2014)
4. Hickey, T., Ju, Q., van Emden, M.H.: Interval arithmetic: from principles to implementation. J. ACM **48**(5), 1038–1068 (2001)
5. Liu, F.: Acceptable consistency analysis of interval reciprocal comparison matrices. Fuzzy Sets Syst. **160**(18), 2686–2700 (2009)
6. Ramanathan, R., Ganesh, L.S.: Group reference aggregation methods in AHP: an evaluation and an intrinsic process for deriving members' weightages. Eur. J. Oper. Res. **79**(2), 249–265 (1994)
7. Saaty, T.L.: The Analytic Hierarchy Process. McGraw Hill, New York (1980)
8. Saaty, T.L.: Decision making with the analytic hierarchy process. Int. J. Serv. Sci. **1**, 83–98 (2008)
9. Saaty, T.L., Vargas, L.G.: Uncertainty and rank order in the analytic hierarchy process. Eur. J. Oper. Res. **32**(1), 107–117 (1987)
10. Saaty, T.L.: Group decision making and the AHP. In: Golden, B.L. et.al.(eds.) The Analytic Hierarchy Process: Applications and Studies, pp. 59–67. McGraw-Hill, New York (1989)
11. Xu, Z.: A direct approach to group decision making with uncertain additive linguistic preference relations. Fuzzy Optim. Decis. Making **5**(1), 21–32 (2006)
12. Yu, J.R., Hsiao, Y.W., Sheiu, H.J.: A multiplicative approach to derive weights in the interval analytic hierarchy process. In: Int. J. Fuzzy Syst. **13**(3) (2011)
13. Zadnik, S.L., Groselj, P.: Estimating priorities in group AHP using interval comparison matrices. Multiple Criteria Decis. Making **8**, 143–159 (2013)

Toward a Conversation Partner Agent for People with Aphasia: Assisting Word Retrieval

Kazuhiro Kuwabara, Takayuki Iwamae, Yudai Wada, Hung-Hsuan Huang and Keisuke Takenaka

Abstract Aphasia is an acquired communication disorder marked by difficulties in understanding the language or expressing oneself using language. In order to support communication of a person with aphasia, a concept of supported communication has been proposed, wherein a conversation partner assists a person with aphasia in communicating. We are designing a conversation partner agent, a software agent that embodies the functions of a conversation partner. In this paper, we focus on the function of assisting a person with aphasia who has difficulty retrieving words. More specifically, multiple choice questions are generated using a domain ontology and presented in appropriate order to a person with aphasia. Through this question and answer process, the system infers the word that a person with aphasia wants to express.

Keywords People with aphasia · Conversation partner · Word retrieval · Information gain

1 Introduction

People with aphasia suffer from a language communication disorder [10]. Aphasia is often the result of a stroke and damage to the part of a brain that plays an important role in using language. Since the probability of having a stroke increases as a person

K. Kuwabara (✉) · Y. Wada · H.-H. Huang
College of Information Science and Engineering, Ritsumeikan University,
Kusatsu, Shiga 525-8577, Japan
e-mail: kuwabara@is.ritsumei.ac.jp

T. Iwamae
Graduate School of Information Science and Engineering,
Ritsumeikan University, Kusatsu, Shiga 525-8577, Japan

K. Takenaka
Abiko City Welfare Center for the Handicapped, Abiko, Chiba 270-1112, Japan

I. Czarnowski et al. (eds.), *Intelligent Decision Technologies 2016*,
Smart Innovation, Systems and Technologies 56,
DOI 10.1007/978-3-319-39630-9_17

gets older, support for aphasia sufferers is becoming much important with the advent of an aging society.

The concept of *Supported Communication* has been proposed to support the daily life of people with aphasia [7], in which a human conversation partner helps aphasia patients understand messages and express thoughts by acting as a kind of interpreter. Helping people with aphasia communicate requires specific skills for which training programs have been proposed and implemented [12].

From this background, we are investigating a way to design a software agent that can act as the conversation partner for people with aphasia. We call such an agent a conversation partner agent. This conversation partner agent is not meant to replace a human conversation partner, but is expected to include a function to assist a human conversation partner. Since a conversation partner must provide a wide range of functions, developing a comprehensive conversation partner agent is difficult. We focus on the function of helping to retrieve a word that a person with aphasia wants to express. In such a case, the human conversation partner often poses a series of multiple-choice questions that may be answered with a simple yes or no. Through this question and answer process, the word the person with aphasia wants to express is identified. In this paper, we present a mechanism to implement such a function as one of the functions of the conversation partner agent.

Our paper is organized as follows: In the next section, we describe related works. Section 3 presents the basic concept of the conversation partner agent and the function of assisting in word retrieval. Section 4 explains the detailed mechanism of word retrieval and Sect. 5 describes the current prototype with its example execution. We conclude our paper in Sect. 6.

2 Related Work

Information Communication Technology (ICT) is being applied in the therapy of people with aphasia [9]. It is also utilized to develop Augmented and Alternative Communication (AAC) applications for people with aphasia to support their communication. Since small portable devices, such as smart phones and tablets, are becoming popular, images or photos in addition to text can easily be deployed.

For example, *PhotoTalk* [2] uses photos to describe the life story of a person with aphasia, and a field study was conducted using a commercial AAC application for iPad, called *TalkingTiles* [6]. In addition, *CoCreation* [1] helps a person with aphasia express daily experiences using photos that are uploaded to the system and automatically saved into clusters. *TalkAbout*, which is a context-aware adaptive AAC system, presents a word list to the user that is adapted to the location and conversation partner [8]. These systems provide a means to communicate with less emphasis on language.

Approaches that put more focus on the language side have also been proposed. For example, a word (vocabulary) list is constructed and utilized to prompt a conversation of a person with aphasia [14]. In this system, words are arranged in a hierarchical

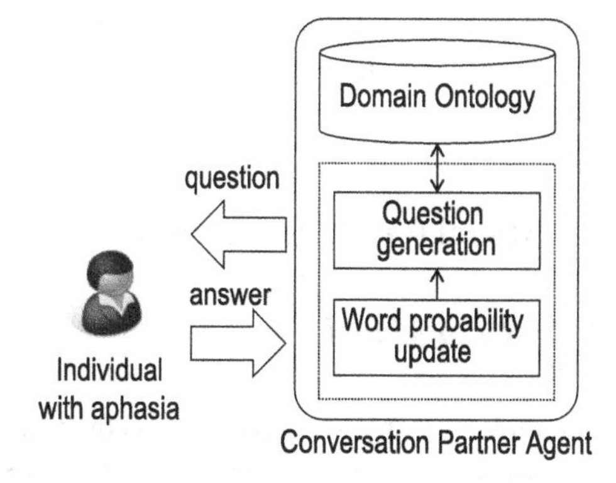

Fig. 1 Assisting word retrieval with a conversation partner agent

order. By following a link from a broader category to a narrower one, the person with aphasia can find the word that he/she wants to talk about.

To assist word retrieval, a system that utilizes a smart phone to identify the name of a food dish has been proposed [3]. This system contains a database that holds various information regarding food and dishes. The system generates a yes or no question and, based on the answers from the person with aphasia, the system narrows down a list of dishes that the person with aphasia has in mind.

In this paper, we focus on word retrieval as in the system proposed by Arima et al. [3]. This is similar in a sense to the popular Akinator game,[1] which guesses a character a player has in mind by asking the player several questions. This game can be thought of as guessing the thing a user has in mind, but the domain of this game is limited to characters both fictitious and real.

3 Assisting Word Retrieval

A conversation partner helps a person with aphasia understand conversations or express thought. As mentioned previously, our aim is to design a software agent that can act as a conversation partner. With the increasing popularity of portable devices, such as smart phones, it is now feasible to have such an agent always at hand.

As a first step toward a conversation partner agent, we focus on the function of assisting a person with aphasia with word retrieval (Fig. 1). Our approach is based on a human conversation partner asking a yes/no question or by selecting from multiple choices to infer the word that the person with aphasia has in mind. We use a domain ontology that holds the information necessary to compose such questions. Words are stored in the domain ontology with their categories and properties.

Since it is unrealistic to guess a word from all of possible words, we let the user select a certain *topic* that limits the words to be guessed from. According to the selected topic, the questions are generated to infer a word from the words that fall

[1] http://en.akinator.com/.

into that selected topic. Starting from the top level categories in the selected topic, the category of the target word is narrowed down. Then, its properties are asked about to pinpoint the word.

4 Model for Word Retrieval

4.1 Data Model

To implement the word guessing method in the software agent, we developed the following model. We assume that there is a set of words D and the system selects a word $w \in D$ that a user has in mind. The system poses a series of questions. We assume that there are n questions q_i $(1 \leq i \leq n)$. Each question effectively divides D into N_i subsets of words, which is denoted by C_{ik} $(k = 1, 2, 3, \ldots, N_i)$. That is,

$$\bigcup_k C_{ik} \equiv D \quad (k = 1, 2, 3 \ldots N_i). \tag{1}$$

To guess the word, we calculate the probability of word w, which is denoted by $p(w)$. Based on the user's answers to the questions posed by the system, the probabilities of words are updated. The word with the highest probability will be presented to the user as a guessed word.

4.2 Database Construction

To generate questions to ask, we use a domain ontology for the particular topic. The domain ontology is constructed using linked data [4] (Fig. 2). The template for the questions is also stored in the domain ontology for categories and properties (Fig. 3).

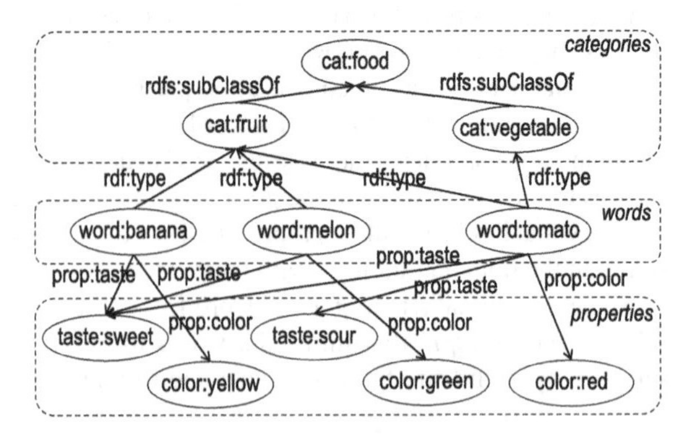

Fig. 2 Part of domain ontology

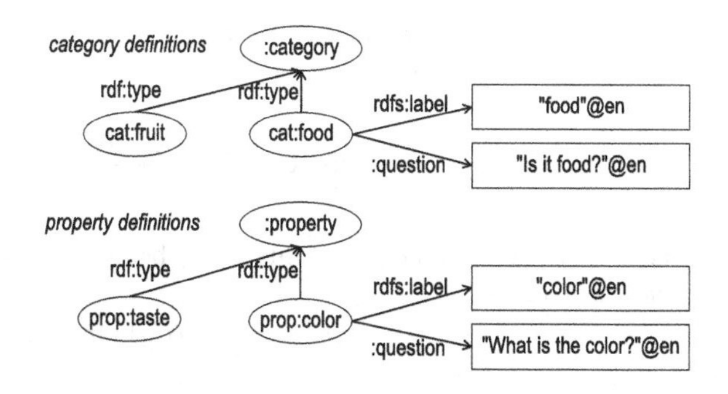

Fig. 3 Question template example

There are two types of questions [11]. The first is to narrow down from a broader category to a narrower one. The second is to narrow down using the property values. For example, let us suppose that a word related to *food* is to be retrieved and, as subcategories of *food*, there are *fruit* and *vegetable* categories. For the first type of question, a question like *Is it a fruit?* or *Is it a vegetable?* will be generated. This question is answered with a simple *yes* or *no*. For a particular word, the answer will be *yes* if the category for the question can be reached following up the `subClass` hierarchy from the category of the word. If not, *no* is the correct answer to that question.

Here, a word can belong to multiple categories. For example, in the topic of *food*, *tomato* may belong to the *vegetable* and *fruit* categories. Note that the domain ontology is meant to be used to generate suitable questions and their multiple choice answers to assist in word retrieval; it is not meant for exact definitions of the concept of words. Thus, the domain ontology is constructed more informally to pursue flexibility.

For the second type of question, properties also have a question template. When a question concerning a property is posed to a user, the possible values are collected from the property values of the current candidate words. If such a property is not defined for a particular word in the domain ontology, it is treated as *don't care*. It is assumed that, for each answer choice, an equal probability is assigned to each possible property value of that particular word.

4.3 Questioning Strategy

In order to achieve a smooth flow of questions, we use questioning strategies based on the information gain as in [3]. First, the entropy H is calculated as follows:

$$H = -\sum_{w \in D} p(w) \log p(w) . \quad (2)$$

The probability of w after the answer to question q_i is known as C_{ik} can be written as $p(w|C_{ik})$. Using the Bayesian theorem, we get:

$$p(w|C_{ik}) = \frac{p(C_{ik}|w)}{p(C_{ik})} p(w) , \tag{3}$$

where $p(C_{ik})$ denotes the probability of C_{ik} being selected as an answer to question q_i and is given as:

$$p(C_{ik}) = \sum_{w \in D} p(C_{ik}|w) p(w) . \tag{4}$$

The entropy after the answer to question q_i is known as C_{ik} is calculated as:

$$H_{ik} = \sum_{w \in D} p(w|C_{ik}) \log p(w|C_{ik}) . \tag{5}$$

The expected average entropy H_i after question q_i is asked is then calculated as:

$$H_i = \sum_{k=1}^{N_i} p(C_{ik}) H_{ik} \tag{6}$$

and the expected information gain IG_i of question q_i is calculated as:

$$IG_i = H - H_i . \tag{7}$$

Basically, the question with the highest information gain should be asked first. However, if the question has a large number of choices, its information gain inherently becomes large. To a human user, a question with a large number of choices may not be appropriate. To take the number of answer choices of a question into consideration, we use the following heuristics to calculate the priority of question q_i, PR_i:

$$PR_i = \frac{1}{N_i - \beta} IG_i , \tag{8}$$

where β denotes a parameter to reflect the effects of the number of answer choices N_i. From the preliminary testing, β is set to 1.5. That is, when $N_i = 2$, the priority of the question is effectively set to the information gain doubled. Thus, the question with fewer choices, such as a yes/no type question, will be given a higher priority.

4.4 Question and Answer Process

First, the probabilities of words in D are equally set if there is no prior knowledge about the word a person with aphasia wants to express. If we have some information

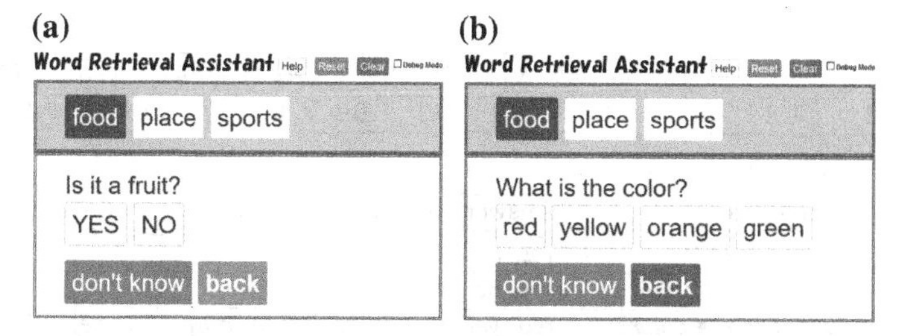

Fig. 4 Screen shots of a prototype implementation. **a** Yes/no type question. **b** Multiple choice question

about the preferences over words, say, from past usage of the system, we can put different weights among the initial probabilities of words.

The question with the highest priority is then selected and presented to the user with a set of possible answers to that question. After the answer is given, the probability of the words is updated. When the probability of words exceeds a given threshold, that word will be shown to the user as a guessed word. If not, based on the updated probability of the words, the priority of remaining questions is recalculated and another question is selected. This loop continues until the word is found or there are no more new questions to ask.

5 Prototype Implementation

5.1 Overview

The prototype system is implemented as a Web application. The domain ontology is stored using the Apache Fuseki server,[2] open source Resource Description Framework (RDF) store. Part of the domain ontology regarding the conversation topics is first converted into an internal representation to generate questions and their answer choices. In addition, we allow a spreadsheet-based data format to be used so that questions and answers can also be prepared using Excel.

The user interface was built using HTML/CSS/JavaScript. By adopting a responsive design, the application can be accessed not only from a PC, but also from a tablet or smart phone. Figure 4 shows screen shots of the current implementation. The upper part of the screen shows a list of possible topics from which the user is to select. The lower part shows a question sentence and its answer choices.

[2]https://jena.apache.org/documentation/serving_data/.

(a)

banana	*broccoli*	*carrot*	*eggplant*	*kiwi*	*lettuce*	*onion*	*orange*	*strawberry*	*tomato*
0.1	0.1	0.1	0.1	0.1	0.1	0.1	0.1	0.1	0.1

(b)

Question	Answer choices
Is it a fruit?	*yes, no*
Is it a vegetable?	*yes, no*
What is the color?	*red, yellow, orange, green, purple, white*
What is the taste like?	*sweet, sour*

Fig. 5 Initial state of the sample execution. **a** Initial probability. **b** List of questions and their answer choices

5.2 Example Session

Let us consider a simple example. Assuming that there are the topics of *food*, *sports*, and *place*, the *food* topic is selected. We also assume, for example's sake, that there are 10 possible words to be selected from. Initially, these words are given equal probabilities (Fig. 5a). The questions are constructed from the subcategories under the *food* topic. In this case, they are *vegetable* and *fruit*. The question sentences are retrieved from the domain ontology and the answer choices are set to *yes* and *no* since they are asking about categories. Questions are also constructed from the properties, *color* and *taste* in this example. Their answer choices are collected from the values of these properties in the domain ontology. The questions and their answer choices are shown in Fig. 5b.

First, the information gains of the questions are calculated using Eq. 7. In this example, the information gain of question *What is the color?* is the highest because it has many answer choices (Fig. 6a). However, it may not be appropriate to ask a question with many choices in the first stage because it forces the user to understand a lot of choices. To give a higher priority to a question with fewer choices, the priority is calculated according to Eq. 8, resulting in the question *Is it a fruit?* being selected because it has the highest priority value.

If the user answers with *yes* to this question, the probability of words are updated, as shown in Fig. 6b. From the updated word probabilities, the information gain and priorities of the questions are recalculated. The question *Is it a vegetable?* will have the highest priority and will be asked next. Please note that the question *Is it a fruit?* is excluded in the list of questions (Fig. 6c) because this question has already been asked. Please also note that, in this example, as *tomato* is assumed to belong to both the *vegetable* and *fruit* categories, this question will eliminate the possibility of *tomato*, if the answer to this question is *no* (only *tomato* belongs to both *vegetable* and *fruit* in this example).

Let us suppose that *no* is given as the answer. Then, the word probabilities are updated as shown in Fig. 6d. The information gain and priorities of the questions are also updated (Fig. 6e). After this calculation, the question *What is the color?* will

(a)

Question	Information gain	Priority
Is it a fruit?	0.693	1.386
Is it a vegetable?	0.673	1.346
What is the color?	1.525	0.339
What is the taste like?	0.162	0.324

(b)

banana	*broccoli*	*carrot*	*eggplant*	*kiwi*	*lettuce*	*onion*	*orange*	*strawberry*	*tomato*
0.2	0	0	0	0.2	0	0	0.2	0.2	0.2

(c)

Question	Information gain	Priority
Is it a vegetable?	0.5	1
What is the color?	1.332	0.533
What is the taste like?	0.223	0.446

(d)

banana	*broccoli*	*carrot*	*eggplant*	*kiwi*	*lettuce*	*onion*	*orange*	*strawberry*	*tomato*
0.25	0	0	0	0.25	0	0	0.25	0.25	0

(e)

Question	Information gain	Priority
What is the color?	1.386	0.555
What is the taste like?	0.203	0.407

Fig. 6 Sample execution session. **a** Information gain and priority of questions. **b** Word probabilities after *yes* is given as an answer to the question *Is it a fruit?* **c** Updated information gain and priority of questions. **d** After *no* is given as an answer to the question *Is it a vegetable?* **e** Updated information gain and priority of questions

have the highest priority value and will be asked next. If the user answers with *red* to this question, the word *strawberry* is inferred. In this way, the word the user has in mind can be identified.

6 Conclusion

This paper presents a conversation partner agent for people with aphasia and a method of assisting them with word retrieval. The current prototype was implemented as a Web application. Through a question and answer process, the word that a person with aphasia has in mind is inferred. The questions are generated from a domain ontology or spreadsheet data. Each question's priority is calculated based on its information gain. By introducing heuristics, a natural order of questions can be achieved.

Much additional work needs to be undertaken. Currently, we do not consider relationships between words. To calculate the information gain more accurately, we need to take into consideration conceptual relationships between words. For example, we can approach this issue using concept vector representation [13]. Moreover, it is also important to take into consideration the characteristics of problems people with aphasia face in understanding semantic information expressed in language [5].

The system's usefulness depends largely on the data it contains. In this sense, accumulating much data into the system is crucial. It is important to make use of resources available on the Web, such as linked open data. In addition, the current prototype focuses on showing text. In a real-world application, images will have greater value. Therefore, we will focus on incorporating images into the system in the future.

Acknowledgments This work was partially supported by JSPS KAKENHI Grant Number 15K00324.

References

1. Al Mahmud, A., Limpens, Y., Martens, J.B.: Expressing through digital photographs: an assistive tool for persons with aphasia. Univ. Access Inf. Soc. **12**(3), 309–326 (2013)
2. Allen, M., McGrenere, J., Purves, B.: The field evaluation of a mobile digital image communication application designed for people with aphasia. ACM Trans. Accessible Comput. **1**(1), 5:1–5:26 (2008)
3. Arima, S., Kuroiwa, S., Horiuchi, Y., Furukawa, D.: Question-asking strategy for people with aphasia to remember food names. J. Technol. Persons Disabil. **3** (2015)
4. Bizer, C., Heath, T., Berners-Lee, T.: Linked data—the story so far. Int. J. Semant. Web Inf. Syst. **5**(3), 1–22 (2009)
5. Hillis, A.E.: Aphasia: progress in the last quarter of a century. Neurology **69**(2), 200–213 (2007)
6. Huijbregts, T., Wallace, J.R.: Talkingtiles: supporting personalization and customization in an AAC app for individuals with aphasia. In: Proceedings of the 2015 International Conference on Interactive Tabletops and Surfaces, pp. 63–72. ITS '15 (2015)
7. Kagan, A.: Supported conversation for adults with aphasia: methods and resources for training conversation partners. Aphasiology **12**(9), 816–830 (1998)
8. Kane, S.K., Linam-Church, B., Althoff, K., McCall, D.: What we talk about: designing a context-aware communication tool for people with aphasia. In: Proceedings of the 14th International ACM SIGACCESS Conference on Computers and Accessibility, pp. 49–56. ASSETS '12 (2012)
9. Kötteritzsch, A., Gerling, K.: Future directions for ICT in aphasia therapy for older adults: enhancing current practices through interdisciplinary perspectives. Stem-, Spraak-en Taalpathologie **20**, 135–150 (2015)
10. National Aphasia Association: Aphasia definitions, http://www.aphasia.org/aphasia-definitions/. Accessed 24 Jan 2016
11. NPO Waon: Let's Talk with People with Aphasia (in Japanese). Chuohoki Publishing (2008)
12. Simmons-Mackie, N., Raymer, A., Armstrong, E., Holland, A., Cherney, L.R.: Communication partner training in aphasia: a systematic review. Arch. Phys. Med. Rehabil. **91**(12), 1814–1837 (2010)

13. Szymański, J., Duch, W.: Context search algorithm for lexical knowledge acquisition. Control Cybern. **41**(1), 81–96 (2012)
14. Yasuda, K., Nemoto, T., Takenaka, K., Mitachi, M., Kuwabara, K.: Effectiveness of a vocabulary data file, encyclopaedia, and Internet homepages in a conversation-support system for people with moderate-to-severe aphasia. Aphasiology **21**(9), 867–882 (2007)

Intelligent Monitoring of Complex Discrete-Event Systems

Gianfranco Lamperti and Giulio Quarenghi

Abstract A complex active system is a discrete-event system exhibiting both horizontal and vertical interaction between components. By inspiration of biological systems in nature, which are organized in a hierarchy of subsystems, the horizontal interaction between components at a given hierarchical level gives rise to an emergent behavior at a superior level, which is unpredictable from a knowledge of the behavior of the interacting components only. Since real critical-systems, such as power networks and nuclear plants, can be conveniently modeled as complex active systems, monitoring and diagnosis of complex active systems is of paramount importance to the safety of society. This is why an intelligent diagnosis framework for complex active systems is presented in this paper. Intelligence means that diagnosis does not require the naive reconstruction of the system behavior as a whole, which would be exponential with the number of components. Experiments show the effectiveness of the diagnosis technique.

1 Introduction

Imagine the operator in the control room of a large and distributed power network. When a misbehavior occurs, like a short circuit on a transmission line, several actions can be triggered by the protection system in order to isolate the shorted line, typically by opening breakers and reconfiguring the power load in order to avoid a blackout, like the India major blackout in 2012, the largest power outage in history [18].

If the reaction of the protection system is abnormal, a possibly large number of alarms and messages are generated. Since the operator is expected to activate specific recovery actions, it is essential that the (possibly overwhelming) stream of information generated by the system, namely the temporal observation, be interpreted correctly in stringent time constraints.

G. Lamperti (✉) · G. Quarenghi
Dipartimento di Ingegneria Dell'Informazione,
Università degli Studi di Brescia, Brescia, Italy
e-mail: gianfranco.lamperti@unibs.it

I. Czarnowski et al. (eds.), *Intelligent Decision Technologies 2016*,
Smart Innovation, Systems and Technologies 56,
DOI 10.1007/978-3-319-39630-9_18

To this end, an automated diagnosis engine is expected to generate diagnosis information online in relatively short time. Specifically, a set of candidate diagnoses are presented to the operator, who is expected to take critical decisions for the safety of the surrounding environment and the involved population. However, a key factor in real applications of automated diagnosis is time response. Hence, it is of paramount importance that the diagnosis engine be not only effective but also efficient.

2 State of the Art and Contribution

Often, real dynamic systems can be modeled as discrete-event systems [3]. Seminal works on diagnosis of discrete-event systems (DES's) maximize offline preprocessing in order to generate an efficient online *diagnoser* [19, 20]. However, these approaches require the generation of the global DES model, which is bound to be impractical for large and distributed systems. Other approaches, like diagnosis of active systems (ACS's) [13], avoid generating the global model of the system by reconstructing online only the behavior of the system which is consistent with the observation. (The behavior is then decorated with candidate diagnoses.) This is achieved by a *naive* technique, where each state of the generated behavior of the ACS is identified by the states of the components, the states of the links, and the state of the observation. Hence, in the worst case, the number of behavior states is still exponential with the number of components of the ACS. This is why more intelligent techniques need to be designed in order to mitigate the explosion of the reconstructed behavior.

This paper presents a novel and intelligent technique for diagnosis of a class of DES's called *complex active systems* (*CAS's*), introduced in [11, 12]. Research on complex systems is increasingly becoming important in several areas of science. However, a consensus concerning a unique definition of a complex system does not yet exist [1, 4, 5, 10, 16, 21]. Intuitively, a complex system is a group or organization which is made up of many interacting components. In a complex system, interactions between components lead to an *emergent behavior*, which is unpredictable from a knowledge of the behavior of the individual components only. Examples of emergent behaviors are climate changes, price fluctuations in markets, building by ants, and the ability of immune systems to distinguish the protected body from external agents.

Inspired by complex systems in nature and society, complexity has been injected into the modeling and diagnosis of active systems [13], a special class of discrete-event systems [3], which are modeled as networks of interacting components, where the behavior of each component is described by a communicating automaton [2]. To this end, the notion of context-sensitive diagnosis was first introduced in [14] and then extended in [15], for active systems that are organized within abstraction hierarchies, so that candidate diagnoses can be generated at different abstraction levels.

Recently, active systems have been equipped with behavior stratification [11, 12], where different networks of components coexist within a hierarchy. This resembles emergent behavior that arises from the vertical interaction of a network with superior levels based on pattern events. Pattern events occur when a network performs strings of component transitions matching patterns which are specific to the application domain.

Pattern events are similar to supervision patterns [9], which are defined by automata rather than regular expressions. However, supervision patterns are defined over DES's without any concern with either complexity or behavior stratification.

CAS's also differs from *Hierarchical Finite State Machines* (HFSM's). The notion of an HFSM was inspired by statecharts [6, 7], a visual formalism for complex systems. HFSM's resemble the idea of class inheritance in object-orientation, thereby providing specific advantages, including factorization and reuse. Diagnosis of HFSMs has been considered in [8, 17]. However, no patterns are involved, events are simple, and diagnosis is context-free.

The contribution of this paper is a novel and intelligent technique for the diagnosis of CAS's, whose time and space complexity is *linear* with the number of components, rather than *exponential* (which is the case for previous papers [11, 12]).

3 Modeling and Diagnosing Complex Active Systems

A complex active system (CAS) is a hierarchy of active systems (ACS's) [13]. In a CAS, the nature of communication is twofold: between components within the same ACS and, secondly, between different ACS's. Within the same ACS, components interact based on events coming: (1) from neighboring components (*internal events*), (2) from neighboring ACS's (*pattern events*), and (3) from outside the CAS (*external events*). Events trigger the transitions of components, which in turn generate new events, thus causing a sequence of component transitions, namely the *trajectory* of the CAS.

Communication between ACS's is achieved based on pattern events [11, 12]. A pattern event is generated in an ACS when the sequence of component transitions within the ACS matches a given pattern. Patterns are domain-dependent and are specified by regular expressions. When a pattern event occurs, it can be consumed by a component c of another ACS, thereby triggering a transition of c. This way, the behavior of an ACS is doomed to influence (although not completely determine) the behavior of another ACS. Since ACS's are connected to one another within the CAS, the occurrence of a pattern event may contribute to a cascade of pattern events in different ACS's.

A CAS C can be either *quiescent* or *reacting*. If quiescent, no event occurs and no transition is performed. C becomes reacting when an external event occurs (e.g. a lightning striking a power transmission line), which can be consumed by a component within C. When reacting, the occurrence of a component transition moves C to a new state, where a state is the composition of all the component states, the occurred (not yet consumed) events, and the states of the pattern-event recognizers.[1] We assume that, sooner or later, C becomes quiescent anew (in other words, the trajectory of C is finite).

Once a real system is modeled as a CAS C, it can be monitored and diagnosed based on the observed behavior. In this paper we focus on *a posteriori diagnosis*. That is, we consider the *observation* generated by a complete trajectory of C, which moves C from the known initial (quiescent) state C_0 to an unknown final (quiescent) state C_f.

The observation $\mathcal{O}$ of C is the tuple of the *local observations* of the ACS's within C, with each local observation being a sequence of *observable labels*. An observable label of an ACS $\mathcal{A}$ is generated by the occurrence of an *observable transition* of a component in $\mathcal{A}$. In other words, $\mathcal{O} = (\mathcal{O}_{\mathcal{A}_1}, \dots, \mathcal{O}_{\mathcal{A}_n})$, where $\forall i \in [1 \mathinner{..} n]$, $\mathcal{O}_{\mathcal{A}_i} = [\ell_1, \dots, \ell_{k_i}]$.

Thus, it is essential to know which are the observable component transitions and their associated observable labels. This is specified by a *viewer* of C, namely $\mathcal{V} = (\mathcal{V}_{\mathcal{A}_1}, \dots, \mathcal{V}_{\mathcal{A}_n})$, which is the composition of the *local viewers* of the ACS's, with each local viewer $\mathcal{V}_{\mathcal{A}_i}$, $i \in [1 \mathinner{..} n]$, being a set of pairs (t, ℓ), where t is an observable transition of a component in $\mathcal{A}_i$ and ℓ an observable label. In other words, $\mathcal{O}$ is the local projection of the trajectory of C on the involved observable component transitions.

However, $\mathcal{O}$ is not sufficient to identify the actual trajectory. Rather, $\mathcal{O}$ is consistent with several (possibly an infinite number of) *candidate trajectories* of C. For each candidate trajectory of C there is a *candidate diagnosis* of C, namely the set of *faulty transitions* in the trajectory. Similarly to a local viewer which specifies observable transitions, faulty transitions are specified by a *local ruler*, a set of pairs (t, f), where t is a component transition and f a *faulty label*. The tuple of all local rulers $\mathcal{R}_{\mathcal{A}_i}$ gives rise to the *ruler* of C, namely $\mathcal{R} = (\mathcal{R}_{\mathcal{A}_1}, \dots, \mathcal{R}_{\mathcal{A}_n})$. A *diagnosis problem* for C is a quadruple $\wp(C) = (C_0, \mathcal{V}, \mathcal{O}, \mathcal{R})$, where C_0 is the initial state of C, $\mathcal{V}$ a viewer of C, $\mathcal{O}$ the observation of C, and $\mathcal{R}$ a ruler of C.

The *solution* of $\wp(C)$, namely $\Delta(\wp(C))$, is the set of candidate diagnoses (based on ruler $\mathcal{R}$), associated with the candidate traces of C (based on C_0, viewer $\mathcal{V}$, and $\mathcal{O}$).

For efficiency reasons, it is convenient to perform some preprocessing on the CAS model before the diagnosis engine is operating. The extent of such offline preprocessing is varying and depends on the performance requirements of the application domain. In particular, in order to detect pattern events, we need to maintain the recognition states of patterns. Since patterns are described by regular expressions, specific automata-based recognizers are to be generated as follows:

1. For each pattern (p, r), a *pattern automaton* P equivalent to r is generated, with final states marked by p;
2. The set $\mathbf{P}$ of pattern automata is partitioned based on ACS and the alphabet of r;
3. For each part $\mathbb{P} = \{P_1, \dots, P_k\}$ in $\mathbf{P}$, the following actions are performed:

[1] Matching a regular expression associated with a pattern event requires maintaining the recognition state within a finite automaton, namely the recognizer.

- A nondeterministic automaton $\mathcal{N}$ is created by generating its initial state n_0 and one empty transition from n_0 to each initial state of P_i, $i \in [1 \,..\, k]$;
- In each P_i, $i \in [1 \,..\, k]$, an empty transition from each non-initial state to n_0 is inserted (this allows for pattern-matching of overlapping strings of transitions);
- $\mathcal{N}$ is determinized into $\mathcal{P}$, where each final state d is marked by the pattern event that is associated with the states in d that are final in the corresponding pattern automaton (in fact, each state d of the deterministic automaton is identified by a subset of the states of the equivalent nondeterministic automaton; besides, we assume that only one pattern event at a time can be generated);
- $\mathcal{P}$ is minimized into the *pattern space* of part $\mathbb{P}$.

Unlike naive behavior reconstruction in diagnosis of ACS's, *intelligent* behavior reconstruction in diagnosis of CAS's (proposed in this paper) avoids materializing the behavior of the CAS (the automaton whose language equals the set of CAS trajectories). Instead, reconstruction is confined to each single ACS based on the local observation *and* the interface constraints on pattern-event occurrences coming from inferior ACS's. Intuitively, the flow of reconstruction in the hierarchy of the CAS is bottom-up.

For an ACS $\mathcal{A}$ with children $\mathcal{A}_1, \ldots, \mathcal{A}_k$ in the hierarchy, the behavior of $\mathcal{A}$, namely $Bhv(\mathcal{A})$, is reconstructed based on the *interfaces* of the children, $Int(\mathcal{A}_1)$, $\ldots$, $Int(\mathcal{A}_k)$, and the local observation of $\mathcal{A}$, namely $\mathcal{O}_{\mathcal{A}}$.

The interface is derived from the behavior. Thus, for any ACS $\mathcal{A}$, both $Bhv(\mathcal{A})$ and $Int(\mathcal{A})$ are needed (with the exception of the root, for which no interface is generated).

The notions of behavior and interface depend on each other. However, such a circularity does not hold for leaf nodes of the CAS: given a leaf node $\mathcal{A}$, the behavior $Bhv(\mathcal{A})$ is reconstructed based on $\mathcal{O}_{\mathcal{A}}$ only, as no interface constraints exist for $\mathcal{A}$. On the other hand, the behavior of the root node needs to be submitted to further decoration-based processing in order to distill the set of candidate diagnoses, namely the *solution* of the diagnosis problem.

In short, four sorts of graphs are required in intelligent reconstruction: *unconstrained behavior* (for leaf nodes), *interface* (for non-root nodes), *constrained behavior* (for non-leaf nodes), and *decorated behavior* (for root node).

A diagnosis δ is a (possibly empty) set of fault labels. In what follows, we make use of the *join* operator between two diagnosis sets Δ_1 and Δ_2, defined as follows:

$$\Delta_1 \bowtie \Delta_2 = \{\delta' \mid \delta' = \delta_1 \cup \delta_2, \delta_1 \in \Delta_1, \delta_2 \in \Delta_2\}. \tag{1}$$

The interface of a behavior is generated as follows:

1. The identifier of a component transition t marking an arc of the behavior and associated with a pattern event is either replaced by the singleton $\{\emptyset\}$, if t is normal, or by the singleton $\{\{f\}\}$, if t is faulty, with f being the fault label associated with t in the ruler.

2. Interpreting transition not associated with pattern events as ϵ-transitions, the obtained nondeterministic automaton (NFA) is determinized so that each state of the resulting deterministic automaton (DFA) contains the ϵ-closure in all its structure (rather than the subset of NFA states only).
3. Within each state d of the DFA, each NFA state n is marked by the diagnosis set generated by all paths starting at the root state in d and ending at n, while identifiers of component transitions are eventually removed.
4. Let $\mathbf{p}$ be the pattern event marking a transition t exiting a state d in the DFA, $\Delta_{\mathbf{p}}$ the diagnosis set associated with $\mathbf{p}$ in step 1, and Δ the diagnosis set associated with the NFA state in d from which t is derived in the determinization process. $\Delta_{\mathbf{p}}$ is replaced by $\Delta \bowtie \Delta_{\mathbf{p}}$, thereby combining the diagnoses in Δ with those in $\Delta_{\mathbf{p}}$.

The behavior of the root node is then decorated by diagnosis sets associated with states, in a way similar to step 3 in marking NFA states within interface states. Eventually, the solution of the diagnosis problem is generated by the union of the diagnosis sets associated with the final states of the decorated behavior of the root node.

4 Reference Application Domain: Power Transmission Networks

In this section we outline the diagnosis technique introduced in Sect. 3 based on a reference example within the domain of power transmission networks.

4.1 Modeling

Displayed on the left of Fig. 1 is a power transmission line. Each side of the line is protected from short circuits by two *breakers*, namely b and r on the left, and b' and r' on the right. Both b and b' are connected to a *sensor* of voltage.

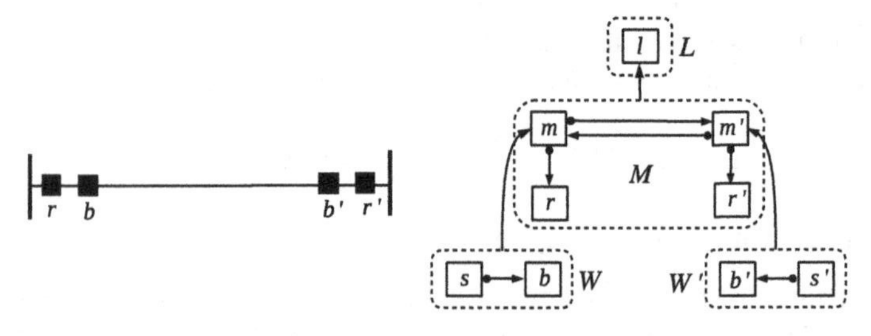

Fig. 1 Power transmission line (*left*) and corresponding CAS Ł (*right*)

If a short circuit (for instance, a lightning) strikes the line, then each sensor will detect the lowering of the voltage and command the associated breaker to open. If both breakers open, then the line will be isolated, thereby causing the short circuit to vanish. If so, the two breakers are commanded to close in order to restore the line.

Still, faulty behavior may occur: either the sensor does not command the breaker to open or the breaker does not open. Such misbehavior is detected by a *monitor* (one for each side of the line). For example, similarly to the sensor, the monitor on the left commands the recovery breaker r to open. In doing so, it also informs the monitor on the right to perform the same action on recovery breaker r'. For safety reasons, once opened, recovery breakers cannot be closed again, thereby leaving the line isolated.

The protected line can be modeled as the CAS outlined on the right of Fig. 1, called Ł, which is composed of four ACS's, namely: W (the protection hardware on the left, including sensor s and breaker b), W' (the protection hardware on the right, including sensor s' and breaker b'), M (the monitoring apparatus, including monitors m and m', and recovery breakers r and r'), and L (including line l).

Arrows within ACS's denote links between components. For instance, W includes a link from s to b, meaning that an event generated by s can be consumed by b. For the sake of simplicity, we assume that, when an event is already present in a link, no transition generating a new event on the same link can be triggered. Links between m and m' allow monitors to communicate to one another. Instead, arrows between ACS's denote links aimed at conveying pattern events. For instance, the link from W to M makes pattern events (occurring in W) available to m in M.

Models of components within Ł are displayed in Fig. 2, namely *sensor*, *breaker*, *monitor*, and *line*. The *topological model* (each one being displayed on top of Fig. 2) consists of the set of *input terminals* and the set of *output terminals*. For instance, *monitor* involves two input terminals, E and I, and two output terminals, O and R. Terminal E is entered by the link exiting the protection hardware (either W or W'), conveying the pattern events occurring in the latter. R is linked to the recovery breaker, while O and I are linked with the other monitor. Displayed under each topological model is a finite automaton specifying the *behavioral model*. In general, each transition is triggered by an *input event* and generates a set of *output events*. Transitions are detailed below (where event e at terminal t is written $e(t)$, while pattern events are in bold):

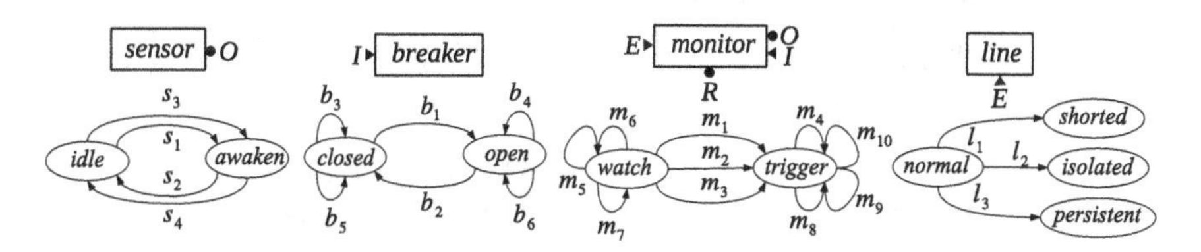

Fig. 2 Component models

- (*Sensor*) s_1 detects low voltage and outputs *op*(*O*); s_2 detects normal voltage and outputs $cl(O)$; s_3 detects low voltage, yet outputs $cl(O)$; s_4 detects normal voltage, yet outputs *op*(*O*).
- (*Breaker*) b_1 consumes *op*(*I*) and opens; b_2 consumes $cl(I)$ and closes; b_3 consumes *op*(*I*), yet keeps being closed; b_4 consumes $cl(I)$, yet keeps being closed; b_5 consumes $cl(I)$; b_6 consumes *op*(*I*).
- (*Monitor*) m_1 consumes **nd**(*E*) and outputs *op*(*R*) and $rc(O)$; m_2 consumes **nd**(*E*) and outputs *op*(*R*) only; m_3 consumes $rc(I)$ and outputs *op*(*R*); m_4 consumes **nd**(*E*); m_5 consumes **di**(*E*); m_6 consumes **co**(*E*); m_7 consumes **nc**(*E*); m_8 consumes **di**(*E*); m_9 consumes **co**(*E*). m_{10} consumes **nc**(*E*).
- (*Line*) l_1: consumes **ni**(*E*); l_2: consumes **nr**(*E*); l_3: consumes **ps**(*E*).

Pattern events have the following meaning. **di**: the protection hardware disconnects the side of the line; **co**: the protection hardware connects the side of the line; **nd**: the protection hardware fails to disconnect the side of the line; **nd**: the protection hardware fails to connect the side of the line; **nr**: the side of the line cannot be reconnected; **ni**: the side of the line cannot be isolated; **ps**: the short circuit persists.

A pattern event p for an ACS $\mathcal{A}$ is specified by a regular expression r on transitions of components in $\mathcal{A}$. The pair (p, r) is a *pattern*. The alphabet of r is a (not necessarily strict) superset of the set of transitions involved in r. We assume the classical operators for regular expressions, namely concatenation, disjunction, optionality, and repetition. Depending on the application domain, other more specific operators can be involved. Patterns for W, W', and L are listed in Table 1. For example, pattern event **nd** occurs either when $s_3(s)$ (the sensor fails to open the breaker) or $s_1(s)\, b_3(b)$ (the sensor commands the breaker to open, yet the breaker fails to open).

For each pattern (p, r) in Table 1, the alphabet of r is defined as follows. For W and W', the alphabet of r is the whole set of transitions of the involved components

Table 1 Specification of patterns by regular expressions

Active system	Pattern event	Regular expression
W	**di**	$b_1(b)$
	co	$b_2(b)$
	nd	$s_3(s) \mid s_1(s)\, b_3(b)$
	nc	$s_4(s) \mid s_2(s)\, b_4(b)$
W'	**di′**	$b_1(b')$
	co′	$b_2(b')$
	nd′	$s_3(s') \mid s_1(s')\, b_3(b')$
	nc′	$s_4(s') \mid s_2(s')\, b_4(b')$
M	**nr**	$m_7(m) \mid m_{10}(m) \mid b_1(r) \mid m_7(m) \mid m_{10}(m') \mid b_1(r')$
	ni	$(m_1(m) \mid m_2(m) \mid m_4(m))\, b_3(r) \mid b_3(r)\, m_4(m) \mid$ $(m_1(m') \mid m_2(m') \mid m_4(m'))\, b_3(r') \mid b_3(r')\, m_4(m')$
	ps	$(m_6(m)\, m_6(m') \mid m_6(m')\, m_6(m))(m_5(m)\, m_5(m') \mid m_5(m')\, m_5(m))$

Table 2 Tabular specification of pattern spaces $\mathcal{P}_W$ (left) and $\mathcal{P}_{\mathbf{ni}}$ (right)

Pattern space $\mathcal{P}_W$										Pattern space $\mathcal{P}_{\mathbf{ni}}$									
	b_1	b_2	b_3	b_4	s_1	s_2	s_3	s_4			$m_1(m)$	$m_2(m)$	$m_4(m)$	$b_3(r)$	$m_1(m')$	$m_2(m')$	$m_4(m')$	$b_3(r')$	
0	1	2			3	4	5	6		0	1	1	1	2	3	3	3	4	
1	1	2			3	4	5	6	***di***	1	1	1	1	5	3	3	3	6	
2	1	2			3	4	5	6	***co***	2	1	1	7	2	3	3	3	4	
3	1	2	5		3	4	5	6		3	1	1	1	2	3	3	3	6	
4	1	2		6	3	4	5	6		4	1	1	1	2	3	3	8	4	
5	1	2			3	4	5	6	***nd***	*5*	1	1	7	2	3	3	3	4	***ni***
6	1	2			3	4	5	6	***nc***	*6*	1	1	1	2	3	3	8	4	***ni***
										7	1	1	1	5	3	3	3	6	***ni***
										8	1	1	1	2	3	3	3	6	***ni***

(breaker and sensor). For M, the alphabet equals the set of transitions involved in r only. Knowing the alphabet of r is essential for the matching of the regular expression against the trajectory of the CAS.

4.2 Diagnosis Problem

We define the diagnosis problem for CAS Ł as $\wp(\text{Ł}) = (\text{Ł}_0, \mathcal{V}, \mathcal{O}, \mathcal{R})$, where:

- In Ł_0, breakers are *closed*, sensors are *idle*, and monitors are *watch* (see Fig. 2);
- $\mathcal{V} = (\mathcal{V}_W, \mathcal{V}_{W'}, \mathcal{V}_M)$, where ($L$ unobservable): $\mathcal{V}_W = \{(b_1(b), opb), (b_2(b), clb), (b_5(b), alr), (b_6(b), alr), (s_1(s), awk), (s_2(s), ide), (s_3(s), awk), (s_4(s), ide)\}$, $\mathcal{V}_{W'} = \{(b_1(b'), opb'), (b_2(b'), clb'), (b_5(b'), alr'), (b_6(b'), alr'), (s_1(s'), awk'), (s_2(s'), ide'), (s_3(s'), awk'), (s_4(s'), ide')\}$, and $\mathcal{V}_M = \{(m_1(m), trg), (m_2(m), trg), (m_3(m), trg), (b_1(r), opr), (b_2(r), clr), (m_1(m'), trg'), (m_2(m'), trg'), (m_3(m'), trg'), (b_1(r'), opr'), (b_2(r'), clr')\}$;
- $\mathcal{O} = (\mathcal{O}_W, \mathcal{O}_{W'}, \mathcal{O}_M)$, $\mathcal{O}_W = [\, awk \,]$, $\mathcal{O}_{W'} = [\, awk', opb' \,]$, $\mathcal{O}_M = [\, trg, opr \,]$;
- $\mathcal{R} = (\mathcal{R}_W, \mathcal{R}_{W'}, \mathcal{R}_M, \mathcal{R}_L)$, where $\mathcal{R}_W = \{(b_3(b), fob), (b_4(b), fcb), (s_3(s), fos), (s_4(s), fcs)\}$, $\mathcal{R}_{W'} = \{(b_3(b'), fob'), (b_4(b'), fcb'), (s_3(s'), fos'), (s_4(s'), fcs')\}$, $\mathcal{R}_M = \{(m_2(m), fm), (m_2(m'), fm'), (b_3(r), for), (b_4(r), fcr), (b_3(r'), for'), (b_4(r'), fcr')\}$, $\mathcal{R}_L = \{(l_1(l), fls), (l_2(l), fli), (l_3(l), flp)\}$.

Specified in Table 2 are pattern spaces $\mathcal{P}_W$ (left), for pattern events **di**, **co**, **nd**, and **nc**, and $\mathcal{P}_{\mathbf{ni}}$ (right), for pattern event **ni**. Listed in first column are the states (0 is the initial state), with final states being shaded. For each pair state-transition, the next state is specified. Listed in the last column are the pattern events associated with final states. Since regular expressions of pattern events for M (**nr**, **ni**, and **ps**) are defined on different alphabets, two additional pattern spaces are generated, namely $\mathcal{P}_{\mathbf{nr}}$ and $\mathcal{P}_{\mathbf{ps}}$.

4.3 Reconstruction

With reference to CAS Ł in Fig. 1 and the diagnosis problem $\wp(\text{Ł}) = (\text{Ł}_0, \mathcal{V}, \mathcal{O}, \mathcal{R})$ defined above, we first need to generate the unconstrained behavior of $\mathcal{W}$ and $\mathcal{W}'$ based on local observations $\mathcal{O}_W = [\, awk \,]$ and $\mathcal{O}_{W'}[\, awk', opb' \,]$, respectively. Displayed in Fig. 3 are $Bhv(W)$ (left) and $Bhv(W')$ (right). For instance, consider the generation of $Bhv(\mathcal{W}')$. As detailed in the right of Table 3, each state is identified by four fields: the state of sensor s', the state of breaker b', the event (if any) ready at terminal $I(b')$, the state of pattern space $\mathcal{P}_{W'}$, and the index of local observation $\mathcal{O}_{W'}$ (from 0 to 2).

The generation of the behavior starts at the initial state 0 and progressively materializes the transition function by applying triggerable transitions to each state created

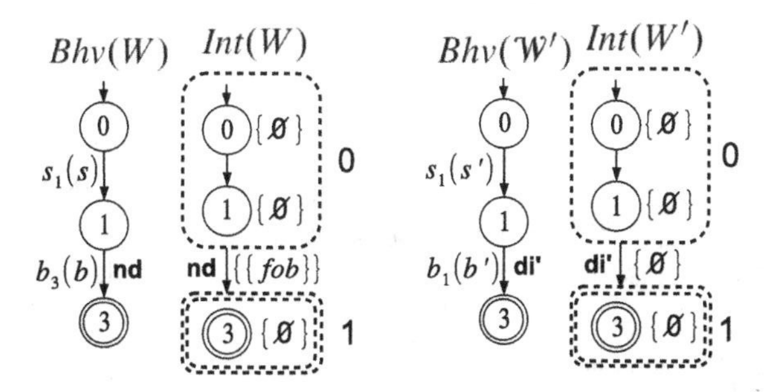

Fig. 3 Unconstrained behaviors $Bhv(W)$ and $Bhv(W')$, along with interfaces $Int(W)$ and $Int(W')$

Table 3 Details on states for behaviors $Bhv(W)$ and $Bhv(W')$

Behavior $Bhv(W)$						Behavior $Bhv(W')$					
State	s	b	$I(b)$	$\mathcal{P}_W$	$\mathcal{O}_W$	State	s'	b'	$I(b')$	$\mathcal{P}_{W'}$	$\mathcal{O}_{W'}$
0	Idle	Closed		0	0	0	Idle	Closed		0	0
1	Awaken	Closed	*op*	3	1	1	Awaken	Closed	*op*	3	1
3	*Awaken*	*Closed*		*5*	*1*	*3*	*Awaken*	*open*		*1*	*2*

so far. However, a part of the reconstructed behavior may be inconsistent. A state (or a transition) is inconsistent when it is not included in any path from initial state 0 to the final state 3. The state is final when all events are consumed and the observation index is complete (equal to the length of the local observation). The transition from 1 to 3 is marked by **di′** pattern event as the state of $\mathcal{P}_{W'}$ becomes final in 3 (left of Table 2).

Shown on the right of each behavior in Fig. 3 are interfaces $Int(W)$ and $Int(W')$. Displayed on the left of Fig. 4 is the constrained behavior $Bhv(M)$. Each state of $Bhv(M)$ includes two sorts of additional information: the pattern events ready (if any) at terminals E, and the pair (I, I') of interfaces states relevant to interfaces $Int(W)$ and $Int(W')$, respectively. A final state needs the additional condition that both states in pair (I, I') be final in the respective interface. When reconstructing a transition t triggered by a pattern event, the latter is required to mark a transition exiting the corresponding state in the interface, otherwise the transition cannot be reconstructed (interface constraints are not met).

The derivation of interface $Int(M)$ shown in Fig. 4 exhibits two peculiarities: step 2 creates a DFA state resulting from two NFA transitions, exiting states 6 and 8 respectively, both marked by pair $(b_1(r), \mathbf{nr})$, and step 3 shall account for diagnosis sets associated with pattern events. The peculiarity in step 3 is essential to the sound and complete generation of candidate diagnoses, as the diagnosis carried by a pattern event contains the diagnosis information relevant to the child ACS (the protection hardware), which is combined with the diagnosis information relevant to the current ACS (the monitoring apparatus).

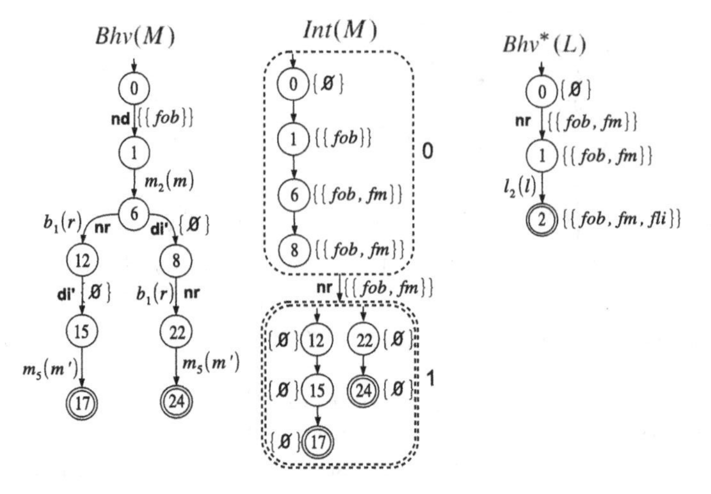

Fig. 4 Constrained behavior $Bhv(M)$, interface $Int(M)$, and decorated behavior $Bhv^*(L)$

4.4 Decoration and Distillation

Outlined on the right of Fig. 4 is the decorated behavior $Bhv^*(L)$. Starting from the singleton $\{\emptyset\}$ marking state 0, the candidate set associated with state 1 is $\Delta(1) = \{\emptyset\} \bowtie \{\{fob, fm\}\} = \{\{fob, fm\}\}$. Since $l_2(l)$ is faulty (associated with label fli), eventually we have $\Delta(2) = \Delta(1) \bowtie \{\{fli\}\} = \{\{fob, fm, fli\}\}$. Hence, the solution of the diagnosis problem $\wp(Ł)$, distilled form the decorated behavior, consists of a single diagnosis involving three faults: breaker b fails to open (fob), monitor m fails to communicate with monitor m' (fm), and line l is isolated (fli).

4.5 Implementation and Results

Intelligent diagnosis of CAS's has been implemented in C++, under Linux Ubuntu 15.10, on a notebook with 4 GB of RAM. The diagnosis framework includes a compiler, for offline preprocessing, and an online diagnosis engine. A specification language was designed, which describes CAS's in terms of architecture, component models, and pattern events, as well as diagnosis problems in terms of viewer, observation, and ruler.

The diagnosis engine can run in either naive or intelligent mode, thereby making it possible to compare the performances of the two techniques. Results of comparison experiments are shown in Fig. 5. The two graphs compare naive vs. intelligent diagnosis, both in processing time (left) and memory allocation (right). The horizontal axis indicates CAS's of increasing size, up to 20 components. The vertical axis indicates, in *logarithmic scale*, either the CPU time (in milliseconds) or the RAM

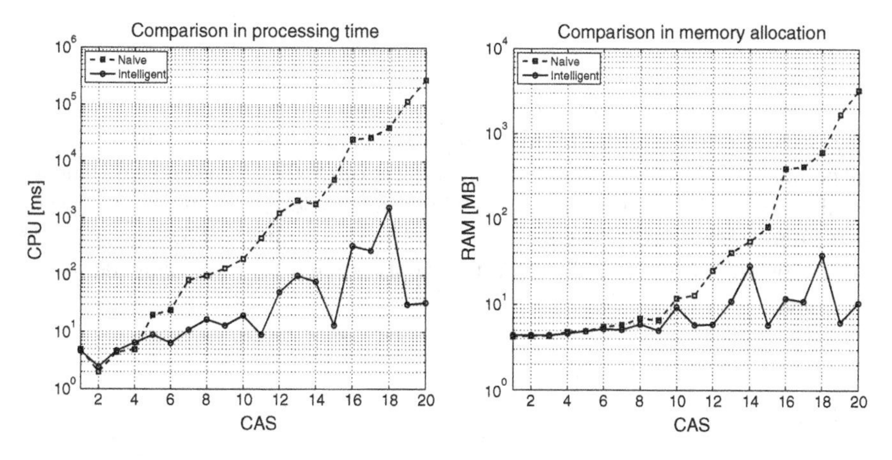

Fig. 5 Comparison in processing time (*left*) and memory allocation (*right*)

allocation (in MB). For instance, for the last CAS, time in naive diagnosis is 270 s, while time in intelligent diagnosis is 0.032 s. For the same experiment, RAM allocation is 3300 MB in naive diagnosis, while it is 10 MB in intelligent diagnosis. In order to analyze the experimental complexity of intelligent diagnosis only, another set of experiments was done, as shown in Fig. 6 (in *normal* scale). Nine CAS's have been considered, from 9 to 640 components. The number of ACS's in each CAS ranges from 4 to 256. Both CPU time and RAM allocation are *linear* (rather than exponential) with the number of components.

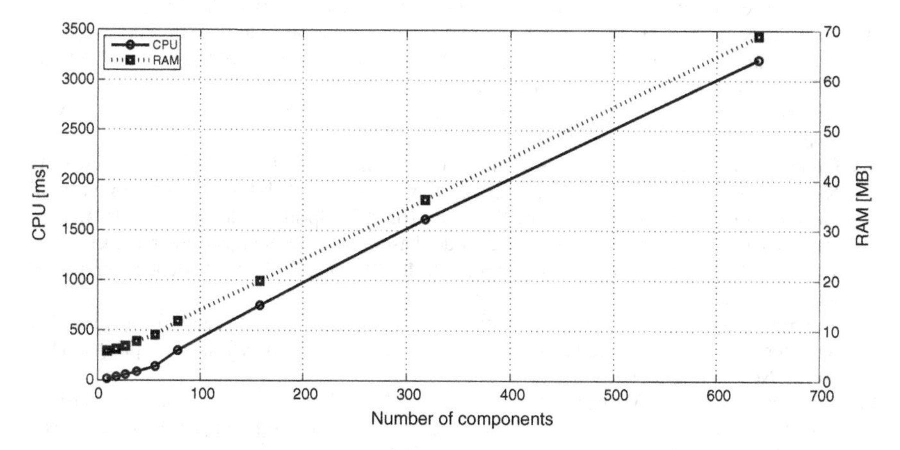

Fig. 6 Experimental complexity for intelligent diagnosis

5 Conclusion

CAS's offer a formalism to capture complexity and abstraction. Most importantly, they can be diagnosed efficiently by intelligent techniques, which avoid the exponential complexity of model-based reasoning typical of DES's. If you choose to model a real system as a DES, it is much more convenient to design the model in terms of interacting modules (ACS's) rather than as a large, monolithic DES. This results in a CAS, for which intelligent diagnosis is not only possible, but also computationally convenient.

References

1. Bossomaier, T., Green, D.: Complex Systems. Cambridge University Press (2007)
2. Brand, D., Zafiropulo, P.: On communicating finite-state machines. J. ACM **30**(2), 323–342 (1983)
3. Cassandras, C., Lafortune, S.: Introduction to Discrete Event Systems, The Kluwer International Series in Discrete Event Dynamic Systems, vol. 11. Kluwer Academic Publishers, Boston (1999)
4. Chu, D.: Complexity: against systems. Theor. Biosci. **130**(3), 229–245 (2011)
5. Goles, E., Martinez, S.: Complex Systems. Springer (2014)
6. Harel, D.: Statecharts: a visual formalism for complex systems. Sci. Comput. Prog. **8**(3), 231–274 (1987)
7. Harel, D., Lachover, H., Naamad, A., Pnueli, A., Politi, M., Sherman, R., Shtull-Trauring, A., Trakhtenbrot, M.: STATEMATE: a working environment for the development of complex reactive systems. IEEE Trans. Softw. Eng. **16**(4), 403–414 (1990)
8. Idghamishi, A., Zad, S.: Fault diagnosis in hierarchical discrete-event systems. In: 43rd IEEE Conference on Decision and Control, pp. 63–68. Paradise Island, Bahamas (2004)
9. Jéron, T., Marchand, H., Pinchinat, S., Cordier, M.: Supervision patterns in discrete event systems diagnosis. In: Seventeenth International Workshop on Principles of Diagnosis DX'06, pp. 117–124. Peñaranda de Duero, Spain (2006)
10. Kaneko, K., Tsuda, I.: Complex Systems: Chaos and Beyond: a Constructive Approach with Applications in Life. Springer (2013)
11. Lamperti, G., Zhao, X.: Diagnosis of higher-order discrete-event systems. In: Cuzzocrea, A., Kittl, C., Simos, D., Weippl, E., Xu, L. (eds.) Availability, Reliability, and Security in Information Systems and HCI, LNCS, vol. 8127, pp. 162–177. Springer, Heidelberg (2013)
12. Lamperti, G., Zhao, X.: Specification and model-based diagnosis of higher-order discrete-event systems. IEEE Int. Conf. Syst. Man Cybern.—SMC 2013, pp. 2342–2347. Manchester, United Kingdom (2013)
13. Lamperti, G., Zanella, M.: Diagnosis of Active Systems—Principles and Techniques, The Springer International Series in Engineering and Computer Science, vol. 741. Springer, Dordrecht, Netherlands (2003)
14. Lamperti, G., Zanella, M.: Context-sensitive diagnosis of discrete-event systems. In: Walsh, T. (ed.) Twenty-Second International Joint Conference on Artificial Intelligence IJCAI'11, vol. 2, pp. 969–975. AAAI Press, Barcelona, Spain (2011)
15. Lamperti, G., Zhao, X.: Diagnosis of active systems by semantic patterns. IEEE Trans. Syst. Man Cybern.: Syst. **44**(8), 1028–1043 (2014)
16. Licata, I., Sakaji, A.: Physics of emergence and organization. World Scientific (2008)
17. Paoli, A., Lafortune, S.: Diagnosability analysis of a class of hierarchical state machines. J. Discrete Event Dyn. Syst.: Theor. Appl. **18**(3), 385–413 (2008)

18. Report of the Enquiry Committee on Grif Disturbance in Northern Region on 30th July 2012 and in Northern, Eastern and North-Eastern Region on 31st July 2012 (2012). http://www.powermin.nic.in/pdf/GRID_ENQ_REP_16_8_12.pdf
19. Sampath, M., Sengupta, R., Lafortune, S., Sinnamohideen, K., Teneketzis, D.: Diagnosability of discrete-event systems. IEEE Trans. Autom. Control **40**(9), 1555–1575 (1995)
20. Sampath, M., Lafortune, S., Teneketzis, D.: Active diagnosis of discrete-event systems. IEEE Trans. Autom. Control **43**(7), 908–929 (1998)
21. Sibani, P., Jensen, H.: Stochastic Dynamics of Complex Systems, Complexity Science, vol. 2. World Scientific (2013)

Anticipation Based on a Bi-Level Bi-Objective Modeling for the Decision-Making in the Car-Following Behavior

Anouer Bennajeh, Fahem Kebair, Lamjed Ben Said and Samir Aknine

Abstract Safety distance models are in a continuous state of improvement due to their important role in the micro-simulation of traffic, intelligent transport systems and safety engineering models. Indeed, many existing models of car-following behavior do not explicitly consider a real link between the increase of the speed of movement and the guarantee of the road safety. In this context, this paper presents a decision-making model for determining velocity and safety distance values basing-on a bi-level bi-objective modeling and that allows simulation parameters anticipation. In fact, our model addresses drivers that circulate in urban zones with normative behaviors. The first objective of the model is to allow agent drivers to have a smooth transition between acceleration and deceleration behaviors according to the leading vehicle actions. Simultaneously, the model intends, as a second objective, to reduce the circulation time by increasing the speed of movement. Agent technology and Tabu search algorithm are used respectively to model drivers and to find the best solution during decision-making. The paper provides first a theoretical background of the research. Then it describes the agent driver decision-making model and the resolution algorithm. Finally it presents and discusses a first simulation and experimentations.

A. Bennajeh (✉) · F. Kebair · L.B. Said
SOIE, Institut Supérieur de Gestion de Tunis - ISGT, Université de Tunis, 41, Avenue de la Liberté, Cité Bouchoucha, 2000 Bardo – Tunis, Tunisie
e-mail: anouer.bennajeh@gmail.com

F. Kebair
e-mail: kebairf@gmail.com

L.B. Said
e-mail: bensaid_lamjed@yahoo.fr

S. Aknine
LIRIS - Université Claude Bernard Lyon 1 - UCBL, 43, Bd du 11 novembre 1918, 69622 Villeurbanne Cedex, France
e-mail: samir.aknine@univ-lyon1.fr

I. Czarnowski et al. (eds.), *Intelligent Decision Technologies 2016*, Smart Innovation, Systems and Technologies 56,
DOI 10.1007/978-3-319-39630-9_19

Keywords Car-following behavior · Safe distance model · Bi-objectives modeling · Bi-level modeling · Anticipation · Making decision · Software agent · Tabu search algorithm

1 Introduction

The car-following driving behavior is an important component in the microscopic traffic simulation, enabling transport engineers to reproduce the dynamic behavior of small discrete intervals. In fact, the car-following microscopic driving behavior describes the longitudinal interactions between the following and the leading vehicle in the same way, where vehicles are described by vectors of state variables (X, V, A), which represent respectively: the spatial location, speed, and acceleration. According to [4], the car-following models can be classified into five categories, which are, the GHR (Gazis-Herman-Rothery) models, safe distance models, linear models, psychophysical models and fuzzy logic based models. Each model is composed of different rules and equations that update these state variables over time. In this paper, we are interested in the safe distance models.

The safe distance model is developed for the first time by Kometani and Sasaki [9]. In fact, this model is based on the calculation of the security distance during the car-following behavior, which uses the equations of physical movement. The principal objective of this model is to avoid the collision between the following and the leading vehicle. One of the widely used safe distance model is [6], which combines a free-flow driving model with a stopping-distance. But, despite the important advantages of this model, where it has been implemented widely in micro-simulation software packages such as SISTM [12], AIMSUN [3] and DRACULA [10], it has a disadvantage resulting in its very strict restriction on the car-following behavior, where, the following vehicle can move only when it has exactly a calculated safety distance with the leading vehicle. Figure 1 illustrates the respect of this strict restriction by seven vehicles.

According to Fig. 1, it is obvious that this restriction is not consistent with the real traffic condition, where it is impossible to have this harmony of velocity between seven vehicles because of this strict restriction of security distance.

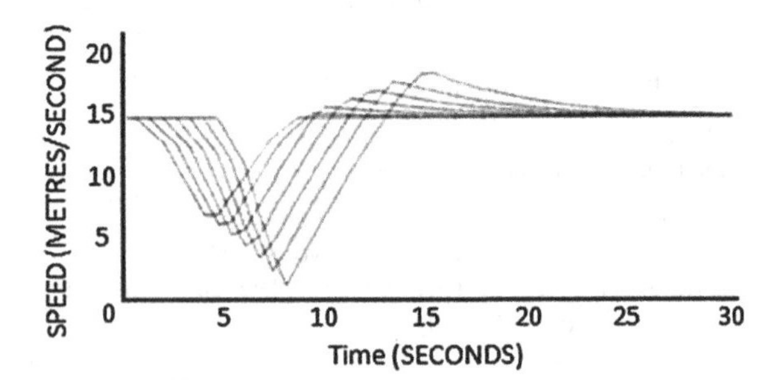

Fig. 1 Speed-time plots for seven successive vehicles [6]

In recent years, there is more research on the safety distance of car-following behavior in order to improve the traditional work [6]; as the research work [11] that proposes a new safety distance model basing on various speed relationships between vehicles in order to improve the calculation of the safety distance for each simulation step. Furthermore, the work of [14] proposed a safe distance model by formulating the safety distance as a function of the velocity difference and safety distance. But most of the research works [6, 11, 14] have the following problems. Firstly, all these research works did not treat the driver behaviors (normative or non normative) during their modeling, which influences considerably on the reality of the simulation, since decision-making differs in these two types of behaviors. Secondly, according to [5], the calculation of the safety distance is based on many factors, which are: the reaction time, the decision time, the action time, the weight of the vehicle, the speed of gravity, the air density, the projection area, the air resistance factor, the efficiency of braking, the friction coefficient, the decay factor and the slope of the road. However, excepting the reaction time, all of the above mentioned factors are not used during the calculation of the safety distance in the research works of [6, 11, 14]. Finally, these research works used a fixed reaction time, which is far from the reality, because the duration of the reaction time is influenced by the driver behaviors. For example, the duration of reaction time for a common driver is 0.6–0.9 s [5].

The simulation is a means to observe and to understand the phenomena that can appear by applying the new transportation strategies in a virtual environment, while reducing the costs of experiments. However, not taking into account, during the simulation, the real behaviors of the road users is against the rules of the Highway Code, which influence and distorts the reality of the simulation. Thus, the overall objective of our research is to integrate the driver behaviors in a computing model which led to the construction of a realistic road traffic simulation. In particular, we are interested in this paper to model drivers with normative behaviors for the car-following behavior. Indeed, this paper intends to construct a new decision-making model to manage the decision-making of the driver, by adopting an approach based on the anticipation concept to estimate the simulation parameters for a next simulation step.

Modeling and implementation to the car-following microscopic driving behavior requires a technology that guarantees the autonomy, the reactivity, the adaptability and the interaction. The software agent technology [13] meets perfectly these criteria and is positioned as an appropriate solution to model the simulation of road traffic with a microscopic approach. Thus, we used this technology to model "driver agents" and consequently to deal and to evaluate their decision making for the velocity and the safety distance that they make.

Basing-on the theoretical background of our research, we present in the next section our bi-level bi-objective modeling with more details. Then, in the following section we present the resolution of our modeling basing on the Tabu search algorithm. Next, in the fourth section we present the simulation results of our model. Finally, we conclude by discussing the first obtained results of our approach and future works.

2 Bi-Level Bi-Objective Modeling

The driver objectives are different according to their behaviors. Indeed, each driver translates its objectives through actions that help him to achieve them. In general, the actions of drivers are categorized under two objectives. The first one is the reduction of the circulation time and the second one is the road safety. To achieve the first objective, the driver increases the velocity of its vehicle, but this action combined with the car-following behavior can reduce the safety distance between the following and the leading vehicle, which influence on the second objective, especially when the leading vehicle reduces its velocity suddenly. However, according to traffic rule, if there is an increase in the velocity, there is an increase in the safety distance, which is not the case with the car-following behavior.

In this context, the two objectives of our following agent driver are totally opposed to the car-following behavior. Moreover, they have an important number of acceleration and deceleration values that we have to test them during simulation, in order to find the best velocity that address the two opposed objectives. Thus, we used a bi-objective modeling to treat this problem while focusing on a normative driver behavior.

Each objective is presented as an objective function, which are: the maximization of the traffic speed and the maximization of the safety distance. At the same time, the following driver agent tries to reduce the circulation time by increasing the velocity of its vehicle; and simultaneously, it tries to take into account the road safety by keeping a safety distance with the leading vehicle. Thus, the objective functions of each following driver agent reflect its choice for the velocity and the safety distance that respond to its need for each simulation step. For the first objective, which is the maximization of the safety distance, we defined the following function (1) for the next simulation step T + 1.

$$\begin{aligned} \text{Max}\, D_{sec\_x}(T+1) &= (T_R + T_D + T_A)*V_x(T+1) + (W/(2*G*\rho*A_f*Cd)) \\ &*\ln\left(1 + \left((\rho*A_f*Cd)/2*(V_x(T+1))^2\right)/((\eta*\mu*W) + (f_r*W*\cos\theta) + (W*\sin\theta))\right) \end{aligned} \quad (1)$$

where, T_R is the reaction time, T_D is the decision time, T_A is the action time, W is the weight of the vehicle, G is the speed of gravity, ρ is the density of air, A_f is the projection area, Cd is the air resistance factor, η is the efficiency of braking, μ is the friction coefficient, fr is the decay factor and θ is the slope of the road. The calculation of the safety distance is based on a combination between the secure stopping distance $(W/(2 * G * \rho * A_f * Cd)) * \ln(1 + ((\rho * A_f * Cd)/2 * (V_x(T + 1))2)/((\eta * \mu * W) + (fr * W * \cos\theta) + (W * \sin\theta)))$ and the distance of reaction $(T_R + T_D + T_A) * V_x(T + 1)$. This objective function based on the safety distance defined by Chen and Wang [5].

The first objective function contains a single variable "$V_x(T + 1)$" that represents the velocity, which will be adopted during the next simulation step T + 1.

In fact, this variable influence directly at the safety distances value, by playing the role of a decision variable in this function.

Concerning the second objective, which is the maximization of the traffic speed for each simulation step, we defined the following function (2), which is based on a single decision variable "a" that presents the acceleration during the increasing speed and the deceleration in the opposite direction.

$$\text{Max}\, V_x(T+1) = V_x(T) + at + m \tag{2}$$

where, "a" is the acceleration or deceleration value, "t" is the duration of the simulation step and "m" is the precision margin that influence on the precision of the velocity of the following vehicle since it is impossible to adopt exactly the calculated velocity by a real driver.

$V_x(T+1)$ is designed as an objective function in Eq. (2), but at the same time it plays the role of a decision variable for the first function (1). Therefore, the modeling of our problematic will be a bi-objective bi-level modeling. Beginning by defining the constraints of lower objective function of our bi-level modeling (the maximization to the traffic speed), the increasing of the velocity relative with the drivers' behaviors, where, according to [1] the drivers with normative behaviors interest firstly to the traffic rules in order to ensure the road security. Thus, we defined eight constraints that ensure the security with this objective function. In fact, these constraints appear depending two simulation parameters, which are, the calculated safety distance "D_{sec_x}" and the real distance "D_{xy}", which presents the real distance between the following vehicle X and the leading vehicle Y. Consequently, there are three possible states that the following agent driver must take into account, which are, "$D_{sec\_x} > D_{xy}$", "$D_{sec\_x} < D_{xy}$" and "$D_{sec\_x} = D_{xy}$".

Starting by the state "$D_{sec\_x} > D_{xy}$", the following vehicle X is in an unsecure situation because the safety distance calculated during the simulation step T is reduced under the pressure of certain factors until it enters in the red zone. Thus, the driver agent of the following vehicle X must avoid colliding with the leading vehicle Y, by ensuring that it does not go reduce more its safety distance during the next simulation step (T + 1). Thus, it reduces the velocity of its vehicle in order that will be lower than the velocity of the leading vehicle Y. The modeling of this constraint translates according to the speed relationships between vehicles. The inequality (3) presents this modeling.

$$a < \left(V_y(T) - V_x(T)\right)/t \tag{3}$$

According to the inequality (3), the following driver agent X decelerates to avoid the collision with the leading vehicle Y. But, in order to present a realistic simulation, the acceleration and deceleration values must be realistic. Therefore, the field to select the variable "a" is based on the work [8]. Thus, in our deceleration state, the value of the decision variable "a" should be between d_{min} and d_{max}.

$$d_{min} <= a <= d_{max} \quad (4)$$

Moving to the state "$D_{sec\_x} < D_{xy}$", the following vehicle X is far to the leading vehicle Y during the simulation step T. In this state, the following driver agent X has two scenarios. First, it can maintain its velocity for the next simulation step T + 1 in order to guarantee its road safety objective. Secondly, it can increase its velocity in order to ensure its objective to reduce the circulation time during the next simulation step T + 1. The modeling of this constraint is expressed by the inequality (5).

$$a >= \left(V_y(T) - V_x(T)\right)/t \quad (5)$$

According to the inequality (5), the following driver agent X may exceed the velocity of the leading vehicle Y, but at the same time, we must guarantee that this increase will take place under the road safety objective. The translation of this warranty is expressed by the velocity relationships between vehicles and the real distance $D_{xy}(T)$ of the simulation step T. This constraint is expressed by the inequality (6).

$$a < \left(D_{xy}(T)/t^2\right) + \left(V_y(T)/t\right) - \left(V_x(T)/t\right) \quad (6)$$

Basing on the realism of our modeling during the simulation and basing on our first objective (the road security), this constraint (inequality (6)) not considered when it give the right to the following driver agent to choose an acceleration value "a" superior to $a_{max,}$ where a_{max} presents the acceleration must not be exceeded to ensure the reality of our simulation.

Our modeling is based on drivers with normative behaviors. According to [1], this type of driver respects the traffic rules first. Hence, even with our objective "the maximization of the traffic speed", the translation of this objective should be achieved by actions that respect the traffic rules. In this context, our modeling should ensure the respect of the maximum speed rule of the traffic zone. This constraint is expressed by the inequality (7).

$$a <= (V_{max} - V_x(T))/t \quad (7)$$

During the acceleration state when the following agent driver chooses to increase the velocity of its vehicle, the value of the decision variable "a" should be between a_{min} and a_{max}, where the field to select the variable "a" is based on [8]. However, when the following driver agent chooses to keep the same velocity of the last simulation step T, the acceleration value should be equal to zero. To choose between the two scenarios, we used a probability "p" for the first scenario and a probability "q" for the second.

$$\begin{cases} a = 0 & \text{with a probability } p \\ a_{min} <= a <= a_{max} & \text{with a probability } q = 1 - p \end{cases} \quad (8)$$

Finally, the state "Dsec_x = Dxy", the following driver agent X tries to keep the perfect real distance of the simulation step T with the leading vehicle Y for the simulation step T + 1. In this state, the following driver agent X may act by two ways. In the first one, it can maintain the velocity of the simulation step T, in order to ensure the objective of the reduction of the circulation time. In the second, it can reduce its velocity to guarantee the road safety objective by taking into account the state, where the leading driver agent Y can suddenly reduce its speed during the next simulation step. The modeling of this constraint translates by the velocity relationships between vehicles. This constraint translates by the inequality (9).

$$a <= \left(V_y(T) - V_x(T)\right)/t \tag{9}$$

The choice of the acceleration value is based on two scenarios. In the first one, the acceleration value should equal to zero. However, in the second one, the acceleration behavior value varies between d_{min} and d_{max} basing on work of literature [8]. To choose between the two scenarios, we used a probability "p" for the first scenario and a probability "q" for the second.

$$\begin{cases} a = 0 & \text{with a probability p} \\ d_{min} <= a <= d_{max} & \text{with a probability q} = 1 - p \end{cases} \tag{10}$$

Turning to present the constraints of the upper objective function Eq. (1), we have one constraint that ensures the road safety by avoiding the longitudinal collision, where the safety distance should be strictly greater than zero.

$$D_{sec\_x} > 0 \tag{11}$$

3 Resolution Based on the Tabu Search Algorithm

Tabu search algorithm was proposed by Fred Glover [7]. Since then, the method has become very popular, thanks to its successes to solve many problems. In fact, the research Tabu algorithm is characterized by the rapidity during research [2]. Thus, this characteristic may help us; especially we need to find the best solution with a research time equal to the decision time of driver agent.

We used the Tabu search in our decision strategy to select the best velocity and safety distance that present the realizable solution and that meets the needs of the driver agent according to the state of its environment. In our research strategy, we used a Tabu list called Tabu list OUT; it contains the acceleration or deceleration values already selected for the calculation of the velocity and the safety distance variables, and the acceleration or deceleration values that do not respect the constraints of our modeling. In this context, the values of the Tabu list OUT should not be selected. At each simulation step, the list will be reset to zero.

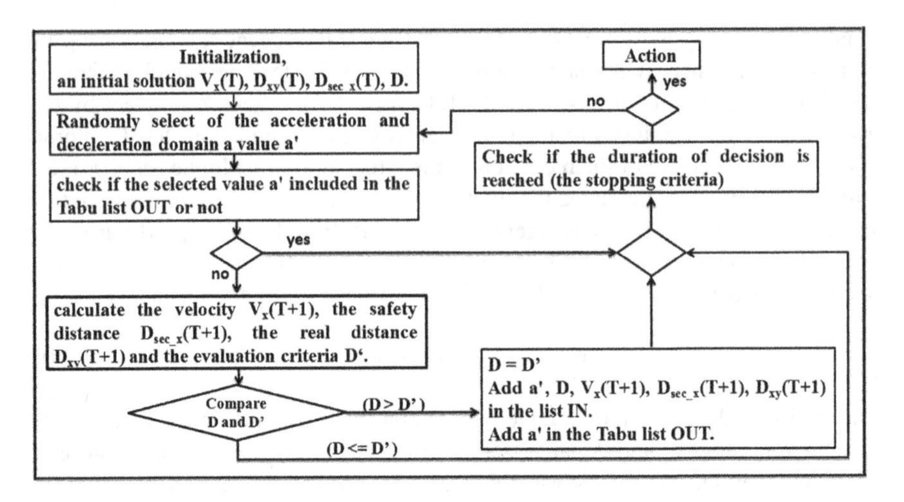

Fig. 2 Application structure of the Tabu search algorithm

The length of the Tabu list OUT is dynamic which adjusts itself during the search. In addition, the stopping criterion of the Tabu search algorithm is the calculation time. Indeed, while we used the Tabu search as a strategy for decision-making by researching the best solution that answers to the needs of the following driver agent, then we chose a random search time between 0.15 and 0.25 s, which is the stopping criterion to select the best realizable solution. This interval represents the decision time for an ordinary behavior driver [5]. Figure 2 presents the application structure of the Tabu search algorithm.

The values of various parameters of the starting solution ($V_x(T)$, $D_{sec\_x}(T)$, $D_{xy}(T)$) are already calculated at the level of the simulation step T. However, we assign a high value to the parameter D that presents the evaluation criteria of the realizable solutions. The list IN is not a Tabu list, just it contains the best realizable solutions, where after research; the driver agent will use the best realizable solution in the list IN.

The selection of the best realizable solution that contains the best velocity and safety distance that correspond to the state of the environment and to the objectives of the driver agent for the simulation step T + 1 is based on the evaluation criteria D. The calculation of the evaluation criteria D is expressed by the Eq. (12).

$$D = \left| D_{xy}(T+1) - D_{sec\_x}(T+1) \right| \tag{12}$$

where, the calculation of the real distance $D_{xy}(T+1)$ is expressed by the Eq. (13).

$$D_{xy}(T+1) = D_{xy}(T) + \left(V_y(T) * t\right) - \left(V_x(T+1) * t\right) \tag{13}$$

4 Simulation

We simulated our model with two cars in an urban zone with bottling conditions, where the leading vehicle circulates with a very slow speed. Therefore, our following agent driver should react with these conditions by basing during the decision-making on the concept of anticipation. We used a cognitive agent to model the driver and we used Jade platform to implement it. For the simulation parameters of Eq. 1, we used the simulation parameters of Mitsubishi Free car 2.0 [5], which are, $W = 1735$ kg, $G = 9.81$ m/s^2, $\rho = 1.25$, $A_f = 2.562$ m^2, $Cd = 0.4$, $\eta = 0.6$, $\mu = 0.8$, $f_r = 0.015$ and $\mu = 0.8$, $T_A = 0.05$–0.15 s, $T_D = 0.15$–0.25 s.

The duration for each simulation step represented by the parameter t is 1 s. Furthermore, we used the acceleration and deceleration interval as defined in [8], where the acceleration values vary between 0.9 and 3.6 m/s^2, and the deceleration values vary between 0.9 and 2.4 m/s^2.

According to the results that we have obtained by the simulation, as it is presented in Fig. 3, we note a rapprochement between the calculated safety distance $D_{sec\_x}(T + 1)$ and the real distance between the two vehicles $D_{xy}(T + 1)$ for each simulation step. Note, that there is not a total domination by the calculated safety distance $D_{sec\_x}(T + 1)$ or the real distance $D_{xy}(T + 1)$ with an acceptable distance margin between 0.036 and 1.584 m. In fact, this distance margin is the evaluation criteria D to select the best realizable solution (see the Fig. 4).

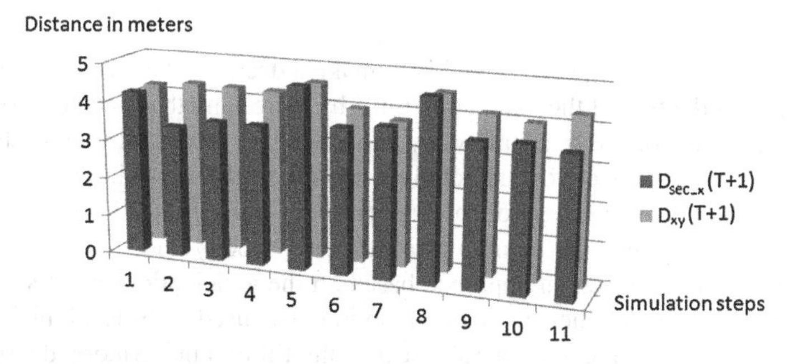

Fig. 3 Comparison between the calculated safety distance and the real distance

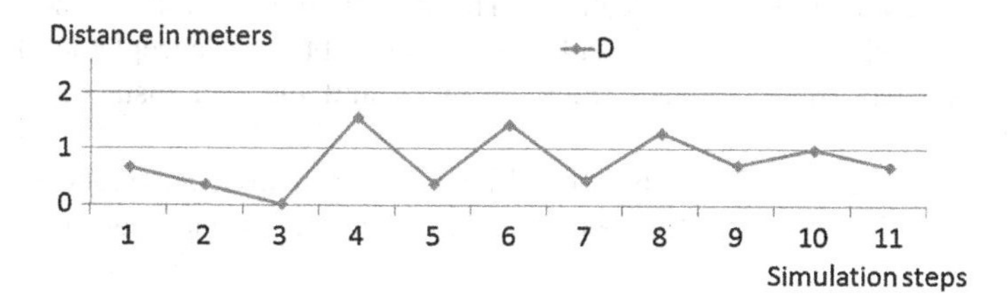

Fig. 4 The values of the evaluation criteria D for each simulation step

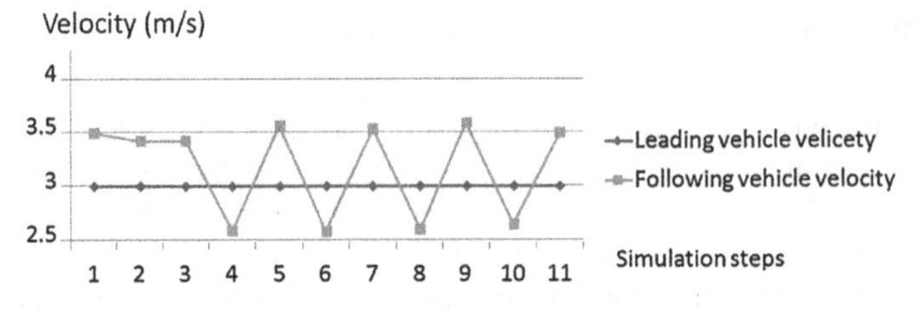

Fig. 5 Comparison of velocity between the leading vehicle and the following vehicle according our approach

In the reality it is impossible to keep a distance between two vehicles strictly equal to the calculated safety distance during circulation [14]. Thus, on Fig. 5 we note that the following driver agent tries to keep a real distance around the calculated distance by controlling the velocity of its vehicle against the velocity of leading vehicle.

Furthermore, according to the results on Fig. 5, we consider the variance of the following vehicle velocity. Thus, this simulation reflects as much as possible the reality of our following driver agent reaction compared with research work [6] (see Fig. 1), where it presents results not real [14].

5 Conclusion

The objective of our proposed model is to make a decision that ensures a real link between the reduction of the circulation time, by increasing the speed of movement and the guarantee of the road safety, by conserving an acceptable safety distance. In fact, these two objectives are totally opposed to the car-following behavior, which requires the consideration of this problem during decision-making. This consists in finding the best compromise between velocity and safety distance. In this context, we proposed an approach based on the anticipation of the simulation parameters. To model the anticipation notion, we used a bi-level bi-objective modeling that we resolve using a Tabu search algorithm. Furthermore, the software agent technology has been used to model and to simulate drivers, in order to make them able to sense and to react according to their environment changes thanks to their autonomy and reactivity features. The model implementation and experimentation provides promising results, since we obtained acceptable distance margins between the calculated safety distance and the real distance for each realizable solution.

The current model concerns only the driver with normative behaviors. Thus, the integration of the non-normative behaviors is the subject of our future work.

References

1. Arnaud, D., René, M., Sylvain, P., Stéphane, E.: A behavioral multi-agent model for road traffic simulation. Eng. Appl. Artif. Intell. **21**, 1443–1454 (2008)
2. Bajeh, A.O., Abolarinwa, K.O.: Optimization: a comparative study of genetic and tabu search algorithms. Int. J. Comput. Appl. (0975–8887) **31**(5) (2011)
3. Barcelo, J., Ferrer, J., Grau, R., Florian, M., Chabini, E.: A route based version of the AIMSUN2 micro-simulation model. 2nd World Congree on ITS, Yokohama. (1995)
4. Brackstone, M., Mcdonald, M.: Car-following: a historical review [J]. Transp. Res. Part F: Traffic Psychol. Behav. **2**(4), 181–196 (1999)
5. Chen, Y.-L., Wang, C.-A.: Vehicle Safety Distance Warning System: A Novel Algorithm for Vehicle Safety Distance Calculating Between Moving Cars. 1550-2252/$25.00 ©16 IEEE. (2007)
6. Gipps, P.G.: A behavioural car following model for computer simulation. Transp. Res. B **15** (2), 105–111 (1981)
7. Glover, F.: Tabu search-part I. ORSA J. Comput. **1**(3) (1989). 0899-1499/89/0103-0190 $01.25.
8. ITE: Transportation and traffic engineering handbook. In: 2nd Edition, Institute of Transportation Engineers. Prentice-Hall, Inc. New Jersey (1982)
9. Kometani, E., Sasaki, T.: Dynamic behaviour of traffic with a nonlinear spacing-speed relationship. In: Proceedings of the Symposium on Theory of Traffic Flow, Research Laboratories, General Motors, pp. 105–119. New York (1959)
10. Liu, R., Van, V.D., Wating, D.P.: DRACULA: dynamic route assignment combining user learning and microsimulation. In: Proceedings of PTRC Summer Annual Conference, Seminar E, pp. 143–152 (1995)
11. Qiang, L., Lunhui, X., Zhihui, C., Yanguo, H.: Simulation analysis and study on car-following safety distance model based on braking process of leading vehicle. In: IEEE Proceedings of the 8th World Congress on Intelligent Control and Automation (2011). 978-1-61284-700-9/11/ $26.00 ©2011
12. Wilson, R.E.: An analysis of Gipps's car-following model of highway traffic. IMA J. Appl. Math. **66**, 509–537 (2001)
13. Wooldridge, M., Jennings, N.R.: Intelligent agents: theory and practice. Knowl. Eng. Rev. **10** (2), 115–152 (1995)
14. Yang, D., Zhu, L.L., Yu, D., Yang, F., Pu, Y.: An enhanced safe distance car-following model. J. Shanghai Jiaotong Univ. (Sci.) **19**(1), 115–122 (2014)

Probabilistic Ontology Definition Meta-Model

Extension of OWL2 Meta-Model for Defining Probabilistic Ontologies

Hlel Emna, Jamoussi Salma, Turki Mohamed and Ben Hamadou Abdelmajid

Abstract In this article, we have proposed an extension of OWL2 meta-model for representing the fundamental elements of probabilistic ontologies (POs). This meta-model, called Probabilistic Ontology Definition Meta-model (PODM), provides support for defining probabilistic ontologies. In addition, we have enriched PODM (by using Object Constraint Language) with a list of constraints specifying invariants that have to be fulfilled by all models that instantiate this meta-model. These constraints allow eliminating ambiguities and inconsistencies that exist in this model.

Keywords Probabilistic ontology · Meta-model · Probabilistic components

1 Introduction

Uncertainty arises from the incorrect or incomplete understanding of the domain. It is a ubiquitous aspect of most real world problems. It exists in many applications, such as natural language disambiguity, diagnosis in medicine, machine learning, etc. Today, there is a very interesting requirement to develop formalisms of knowledge representation allowing to deal with uncertainty. Despite that the

H. Emna (✉) · J. Salma · T. Mohamed · B.H. Abdelmajid
Miracl Laboratory, Technology Center of Sfax, BP 242–3021 Sakiet Ezzit, Sfax, Tunisia
e-mail: emnahlel@gmail.com

J. Salma
e-mail: jamoussi@gmail.com

T. Mohamed
e-mail: med.turki@gmail.com

B.H. Abdelmajid
e-mail: abdelmajid.benhamadou@gmail.com

I. Czarnowski et al. (eds.), *Intelligent Decision Technologies 2016*,
Smart Innovation, Systems and Technologies 56,
DOI 10.1007/978-3-319-39630-9_20

ontologies have become standard for knowledge representation of particular domain, but they are not able to represent and reason with uncertainty. In literature, several works have been proposed for representing probabilistic ontologies (see the survey [1]). However, there is no a standard for doing so. Moreover, these works have not focused on the proposal of a meta-model for defining the fundamental components of POs.

On the other hand, a model is an abstraction of phenomena in the real world. A meta-model is a model of a model, and meta-modeling is the process of generating such meta-models. In Object Management Group (OMG) terminology an ontology is a kind of a data model. In literature, there are various meta-models for defining classical ontologies such as the ODM (Ontology Definition Meta-model) for OWL2 [2]. This meta-model should be designed to represent the fundamental components of all ontologies that are expressed in OWL2. However, the components of probabilistic ontologies are not taken into account. So, how we can extend the OWL2 meta-model for defining POs? What are the fundamental components of probabilistic ontologies? And what are its definitions? This paper falls within this context. The objective of this paper is to propose an extension of OWL2 meta-model for supporting uncertainty.

The remainder of this paper is organized as follows: Sect. 1 presents some related works, Sect. 2 introduces MOF and OCL, Sect. 3 introduces the OWL2 meta-model. In Sect. 4, we present our probabilistic ontology definition meta-model (PODM). Finally, we present a conclusion and future work.

2 Related Work

An ontology is a model that is defined by a machine-interpretable language. It defines a common set of concepts, properties, instances, etc. that are used to describe and represent knowledge of a particular domain. Thanks to these components, ontologies are used to model the reality (real world applications). But this world includes inaccuracies and imperfections which cannot be represented by classical or traditional ontologies (CO). Probabilistic Ontology has come to remedy this defect. According to [3], PO is "an explicit, formal knowledge representation that expresses knowledge about a domain *of application. This includes: (1) Types of entities that exist in the domain; (2) Properties of those entities; (3) Relationships among entities; (4) Processes and events that happen with those entities; (5) Statistical regularities that characterize the domain; (6) Inconclusive, ambiguous, incomplete, unreliable, and dissonant knowledge related to entities of the domain; and (7) Uncertainty about all the above forms of knowledge*". We can define the PO simply as a CO enriched with uncertain knowledge. In other words, an ontology is a PO only if it contains at least one probabilistic component [4, 5]. In literature, various researchers have presented the requirements to model the uncertain knowledge on the semantic web. The authors of [1] have described some areas where probabilistic information plays a role in the context of the semantic web such as representation of uncertain information, ontology learning, document classification and etc. So, it is necessary to model the uncertain knowledge of a particular domain. However, there is no a standard for defining POs.

In addition, various researchers have proposed probabilistic extensions of ontology languages, especially OWL, for representing POs such as OntoBayes, BayesOWL, BeliefOWL, etc. Each of these languages contains a set of classes and properties to represent uncertain knowledge. In [6], the authors have proposed a probabilistic extension of OWL called BayesOWL by using the Bayesian network (BN) [7, 8]. For adding uncertainty to existing ontologies, they have defined a list of new OWL classes such as "*PriorProbObj*". This class allows to specify the prior probability (P(A), where A is a class). Similarly, the authors of [9] have integrated the BN in OWL to retain the benefits of both. To represent the uncertain information, they have defined some new OWL classes for example "*PriorProb*" and "*CondProb*". These classes are defined to identify the prior and conditional probability. The authors of [10] have proposed a probabilistic extension of OWL, called BeliefOWL. The latter is able to support the uncertainty thanks to new OWL classes such as "*beliefDistribution*", "*priorBelief*" and "*condBelief*" which allow expressing the prior and conditional evidence.

Description Logics (DLs) [11] are a family of ontological knowledge representation languages. They represent knowledge in terms of objects, concepts, and roles. To encode uncertainty, probabilistic description logics (PDLs) must be contemplated. The literature contains a number of proposals for PDLs [12–14]. P-SHOQ is a probabilistic DL [13], extension of the description logic SHOQ [15]. It adds to the syntax for SHOQ a list of conditional constraints that are defined as expressions $P(D|C)$ $[l, u]$ with C, D are classes and $[l, u]$ is an interval between 0 and 1. These constraints can be used to represent different kinds of probabilistic knowledge, for example $(D|\{o\})[l; u]$ means "*o is an instance of the concept D with a probability in [l; u]*". CRALC [14] is a probabilistic DL, extension of the description logic ALC [16]. It retains all constructors offered by ALC (conjunction, disjunction, etc.) by adding probabilistic inclusion such that $P(C \mid D) = \alpha$ or $P(r) = \beta$, with C and D are two concepts and r is a role. For example, the sentence P(Professor|Researcher) = 0.4 indicates the probability that an element of the domain is a Professor given that it is a Researcher. This sentence is called probabilistic inclusion.

As previously mentioned, various works have been proposed in literature for representing POs (see the survey [1]). However, currently there is no established foundation or no standard for doing so. Moreover, these works have not focused on the proposal of a meta-model for defining POs by specifying the new probabilistic components of ontology, which allow to support the uncertainty, and the constraints associated with these components. This paper falls within this context. We think that future standard OWL2 versions (the recent version of the W3C is OWL 2.0) should be extended in a way to allow the creation of the POs. In this work, we have proposed an extension of ODM for OWL2 for defining POs.

3 MOF and OCL

MOF (Meta-Object Facility), maintained by OMG, defines an abstract language and framework for specifying, constructing and managing technology neutral meta-models [17]. It plays a crucial role in OMG's four-layer metadata architecture which provides a solid basis for defining meta-models. In many cases, a graphical model like a class diagram UML is not enough for unambiguous constraints [18]. Also, the set of constraints concerning the objects in the model need to be described. In the world MOF, these constraints are expressed using OCL that started as a complement of the UML notation with the goal to overcome the limitations of UML in terms of precisely specifying detailed aspects of a system design. OCL is a textual specification language for describing rules that apply to UML. It is a formal language which allows to specify constraints on models in a more precise and concise way [19]. In literature, several researchers have added to models (or meta-models) a set of constraints using this language to refine its syntax such as [20].

4 Ontology Definition Meta-Model

The Web Ontology Language (OWL) is the standardized ontology language for the web. It is a semantic markup for publishing and sharing ontologies on the World Wide Web. There exists a number of related works for defining meta-models for OWL [21]. The most relevant works are OMG OWL meta-model [22] and W3C OWL2 meta-model [2]. In this article, we are interested only in OWL2 meta-model (the recent version of the W3C is OWL 2.0). The authors of [2] use for defining this meta-model a very simple form of UML class diagrams. This meta-model is composed of several UML class diagrams, a class diagram per element of the OWL2 meta-model. A fragment of this meta-model is shown in Fig. 1. In this paper, we concentrate only on these elements: classes, individual and properties.

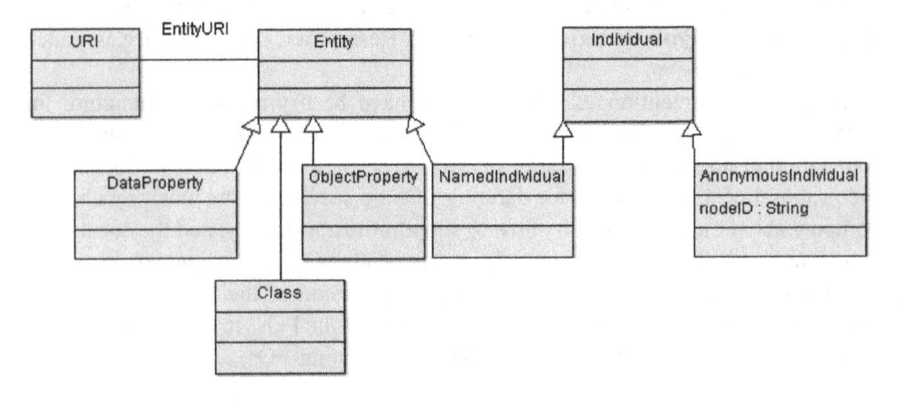

Fig. 1 Fragment of meta-model for OWL2

- **Classes**: They are identified using a URI. They represent a category of individuals which share the same properties [23].
- **Individuals**: They represent objects in the domain in which we are interested. They consist of named individuals and anonymous individuals. Named individuals can be used to represent actual objects from the domain of interest. They are identified using a URI. For example, the instance "*Jhon*" can be used to represent a particular person. Anonymous individuals are analogous to blank nodes in RDF. They are identified by a local node ID. For example, "*John*" lives at some address which is represented using the anonymous individual "*a1*".
- **Properties**: They can be used to represent relationships in the domain of interest. OWL2 distinguishes two kinds of properties: Object properties and Data properties. Data properties connect individuals with literals and Object properties connect two pairs of individuals.

5 Probabilistic Ontology Definition Meta-Model

In literature, there are no approaches to offer metadata about the defined classes or properties in probabilistic ontologies. In meta-model for OWL2 [2], there is no defined way to represent the uncertain or probabilistic components of an ontology. This requires the proposal of a new approach in terms of a meta-model to specify the fundamental components of a PO. In this section, we have defined an extension of ODM for OWL2 [2], named Probabilistic Ontology Definition Meta-model (PODM) which provides support for defining POs (see Fig. 2). The latter is a structural model represented by UML class diagram. It includes a list of new elements (attributes, relationships and concepts) that are not supported by meta-model for OWL2 for representing the structure of POs. Indeed, we have extended the meta-model for OWL2 by a list of components specified to support

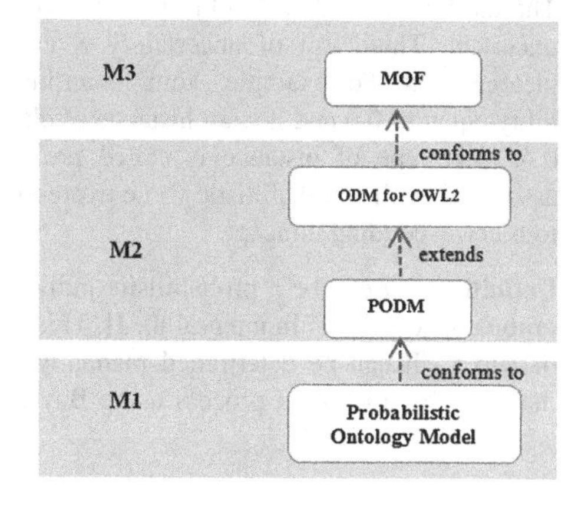

Fig. 2 From a MOF Model to a probabilistic ontology model

uncertainty like *ProbabilisticObjectProperty*, *ProbabilisticClass*, *ProbabilisticDataProperty*, *ProbabilisticIndividual*, etc. These components are the basis for defining POs. In the next subsection, we will present the meta-model PODM and its components.

5.1 Components of Probabilistic Ontology

As previously mentioned, we can define the probabilistic ontology simply as an extension of the classical ontology by integrating uncertain knowledge. It can describe the knowledge of a particular domain ideally in a format that can be read and processed by a computer and incorporate uncertainty to this knowledge [4]. We can divide the fundamental components of probabilistic ontology in precise (or classical) components and probabilistic (or uncertain) components:

- Classical components: They consist of classical individual (classical anonymous individual and classical named individual), classical class and classical property (classical data property and classical object property). These components have the same definition and play the same role of the basic elements of meta-model for OWL2 [2] which are Anonymous individual, Named Individual, Class, Data Property, Object Property. So, these classical components are supported by ODM for OWL2 (see Fig. 1).
- Probabilistic components: They consist of Probabilistic Individual (Probabilistic Named Individual and Probabilistic Anonymous Individual), Probabilistic Class, Probabilistic Object Property and Probabilistic Data Property. They are not supported by ODM for OWL2. So, this paper aims to extend this meta-model by these probabilistic components for obtaining a meta-model which is able to support uncertainty (PODM). In the rest of this section, we will present a description of these new components.

Probabilistic Individual
The attribution of data or objects to the corresponding concept (or class) may be uncertain. This form of uncertainty is caused by: lack of information, incompleteness, etc. For example, "tom" is an instance of class "Animal" with a probability equal to 0.6 and it is an instance of class "Person" with a probability equal to 0.4. This type of instance is called probabilistic or uncertain individual. It is associated with a probabilistic value expressing the belonging degree of an instance to a corresponding concept.

Definition 1 Let I be a probabilistic individual of class C. It is associated with a probabilistic value P, in interval]0, 1[. This value expresses the degree of belonging of I to C. It can be determined manually with the aid of an ontologist or automatically by a learning process using Bayesian Network, probabilistic model, etc.

Similarity to Web Ontology Language (OWL2), we can distinguish two kinds of probabilistic individual: probabilistic named individual (identified with URI) and probabilistic anonymous individual.

Probabilistic class

The classes (or concepts) of OWL ontology describe a collection of objects for a particular domain. If this collection includes one or more probabilistic instances then the type of this class becomes a probabilistic class. Let *C* be a class of an OWL ontology and $I = \{I_1, \ldots, I_i, \ldots, I_n\}$ be a list of instances of this concept. In OWL ontology, we can distinguish two types of concepts: if all elements of *I* are classical instances then *C* is a classical concept and if *I* contains at least one probabilistic instance then *C* is a probabilistic concept. Assuming that *C* is a probabilistic concept, *N* is the total number of instances of this concept and *NP* is the number of probabilistic instances of *C*. This concept is attached with a probabilistic value *ProbV* which expresses uncertainty. This value is calculated with the aid of this formula:

$$\mathrm{ProbV} = \frac{\mathrm{NP}}{\mathrm{N}} \in [0, 1] \tag{1}$$

For example, let *C* be a probabilistic concept of an ontology *O*, I_1, I_2 and I_3 be the list of instances of *C* and I_2 be a probabilistic instance. Then *ProbV* is equal to 0.33.

Definition 2 C is probabilistic class if and only if one of its instances is a probabilistic instance. This class is defined as a collection of uncertain objects and/or certain objects. It is associated with a probability which represents the degree of certitude of this concept. This degree can be calculated with the formula previously mentioned.

Probabilistic Object Property

In the real world, it is often the case that the relationships between resources hold probabilistically. Some examples of this relation are represented by: "*Imagery*" (Theme) is connected to "*Data-Mining*" (Theme) with a probability of 0.7, "*John*" (Person) is interested in "*Imagery*" (Theme) with a probability of 0.8. In probabilistic ontology, these relations "*be-connected*" and "*be-interested*" are considered as probabilistic object properties. These properties are attached with probabilistic values which represent uncertainty.

Definition 3 R is probabilistic object property between two instances if and only if it represents an imprecise interaction between these two components. It is associated with a probability which represents the uncertain appearance and can be determined manually by an ontologist or automatically with a learning process (Bayesian Network, probabilistic model, etc.).

Probabilistic Data Property

The probabilistic data properties are classical properties that are attached with probabilities which represent the uncertain appearance. Generally, the extraction of knowledge in an automatic way provides us uncertain and undetermined knowledge, because the knowledge extracted by using automatic or semi-automatic systems is uncertain and probabilistic. For example, the extraction of hobby for each person can be realized automatically or semi-automatically from social networks (Facebook, Twitter, etc.). The result of this task is a list of uncertain knowledge (list of hobbies for each person). In an ontology, the concept "Person" can be used to model the set of persons. The data property "name" can be used to represent the name for each person. The data property "hobby" can be used to model the hobbies for each person. The first property is a precise element of this ontology. However, the second property is a probabilistic or uncertain element of this ontology (probabilistic data property). It is attached with probabilistic value that expresses the degree of certitude of this knowledge. For example, the hobby of "John" (instance of Person) is "music" (value of the probabilistic data property "hobby") with a degree equal to 0.5.

Definition 4 Probabilistic data property is defined as a classical data property that is associated with a degree of certitude (probabilistic value in interval]0, 1[). As mentioned previously, this degree can be determined manually with the aid of an ontologist or automatically by using a learning process (Bayesian Network, probabilistic model, etc.).

5.2 Probabilistic Ontology Definition Meta-Model (PODM)

The Fig. 3 shows the components of PODM, an extension of standard ODM of OWL2. Indeed, this meta-model contains a list of new components (represented in Fig. 3 with the color white) which are not supported by ODM for OWL2. These

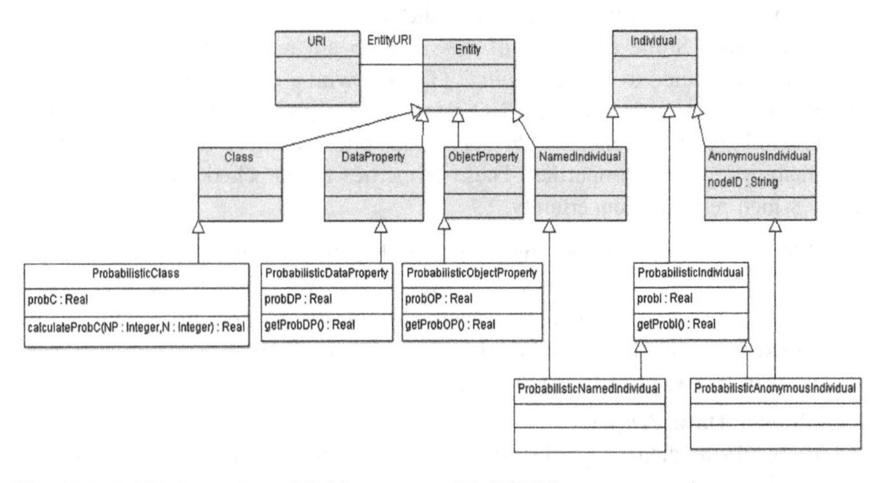

Fig. 3 Probabilistic ontology definition meta-model (PODM)

components are *ProbabilisticDataProperty, ProbabilisticIndividual, ProbabilisticObjectProperty, ProbabilisticClass, ProbabilisticNamedIndividual* and *ProbabilisticAnonymousIndividual*. Also, PODM contains a list of components which are supported by ODM for OWL2 (represented in Fig. 3 with the color yellow) such as *Class, ObjectProperty, DataProperty, AnonymousIndividual, Individual* and *NamedIndividual*, etc. The central class in PODM is the class *Entity* which describes the basic components of POs which are *NamedIndividual, ObjectProperty, Class* and *DataProperty*. The association between the class *Entity* and the class *URI* called *EntityURI* indicates that each entity is identified with URI. The class *ProbabilisticObjectProperty* is subclass of *ObjectProperty*. It has an attribute named *probOP* which represent uncertainty and a method called *getProbOP()*. The latter allows returning the value of *probOP*. The class *ProbabilisticDataProperty* is subclass of *DataProperty*: this states that *ProbabilisticDataProperty* is more specific than *DataProperty*. It has an attribute named *probDP* which represent uncertainty and a method called *getProbDP()* which return the value of this attribute. These two classes (identified with URI) allow to represent the probabilistic properties of probabilistic ontologies. The class *ProbabilisticIndividual* is subclass of Individual. It has an attribute named *probI* and a method called *getProbI()*. *probI* represent the degree of belonging of an instance to a corresponding concept and *getProbI()* allows to return the value of this attribute. Similarity to OWL2, we can distinguish two kinds of probabilistic individual: probabilistic named individual (identified with URI) and probabilistic anonymous individual. So, the superclass of *ProbabilisticNamedIndividual* and *ProbabilisticAnonymousIndividual* is *ProbabilisticIndividual*. The class *ProbabilisticClass* (identified by URI) is subsumed by the class *Class*. Roughly speaking, this states that *ProbabilisticClass* is more specific than *Class*. It has an attribute named *probC* and a method called *calculateProbC()*. The latter allows calculating the value of this attribute by using the formula 1.

5.3 Augmenting PODM with OCL Constraints

OCL is a formal language used to formally specify constraints on models (or meta-models) in a more precise and concise way than it is possible to do with diagrams only. In this work, we have enriched PODM with a list of OCL constraints that have neither the ambiguities of natural language nor the difficulty of using complex mathematics. We have used ArgoUML [24] for creating these constraints (see the Table 1).

Table 1 The list of constraints OCL and their significations

Probabilistic components	Constraints and its significations

ProbabilisticObjectProperty	[1] For all instances of the class *ProbabilisticObjectProperty*, the *probOP* attribute must always be positive and less than 1
	`context: ProbabilisticObjectProperty`
	`Inv:ProbValueOP: (self.probOP > 0) and (self.`
	`probOP < 1)`
	[2] The result of *getProbOP()* must equal to the value of *probOP*
	`context ProbabilisticObjectProperty::`
	`getProbOP(): Real`
	`post POP: result1 = probOP`

ProbabilisticDataProperty	[3] For all instances of the class *ProbabilisticDataProperty*, the *probDP* attribute must always be positive and less than 1
	`context: ProbabilisticDataProperty`
	`Inv:ProbValueDP: (self.probDP > 0) and (self.`
	`ProbDP < 1)`
	[4] The result of *getProbDP()* must equal to the value of *probDP*
	`context ProbabilisticDataProperty::getProbDP`
	`(): Real post PDP: result2 = probDP`

ProbabilisticIndividual	[5] For all instances of *ProbabilisticIndividual*, the *probI* attribute must always be positive and less than 1
	`context: ProbabilisticIndividual`
	`inv:ProbValueInst: (self.probI > 0) and`
	`(self.probI < 1)`
	[6] The result of *getProbI()* must equal to the value of *probI*
	`context ProbabilisticIndividual::getProbI():`
	`Real post PI: result3 = probI`

ProbabilisticClass	[7] For all instances of *ProbabilisticClass*, the *probC* attribute must always in this interval]0, 1]
	`context: ProbabilisticClass`
	`inv:ProbValueClass: (self.probC > 0) and`
	`(self.probC <=1)`
	[8] The result of *calculateProbC(NP, P)* should be equal to NP (the number of probabilistic instances of ProbabilisticClass) divided by N (the total number of instances of ProbabilisticClass)
	`context ProbabilisticClass::calculateProbC`
	`(NP: Integer, N: Integer): Real`
	`post PC: self.probC = NP/N`

6 Conclusion and Perspectives

In this paper, firstly we have briefly presented MOF and OCL, ODM for OWL2, etc. Secondly, we have proposed an extension of ODM for OWL2, called PODM (Probabilistic Ontology Definition Meta-model), for supporting uncertainty. To enrich this meta-model with several constraints, we have used the language OCL. So, PODM can be considered as a part of the effort to specify standard meta-model for defining probabilistic ontologies. To our knowledge, our work is the first one to propose a meta-model which provides support for defining probabilistic ontologies.

In the future work, we will try to extend Axioms and Expressions of OWL2 for supporting uncertainty. In addition, we will try to create a probabilistic ontology of a particular domain based on this proposed meta-model.

References

1. Predoiu, L., Stuckenschmidt, H.: Probabilistic models for the semantic web: survey. In: Web Technologies: Concepts, Methodologies, Tools, and Applications, pp. 1896–1928. University of Victoria, Germany (2010)
2. Motik, B., Patel-Schneider, P.F., Parsia, B.: OWL2 web ontology language: structural specification and functional-style syntax (2012)
3. da Costa, P.C.G., Laskey, K.B.: PR-OWL: A Bayesian ontology language for the Semantic Web. In: Proceedings URSW, pp. 23–33 (2005)
4. Hlel, E., Jamoussi, S., Ben Hamadou, A.: A probabilistic ontology for the prediction of author's Interests. In: International Conference ICCCI (Springer), pp. 492–501 (2015)
5. Hlel, E., Jamoussi, S., Ben Hamadou, A.: Intégration d'un réseau bayésien dans une ontologie. In: Actes des 25 journées francophones IC2014, pp. 295–297 (2014)
6. Ding, Z., Peng, Y.: A probabilistic extension to ontology language OWL. In: Proceedings HICSS, pp. 5–8 (2004)
7. Holmes, D.E., Jain, L.C.: Innovations in Bayesian networks: theory and applications. Comput. Intell. (2008). Springer
8. Ben Mrad, A., Delcroix, V., Piechowiak, S., Leicester, P., Abid, M.: An explication of uncertain evidence in Bayesian networks: likelihood evidence and probabilistic evidence—uncertain evidence in Bayesian networks. Appl. Intell. 802–824 (2015)
9. Yang, Y., Calmet, J.: Ontobayes: an ontology-driven uncertainty model. In: Conference on Intelligent Agents, Web Technologies and Internet Commerce, pp. 457–464 (2005)
10. Essaid, A., Ben Yaghlane, B.: BeliefOWL: an evidential representation in OWL ontology. In: International Workshop on URSW, Washington DC, USA, pp. 77–80 (2009)
11. Baader, F., Calvanese, D., McGuinness, D., Nardi, D., Patel-Schneider, P.F.: The Description Logic Handbook: Theory, Implementation and Applications. Cambridge University Press (2003)
12. Nottelmann, H., Fuhr, N.: pDAML + OIL: a probabilistic extension to DAML + OIL based on probabilistic datalog. In :Information Processing and Management of Uncertainty in Knowledge-Based Systems (2004)
13. Giugno, R., Lukasiewicz, T.: P-SHOQ(D): a probabilistic extension of SHOQ(D) for probabilistic ontologies in the semantic web. In: INFSYS, Austria, Research Report (2002)
14. Fabio, G., Rodrigo, B.P., Takiyamaa, F.I., Kate, R.C.: Computing inferences for credal ALC terminologies. In: URSW, pp. 94–97 (2011)
15. Horrocks, I., Sattler, U.: Ontology reasoning in the SHOQ(D) description logic. In: The 17th Conference on Artificial Intelligence (IJCAI) (2001)
16. Schmidt-Schauss, M., Smolka, G.: Attributive concept descriptions with complements. In: Artif. Intell. 1–26 (1991)
17. OMG-MOF, Object Management Group, MOF 2.0 Core Specification (2009)
18. OMG-UML, Unified Modeling Language, OMG UML (2015)
19. OMG-OCL, Object Constraint Language OCL, version 2.4 (2014)
20. Brockmans, S., Volz, R., Eberhart, A., Löffler, P.: Visual modeling of OWL DL ontologies using UML. In: The Third International SW, pp. 198–213, Japan (2004)
21. Parreiras, F.S.: Semantic web and model-driven engineering, 1st edn. Institute of Electrical and Electronics Engineers. IEEE Press (2012)
22. OMG-ODM, Ontology Definition Metamodel Version 1.0 (2009)

23. Horridge, M.: A Practical Guide to Building OWL Ontologies Using Protégé 4 and CO-ODE Tools. University of Manchester, Manchester (2009)
24. Ramirez, A., Vanpeperstraete, P., Rueckert, A., Odutola, K., Bennett, J., Tolke, L., van der Wulp, M.: ArgoUML user manual a tutorial and reference description (2010)

Development Aid Decision Making Framework Based on Hybrid MCDM

Eric Afful-Dadzie, Zuzana Komínková Oplatková, Stephen Nabareseh and Michael Adu-Kwarteng

Abstract Developmental Aid Programs or what is normally referred to as Official Development Assistance (ODA), plays a key role in the growth process of many underdeveloped and developing countries. However, with a recent growing scarcity of resources and a general call for strict accountability from aid-recipient countries, there has been some interests in the design of aid performance evaluation methodologies or tools. To deepen such interests, this paper proposes a hybrid Fuzzy AHP Synthetic Extent—PROMETHEE framework of how aid-recipient countries could be evaluated and selected in developmental aid programs. With the use of the Organization for Economic Co-operation and Development (OECD) set of subjective criteria for evaluating aid programs, a decision framework pre-defined by linguistic terms of triangular fuzzy number format is provided for such selection decision problem. The framework serves to deepen transparency, fairness, value for money and sustainability of development aid programs.

Keywords Development aid · Fuzzy AHP · Fuzzy PROMETHEE · Performance evaluation · Poor countries

E. Afful-Dadzie (✉) · Z.K. Oplatková
Faculty of Applied Informatics, Tomas Bata University, Zlin, Czech Republic
e-mail: afful@fai.utb.cz

Z.K. Oplatková
e-mail: kominkovaoplatkova@fai.utb.cz

S. Nabareseh · M. Adu-Kwarteng
Faculty of Management and Economics, Tomas Bata University, Zlin, Czech Republic
e-mail: nabareseh@fame.utb.cz

M. Adu-Kwarteng
e-mail: kwarteng@fame.utb.cz

I. Czarnowski et al. (eds.), *Intelligent Decision Technologies 2016*,
Smart Innovation, Systems and Technologies 56,
DOI 10.1007/978-3-319-39630-9_21

1 Introduction

Growth and development agenda for most developing countries still depend heavily on development assistance programs. Billions in various aids go to developing countries and poor nations every year. Historically, most aid programs are based on bilateral agreements between the donor and the recipient countries and therefore are often not necessarily competitive. However, there are other aid programs that are highly competitive where applicants are required to meet some set criteria. One of such competitive aid programs is the Millennium Challenge Corporation funds advanced to countries to fight global poverty. Development assistances typically come in the form of various aid instruments, such as budgetary supports, social funds, humanitarian aid and pooled funding among others—and go a long way to alleviate poverty, build and repair institutions [1]. Some of the organizations and nations at the forefront of aid donations include the Organization for Economic Cooperation and Development (OECD), UK's Department for International Development (DFID), the United States Agency for International Development (USAID), Canada's Country Indicators for foreign Policy (CIFP), the African Development Bank (AfDB) and the World Bank among others. Recent shortfalls in aid allocation have called for stringent measures to ensure that aid money are efficiently utilized by recipient countries. This phenomenon is evidenced by recent review of criteria sets used by most donor agencies and countries as far as selection and performance evaluation of recipients are concerned [2, 3]. In furtherance to changes in some of the criteria used in selecting countries for development assistance programs, an issue of utmost importance is improving the methodologies used in such selection decision problems [3, 4].

Evaluation of aid programs is generally seen as a complex task because of several non-aid factors that tend to affect the overall outcomes [3]. Adding to the complexity are also the subjective sets of criteria used in most aid evaluation programs. To deal with such complex evaluations, robust and efficient methods that cater for the uncertainties are needed. In recent times, there seems to be a gradual interest in the design of tools and methodologies used in measuring aid efficiency [5] and overall performance of development aid programs. Palenberg [6], reviewed and proposed a number of such tools in the most comprehensive review of tools and methodologies appropriate for aid efficiency measurement. Though the review by [6] mentions multi-attribute decision making methods (MADM), it fails to capture instances where uncertainty modelling in aid allocation selection programs can be carried out. Giving that most criteria for appraising development aid programs are often subjective, using a deterministic MADM method in this instance would not be methodologically appropriate.

This paper therefore proposes a hybrid fuzzy MCDM framework composed of the Fuzzy Analytical Hierarchy Process (AHP) Synthetic Extent and Preference Ranking Organization METHod for Enrichment of Evaluations (PROMETHEE) methods to evaluate countries vying for development aid programs. The proposed method helps to keep track of progress of countries whiles ensuring that future aid allocations are based on performance of previous aid programs. The fuzzy AHP Synthetic Extent method is used to set criteria weights whiles the fuzzy PROMETHEE method is used in the ranking and selection of deserving countries. The two methods are ideal for such problem because of their ability to weigh individual preferences and compare pairs of alternatives to select one which is superior in terms of a criteria. With the aid of the Organisation for Economic Co-operation and Development (OECD) set of subjective criteria for evaluating aid programs, pre-defined by linguistic terms of triangular fuzzy number format, a numerical example is provided of how participating developing countries can be ranked in such selection decision problem. The framework helps to deepen transparency, fairness, value for money and sustainability of development aid programs.

The rest of the paper is presented as follows: Uncertainty modelling with fuzzy set theory is briefly explained. The hybrid framework is introduced where the methodological steps of fuzzy AHP Synthetic Extent and fuzzy PROMETHEE are outlined. Finally, a numerical example of how the hybrid MCDM decision framework works is presented to demonstrate ranking and selection of participating developing countries in developmental assistance programs.

2 Modelling Subjectivity with Fuzzy Sets

Zadeh [7] introduced the fuzzy set theory to tackle issues of uncertainty, imprecision and vagueness in information that are not statistical in nature. The fuzzy sets concept has been applied extensively in multi-criteria decision making (MCDM). The following presents the definitions and basic operations of the fuzzy set theory.

A. *Fuzzy Set*

Let X be a nonempty set, the universe of discourse $X = \{x_1, x_2, \ldots, x_n\}$. A fuzzy set A of X is a set of ordered pairs $\{(x_1, f_A(x_1)), (x_2, f_A(x_2)), \ldots, (x_n, f_A(x_n))\}$, characterized by a membership function $f_A(x)$ that maps each element x in X to a real number in the interval [0,1]. The function value $f_A(x)$ stands for the membership degree of x in A. This paper uses the Triangular Fuzzy Number (TFN) defined below for evaluation.

B. *Triangular Fuzzy Number*

In triangular fuzzy number (TFN), the membership function is expressed as a triplet (t, u, v). The membership function $f_A(x)$ of the triangular fuzzy number is shown in Eq. (1):

$$f_A(x) = \begin{cases} 0 & x < t \\ \frac{x-t}{u-t}, & t \leq x \leq u \\ \frac{v-x}{v-u}, & u \leq x \leq v \end{cases} \quad (1)$$

The value of x at u gives the maximal value of $f_A(x)$, that is $f_A(x) = 1$. The value of x at t represents the minimal grade of $f_A(x)$, i.e. $f_A(x) = 0$. The constants t and v stand for the lower and upper bounds of the available area data respectively. According to [8], fuzzy models using TFNs are effective for solving decision-making problems with subjective and vague available information. The TFNs are used in very practical applications because of the computational efficiency and its simplicity.

C. *Basic fuzzy sets operations*

Supposing $M_1 = (t_1, u_1, v_1)$ and $M_2 = (t_2, u_2, v_2)$ are two TFNs, then the following are some relevant basic operations on these two TFNs used in this study.

$$M_1 \oplus M_2 = (t_1, u_1, v_1) \oplus (t_2, u_2, v_2) = (t_1 + t_2, u_1 + u_2, v_1 + v_2) \tag{2}$$

$$M_1 \ominus M_2 = (t_1, u_1, v_1) \ominus (t_2, u_2, v_2) = (t_1 - t_2, u_1 - u_2, v_1 - v_2) \tag{3}$$

$$M_1 \otimes M_2 = (t_1, u_1, v_1) \otimes (t_2, u_2, v_2) = (t_1 t_2, u_1 u_2, v_1 v_2) \tag{4}$$

$$M_1 \ominus M_2 = (t_1, u_1, v_1) \ominus (t_2, u_2, v_2) = (t_1 / v_2, u_1 / u_2, v_1 / t_2) \tag{5}$$

A reciprocal of a fuzzy number is also as expressed in Eq. (6)

$$M_1^{-1} = (t_1, u_1, v_1)^{-1} = (1/t_1, 1/u_1, 1/v_1) \tag{6}$$

3 Hybrid MCDM Framework: Fuzzy AHP—PROMETHEE

Hybrid approaches in MCDM are utilized mostly to harness the strengths of various MCDM methods at different stages in the decision making process. This approach helps to get the most out of each decision making stage and improves the final score. In this paper the combined strength of Fuzzy AHP Synthetic Extent and fuzzy PROMETHEE method are used to rank and select deserving countries competing for development aid. In the following section, the fuzzy AHP Synthetic Extent method is introduced together with all relevant theories and mathematical expressions. This is followed by an introduction to fuzzy PROMETHEE and its relevant methodological steps.

3.1 Weight Setting with Fuzzy AHP Synthetic Extent

Chang's fuzzy AHP Synthetic Extent [9], basically provides an extension to Saaty's [10] widely used AHP method. In this paper, the method is used to set criteria weights in the hybrid framework. Mathematically, Let = $\{x_1, x_2, \ldots, x_n\}$ be an object set, and $G = \{g_1, g_2, \ldots, g_n\}$ be a set of goals. According to Chang [9], the extent analysis method is essentially applied on each object x_i for each goal respectively. Thus for an m extent analysis performed on each object and goal respectively, Eq. (7) expresses the concept.

$$M_{gi}^{1}, M_{gi}^{2}, \ldots, M_{gi}^{m}, i = 1, 2, \ldots, n. \tag{7}$$

where all the $M_{gi}^{j}(j = 1, 2, \ldots, m)$ are triangular numbers.

In the following steps, the fuzzy extent analysis method is outlined.

Step 1 Computing Fuzzy synthetic extent with respect to ith object.
Let Eq. (7) be the values of extent analysis of an ith object for m goals. Then the value of the fuzzy synthetic extent is computed as shown in Eq. (8).

$$S_i = \sum_{j=1}^{m} M_{g_i}^{j} \otimes \left[\sum_{i=1}^{n} \sum_{j=1}^{m} M_{g_i}^{j} \right]^{-1} \tag{8}$$

where $\sum_{j=1}^{m} M_{g_i}^{j}$ is derived using the fuzzy addition operation as expressed in Eq. (2) for each matrix such that $\sum_{j=1}^{m} M_{g_i}^{j} = \left(\sum_{j=1}^{m} t_j, \sum_{j=1}^{m} u_j, \sum_{j=1}^{m} v_j \right)$.

Step 2 Computing degree of possibility of $S_2 . \geq S_1$
Let $S_1 = (t_1, u_1, v_1)$ and $S_2 = (t_2, u_2, v_2)$ be two convex triangular fuzzy numbers. The degree of possibility of $S_2 \geq S_1$ is defined as:

$$V(S_2 \geq S_1) = sup_{x \geq y} \left[\min(\mu_{S_2}(x), \mu_{S_1}(y)) \right] \tag{9}$$

Since S_1 and S_2 are convex fuzzy numbers, the following expressions in Eqs. (10) and (11) hold.

$$V(S_2 \geq S_1) = 1 \quad iff \quad s_2 \geq s_1 \tag{10}$$

$$V(S_2 \geq S_1) = hgt(S_1 \cap S_2) = \mu_{S_2}(d) \tag{11}$$

The ordinate of D is fully expressed as in Eq. (12)

$$V(S_2 \geq S_1) = hgt(S_1 \cap S_2) = \frac{u_2 - v_2 - v_2}{(u_2 - v_2) - (u_1 - t_1)} \tag{12}$$

Step 3 This stage computes the degree of possibility for a convex fuzzy number to be greater than k convex fuzzy numbers $S_i(i = 1, 2, \ldots, k)$. In the following expressions, Eq. (13) computes this notion.

$$\begin{aligned} &V(S \geq S_1, S_2, \ldots, M_k) = V[(S \geq S_1)\, and\, (S \geq S_2)\, and \ldots (S \geq S_k)] \\ &min V(S \geq S_i), i = 1, 2, \ldots, k \end{aligned} \tag{13}$$

Step 4 Computing the normalized weight W.

$$\text{Assuming that } d'(A_i) = min\, V(S_i \geq S_k) \text{ for } k = 1, 2, \ldots, n, k \neq i. \tag{14}$$

Then the normalized weight vector is expressed as in Eq. (15):

$$W' = \left(d'(A_1), d'(A_2), \ldots, d'(A_n)\right)^T \tag{15}$$

where $A_i (i = 1, 2, \ldots, n)$ are n elements.

4 Aggregation and Ranking with Fuzzy PROMETHEE

The PROMETHEE method known fully as the Preference Ranking Organization Method for Enrichment was introduced by Brans and Vincke [8]. The method works by performing a pairwise comparison of pairs of alternatives that are scored between an interval of [0, 1] by the help of a preference function. In this paper, the PROMETHEE I, which deals with partial ranking is combined effectively with PROMETHEE II that offers a full ranking approach [11, 12]. The partial ranking offered by PROMETHEE I, first uses the sum of indices, $\pi(m, l)$ to determine the preference of an alternative m to the rest of the alternatives, l. This is known as the 'leaving flow' $\varnothing^+(m)$, signifying the relative better performance of m over the other alternatives, l. In this instance, the alternative with the highest 'leaving flow' is adjudged the best. In similar manner, the sum of indices, $\pi(l, m)$ known as the 'entering flow' $\varnothing^-(m)$ signifies the superiority of the rest of the alternatives l over alternative m been considered. PROMETHEE II on the other hand introduces a construct called the net flow $\varnothing(m)$ which computes the difference between the leaving and the entering flows to derive a complete or full ranking. In view of this, the alternative with the highest net flow becomes the ideal or best alternative. The following section presents a systematic outline of the fuzzy PROMETHEE method culled from [11, 12].

Step 1a *Construction of the fuzzy decision matrix.*
Let $A = \{A_1, A_2, \ldots A_m\}$ be the set of alternatives to be ranked, $C = \{C_1, C_2, \ldots C_n\}$, the set of criteria and, $k = \{D_1, D_2, \ldots D_d\}$ the sets of decision makers. Equation (16), shows a decision matrix for decision maker, k = 1, 2, ..., d.

$$\tilde{k} = \begin{matrix} & C_1 & C_2 & & C_n \\ A_1 & \tilde{x}_{11} & \tilde{x}_{12} & \cdots & \tilde{x}_{1n} \\ A_2 & \tilde{x}_{21} & \tilde{x}_{22} & \cdots & \tilde{x}_{2n} \\ \vdots & \vdots & \vdots & \ddots & \vdots \\ A_m & \tilde{x}_{m1} & \tilde{x}_{m2} & \cdots & \tilde{x}_{nm} \end{matrix}, i = 1, 2, \ldots, m, j = 1, 2, \ldots, n \quad (16)$$

where x_{ij} is the rating of alternative A_i with respect to criterion C_j, expressed in Triangular fuzzy number format (TFN). This implies that the rating of a decision maker k is $\tilde{x}_{ij} = \left\{ t_{ij}^k, u_{ij}^k, v_{ij}^k \right\}$.

Step 1b *Aggregation of decisions.*

The ratings of the alternatives are aggregated at this stage. This is carried out using the interval valued technique as illustrated in Eq. (17).

$$\tilde{x}_{ij} = \frac{1}{n}\left[\tilde{x}_{ij}^1 + \tilde{x}_{ij}^2 + \cdots, + \tilde{x}_{ij}^n\right] \quad (17)$$

Step 2 *Normalization of the decision matrix.*

The aggregated fuzzy decision matrix derived in step 1b above is normalized to ensure uniformity in the units. The normalized fuzzy decision matrix is defined in Eq. 18 and computed using Eq. 19. It must be noted that the result of the normalized matrix still remains a TFN.

$$\tilde{S} = \left[\tilde{s}_{ij}\right]_{mxn}, i = 1, 2, \ldots, m; \mathrm{j} = 1, 2, \ldots, \mathrm{n} \quad (18)$$

$$\tilde{S}_{ij} = \left(\frac{\tilde{t}_{ij}}{v_j^+}, \frac{\tilde{u}_{ij}}{v_j^+}, \frac{\tilde{v}_{ij}}{v_j^+}\right) v_j^+ = \max_i v_{ij} \quad (19)$$

Step 4 *Construction of the fuzzy preference function.*

The fuzzy preference function $\tilde{P}_j(m,n)$ expresses the utility of the decision-makers' regarding the pairs of alternatives. The usual-criterion function (Type I) is used in this paper. However there are other functions such as the quasi-criterion, criterion with linear preference, level-criterion, Gaussian-criteria among others [12].

$$\tilde{P}_j(m,n) = \begin{cases} 0 & \tilde{s}_{mj} \leq \tilde{s}_{nj} \\ 1 & \tilde{s}_{mj} > \tilde{s}_{nj} \end{cases}, j = 1, 2, \ldots k \quad (20)$$

Step 5 *Computation of weighted aggregated preference function.*

The weighted aggregated preference function is computed using Eq. 21.

$$\tilde{\pi}(m,l) = \sum_{j=1}^{k} \tilde{P}_j(m,n)\tilde{w}_j \quad (21)$$

where $\tilde{w}_j$ is the importance weight of the criteria which is computed using the fuzzy AHP synthetic extent as shown in Sect. 3.1.

Step 6 *Computation of the leaving, entering and net flows.*
In this step, each alternative is related to $(n-1)$ alternatives that results in either a positive or a negative flow [11, 12]. The approach calculates the leaving, entering and net flows using Eqs. 22–24 respectively.
Leaving flow:

$$\varnothing^{+}(m) = \frac{1}{n-1}\sum_{m \neq 1} \tilde{\pi}(m, l), \forall m, l \in A \tag{22}$$

Entering flow:

$$\varnothing^{-}(m) = \frac{1}{n-1}\sum_{m \neq 1} \tilde{\pi}(l, m), \forall m, l \in A \tag{23}$$

where n is the number of alternatives.

Step 7 *Establishing ranking.*
This step uses PROMETHEE II for a full ranking using the net flow in Eq. 24.
Net flow:

$$\varnothing(m) = \varnothing^{+}(m) - \varnothing^{-}(m), \forall m \in A \tag{24}$$

5 Numerical Example

The hybrid fuzzy AHP Synthetic Extent—fuzzy PROMETHEE method is demonstrated in this numerical example of selecting deserving countries in developmental aid programmes. The OECD criteria [13] adopted in this example are as following: Sustainability (C1), Efficiency (C2), Impact (C3), Effectiveness (C4), and Relevance (C5). These 5 criteria set are used by 3 decision makers to evaluate 5 countries within the lower middle income bracket, a category that normally qualifies for such assistances. In this paper Afghanistan (A1), Ethiopia (A2), Haiti (A3), Somalia (A4) and Bangladesh (A5) are used in no particular order.

Table 1 is obtained through a pairwise comparison of each of the criteria per the ratings of the decision makers regarding their importance. The result is the fuzzy evaluation matrix in Table 1. Applying Eq. (8), the values of the fuzzy synthetic extent for each of the criteria are computed. For instance the value for S_1 is computed as below.

Table 1 Aggregated criteria weight by synthetic extent

	C_1	C_2	C_3	C_4	C_5
C_1	(1, 1, 1)	(0.87, 0.57, 0.98)	(0.33, 0.39, 0.49)	(0.04, 0.2, 0.45)	(0.7, 0.87, 1.0)
C_2	(0.29, 0.33, 0.40)	(1, 1, 1)	(0.27, 0.41, 0.55)	(0.22, 0.5, 0.67)	(0.5, 0.8, 0.97)
C_3	(0.55, 0.75, 0.87)	(0.4, 0.5, 0.67)	(1, 1, 1)	(0.33, 0.53, 0.7)	(0.66, 0.88, 0.99)
C_4	(0.33, 0.53, 0.7)	(0.32, 0.6, 0.77)	(0.13, 0.33, 0.41)	(1, 1, 1)	(0.52, 0.63, 0.88)
C_5	(0.67, 0.88, 0.96)	(0.67, 0.86, 1.0)	(0.57, 0.87, 0.98)	(0.4, 0.5, 0.67)	(1, 1, 1)

$$S_1 = (2.94, 3.03, 3.92) \otimes \left(\frac{1}{20.11}, \frac{1}{16.93}, \frac{1}{13.77}\right) = (0.15, 0.18, 0.28)$$

$$S_2 = (2.28, 3.04, 3.59) \otimes \left(\frac{1}{20.11}, \frac{1}{16.93}, \frac{1}{13.77}\right) = (0.11, 0.18, 0.26)$$

This is replicated to obtain the fuzzy synthetic extent for all the criteria. Now by using Eqs. (10) and (11), the various values are compared as done in the following.

$$V(S_1 \geq S_2) = 1.$$

However, for example $V(S_1 \geq S_5)$, since $S_5 \geq S_1$,

$$V(S_1 \geq S_5) = \frac{(0.16 - 0.28)}{(0.18 - 0.28) - (0.24 - 0.16)} = 0.67$$

Finally, by using Eq. (14), the weight vectors are obtained as follows:

$$d'(C1) = V(S_1 \geq S_2, S_3, S_4, S_5) = \min(1, 0.76, 1, 0.67) = 0.67$$
$$d'(C2) = V(S_2 \geq S_1, S_3, S_4, S_5) = \min(1, 0.73, 1.07, 0.63) = 0.63$$
$$d'(C3) = V(S_3 \geq S_1, S_2, S_4, S_5) = \min(1, 1, 1, 0.88) = 0.88$$
$$d'(C4) = V(S_4 \geq S_1, S_2, S_3, S_5) = \min(1, 1, 0.75, 0.65) = 0.65$$
$$d'(C5) = V(S_5 \geq S_1, S_2, S_3, S_4) = \min(1, 1, 1, 1) = 1$$

Then with the aid of Eq. (15), the normalized weight of W' = (0.67, 0.63, 0.88, 0.65, 1) is also obtained as follows as W = (0.17, 0.16, 0.23, 0.17, 0.26) where W is a non-fuzzy number.

5.1 *Fuzzy PROMETHEE*

The second part of the decision framework now factors the weight obtained through the fuzzy AHP synthetic extent in decision makers' aggregated ratings. In Table 3

Table 2 Linguistic terms for the alternative ratings

Linguistic terms	Triangular fuzzy number
Very low (VL)	(0.0, 0.0, 2.5)
Low (L)	(0.0, 2.5, 5.0)
High (H)	(2.5, 5.0, 7.5)
Very high (VH)	(5.0, 7.5, 10.0)
Extremely high (EH)	(7.5, 10.0, 10.0)

are the ratings by 3 decision makers. Table 2 shows the linguistic scale used by decision makers in their assessment of the candidate countries vying for development aid support.

The next step after the aggregation of decision makers' ratings is to compute preference functions which describe decision-makers' aggregated utility between pairs of alternatives. This paper used the 'usual criterion' as expressed in Eq. 20.

Subsequently, the preference functions are weighted using the weights obtained through the fuzzy AHP synthetic extent analysis shown in the first part of the paper. Using Eq. (21) for this task, the result of the weighted aggregated preference functions are as shown in Table 4.

Table 3 Alternatives' (Country) ratings by decision makers

		D_1	D_2	D_3			D_1	D_2	D_3			D_1	D_2	D_3
C_1	A_1	VL	VH	EH	C_3	A_1	L	EH	H	C_5	A_1	VL	EH	VL
	A_2	L	H	H		A_2	VL	H	EH		A_2	VH	H	L
	A_3	H	VH	H		A_3	L	L	VL		A_3	L	L	VH
	A_4	VH	L	VL		A_4	H	VL	L		A_4	H	VH	EH
	A_5	H	VL	L		A_5	H	H	VH		A_5	L	EH	L
C_2	A_1	H	VL	H	C_4	A_1	H	VL	H					
	A_2	H	EH	VH		A_2	VL	H	VL					
	A_3	VH	H	H		A_3	H	L	H					
	A_4	H	VH	L		A_4	H	H	H					
	A_5	VL	L	L		A_5	EH	H	EH					

Table 4 Weighted aggregated preference function

	A1	A2	A3	A4
A1		0.51	0.69	0.69
A2	0.99		1.2	0.69
A3	2.97	2.46		0
A4	0.99	0.51	1.2	
A5	0.51	0	1.2	*0.69*

Table 5 The leaving, entering, net flows and the resultant ranking of alternatives

	Leaving flow $\varnothing^{+}(m)$	Entering flow $\varnothing^{-}(m)$	Net flow $\varnothing(m)$	Ranking
A1	0.411	0.780	−0.368	*5*
A2	0.480	0.497	−0.0171	*3*
A3	0.920	0.613	0.304	*1*
A4	0.530	0.296	0.231	*2*
A5	0.343	0.493	−0.15	*4*

The last stage in the fuzzy PROMETHEE computes the leaving, entering and net flows respectively using Eqs. 22–24. This three constructs are then used in determining the final ranking of the alternatives. As demonstrated in Table 5 and per the ranking, alternative A3 (Haiti) would emerge as the best lower middle income country to receive the allocated development aid.

6 Conclusion

The research focused on developing a hybrid MCDM framework appropriate for selection of countries vying for competitive developmental aid support. The framework uses a numerical example to demonstrate how in the face of scarce resources, such developmental aid support can be given based on performance. The criteria set used is based on what OECD (one of the largest source of development aid) uses in selecting deserving countries. The proposed hybrid framework draws on the strength of fuzzy AHP Synthetic Extent and fuzzy PROMETHEE to ensure that the final selection is fair and reliable. Further, the decision framework demonstrates how in a highly uncertain area such as developmental aid support, the proposed hybrid framework could be used to realize the selection of deserving countries. In addition, the proposed method helps to keep track of progress of countries whiles ensuring that future aid allocations are based on performance of previous aid programs. The two methods in the hybrid framework are ideal for such decision problem because of their ability to pitch alternatives in a head-on pair wise comparison to determine one which is superior in relation to the underlying criteria. The proposed hybrid framework can be utilized in other similar decision problems where competition is keen and a deserving alternative has to be selected.

Acknowledgement This work was supported by Grant Agency of the Czech Republic—GACR P103/15/06700S, further by financial support of research project NPU I no. MSMT-7778/2014 by the Ministry of Education of the Czech Republic and also by the European Regional Development Fund under the Project CEBIA-Tech No. CZ.1.05/2.1.00/03.0089. Further, this work was supported by Internal Grant Agency of Tomas Bata University under the project no. IGA/CebiaTech/2016/007.

References

1. Kharas, K.: Measuring Aid Effectiveness Effectively: A Quality of Official Development Assistance Index. Brookings Institution Press (2011). http://www.brookings.edu/research/opinions/2011/07/26-aid-effectiveness-kharas. Accessed 10 Dec 2014
2. Morrison, K.M.: As the world bank turns: determinants of IDA lending in the cold war and after. Bus. Politics **13** (2011)
3. Claessens, S., Cassimon, D., Van Campenhout, B.: Evidence on changes in aid allocation criteria. World Bank Econ. Rev. **23**(2), 185–208 (2009)
4. Afful-Dadzie, E., Nabareseh, S., Afful-Dadzie, A., Oplatková, Z.K.: A fuzzy TOPSIS framework for selecting fragile states for support facility. Qual. Quant. **49**(5), 1835–1855 (2014)
5. Dalgaard, C.J., Hansen, H.: Evaluating Aid Effectiveness in the Aggregate: A Critical Assessment of the Evidence. University Library of Munich, Germany (2010)
6. Palenberg, M.: Tools and Methods for Evaluating the Efficiency of Development Interventions. Evaluation Working Papers (2011)
7. Zadeh, A.: Information and control. Fuzzy Sets **8**(3), 338–353 (1965)
8. Vincke, P., Brans, J.: A preference ranking organization method. the PROMETHEE method for MCDM. Manage. Sci. **31**(6), 647–656 (1985)
9. Chang, D.Y.: Extent Analysis and Synthetic Decision, Optimization Techniques and Applications, vol. 1, pp. 352. World Scientific, Singapore (1992)
10. Saaty, T.L., Thomas L.: How to make a decision: the analytic hierarchy process. Eur. J. Oper. Res. **48**(1), 9–26 (1990)
11. Amaral, T.M., Costa, A.P.: Improving decision-making and management of hospital resources: an application of the PROMETHEE II method in an Emergency Department. Oper. Res. Health Care **3**(1), 1–6 (2014)
12. Xiaojuan, T., Liu, X., Wang, L.: An improved PROMETHEE II method based on axiomatic fuzzy sets. Neural Comput. Appl. **25**, 1675–1683 (2014)
13. OECD: International Development Statistics. OECD Press (2014)

Specialized Decision Techniques for Data Mining, Transportation and Project Management

Measuring Quality of Decision Rules Through Ranking of Conditional Attributes

Urszula Stańczyk

Abstract One of the reasons for the wide popularity of rule classification systems is their ability to enhance understanding of mined data, and structures present in it. The discovered patterns are stated explicitly, which allows for more transparent descriptions of learned knowledge. To reach this goal of good descriptive and generalisation properties, induced rules need to be of a certain quality, which is typically measured by the predictive accuracy of the rule classifier. The paper presents research dedicated to measuring qualities of the inferred rules by taking into account a ranking of considered conditional attributes. Calculated quality measures along with supports of rules lead to construction of new classifiers, with improved parameters. The process is illustrated by a case of binary authorship attribution based on recognition of writing styles.

Keywords Decision rule · Quality measure · Conditional attribute · Ranking of attributes · Weights of attributes

1 Introduction

In order to obtain a rule classification system with high predictive accuracy, some measure of quality for constituent decision rules needs to be considered and exploited. Standard approaches refer to the direct parameters of rules, such as supports or strengths, lengths, and basing on them rule interestingness measures are constructed [6, 7]. An analysis of rule qualities leads to their filtering and optimisation of decision algorithms [13]. The process of rule selection can also be driven by some obtained ranking of the considered conditional attributes [18].

A ranking of variables is a result of assigning a value of some score function to the considered features, which defines their ordering. Typically the highest ranking attribute receives the highest value from the scoring function, and these values

U. Stańczyk (✉)
Silesian University of Technology, Akademicka 16, 44-100 Gliwice, Poland
e-mail: urszula.stanczyk@polsl.pl

I. Czarnowski et al. (eds.), *Intelligent Decision Technologies 2016*,
Smart Innovation, Systems and Technologies 56,
DOI 10.1007/978-3-319-39630-9_22

decrease further down the list, with the smallest assigned for the least ranking element. Rankings can be obtained through some statistical measures [11], dedicated algorithms, machine learning approaches, or procedures [4, 17].

The paper presents a research framework within which firstly for all conditional attributes their ranking by Relief algorithm is obtained [10], and next, basing on the resulting ordering of variables, three sets of weights are assigned to them. These weights are then exploited to evaluate a defined quality measure for all induced decision rules, depending on attributes they include. The values of the quality measure, with the help of supports of rules, are used in the process of constructing rule classifiers in search of optimised solutions, where predictive accuracy and numbers of rules included are treated as two optimisation criteria. The procedures of rule selection are illustrated with the binary classification task of authorship attribution, based on recognising writing styles [9].

The decision rules studied were induced within Dominance-based Rough Set Approach (DRSA), which is a modified version of rough set processing as defined by Pawlak [12], invented to support ordinal properties in data sets and enable not only nominal but also ordinal classification [14].

The paper is organised as follows. Section 2 describes related research areas, Sect. 3 explains the research framework, Sects. 4 and 5 present respectively pre-processing works and obtained results, while Sect. 6 concludes the paper.

2 Related Research Areas

The research described in this paper involved feature selection, rule classifiers, and filtering of decision rules, which are presented in the following subsections.

2.1 Attribute Ranking

Feature selection process can be based on their ranking [8]. Each ranking produces some kind of a score function that leads to an ordering of the considered variables. The values assigned by the score function can be interpreted as the direct weights of conditional attributes [17], or just treated as means for ordering, without looking for meaning or further application.

Rankings can be obtained by statistical measures [11], such as entropy, information gain, gain ratio, etc., or through dedicated algorithms [18] such as Relief, used in the presented research (pseudo-code shown as Algorithm 1), which iteratively calculates the differences between pairs of instances, to measure their suitability for discerning decision classes. The difference function *diff* takes into account values of attributes and averages results from all considered nearest hits (the closest neighbours in the same class) and nearest misses (the closest neighbours in the other class or classes) weighted by prior probabilities for each class.

Algorithm 1 Pseudo-code for RELIEF

Input: set of learning instances X, set A of N attributes,
set of classes $\boldsymbol{Cl}$, probabilities of classes $P(Cl)$,
number of iterations m,
number k of considered nearest instances from each class;
Output: vector of weights $\boldsymbol{w}$ for all attributes;
begin
for i = 1 **to** N **do** $w(i) = 0$ **endfor**
for i = 1 **to** m **do**
choose randomly an instance $x \in X$
find k nearest hits H_j
for each class $Cl \neq class(x)$ **do** find k nearest misses $M_j(Cl)$ **endfor**
for $l = 1$ **to** N **do**

$$w(l) = w(l) - \sum_{j=1}^{k} \frac{diff(l,x,H_j)}{m \times k} + \sum_{Cl \neq class(x)} \frac{\frac{P(Cl)}{1-P(class(x))} \sum_{j=1}^{k} diff(l,x,M_j(C))}{m \times k}$$

endfor
endfor
end {algorithm}

Ordering of variables can also be an outcome of some systematic procedure such as sequential forward or backward search, wrapped around the performance of the employed classification system [16]. With this approach there are no direct numerical values or weights assigned to attributes, only their relative (as the search is rarely exhaustive) ordering is discovered.

2.2 *DRSA Classifiers*

Classical Rough Set Approach (CRSA) defined by Pawlak [12] enables to perceive granules of knowledge as equivalence classes of objects that cannot be discerned basing on values of the describing them attributes. Such perspective allows only for nominal classification as ordinal properties are disregarded. To prevent that and support multi-criteria decision making Dominance-based Rough Set Approach (DRSA) [14] observes that the objects and the values of their attributes are less or more *preferred*. Defined preferences for all attributes (including decision attribute) lead to dominance cones—granules of knowledge corresponding to upward and downward unions of classes, and classification to *at most* some class, or *at least* some class.

With such definitions DRSA is well suited to data mining in cases of continuous values of attributes [15], as they are naturally ordered, and only specific preferences need to be chosen or discovered for all variables.

2.3 Quality of Decision Rules

One of elements differentiating rule inferring algorithms is the fact whether they generate the complete set of rules that can possibly be constructed, or just some subset that is considered as interesting or sufficient [3, 13]. Both approaches have advantages and disadvantages. The latter can significantly shorten processing time needed, yet it can result in omitting some important rules, while the former means more time required and access to such high numbers of rules that hinders recognition and detection of those rules that are the best for an intended purpose.

Whichever approach to induction is employed, once the rules are available, they can be further mined and filtered to find some sought out solution. Discovery of rule quality can be governed by their explicit parameters, such as length, strength, support, included conditions on variables, or coverage [19].

It is generally acknowledged that quite often short decision rules, with fewer conditions included, produce better results, that is give better recognition and higher classification accuracy [1]. High support of a rule means that is describes some pattern detectable in a high number of learning instances, which makes it likely to appear also within the testing samples [16].

Quality of decision rules can also be evaluated by some defined measures [5], as described in the next section of the paper.

3 Research Framework

The framework for research works presented in this paper consisted of the steps:

1. Pre-processing: (a) preparation of input datasets, (b) obtaining a ranking of all considered features, (c) generating decision rules, (d) mining induced rules to establish their parameters, (e) filtering rules by imposing hard constraints on their parameters to find the best algorithm,
2. Evaluation of rule quality: (a) assigning weights to all attributes, (b) calculating defined measures for all inferred rules, using defined weights of included attributes,
3. Construction of modified classifiers: (a) ordering rules by evaluated measures and rule supports, (b) filtering rules while following their rankings,
4. Tests and analysis of the obtained results, detailed in the next subsections.

3.1 Ranking-Based Weights of Attributes

If a ranking algorithm calculates and assigns numerical values to attributes, these values can be used as weights in calculations of quality measures for rules including these attributes. Yet in cases of rankings obtained through some search procedure, i.e. sequential selection, these weights need to be assigned arbitrarily.

When weights are assigned in the arbitrary manner, one of important decisions to make is whether to divide the space between the maximal and minimal values into sub-intervals with equal or different widths. The former approach focuses on observing the order of variables and not on how far apart they are, while the latter can take into account the individual or relative relevance of attributes, or the fact that in some circumstances the differences in relevance are higher and the process of ordering rather simple, while in others the variables are very similar and detailed analysis is required to pick one out of all remaining.

In the research three different sets of weights were employed for attributes:

- WI1-Relief—weights calculated in Relief algorithm applied to all variables,
- WI2-RED—arbitrarily assigned equidistant weights calculated according to the formula for the weight at *i*th position:

$$\forall_{i\in\{1,\ldots,N\}} w_i = 1 - \frac{i-1}{N} = \frac{N+1-i}{N}. \tag{1}$$

  The interval (0,1> is divided into N equal parts, N being the number of variables, and the first variable in the ranking gets the weight 1, the second $(N-1)/N$, and so on, to the least ranking variable with the weight $1/N$.
- WI3-RDD—arbitrarily assigned weights with decreasing distances between them, according to the equation:

$$\forall_{i\in\{1,\ldots,N\}} w_i = \frac{1}{i}. \tag{2}$$

  The highest ranking attribute has the weight of 1, the second in ranking $1/2$, and so on, to the lowest ranking variable with the weight $1/N$.

3.2 *Defined Rule Quality Measures*

With the assumption of previously obtained ranking of conditional attributes, there was defined a quality measure for decision rules in the following manner:

$$QM(r_i) = \prod_{j=1}^{K_{r_i}} w(a_j), \tag{3}$$

where r_i is a generated rule, K_{r_i} the number of its conditions, and $w(a_j)$ is a weight assigned to the attribute a_j. In short, the quality measure for the rule corresponds to the product of weights of all conditional attributes included in this rule. With such way of constructing measures and weights being fractional numbers, the more conditions a rule contains, the lower the value of its quality measure, thus the quality indicator is in fact inversely proportional to its length.

Since to calculate a measure only a subset of involved variables matters, and not particular conditions on them, it means that all rules with the conditions on the same subsets of attributes will get the same value of the quality measure. Yet even if the rules have the same conditional attributes included, they can still vary in quality, which is then indicated by their support values (and other parameters as well). This fact can be included in equation, by modifying the definition for the measure:

$$QMS(r_i) = S_{r_i} \prod_{j=1}^{K_{r_i}} w(a_j) = S_{r_i} \cdot QM(r_i) , \qquad (4)$$

where S_{r_i} denotes the value of support for rule r_i.

4 Pre-processing Step

To minimise influence of other factors on the studied problem, as a classification task there was selected binary authorship recognition with balanced classes.

4.1 Datasets for Style Mining

A writing style is independent on the form of presentation as its definition refers to quantitative descriptors that remain unique and characteristic for the author, making them distinguishable from others, and the task of authorship attribution suitable for execution by either statistic or machine learning approaches [9].

In recognition of authorship as characteristic features there are typically analysed lexical and syntactic markers, referring to the frequencies of usage of words, phrases, characters, patterns of sentences indicated by punctuation marks. These features can be studied separately from the employed inducer with the filtering approach, or with taking into account properties of the classifier.

In the research there was considered recognition of authorship between two writers, Jane Austen and Edith Wharton. The learning and testing samples corresponded to parts of longer works, to provide more information on style variations. As characteristic features 25 lexical and syntactic markers were selected, giving frequencies of occurrence for some function words (from the list of the most commonly used English words published by Oxford University Press) and punctuation marks that reflect syntactic structures of sentences. Frequencies are quite clearly continuous numbers, and in such form the values were processed, that is no discretisation was executed [2].

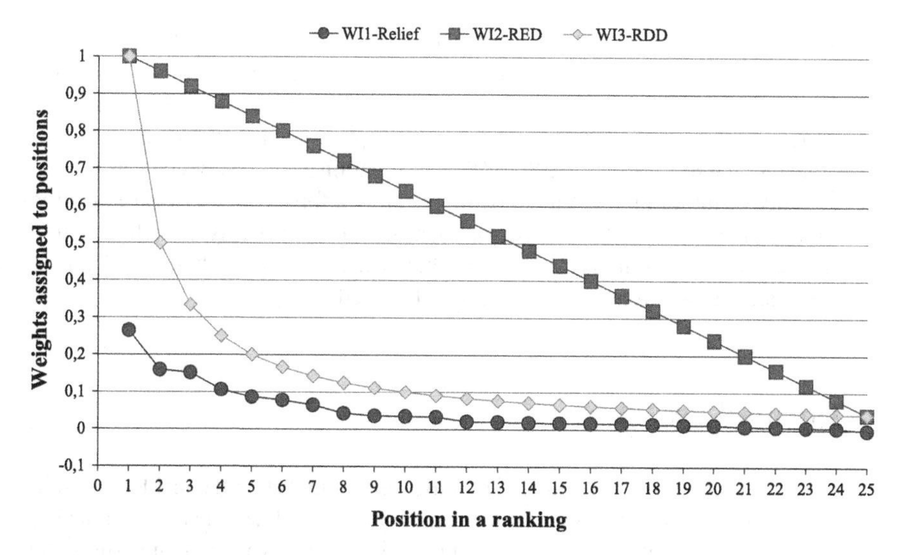

Fig. 1 The three sets of weights assigned to positions in a ranking

4.2 *Sets of Attribute Weights*

As described in the previous section, three sets of weights were assigned to attributes, constituting series shown in Fig. 1. WI1-Relief weights were calculated through Relief algorithm, and WI2-RED and WI3-RDD assigned arbitrarily.

In the first case the differences between subsequent values of weights are irregular, some close, others further apart, since they are calculated in such way in Relief algorithm. In the second and third cases the maximal and minimal values of weights are the same, yet for the former the difference between neighbouring values is constant, and for the latter, the distances are systematically decreasing.

4.3 *DRSA Decision Algorithm*

Within DRSA approach two decision algorithms were generated, firstly only with providing a minimal cover, and with testing cost preference for all attributes (the lower values of attributes the higher class they should be attributed to), and then gain preference (higher values point to the higher class). The latter preference resulted in higher predictive accuracy thus that was the one chosen for the generation of all rules on examples algorithm.

The complete algorithm contains 62,383 constituent rules, with lengths varying from 1 to 10, and number of supporting training instances from 1 to 77. Without any hard constraints on rules the complete algorithm gives zero recognition as the high number of rules causes all decisions to be ambiguous.

Associating quality of rules with their support, revealed by the number of supporting them training samples, means looking for such values of support that, when required of rules, leads to this selection of rules which results in the highest correct recognition ratio. This point in the search space was found for support equal 66 or greater, where 17 rules gave 86.67 % classification accuracy, which was used as the reference point for all comparisons. In this and all following cases classification accuracy is given as the total number of unambiguously correctly classified testing samples, while all ambiguous cases of matching rules with contradicting verdicts or no rules matching are treated as incorrectly classified.

5 Obtained Results

Once decision rules were inferred, and rankings and weights for all considered conditional attributes defined, it was possible to employ newly defined measures of quality for all rules, which resulted in their ordering. As there were two quality measures and three sets of weights, six different orderings of decision rules were established, basing on which in the next step subsets of rules were chosen for decision algorithms, to be compared against the previously found best algorithm.

For all three sets of attribute weights using *QM* measure meant ordering and grouping rules at the same time as many rules had the same values of the measure. Hence with each step of rule selection typically not single rules but their groups were added to the subset.

Application of *QMS* measure resulted in highly differentiated values for rules, in fact hardly any rules had the same values, so some decision had to be made with respect to the number of rules selected in each step—if we retrieve decision rules strictly within the same value of the measure, then in many cases the selection gives just single rules, and very high numbers of constructed decision algorithms, varying just by single rules. Yet even so small differences can mean a significant change in classification accuracy.

For all constructed decision algorithms, even for those based on *QMS* measure values, still the supports of rules were taken into consideration by imposing hard constraints on rules, which led to their further reduction and shortened algorithms. In each of the six search paths there were found several algorithms with improved parameters, either with the same predictive accuracy as the previously found the best algorithm, but with fewer rules, or with more rules included but higher correct classification ratio. The parameters of these enhanced rule classifiers are displayed in Table 1, where there is shown the value of the maximal classification accuracy obtained for certain number of rules with supports equal or higher than the specified values, separately for each search path.

An analysis of best results from all six orderings of rules, given in Fig. 2, leads to the conclusion that in the optimisation space with two parameters, a maximal classification accuracy and a minimal length of the algorithm (understood as a number of included rules), there are two Pareto points. While keeping the recognition at the

Table 1 The best results in classification accuracy [%] with respect to minimal support required of rules and their number denoted as NR, for three sets of weights for attributes and both quality measures

		WI1-Relief					WI2-RED			WI3-RDD			
		Support											
QM	NR	44	46	52	55	66	52	55	66	46	52	55	66
	29		90.00							90.00			
	27				86.67			86.67					
	25			87.78							87.78		
	19	86.67											
	17					86.67			86.67				86.67
	16					86.67							86.67
	14					86.67			86.67				86.67
	13					86.67			86.67				86.67
	11					86.67							86.67
		WI1-Relief					WI2-RED			WI3-RDD			
		Support											
QMS	NR	44	46	52	55	66	52	55	66	46	52	55	66
	27				86.67								
	25			87.78							87.78		
	24							86.67					
	23											86.67	
	22				86.67		86.67					86.67	
	21							86.67					
	20				86.67		86.67					86.67	
	19							86.67				86.67	
	18	86.67			86.67		86.67					86.67	
	17				86.67	86.67		86.67	86.67			86.67	86.67
	16				86.67								86.67
	15	86.67			86.67							86.67	86.67
	14				86.67	86.67			86.67			86.67	86.67
	13								86.67				86.67
	12								86.67				86.67
	11					86.67			86.67				86.67

reference level of 86.67 % the algorithm can be reduced from 17 to 11 rules with supports of these rules equal to 66 or higher, or, we can have prediction increased to 90 % but at the cost of increasing the length of the algorithm to 29 rules and lowering the constraint on minimal rule support to 46.

The first of the two optimised algorithms, with the reduced number of rules, was found in all but one search paths, while the other, with increased recognition, only in two, for employing *QM* measure and WI1-Relief or WI2-RDD weights, both sets characterised by unequal distances between subsequent values of weights, decreasing while following down the ranking.

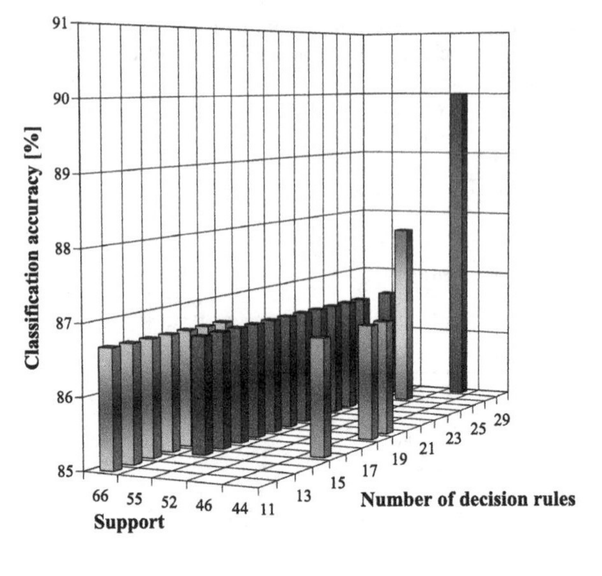

Fig. 2 The best results in classification accuracy with respect to minimal support required of rules and their number, accumulated from all search paths

6 Conclusions

The paper presents research on quality of decision rules, measured while taking into account positions within a ranking for included conditional attributes. Basing on these positions, three sets of weights were assigned to all variables and these weights were used in evaluation of defined quality measures for induced decision rules. Values of measures along with information on rule supports helped to filter decision rules in search of optimised decision algorithms, while treating classification accuracy and lengths of algorithms as optimality criteria. The described procedure led to improved solutions, which was illustrated by a case of binary authorship attribution through recognising writing styles.

Acknowledgments 4eMka Software used for DRSA processing was developed at the Poznan University of Technology (http://www-idss.cs.put.poznan.pl/). The research works presented in the paper were performed at the Silesian University of Technology, Gliwice, Poland within the project BK/RAu2/2016.

References

1. Amin, T., Chikalov, I., Moshkov, M., Zielosko, B.: Relationships between length and coverage of decision rules. Fundam. Informaticae **129**, 1–13 (2014)
2. Baron, G.: Influence of data discretization on efficiency of Bayesian classifier for authorship attribution. Procedia Comput. Sci. **35**, 1112–1121 (2014)
3. Bayardo, Jr., R., Agrawal, R.: Mining the most interesting rules. In: Proceedings of the 5th ACM SIGKDD International Conference on Knowledge Discovery and Data Mining. pp. 145–154 (1999)

4. Ferreira, A., Figueiredo, M.: Incremental filter and wrapper approaches for feature discretization. Neurocomputing **123**, 60–74 (2014)
5. Freitas, A.: On rule interestingness measures. Knowl.-Based Syst. **12**(5–6), 309–315 (1999)
6. Greco, S., Słowiński, R., Szczęch, I.: Analysis of monotonicity properties of some rule interestingness measures. Control Cybern. **38**(1), 9–25 (2009)
7. Gruca, A., Sikora, M.: Rule based functional description of genes—estimation of the multicriteria rule interestingness measure by the UTA method. Biocybern. Biomed. Eng. **33**, 222–234 (2013)
8. Jensen, R., Shen, Q.: Computational Intelligence and Feature Selection. Wiley, Hoboken (2008)
9. Jockers, M., Witten, D.: A comparative study of machine learning methods for authorship attribution. Literary Linguistic Comput. **25**(2), 215–223 (2010)
10. Kononenko, I.: Estimating attributes: analysis and extensions of RELIEF. In: Bergadano, F., De Raedt, L. (eds.) Machine Learning: ECML-94. LNCS, vol. 784, pp. 171–182. Springer, Berlin (1994)
11. Mansoori, E.: Using statistical measures for feature ranking. Int. J. Pattern Recogn. Artif. Intell. **27**(1), 1350003–14 (2013)
12. Pawlak, Z.: Rough sets and intelligent data analysis. Inform. Sci. **147**, 1–12 (2002)
13. Sikora, M., Wróbel, L.: Data-driven adaptive selection of rule quality measures for improving the rule induction algorithm. LNCS **6743**, 279–287 (2011)
14. Słowiński, R., Greco, S., Matarazzo, B.: Dominance-based rough set approach to reasoning about ordinal data. In: Kryszkiewicz, M., Peters, J., Rybiński, H., Skowron, A. (eds.) Rough Sets and Intelligent Systems Paradigms. LNCS (LNAI), vol. 4585, pp. 5–11. Springer, Berlin (2007)
15. Stańczyk, U.: Dominance-based rough set approach employed in search of authorial invariants. In: Kurzyński, M., Woźniak, M. (eds.) Computer Recognition Systems 3, AISC, vol. 57, pp. 315–323. Springer, Berlin (2009)
16. Stańczyk, U.: On performance of DRSA-ANN classifier. In: Corchado, E., Kurzyński, M., Woźniak, M. (eds.) Hybrid Artificial Intelligence Systems. Part 2, LNCS (LNAI), vol. 6679, pp. 172–179. Springer, Berlin (2011)
17. Stańczyk, U.: Weighting of attributes in an embedded rough approach. In: Gruca, A., Czachórski, T., Kozielski, S. (eds.) Man-Machine Interactions 3, AISC, vol. 242, pp. 475–483. Springer, Berlin (2013)
18. Stańczyk, U.: Selection of decision rules based on attribute ranking. J. Intell. Fuzzy Syst. **29**(2), 899–915 (2015)
19. Zielosko, B.: Optimization of decision rules relative to coverage—comparative study. In: Kryszkiewicz, M., Cornelis, C., Ciucci, D., Medina-Moreno, J., Motoda, H., Raś, Z. (eds.) Rough Sets and Intelligent Systems Paradigms, LNCS, vol. 8537, pp. 237–247. Springer, Berlin (2014)

Greedy Algorithm for Optimization of Association Rules Relative to Length

Beata Zielosko and Marek Robaszkiewicz

Abstract In the paper, an optimization of α-Association rules constructed by greedy algorithm is proposed. It allows us to decrease the number of rules and obtain short rules, what is important from the point of view of knowledge representation. Experimental results for data sets from UCI Machine Learning Respository are presented.

Keywords Rough sets · Length · Greedy algorithm · Association rules

1 Introduction

Association rule mining is one of the popular field of data mining and knowledge discovery. It aims to extract interesting associations or frequent patterns among sets of items in data repository. Association rules are widely used in various areas such as business field for decision making and effective marketing, telecommunication networks, etc. Even, when data are initially incomplete due to the way they were acquired or merged, they can be preprocessed [22] to obtain suitable data set for association rule mining.

There are many approaches for construction of association rules. The most popular, is Apriori algorithm based on frequent itemsets [2], and frequent pattern growth algorithm [6] that adopts divide and conquer strategy. During years, many new algorithms were designed which are based on, e.g., hash based technique [14], transaction reduction [1], partitioning the data [17], and others [3, 7, 13].

B. Zielosko (✉)
Institute of Computer Science, University of Silesia, 39, Będzińska St., 41-200 Sosnowiec, Poland
e-mail: beata.zielosko@us.edu.pl

M. Robaszkiewicz
EL-PLUS, 11, Inwalidzka St., 41-506 Chorzów, Poland
e-mail: marek.robaszkiewicz@gmail.com

I. Czarnowski et al. (eds.), *Intelligent Decision Technologies 2016*, Smart Innovation, Systems and Technologies 56,
DOI 10.1007/978-3-319-39630-9_23

Association rules can be definded in many ways. In the paper, a special kind of association rules is studied, i.e., they relate to decision rules. Similar approach was considered in [11, 12, 24, 25]. Application of rough sets theory to the construction of rules for knowledge representation or classification tasks is usually connected with the usage of decision table [15] as form of input data representation. In such a table one attribute is distinguished as a decision attribute and it relates to a rule's consequence. However, in the last years, associative mechanism of rules construction, where all attributes can occur as premises or consequences of particular rules, is popular [13, 23].

There are different rule quality measures that are used for induction or classification tasks [8, 20, 21]. The most popular, for association rules mining, are support (sometimes called as coverage) and confidence.

In the paper, the length is considered as a rule's evaluation measure. The choice of this measure is connected with the Minimum Description Length principle introduced by Rissanen [16]: the best hypothesis for a given set of data is the one that leads to the largest compression of the data. Construction and optimization of rules relative to length can be considered as important task for knowledge representation. Short rules are easier for understanding and interpreting by experts. Unfortunately, the problem of construction of rules with minimum length is *NP*-hard [12, 13]. The most part of approaches, with the exception of brute-force and dynamic programming cannot guarantee the construction of rules with the minimum length. So, approximate polynomial algorithms for this problem should be studied.

The paper, based on theoretical results from [11, 12], presents a greedy algorithm for construction of association rules. Then, optimization of rules relative to length is proposed, i.e., the shortest rules, for each row of information system [15], are selected. The aim of the paper is to show that proposed optimization allows one to obtain short association rules and reduce the number of rules.

In [10] it was shown that for wide classes of binary information systems the number of irreducible association rules for almost all information systems is not polynomial in the number of attributes. It means that efficient approaches for construction of relatively small sets of "important" rules should be investigated. In [11, 12] it was shown that under the assumption on the class *NP*, the greedy algorithm is close to the best polynomial approximate algorithms for partial association rule minimization. The approach presented in this paper is different from the known approaches based on the frequent itemsets, but allows one to construct "important" in some sense rules in a reasonable time.

In the paper, not only exact but also approximate association rules are considered. Exact rules can be overfitted, i.e., dependent essentially on the noise or adjusted too much to the existing examples. If rules are considered as a way of knowledge representation [18], then instead of an exact rule with many attributes it is more appropriate to work with a partial rule containing smaller number of attributes and having relatively high accuracy.

Possible applications of association rules obtained using presented approach are knowledge representation, construction of classifiers, inference processes in knowledge base systems, filling missing values of attributes.

The paper contains experimental results for data sets from UCI Machine Learning Repository [9], mainly, for the length of optimized rules and the number of rules.

The paper consists of six sections. Section 2 contains main notions. In Sect. 3, algorithm for construction and optimization of rules is presented. In Sect. 4 some theorems about precision of greedy algorithm are presented. Section 5 contains experimental results with data sets from UCI Machine Learning Repository. Section 6 contains conclusions.

2 Main Notions

An *information system* I is a table with n rows and m columns labeled with attributes $a_1, \dots, a_m$. This table is filled by nonnegative integers (values of attributes). Formally, information system is defined as a pair $I = (U, A)$, where $U = \{r_1, \dots, r_n\}$ is nonempty, finite set of objects (rows), $A = \{a_1, \dots, a_m\}$ is nonempty, finite set of attributes, i.e., $a : U \rightarrow V_a$ is a function, for any $a \in A$, V_a is the set of values of attribute a, called the domain of a. From information system, a decision table can be obtained. Formally, decision table is defined as $T = (U, A \cup \{d\})$, where $U = \{r_1, \dots, r_n\}$ is nonempty, finite set of objects (rows), $A = \{a_1, \dots, a_{m-1}\}$ is nonempty, finite set of attributes. Elements of the set A are called conditional attributes, $d \notin A$ is a distinguished attribute, called a decision attribute.

Let $r = (b_1, \dots, b_m)$ be a row of I, and a_p be an attribute from the set $\{a_1, \dots, a_m\}$. Let us denote by $U(I, r, a_p)$ the set of rows from I which are different from r in the column a_p and in at least one column a_j such that $j \in \{1, \dots, m\} \backslash \{p\}$. We will say that an attribute a_i *separates* a row $r' \in U(I, r, a_p)$ from the row r if the rows r and r' have different values at the intersection with the column a_i. The triple (I, r, a_p) will be called an *association rule problem.*

To define a notion of an approximate rule, a parameter α and the value $|U(I, r, a_p)|$ are used. Let $\alpha \in \mathbf{R}$ and $0 \leq \alpha < 1$. A rule

$$(a_{i_1} = b_{i_1}) \wedge \cdots \wedge (a_{i_t} = b_{i_t}) \rightarrow a_p = b_p \tag{1}$$

is called an *α-association rule* for (I, r, a_p) if $i_1, \dots, i_t \in \{1, \dots, m\} \setminus \{p\}$, and attributes $a_{i_1}, \dots, a_{i_t}$ separate from r at least $(1 - \alpha)|U(I, r, a_p)|$ rows from $U(I, r, a_p)$ (such rules are also called *partial* association rules). For example, for α=0.1, 0.1-association rule means that at least 90 % of rows from $U(I, r, a_p)$ should be separated from the row r. If $\alpha = 0$ then α-association rule is an exact association rule (100 % of rows from $U(I, r, a_p)$ should be separated from the row r).

The length of α-association rule it is the number of descriptors (pairs "attribute= value") from the left-hand side of rule. The coverage of α-association rule it is the number of rows in I, for which the left-hand side and the right-hand side of α-association rule is true. The coverage of rules is also known as a support of association rules, especially when it is presented as a percentage of rows from data set that the given rule satisfies. Coverage allows us to discover major patterns in the data and the choice of this rules' evaluation measure is important from the point of view of knowledge representation.

3 Greedy Algorithm for Construction and Optimization of α-association Rules

Algorithm 3.1 pretenses a greedy algorithm for construction of α-association rules. This algorithm is applied sequentially to each row of information system I.

Algorithm 3.1: Greedy algorithm for α-association rule construction

Input : Information system I with attributes $a_1, \ldots, a_m$, row $r = (b_1, \ldots, b_m)$ of I, and real number α, $0 \leq \alpha < 1$.
Output: α-association rules for triples (I, r, a_p), $p = 1, \ldots, m$.

```
for i = 1 to m do
    a_p = a_i;
    Q ← ∅;
    while attributes from Q separate from r less than (1 − α)|U(I, r, a_p)| rows
    from U(I, r, a_p) do
        select a_i ∈ {a_1, ..., a_m} \ {a_p} with minimum index i such that a_i
        separates from r the maximum number of rows from U(I, r, a_p)
        unseparated by attributes from Q
        Q ← Q ∪ {a_i};
    end
    return ∧_{a_i∈Q} (a_i = b_i) → a_p = b_p;
end
```

By $L_{greedy}(\alpha) = L_{greedy}(\alpha, I, r, a_p)$ is denoted the length of constructed α association rule for (I, r, a_p). By $L_{min}(\alpha) = L_{min}(\alpha, I, r, a_p)$ is denoted the minimum length of α-association rule for (I, r, a_p).

The union of α-association rules obtained for a given row r, for triples (I, r, a_p), $p = 1, \ldots, m$, is considered as a set of α-association rules $Rul(\alpha, I, r)$ for information system I, row r and α.

Now, an optimization of α-association rules, constructed by the greedy algorithm, relative to length is presented. The aim of the proposed optimization is to select, for each row of I, the shortest rules.

The minimum length of α-association rule for the row r of information system I and α, is denoted by $Opt^l(\alpha, I, r)$.

$$Opt^l(\alpha, I, r) = \min\{L_{greedy}(\alpha, I, r, a_p) \; : \; p = 1, \ldots, m\}.$$

By $Rul^l(\alpha, I, r)$ is denoted the set of α-association rules with minimum length, for the row r of information system I and α. During optimization, among all numbers $L_{greedy}(\alpha, I, r, a_p), p = 1, \ldots, m$, only these are selected where

$$L_{\text{greedy}}(\alpha, I, r, a_p) = Opt^l(\alpha, I, r).$$

Then the set $Rul^l(\alpha, I, r)$ is the union of rules for which the last equality holds.

4 Precision on Greedy Algorithm

In this section, based on results from [5, 11–13, 19] some theorems are presented, to show relevancy of greedy algorithm from the point of view of minimization of rule length.

Let I be the information system with m columns labeled with attributes $a_1, \ldots, a_m$, r be a row of I, and $a_p \in \{a_1, \ldots, a_m\}$.

Theorem 1 *Let* $0 \leq \alpha < 1$ *and* $\lceil (1-\alpha)|U(I, r, a_p)| \rceil \geq 2$. *Then*

$$L_{\text{greedy}}(\alpha) < L_{\min}(\alpha)(\ln \lceil (1-\alpha)|U(I, r, a_p)| \rceil - \ln\ln \lceil (1-\alpha)|U(I, r, a_p)| \rceil + 0.78).$$

Theorem 2 *Let* $0 \leq \alpha < 1$. *Then for any natural* $t \geq 2$ *there exists an association rule problem* (I, r, a_p) *such that* $\lceil (1-\alpha)|U(I, r, a_p)| \rceil = t$ *and* $L_{\text{greedy}}(\alpha) > L_{\min}(\alpha)$ $(\ln \lceil (1-\alpha)|U(I, r, a_p)| \rceil - \ln\ln \lceil (1-\alpha)|U(I, r, a_p)| \rceil - 0.31)$.

Theorem 3 *Let* $0 \leq \alpha < 1$ *and* $U(I, r, a_p) \neq \emptyset$. *Then*

$$L_{\text{greedy}}(\alpha) \leq L_{\min}(\alpha) \left(1 + \ln \left(\max_{j \in \{1,\ldots,m\} \setminus \{p\}} |U(I, r, a_p, a_j)| \right) \right).$$

Theorem 4 *Let* $0 \leq \alpha < 1$. *Then the problem of construction of* α*-association rule with minimum length is NP-hard.*

Theorem 5 *Let* $\alpha \in \mathbf{R}$ *and* $0 \leq \alpha < 1$. *If* $NP \not\subseteq DTIME(n^{O(\log \log n)})$, *then for any* ε, $0 < \varepsilon < 1$, *there is no polynomial algorithm that for a given association rule problem* (I, r, a_p) *with* $U(I, r, a_p) \neq \emptyset$ *constructs an* α*-association rule for* (I, r, a_p) *which length is at most*

$$(1-\varepsilon)L_{\min}(\alpha, I, r, a_p) \ln |U(I, r, a_p)|.$$

From Theorem 3 it follows that $L_{\text{greedy}}(\alpha) \leq L_{\min}(\alpha)(1 + \ln |U(I, r, a_p)|)$. From this inequality and from Theorem 5 it follows that, under the assumption $NP \not\subseteq DTIME(n^{O(\log \log n)})$, the greedy algorithm is close to the best polynomial approximate algorithms for partial association rule minimization.

5 Experimental Results

The aim of the experiments is to show, mainly, the relevancy of the optimization (i) relative to length of α-association rules constructed by the greedy algorithm and (ii) relative to the number of rules.

Experiments were made on data sets from UCL Machine Learning Repository [9]. Each data set was considered as the information system I and for each triple (I, r, a_p), $p = 1, \ldots, m$, α-association rules were constructed by the greedy algorithm. Then, optimization of α-association rules relative to length was preformed.

Table 1 presents the length of α-association rules after optimization, for $\alpha = \{0.0, 0.1, 0.2, 0.3\}$. For each row r of I, the minimum length of α-association rule for I, r and α (value $Opt^l(\alpha, I, r)$) was obtained. After that, for rows of I, the minimum length of α-association rule (column "Min"), the maximum length of such rule (column "Max") and the average length of α-association rules (column "Avg") were obtained. Column "Rows" contains the number of rows in I, column "Attr" contains the number of attributes in I. We can see that the length of α-association rules is not decreasing when the value of α is increasing. For $\alpha = 0.3$, the minimum, average and maximum length of rules are equal to 1. In the case of exact rules ($\alpha = 0.0$), the maximum length of 0-association rules, for all data sets, is smaller than the number of attributes, in particular, for "Zoo-data" (17 attributes, the maximum length is equal to 1), "Teeth" (9 attributes, the maximum length is equal to 1), "Soybean-small" (36 attributes, the maximum length is equal to 1), "House-votes" (17 attributes, the maximum length is equal to 2), "Breast-cancer" (10 attributes, the maximum length is equal to 3), such values are in bold.

The length of α-association rules before optimization, for $\alpha = \{0.0, 0.1, 0.2, 0.3\}$, was calculated also. Table 2 presents comparison of the average length of α-association rules before and after optimization. Each input of this table is equal to the average length of α-association rules before optimization divided by the average length of α-association rules after optimization. The biggest difference (more than two times) is noticable, for $\alpha = 0.0$ ("Adult-stretch", "Monks-1-test", "Monks-3-test", "Nursery"), and for $\alpha = 0.1$ ("Adult-stretch"), such values are in bold. It means that proposed optimization allows us to obtain short rules, in particular, for small values of α. The same average length (before and after optimization) exists only of $\alpha = 0.2$ ("Balance-scale") and $\alpha = 0.3$ ("Balance-scale", "Cars", "Nursery", "Soybean-small", "Teeth").

The number of α-association rules, for $\alpha = \{0.0, 0.1, 0.2, 0.3\}$, before and after optimization was calculated. Table 3 (column "diff") presents comparison of the number of rules before and after optimization (column "nr"), for $\alpha = \{0.0, 0.1, 0.2, 0.3\}$. Each input of this table is equal to the number of α-association rules before optimization divided by the number of α-association rules after optimization. The number of rules before optimization is equal to $n \times m$, where n is the number of rows (column "Rows"), m is the number of attributes (column "Attr"). The differences (more than two times) are noticable, the biggest one is equal to seven, such

Table 1 Length of α-association rules after optimization, $\alpha = \{0.0, 0.1, 0.2, 0.3\}$

Name of data set	Rows	Attr	$\alpha = 0.0$			$\alpha = 0.1$			$\alpha = 0.2$			$\alpha = 0.3$		
			Min	Avg	Max	Min	Avg	Max	Min	Avg	Max	Min	Avg	Max
Adult-stretch	16	5	1	1.00	1	1	1.00	1	1	1.00	1	1	1.00	1
Balance-scale	625	5	3	3.21	4	1	1.68	2	1	1.00	1	1	1.00	1
Breast-cancer	266	10	1	1.97	3	1	1.10	2	1	1.00	1	1	1.00	1
Cars	1728	7	1	2.91	6	1	1.72	2	1	1.13	2	1	1.00	1
Hayes-roth-data	69	5	1	2.13	3	1	1.87	2	1	1.06	2	1	1.00	1
House-votes	279	17	2	2.00	2	1	1.00	1	1	1.00	1	1	1.00	1
Lenses	24	5	1	1.32	3	1	1.07	2	1	1.22	2	1	1.00	1
Monks-1-test	432	7	1	2.25	3	1	1.95	2	1	1.70	2	1	1.00	1
Monks-3-test	432	7	1	1.80	2	1	1.87	2	1	1.00	1	1	1.00	1
Nursery	12960	9	1	2.56	6	1	1.87	2	1	1.00	1	1	1.00	1
Shuttle-landing	15	7	1	1.00	1	1	1.00	1	1	1.00	1	1	1.00	1
Soybean-small	47	36	1	1.00	1	1	1.00	1	1	1.00	1	1	1.00	1
Teeth	23	9	1	1.00	1	1	1.00	1	1	1.00	1	1	1.00	1
Tic-tac-toe	958	10	3	3.49	4	2	2.00	2	1	1.53	2	1	1.00	1
Zoo-data	59	17	1	1.00	1	1	1.00	1	1	1.00	1	1	1.00	1

Table 2 Comparison of the average length of α-association rules, $\alpha = \{0.0, 0.1, 0.2, 0.3\}$

Data set	$\alpha = 0.0$	$\alpha = 0.1$	$\alpha = 0.2$	$\alpha = 0.3$
Adult-stretch	**2.45**	**2.45**	1.75	1.65
Balance-scale	1.08	1.09	1.00	1.00
Breast-cancer	1.53	1.29	1.08	1.01
Cars	1.64	1.08	1.57	1.00
Hayes-roth-data	1.28	1.02	1.21	1.03
House-votes	1.82	1.79	1.35	1.15
Lenses	1.80	1.80	1.30	1.24
Monks-1-test	**2.06**	1.08	1.11	1.18
Monks-3-test	**2.50**	1.09	1.86	1.11
Nursery	**2.46**	1.02	1.12	1.00
Shuttle-landing	1.46	1.33	1.20	1.14
Soybean-small	1.47	1.15	1.02	1.00
Teeth	1.14	1.06	1.02	1.00
Tic-tac-toe	1.34	1.01	1.12	1.06
Zoo-data	1.57	1.20	1.07	1.01

Table 3 Comparison of the number of α-association rules, $\alpha = \{0.0, 0.1, 0.2, 0.3\}$

Data set	Rows	Attr	$\alpha = 0.0$		$\alpha = 0.1$		$\alpha = 0.2$		$\alpha = 0.3$	
			nr	diff	nr	diff	nr	diff	nr	diff
Adult-stretch	16	5	20	**4.00**	20	**4.00**	20	**4.00**	28	**2.86**
Balance-scale	625	5	2059	1.52	1700	1.84	3119	1.00	3125	1.00
Breast-cancer	266	10	798	**3.33**	1756	1.51	2438	1.09	2624	1.01
Cars	1728	7	2309	**5.24**	6202	1.95	3130	**3.86**	12096	1.00
Hayes-roth-data	69	5	106	**3.25**	223	1.55	286	1.21	328	1.05
House-votes	279	17	1064	**4.46**	1551	**3.06**	3121	1.52	4027	1.18
Lenses	24	5	36	**3.33**	45	**2.67**	56	**2.14**	69	1.74
Monks-1-test	432	7	432	**7.00**	2052	1.47	1080	**2.80**	2484	1.22
Monks-3-test	432	7	528	**5.73**	1572	1.92	432	**7.00**	2700	1.12
Nursery	12960	9	18598	**6.27**	86070	1.36	102074	1.14	116640	1.00
Shuttle-landing	15	7	74	1.42	74	1.42	86	1.22	89	1.18
Soybean-small	47	36	622	**2.72**	880	1.92	1015	1.67	1033	1.64
Teeth	23	9	185	1.12	195	1.06	203	1.02	207	1.00
Tic-tac-toe	958	10	2288	**4.19**	9517	1.01	5684	1.69	9006	1.06
Zoo-data	59	17	591	1.70	800	1.25	930	1.08	988	1.02

Table 4 Comparison of coverage of α-association rules, $\alpha = \{0.0, 0.1, 0.2, 0.3\}$

Name of data set	$\alpha = 0.0$			$\alpha = 0.1$			$\alpha = 0.2$			$\alpha = 0.3$		
	cov1	cov2	cov	cov1	cov2	cov	cov1	cov2	cov	cov1	cov2	cov
Adult-stretch	2.50	6.40	**2.56**	2.50	6.40	**2.56**	3.10	5.14	1.66	3.20	5.14	1.61
Balance-scale	1.66	2.01	1.21	16.90	26.60	1.57	36.21	36.27	1.00	36.22	36.22	1.00
Breast-cancer	4.73	9.42	1.99	12.74	15.95	1.25	17.93	18.84	1.05	18.87	19.03	1.01
Cars	49.03	250.45	**5.11**	77.13	120.78	1.57	90.34	260.02	**2.88**	160.40	160.40	1.00
Hayes-roth-data	2.39	4.93	**2.06**	3.21	3.94	1.23	5.02	5.63	1.12	6.41	6.49	1.01
House-votes	20.33	56.25	**2.77**	47.11	92.44	1.96	64.45	86.52	1.34	73.59	82.52	1.12
Lenses	2.97	6.32	**2.13**	3.39	5.93	1.75	3.83	5.09	1.33	4.90	5.85	1.19
Monks-1-test	8.33	45.00	**5.40**	18.33	22.11	1.21	22.71	38.40	1.69	43.29	48.52	1.12
Monks-3-test	10.75	51.82	**4.82**	23.30	31.15	1.34	28.92	108.67	**3.76**	51.52	56.27	1.09
Nursery	331.52	2064.42	**6.23**	552.75	675.73	1.22	1084.42	1202.41	1.11	1160.38	1160.38	1.00
Shuttle-landing	1.98	2.19	1.11	1.98	2.13	1.08	2.02	2.09	1.03	2.02	2.08	1.03
Soybean-small	6.85	9.96	1.45	7.33	8.26	1.13	7.62	7.71	1.01	7.65	7.66	1.00
Teeth	3.23	3.49	1.08	3.23	3.36	1.04	3.23	3.27	1.01	3.23	3.23	1.00
Tic-tac-toe	5.18	14.49	**2.80**	29.47	29.60	1.00	45.41	62.10	1.37	84.39	87.15	1.03
Zoo-data	10.50	13.79	1.31	11.51	12.97	1.13	12.22	12.73	1.04	12.52	12.60	1.01

Table 5 Length and number of rules constructed by Apriori algorithm [4]

Name of data set	Number of rules	Length		
		Min	Avg	Max
Adult-stretch	584	1	2.53	4
Balance-scale	324	1	2.48	4
Breast-cancer	403356	1	4.92	9
Cars	11216	1	4.00	6
Hayes-roth-data	28	1	**1.71**	3
Lenses	9	1	1.33	2
Monks-1-test	75	1	2.13	4
Monks-3-test	75	1	**2.13**	4
Shuttle-landing	28	1	1.71	3
Teeth	1439	1	2.44	5
Tic-tac-toe	47	1	**1.79**	3
Zoo-data	123	1	1.85	3

values are in bold. It confirms that proposed optimization decreses the number of rules significantly, in particular, for small values of α.

Table 4 (column "cov") presents comparison of the average coverage of α-association rules before (column "cov1") and after optimization (column "cov2") relative to length, for $\alpha = \{0.0, 0.1, 0.2, 0.3\}$. Each input of this table is equal to the average coverage of rules before optimization divided by the average coverage of rules after optimization. The differences are noticable, especially for $\alpha = 0.0$, the biggest one is more than six times, for data set "Nursery". In general, α-association rules after optimization relative to length have greater average coverage than before optimization. In [24], the number of α-association rules before optimization depending on the minimum coverage (support) was presented.

Table 5 presents the number of rules and minimum (column "min"), average (column "avg"), and maximum length (column "max") of association rules constructed by Apriori algorithm implemented by Borglet [4] (http://www.borgelt.net/apriori.html). The approach presented in this paper is totally different, so it is difficult to make comparison. However, regarding of average length of constructed rules, proposed optimization allows us to obtain often shorter rules. Bold values in column "avg" denotes better results for the Apriori algorithm.

Figure 1 presents, for data set "Cars", comparison of rules construed by the greedy algorithm ($\alpha = 0$) and optimized relative to length, and constructed by the Apriori algorithm [4]. The left-hand side of Fig. 1 presents the number of rules for a given length. The right-hand side of Fig. 1 presents the average coverage of rules for a given length (the values on the y axis are presented in the logarithmic scale). We can see that for "Cars", in general, Apriori algorithm constructs more rules than the greedy one, however the greedy algorithm constructs much more short rules with length equal to 1 than the Apriori algorithm. Regarding of the average coverage,

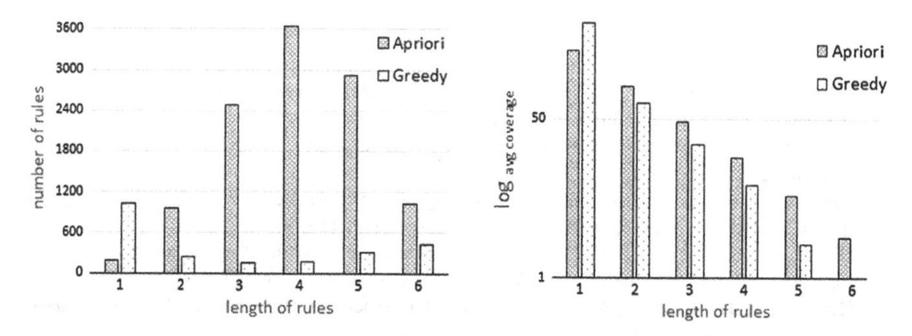

Fig. 1 Comparison for the data set "Cars"

in general, Apriori algorithm constructs rules with higher coverage than the greedy one. Only for rules with length equal to 1 the average coverage of rules constructed by the greedy algorithm is grater than the average coverage of rules constructed by the Apriori algorithm. This is due to the number of rules with length equals 1 is much greater.

6 Conclusions

In the paper, an optimization of α-association rules constructed by the greedy algorithm was proposed. The presented approach is different from the known approaches based on the frequent itemsets, but allows one to construct "important" in some sense rules in a reasonable time. Construction and optimization of rules relative to length can be considered as important task for knowledge representation. Short rules are easier for understanding and interpreting by experts.

Experimental results show that α-association rules after optimization relative to length are short and the number of rules is reduced, often, more than two times. It should be noted also, that optimization relative to length allows us to obtain α-association rules which average coverage is usually greater than before optimization.

Future works will be connected with the construction of classifiers for association rules.

References

1. Agrawal, R., Srikant, R.: Fast algorithms for mining association rules in large databases. In: Bocca, J.B., Jarke, M., Zaniolo, C. (eds.) VLDB, pp. 487–499. Morgan Kaufmann (1994)
2. Agrawal, R., Imieliński, T., Swami, A.: Mining association rules between sets of items in large databases. In: SIGMOD '93, pp. 207–216. ACM (1993)

3. Borgelt, C.: Simple algorithms for frequent item set mining. In: Koronacki, J., Raś, Z.W., Wierzchoń, S.T., Kacprzyk, J. (eds.) Advances in Machine Learning II, Studies in Computational Intelligence, vol. 263, pp. 351–369. Springer, Berlin Heidelberg (2010)
4. Borgelt, C., Kruse, R.: Induction of association rules: Apriori implementation. 15th Conference on Computational Statistics (Compstat 2002. Berlin, Germany), pp. 395–400. Physica Verlag, Heidelberg (2002)
5. Feige, U.: A threshold of $\ln n$ for approximating set cover. In: Leighton, F.T. (ed.) Journal of the ACM (JACM), vol. 45, pp. 634–652. ACM New York (1998)
6. Han, J., Pei, J., Yin, Y., Mao, R.: Mining frequent patterns without candidate generation: a frequent-pattern tree approach. Data Min. Knowl. Discov. **8**(1), 53–87 (2004)
7. Herawan, T., Deris, M.M.: A soft set approach for association rules mining. Knowled.-Based Syst. **24**(1), 186–195 (2011)
8. Kozak, J., Boryczka, U.: Multiple boosting in the ant colony decision forest meta-classifier. Knowled.-Based Syst. **75**, 141–151 (2015)
9. Lichman, M.: UCI Machine Learning Repository. http://archive.ics.uci.edu/ml/. Accessed Feb 2016
10. Moshkov, M.J., Skowron, A., Suraj, Z.: On minimal rule sets for almost all binary information systems. Fundam. Inform. **80**(1–3), 247–258 (2007)
11. Moshkov, M.J., Piliszczuk, M., Zielosko, B.: On construction of partial association rules. In: Wen, P., Li, Y., Polkowski, L., Yao, Y., Tsumoto, S., Wang, G. (eds.) RSKT, LNCS, vol. 5589, pp. 176–183. Springer (2009)
12. Moshkov, M.J., Piliszczuk, M., Zielosko, B.: Greedy algorithm for construction of partial association rules. Fundam. Inform. **92**(3), 259–277 (2009)
13. Nguyen, H.S., Ślęzak, D.: Approximate reducts and association rules - correspondence and complexity results. In: Zhong, N., Skowron, A., Ohsuga, S. (eds.) RSFDGrC, LNCS, vol. 1711, pp. 137–145. Springer (1999)
14. Park, J.S., Chen, M.S., Yu, P.S.: An effective hash based algorithm for mining association rules. In: Carey, M.J., Schneider, D.A. (eds.) SIGMOD Conference, pp. 175–186. ACM Press (1995)
15. Pawlak, Z., Skowron, A.: Rudiments of rough sets. Inf. Sci. **177**(1), 3–27 (2007)
16. Rissanen, J.: Modeling by shortest data description. Automatica **14**(5), 465–471 (1978)
17. Savasere, A., Omiecinski, E., Navathe, S.B.: An efficient algorithm for mining association rules in large databases. In: Dayal, U., Gray, P.M.D., Nishio, S. (eds.) VLDB, pp. 432–444. Morgan Kaufmann (1995)
18. Skowron, A.: Rough sets in KDD - plenary talk. In: Shi, Z., Faltings, B., Musen, M. (eds.) Proceedings of the 16th IFIP, pp. 1–14. World Computer Congress, Publishing House of Electronic Industry (2000)
19. Slavík, P.: A tight analysis of the greedy algorithm for set cover. In: Proceedings of the Twenty-Eighth Annual ACM Symposium on Theory of Computing. pp. 435–441. ACM New York (1996)
20. Stańczyk, U.: Selection of decision rules based on attribute ranking. J. Intell. Fuzzy Syst. **29**(2), 899–915 (2015)
21. Stefanowski, J., Vanderpooten, D.: Induction of decision rules in classification and discovery-oriented perspectives. Int. J. Intell. Syst. **16**(1), 13–27 (2001)
22. Tkacz, M.A.: Artificial neural networks in incomplete data sets processing. In: Kopotek, M.A., Wierzchoń, S.T., Trojanowski, K. (eds.) IIS: IIPWM'05. pp. 577–584. Advances in Soft Computing, Springer (2005)
23. Wieczorek, A., Słowiński, R.: Generating a set of association and decision rules with statistically representative support and anti-support. Inf. Sci. **277**, 56–70 (2014)
24. Zielosko, B.: Greedy algorithm for construction of partial association rules. Studia Inform. **31**(2A), 225–236 (2010) (in Polish)
25. Zielosko, B.: Global optimization of exact association rules relative to coverage. In: Kryszkiewicz, M., Bandyopadhyay, S., Rybiński, H., Pal, S.K. (eds.) PReMI 2015. LNCS, vol. 9124, pp. 428–437. Springer (2015)

Decision Rules with Collinearity Models

Leon Bobrowski

Abstract Data mining algorithms are used for discovering general regularities based on the observed patterns in data sets. Flat (multicollinear) patterns can be observed in data sets when many feature vectors are located on a planes in the multidimensional feature space. Collinear patterns can be useful in modeling linear interactions between multiple variables (features) and can be used also in a decision support process. Flat patterns can be efficiently discovered in large, multivariate data sets through minimization of the convex and piecewise linear (*CPL*) criterion functions.

Keywords Data mining • Decision rules • Collinear models • *CPL* criterion functions

1 Introduction

Patterns (regularities) can be discovered in large data sets by using variety of data mining tools [1]. Observed patterns could allow for extraction an useful information from a given data set. Extracted regularities are used in solving many practical problems linked, for example, to decision support, stock forecasting, or to scientific discoveries.

We assume that data sets are structuralized as feature vectors of the same dimensionality representing particular objects (patients, events, dynamical processes) [2]. Different patterns composed of numerous feature vectors can be extracted from the same data set by using different computational tools serving, for example, in cluster analysis, anomaly detection or mining of interaction models.

L. Bobrowski (✉)
Faculty of Computer Science, Białystok University of Technology, Białystok, Poland
e-mail: l.bobrowski@pb.edu.pl

L. Bobrowski
Institute of Biocybernetics and Biomedical Engineering, PAS, Warsaw, Poland

I. Czarnowski et al. (eds.), *Intelligent Decision Technologies 2016*, Smart Innovation, Systems and Technologies 56,
DOI 10.1007/978-3-319-39630-9_24

A collinear (flat) patterns mean clusters around central planes in multidimensional feature space [3]. Central planes can be represented as the *vertexical planes* which are based on degenerated vertices in parameter space [4]. The considered representation of the flat patterns can be compared with the techniques based on the Hough transform [5, 6]. Feature vectors subsets located around the *vertexical planes* can be discovered in large data sets through minimization of the convex and piecewise linear (*CPL*) criterion functions [7]. Theoretical properties of this method of flat patterns discovering are analyzed in the presented article.

2 Vertices in the Parameter Space

Let us assume that the data set C contains m feature vectors $\mathbf{x}_j[n] = [x_{j,1},\ldots,x_{j,n}]^T$ belonging to a given n-dimensional feature space $F[n]$ ($\mathbf{x}_j[n] \in F[n]$):

$$C = \{x_j[n]\}, \text{ where } j = 1, \ldots, m \tag{1}$$

Components $x_{j,i}$ of the feature vector $\mathbf{x}_j[n]$ can be treated as the numerical results of n standardized examinations of the j-th object O_j, where $x_{ji} \in \{0, 1\}$ or $x_{ji} \in R$.

Each of m feature vector $\mathbf{x}_j[n]$ from the set C (1) can define the following (*dual*) hyperplane h_j in the parameter space R^n ($\mathbf{w}[n] \in R^n$):

$$(\forall j \in \{1, \ldots, m\}) \quad h_j = \left\{ w[n]: x_j[n]^T w[n] = 1 \right\} \tag{2}$$

where $\mathbf{w}[n] = [w_1, \ldots, w_n]^T$ is the parameter (*weight*) vector ($\mathbf{w}[n] \in R^n$).

Each of n unit vectors $\mathbf{e}_i[n] = [0, \ldots, 1, \ldots, 0]^T$ defines the following hyperplane h_i^0 in the parameter space R^n:

$$(\forall i \in \{1, \ldots, n\}) \quad h_i^0 = \left\{ \mathbf{w}[n]: \mathbf{e}_i[n]^T \mathbf{w}[n] = 0 \right\} = \{ \mathbf{w}[n]: w_i = 0 \} \tag{3}$$

Let us consider a set S_k of n linearly independent vectors which is composed of n' feature vectors $\mathbf{x}_j[n]$ ($j \in J_k$) and $n - n'$ unit vectors $\mathbf{e}_i[n]$ ($i \in I_k$).

$$S_k = \left\{ \mathbf{x}_j[n]: j \in J_k \right\} \cup \left\{ \mathbf{e}_i[n]: i \in I_k \right\} \tag{4}$$

The k-th *vertex* $\mathbf{w}_k[n]$ in the parameter space R^n is the intersection point of n hyperplanes h_j (2) or h_i^0 (3) defined by the linearly independent vectors $\mathbf{x}_j[n]$ ($j \in J_k$) and $\mathbf{e}_i[n]$ ($i \in I_k$) from the set S_k (4). The intersection point $\mathbf{w}_k[n] = [w_{k,1}, \ldots, w_{k,n}]^T$ (9) is defined by the following linear equations:

$$(\forall j \in J_k) \quad \mathbf{w}_k[n]^T \mathbf{x}_j[n] = 1 \tag{5}$$

and

$$(\forall i \in I_k) \quad \mathbf{w}_k[n]^T \mathbf{e}_i[n] = 0 \ or \ w_{k,i} = 0 \tag{6}$$

The Eqs. (5) and (6) can be represented in the following matrix form:

$$\mathbf{B}_k[n] \ \mathbf{w}_k[n] = \mathbf{1}'[n] = [1, \ldots, 1, 0, \ldots, 0]^T \tag{7}$$

The square, nonsingular matrix $\mathbf{B}_k[n]$ is the k-th *basis* linked to the vertex $\mathbf{w}_k[n]$:

$$\mathbf{B}_k[n] = \left[\mathbf{x}_{j(1)}[n], \ldots, \mathbf{x}_{j(n')}[n], \mathbf{e}_{i(n'+1)}[n], \ldots, \mathbf{e}_{i(n)}[n]\right]^T \tag{8}$$

and

$$\mathbf{w}_k[n] = \mathbf{B}_k[n]^{-1}\mathbf{1}'[n] = [\mathbf{r}_1[n]], \ldots, \mathbf{r}_n[n]\mathbf{1}'[n] \tag{9}$$

where the symbol $\mathbf{r}_i[n]$ means the i-th column of the reverse matrix $\mathbf{B}_k[n]^{-1}$ ($\mathbf{B}_k[n]^{-1} = [\mathbf{r}_1[n], \ldots., \mathbf{r}_n[n]]$).

Definition 1 The *rank* r_k ($1 \leq r_k \leq n$) of the vertex $\mathbf{w}_k[n] = [w_{k,1},\ldots,w_{k,n}]^T$ (9) is defined as the number of nonzero components $w_{k,i}$ ($w_{k,i} \neq 0$) of the vector $\mathbf{w}_k[n]$.

In accordance with the Eq. (6), if the i-th unit vectors $\mathbf{e}_i[n]$ is contained in the basis $\mathbf{B}_k[n]$ (8), then the i-th component $w_{k,i}$ of the vertex $\mathbf{w}_k[n] = [w_{k,1},\ldots,w_{k,n}]^T$ (9) is equal to zero ($w_{k,i} = 0$). The reverse sentence is not necessarily true. It can be given examples of such matrices $\mathbf{B}_k[n]$ (8) that the vector $\mathbf{x}_{j(i)}[n]$ from the set C (1) constituting the i-th row of the matrix $\mathbf{B}_k[n]$ results in the i-th component $w_{k,i}$ (9) equal to zero ($w_{k,i} = 0$).

Definition 2 The *standard basis* $\mathbf{B}_k[n]$ (8) linked to the vertex $\mathbf{w}_k[n]$ of the *rank* r_k is composed of r_k feature vectors $\mathbf{x}_{j(i)}[n]$ from the set C (1) and $n - r_k$ unit vectors $\mathbf{e}_i[n]$.

Definition 3 The *degree of degeneration* of the vertex $\mathbf{w}_k[n]$ (9) of the rank r_k is defined as the number $d_k = m_k - r_k$, where m_k is the number of such feature vectors $\mathbf{x}_j[n]$ ($\mathbf{x}_j[n] \in C$) from the set C (1), which define the hyperplanes h_j (2) passing through this vertex ($\mathbf{w}_k[n]^T\mathbf{x}_j[n] = 1$).

3 Vertexical Planes in the Feature Space

The hyperplane $H(\mathbf{w}[n], \theta)$ in the n-dimensional feature space $F[n]$ is defined in the following manner:

$$H(\mathbf{w}[n],\theta) = \left\{\mathbf{x}[n]: \mathbf{w}[n]^{\mathrm{T}}\mathbf{x}[n] = \theta\right\} \tag{10}$$

where $\mathbf{w}[n]$ is the *weight vector* ($\mathbf{w}[n] \in R^n$) and θ is the *threshold* ($\theta \in R^1$).

The ($r_k - 1$)-dimensional *vertexical plane* $P_k(\mathbf{x}_{j(1)}[n], \ldots, \mathbf{x}_{j(rk)}[n])$ in the feature space $F[n]$ is defined as the linear combination of r_k ($r_k > 1$) *supporting vectors* $\mathbf{x}_{j(i)}[n]$ ($j \in J_k$) (5) belonging to the basis $\mathbf{B}_k[n]$ (8) [4]:

$$P_k\left(\mathbf{x}_{j(1)}[n], \ldots, \mathbf{x}_{j(rk)}[n]\right) = \left\{\mathbf{x}[n]: \mathbf{x}[n] = \alpha_1\mathbf{x}_{j(1)}[n] + \cdots + \alpha_k\mathbf{x}_{j(rk)}[n]\right\} \tag{11}$$

where $j(i) \in J_k$ (5) and the parameters α_i ($\alpha_i \in R^1$) fulfill the following equation:

$$\alpha_1 + \cdots + \alpha_{rk} = 1 \tag{12}$$

Two linearly independent supporting vectors $\mathbf{x}_{j(1)}[n]$ and $\mathbf{x}_{j(2)}[n]$ from the set C (1) *span* the line $l(\mathbf{x}_{j(1)}[n], \mathbf{x}_{j(2)}[n])$ in the feature space $F[n]$ ($\mathbf{x}[n] \in F[n]$):

$$l\left(\mathbf{x}_{j(1)}[n], \mathbf{x}_{j(2)}[n]\right) = \left\{\mathbf{x}[n]: \mathbf{x}[n] = \mathbf{x}_{j(1)}[n] + \alpha\left(\mathbf{x}_{j(2)}[n] - \mathbf{x}_{j(1)}[n]\right)\right\} \tag{13}$$

where $\alpha \in R^1$. The Eq. (13) can be given in the below form:

$$l\left(\mathbf{x}_{j(1)}[n], \mathbf{x}_{j(2)}[n]\right) = \left\{\mathbf{x}[n]: \mathbf{x}[n] = (1-\alpha)\mathbf{x}_{j(1)}[n] + \alpha\mathbf{x}_{j(2)}[n]\right\} \tag{14}$$

Remark 1 The line $l(\mathbf{x}_{j(1)}[n], \mathbf{x}_{j(2)}[n])$ (14) in the feature space $F[n]$ is the *vertexical plane* $P_k(\mathbf{x}_{j(1)}[n], \mathbf{x}_{j(2)}[n])$ (11) spanned by two supporting vectors $\mathbf{x}_{j(1)}[n]$ and $\mathbf{x}_{j(2)}[n]$ with $\alpha_1 = 1 - \alpha$ and $\alpha_2 = \alpha$.

Theorem 1 *If the vector* $\mathbf{x}_j[n]$ *is situated on the* (r_k − 1)*-dimensional vertexical plane* $P_k(\mathbf{x}_{j(1)}[n], \ldots, \mathbf{x}_{j(rk)}[n])$ (11), *where* $j(i) \in J_k$ (5), *then the feature vector* $\mathbf{x}_j[n]$ *defines the hyperplane* h_j (2) *which passes through the vertex* $\mathbf{w}_k[n]$ (9).

Proof If the vector $\mathbf{x}_j[n]$ is situated on the (r_k − 1)-dimensional vertexical plane $P_k(\mathbf{x}_{j(1)}[n], \ldots, \mathbf{x}_{j(rk)}[n])$ (11), then the following equation is fulfilled:

$$\mathbf{x}_j[n] = \alpha_{j,1}\mathbf{x}_{j(1)}[n] + \cdots + \alpha_{j,rk}\mathbf{x}_{j(rk)}[n] \tag{15}$$

where the parameters $\alpha_{j,i}$ satisfy the equation $\alpha_{j,1} + \cdots + \alpha_{j,rk} = 1$ (12). Thus (5)

$$\mathbf{w}_k[n]^{\mathrm{T}}\mathbf{x}_j[n] = \mathbf{w}_k[n]^{\mathrm{T}}\left(\alpha_{j,1}\mathbf{x}_{j(1)}[n] + \cdots + \alpha_{j,rk}\mathbf{x}_{j(rk)}[n]\right) = \alpha_{j,1} + \cdots + \alpha_{j,rk} = 1 \tag{16}$$

It means, that the hyperplane h_j (2) passes through the vertex $\mathbf{w}_k[n]$ (9). □

Each feature vector $\mathbf{x}_j[n]$ (1) can be represented as the linear combination of the basis vectors $\mathbf{x}_{j(i)}[n]$ and $\mathbf{e}_{i(l)}[n]$ (8) in the vertex $\mathbf{w}_k[n]$ (9), where $\alpha_{j,i} \in R^1$:

$$\mathbf{x}_j[n] = \alpha_{j,1}\mathbf{x}_{j(1)}[n] + \cdots + \alpha_{j,rk}\mathbf{x}_{j(rk)}[n] + \alpha_{j,rk+1}\mathbf{e}_{i(rk+1)}[n] + \cdots + \alpha_{j,n}\mathbf{e}_{i(n)}[n] \tag{17}$$

Because $(\forall j(i) \in J_k)$ $(\mathbf{w}_k[n])^T\mathbf{x}_{j(i)}[n] = 1$ (5) and $(\forall i(l) \in I_k)$ $(\mathbf{w}_k[n])^T\mathbf{e}_{i(l)}[n] = 0$ (6), then (16)

$$\mathbf{w}_k[n]^T\mathbf{x}_j[n] = \alpha_{j,1} + \cdots + \alpha_{j,rk} \tag{18}$$

So, the below rule holds:

$$\textit{if } \alpha_{j,1} + \cdots + \alpha_{j,rk} = 1, \textit{ then } (\mathbf{w}_k[n])^T\mathbf{x}_j[n] = 1 \tag{19}$$

The condition $\mathbf{w}_k[n]^T\mathbf{x}_j[n] = 1$ means, that the hyperplane h_j (2) defined by the feature vector $\mathbf{x}_j[n]$ (17) with the condition $\alpha_{j,1} + \cdots + \alpha_{j,rk} = 1$ passes through the vertex $\mathbf{w}_k[n]$ (9) of the rank r_k.

Theorem 2 *If the vertex* $\mathbf{w}_k[n] = [w_{k,1}, \ldots, w_{k,n}]^T$ (9) *has the maximal rank* $r_k = n$, *then each hyperplane* h_j (2) *which passes through this vertex* $(\mathbf{w}_k[n]^T\mathbf{x}_j[n] = 1)$ *is defined by the feature vector* $\mathbf{x}_j[n]$ *which is located on the* $(n - 1)$*-dimensional vertexical plane* $P_k(\mathbf{x}_{j(1)}[n], \ldots, \mathbf{x}_{j(n)}[n])$ (11) *equal to the hyperplane* $H(\mathbf{w}_k[n], 1)$ (10).

Proof If the vertex $\mathbf{w}_k[n] = [w_{k,1}, \ldots, w_{k,n}]^T$ (9) has the rank $r_k = n$, then the basis $\mathbf{B}_k[n]$ (7) is composed from n feature vectors $\mathbf{x}_{j(i)}[n]$ and contains no unit vectors $\mathbf{e}_i[n]$ $(\mathbf{B}_k[n] = [\mathbf{x}_{j(1)}[n], \ldots, \mathbf{x}_{j(n)}[n]]^T)$. In these circumstances the Eqs. (15) and (16) are fulfilled and the thesis of the Theorem 2 holds. □

4 Convex and Piecewise Linear (*CPL*) Criterion Functions

The convex and piecewise linear (*CPL*) penalty functions $\varphi_j(w[n])$ are defined on the feature vectors $x_j[n]$ from the C (1) [3]:

$$\begin{aligned} &(\forall \mathbf{x}_j[n] \in C)\, 1 - \mathbf{w}[n]^T\mathbf{x}_j[n] \quad \textit{if} \quad \mathbf{w}[n]^T\mathbf{x}_j[n] \le 1 \\ &\varphi_j(\mathbf{w}[n]) = \left|1 - \mathbf{w}[n]^T\mathbf{x}_j[n]\right| = \mathbf{w}[n]^T\mathbf{x}_j[n] - 1 \quad \textit{if} \quad \mathbf{w}[n]^T\mathbf{x}_j[n] > 1 \end{aligned} \tag{20}$$

The penalty functions $\varphi_j(\mathbf{w}[n])$ are equal to the absolute values $|1 - \mathbf{w}[n]^T\mathbf{x}_j[n]|$.

The criterion function $\Phi_k(\mathbf{w}[n])$ is defined as the weighted sum of the penalty functions $\varphi_j(\mathbf{w}[n])$ (20) linked to the feature vectors $\mathbf{x}_j[n]$ from the data subset C_k $(C_k \subset C)$ (1) [3]:

$$\Phi_k(\mathbf{w}[n]) = \sum_{j \in J_k} \beta_j \varphi_j(\mathbf{w}[n]) \tag{21}$$

where $J_k = \{j: \mathbf{x}_j[n] \in C_k\}$ is the set of indices j of the feature vectors $\mathbf{x}_j[n]$ from the subset C_k and the positive parameters β_j ($\beta_j > 0$) can be treated as the *prices* of particular feature vectors $\mathbf{x}_j[n]$. The standard value of the parameters β_j one ($\beta_j = 1.0$).

The criterion function $\Phi_k(\mathbf{w}[n])$ (21) is convex and piecewise linear because is the sum of such type of functions $\beta_j\, \varphi_j(\mathbf{w}[n])$. The minimal value Φ_k^* of the criterion function $\Phi_k(\mathbf{w}[n])$ can be found in one of the vertices $\mathbf{w}_k[n]$ (9) [7]. The optimal vertex $\mathbf{w}_k^*\,[n]$ (9) constitutes the minimal value Φ_k^* of the function $\Phi_k(\mathbf{w}[n])$ (21):

$$\left(\exists \mathbf{w}_k^*[n]\ (9)\right)(\forall \mathbf{w}[n])\Phi_k(\mathbf{w}[n]) \geq \Phi_k\left(\mathbf{w}_k^*[n]\right) = \Phi_k^* \geq 0 \tag{22}$$

The basis exchange algorithms which are similar to the linear programming allow to find efficiently an optimal vertex $\mathbf{w}_k^*[n]$ (22) even in the case of large, multidimensional data subsets C_k ($C_k \subset C$) (1) [8].

The optimal weight vector $\mathbf{w}_k^*[n]$ (22) can be used in the definition of the below hyperplane $H(\mathbf{w}_k^*[n],1)$ (10) with the threshold θ equal to one ($\theta = 1$):

$$H\left(\mathbf{w}_k^*[n], 1\right) = \left\{\mathbf{x}[n]: \mathbf{w}_k^*[n]^T \mathbf{x}[n] = 1\right\} \tag{23}$$

Theorem 3 *The minimal value* $\Phi_k(\mathbf{w}_k^*[n])$ (22) *of the criterion function* $\Phi_k(\mathbf{w}[n])$ (21) *defined on elements* $\mathbf{x}_j[n]$ *of the subset* $C_k \subset C$ (1) *is equal to zero* ($\Phi_k(\mathbf{w}_k^*[n]) = 0$), *if and only if all the feature vectors* $\mathbf{x}_j[n]$ *from the subset* C_k *are situated on some hyperplane* $H(\mathbf{w}, \theta)$ (10) with $\theta \neq 0$.

Proof Let us suppose that all the feature vectors $\mathbf{x}_j[n]$ from the subset C_k are situated on some hyperplane $H(\mathbf{w}'[n], \theta')$ (2) with $\theta' \neq 0$:

$$\left(\forall \mathbf{x}_j[n] \in C_k\right)\ \mathbf{w}'[n]^T \mathbf{x}_j[n] = \theta' \tag{24}$$

From this

$$\left(\forall \mathbf{x}_j[n] \in C_k\right)\ \left(\mathbf{w}'[n]/\theta'\right)^T \mathbf{x}_j[n] = 1 \tag{25}$$

The above equations mean that the penalty functions $\varphi_j(\mathbf{w}[n])$ (20) are equal to zero in the point $\mathbf{w}'[n]/\theta'$:

$$\left(\forall \mathbf{x}_j[n] \in C_k\right)\ \varphi_j\left(\mathbf{w}'[n]/\theta'\right) = 0 \tag{26}$$

or (21)

$$\Phi_k\left(\mathbf{w}'[n]/\theta'\right) = 0 \tag{27}$$

On the other hand, if the criterion function $\Phi_k(\mathbf{w}[n])$ (21) is equal to zero in some point $\mathbf{w}'[n]$, then each of the penalty functions $\varphi_j(\mathbf{w}'[n])$ (20) has to be equal to zero:

$$(\forall \mathbf{x}_j[n] \in C_k)\ \varphi_j(\mathbf{w}'[n]) = 0 \tag{28}$$

or

$$(\forall \mathbf{x}_j[n] \in C_k)\ \mathbf{w}'[n]^T \mathbf{x}_j[n] = 1 \tag{29}$$

The above equations mean that each feature vector $\mathbf{x}_j[n]$ from the subset C_k is located on the hyperplane $H(\mathbf{w}'[n], 1)$ (23). □

It has been also proved that the minimal value $\Phi_k(\mathbf{w}_k^*[n])$ (22) of the criterion function $\Phi_k(\mathbf{w}[n])$ (21) does not depend on linear, non-singular data transformations (the *invariance property*) [7]:

$$\Phi_k'(\mathbf{w}_k'[n]) = \Phi_k(\mathbf{w}_k^*[n]) \tag{30}$$

where $\Phi_k'(\mathbf{w}_k'[n])$ is the minimal value of the criterion functions $\Phi_k(\mathbf{w}[n])$ (21) defined on the transformed feature vectors $\mathbf{x}_j'[n]$:

$$(\forall \mathbf{x}_j[n] \in C_k)\ \mathbf{x}_j'[n] = A\,\mathbf{x}_j[n] \tag{31}$$

and A is a non-singular matrix of the dimension $(n \times n)$ (the matrix A^{-1} exists).

The minimal value Φ_k^* (22) of the criterion function $\Phi_k(\mathbf{w}[n])$ (21) is characterized by two below *monotonicity properties*:

The *positive monotonicity property due to reduction of feature vectors* $\mathbf{x}_j[n]$: The reducing of the subset C_k to $C_{k'}$ by neglecting some feature vectors $\mathbf{x}_j[n]$ cannot cause an increase of the minimal value Φ_k^* (22) of the criterion function $\Phi_k(\mathbf{w}[n])$ (21):

$$(C_{k'} \subset C_k) \Rightarrow (\Phi_{k'}^* \le \Phi_k^*) \tag{32}$$

where the symbol $\Phi_{k'}^*$ stands for the minimal value (22) of the criterion function $\Phi_{k'}(\mathbf{w}[n])$ (21) defined on the elements $\mathbf{x}_j[n]$ of the subset $C_{k'}$ ($\mathbf{x}_j[n] \in C_{k'}$).

The implication (30) can be justified by the remark that neglecting some feature vectors $\mathbf{x}_j[n]$ results in neglecting some non-negative components $\varphi_j(\mathbf{w}[n])$ (20) in the criterion function $\Phi_k(\mathbf{w}[n])$ (21).

The *negative monotonicity property due to reduction of features* x_i: The reduction of the feature space $F[n]$ to $F'[n']$ by neglecting some features x_i cannot result in a decrease of the minimal value $\Phi_k(\mathbf{w}_k^*[n])$ (22) of the criterion function $\Phi_k(\mathbf{w}[n])$ (21):

$$\left(F'[n'] \subset F[n]\right) \Rightarrow \left(\Phi_{k'}^{*} \geq \Phi_{k}^{*}\right) \quad (33)$$

where the symbol $\Phi_{k'}^{*}$ stands for the minimal value (22) of the criterion function Φ_k' $(\mathbf{w}[n'])$ (21)) defined on the vectors $\mathbf{x}_j[n']$ from the reduced feature space $F'[n']$ ($\mathbf{x}_j[n'] \in F'[n']$, $n' < n$). The relation (33) results from the fact that neglecting some features x_i is equivalent to imposing additional constraints in the form of the condition $w_i = 0$ (6) on the parameter space R^n.

5 The Vertexical Feature Subspace

The criterion function $\Phi_k(\mathbf{w}[n])$ is defined (21) on the n-dimensional feature vectors $\mathbf{x}_j[n]$ ($\mathbf{x}_j[n] \in F[n]$) from the data subset C_k ($C_k \subset C$) (1). Because the function $\Phi_k(\mathbf{w}[n])$ (21) is convex and piecewise linear (*CPL*), their minimal value can be found in one of the vertices $\mathbf{w}_k[n]$ (9). This property is linked to the fundamental theory of linear programming [9]. The optimal vertex $\mathbf{w}_k^*[n]$ (9) constitutes the minimal value $\Phi_k(\mathbf{w}_k^*[n])$ (22) of the function $\Phi_k(\mathbf{w}[n])$. The basis exchange algorithm allows to find efficiently the optimal vertex $\mathbf{w}_k^*[n]$ (22) [8].

In accordance with the Theorem 2, the minimal value $\Phi_k(\mathbf{w}_k^*[n])$ (22) of the funtion $\Phi_k(\mathbf{w}[n])$ is equal to zero ($\Phi_k(\mathbf{w}_k^*[n]) = 0$) if all feature vectors $\mathbf{x}_j[n]$ from the subset C_k are situated on the hyperplane $H(\mathbf{w}_k^*[n],1)$ (23). But usually, we meet the situation, when $\Phi_k(\mathbf{w}_k^*[n]) > 0$. As it results from the property (32), the minimal value $\Phi_k(\mathbf{w}_k^*[n])$ (22) can be always reduced to zero through neglecting sufficient number of feature vectors $\mathbf{x}_j[n]$ with the largest values of the penalty functions $\varphi_j(\mathbf{w}_k^*[n])$ (20) in the optimal vertex $\mathbf{w}_k^*[n]$ (22). In result of neglecting of the vector $\mathbf{x}_j[n]$ with the largest value $\varphi_j(\mathbf{w}_k^*[n])$ (20), the set C_k, is replaced by the reduced data set C_k^{II}:

$$C_k \rightarrow C_k^{II} \quad (34)$$

The reduced criterion funtion $\Phi_k^{II}(\mathbf{w}[n])$ (21) is defined on elements $\mathbf{x}_j[n]$ of the set C_k^{II}. The minimal value $\Phi_k^{II}(\mathbf{w}_k^{II}[n])$ of the funtion $\Phi_k^{II}(\mathbf{w}[n])$ (21) can be located in a new optimal vertex $\Phi_k^{II}[n]$ by using the basis exchange algorithm [8]. If $\Phi_k^{II}(\mathbf{w}_k^{II}[n]) > 0$, then the next vector $\mathbf{x}_j[n]$ is reduced from the set C_k, and so on.

$$(\forall \mathbf{w}[n]) \; \Phi_k^{II}(\mathbf{w}[n]) \geq \Phi_k^{II}\left(\mathbf{w}_k^{II}[n]\right) \quad (35)$$

Designing postulate: The reduced data set C_k^{II} (34) should contain the maximal number of feature vectors $\mathbf{x}_j[n]$ while assuring the below condition:

$$\Phi_k^{II}\left(\mathbf{w}_k^{II}[n]\right) = 0 \tag{36}$$

We can infer on the basis of the Theorem 2, that the above designing postulate is aimed at location of a large number of feature vectors $\mathbf{x}_j[n]$ on the hyperplane H ($\Phi_k^{II}[n]$, 1) (23). In other words, the optimal vertex $\Phi_k^{II}[n]$ (35) should have a high degree of degeneration $d_k = m_k - r_k$, where r_k is the rank of the vertex $\Phi_k^{II}[n]$ and m_k is the number of the hyperplanes h_j (2) passing through this vertex. The designing postulate may allow to find in an unique manner the optimal vertex $\Phi_k^{II}[n]$ (35).

If the rank r_k of the vertex $\Phi_k^{II}[n] = [w_{k,1}, \ldots, w_{k,n}]^T$ (35) is less than n ($r_k < n$) then the basis $\mathbf{B}_k[n]$ (8) contains $n - r_k$ unit vectors $\mathbf{e}_i[n]$ and $n - r_k$ components $w_{k,i}$ have to be equal to zero ($w_{k,i} = 0$). The vertexical feature subspace $F_k[r_k]$ ($F_k[r_k] \subset F[n]$) can be defined on the basis of the optimal vertex $\Phi_k^{II}[n]$ (35).

Definition 4 The k-th vertexical subspace $F_k[r_k]$ ($F_k[r_k] \subset F[n]$) is composed of such r_k features x_i from the subset J_k ($x_i \in J_k$) which are linked to the non-zero components $w_{k,i}$ ($w_{k,i} \neq 0$) of the optimal vector $\Phi_k^{II}[n] = [w_{k,1}, \ldots, w_{k,n}]^T$ (35) and (36):

$$(\forall i \in \{1, \ldots, n\}) \textit{ if } w_{k,i} \neq 0, \textit{ then } x_i \in J_k \tag{37}$$

The reduced vectors $\mathbf{y}_j[r_k]$ ($\mathbf{y}_j[r_k] \in F_k[r_k]$) are obtained from the feature vectors $\mathbf{x}_j[n] = [x_{j,1}, \ldots, x_{j,n}]^T$ (1) by neglecting such $n - r_{k(i)}$ components $x_{j,i}$ ($x_i \notin J_k$) which are linked to the weights $w_{k,i}$ equal to zero (37).

Each of m reduced vectors $\mathbf{y}_j[r_k]$ (37) defines the below hyperplane h_j' (2) in the r_k-dimensional parameter space R^{rk} ($\mathbf{v}[r_k] \in R^{rk}$):

$$(\forall j \in \{1, \ldots, m\})\ h_j' = \left\{\mathbf{v}[r_k] : \mathbf{y}_j[r_k]^T \mathbf{v}[r_k] = 1\right\} \tag{38}$$

where $\mathbf{v}[r_k] = [v_1, \ldots, v_{rk}]^T$ is the parameter (*weight*) vector ($\mathbf{v}[r_k] \in R^{rk}$).

The k-th *vertex* $\mathbf{v}_k[r_k]$ in the parameter space R^{rk} is the intersection point of r_k hyperplanes h_j' (38) defined by the linearly independent vectors $\mathbf{y}_j[r_k]$ ($\mathbf{y}_j[r_k] \in F_k[r_k]$). The vertex $\mathbf{v}_k[r_k] = [v_{k,1}, \ldots, v_{k,rk}]^T$ is defined by the below linear equations (5):

$$\left(\forall j \in J_k'\right)\ \mathbf{v}_k[r_k]^T \mathbf{y}_j[r_k] = 1 \tag{39}$$

or (9)

$$\mathbf{v}_k[r_k] = \mathbf{B}_k'[r_k]^{-1} \mathbf{1}[r_k] = \mathbf{r}_1[r_k] + \cdots + \mathbf{r}_n[r_k] \tag{40}$$

where $\mathbf{B}_k'[r_k] = [\mathbf{y}_{j(1)}[r_k], \ldots, \mathbf{y}_{j(rk')}[r_k]]^T$ and $(\mathbf{B}_k'[r_k])^{-1} = [\mathbf{r}_1[r_k], \ldots, \mathbf{r}_{rk}[r_k]]$).

Theorem 4 *The hyperplane* h_j' (38) *passes through the vertex* $\mathbf{v}_k[r_k]$ (40) ($\mathbf{v}_k[r_k]^T \mathbf{y}_j[r_k] = 1$) *if and only if the reduced vector* $\mathbf{y}_j[r_k]$ *is located on the*

($r_k - 1$)-dimensional vertexical plane $P_k(\mathbf{y}_{j(1)}[r_k],\ldots,\mathbf{y}_{j(rk)}[r_k])$ (11) *equal to the hyperplane* $H_k(\mathbf{v}_k[r_k],1)$ (10) *in the* r_k*-dimensional vertexical subspace* $F_k[r_k]$.

Proof The vertex $\mathbf{v}[r_k]$ (40) in the r_k-dimensional parameter space R^{rk} has the maximal rank r_k. It means that the basis $\mathbf{B}_k^{'}[r_k]$ contains no unit vectors $\mathbf{e}_i[r_k]$. It means also that the assumptions of the Theorem 2 are fulfilled. So, the thesis of the above theorem holds. □

6 Multistage Procedure for Seeking Vertexical Planes

Large, multivariate data set C (1) may contain a lot of degenerated vertices $\mathbf{w}_k[n]$ (9). The multistage procedure sketched below allows to identify some vertices $\mathbf{w}_k[n]$ with the greatest degree of degeneration d_k (*Definition* 3).

The data set C_1 ($C_1 \subset C$) (1) is used during the first stage of the procedure ($k = 1$) and allows to find the optimal vertex $\mathbf{w}_1^{II}[n]$ (35) fulfilling the condition (36). During the k-th stage ($k > 1$), the minimization of the criterion funtion $\Phi_k^{III}(\mathbf{w}[n])$ (21) combined with the gradual reduction of the data set C_k to C_k^{II} (34) allows to identify such optimal vertex $\Phi_k^{II}[n]$ that the minimal value $\Phi_k^{II}(\mathbf{w}_k^{II}[n])$ (35) has become equal to zero ($\Phi_k(\mathbf{w}_k^{II}[n]) = 0$). The next ($k$ +1) stage begins by the removing from the set C_k of all such feature vectors $\mathbf{x}_j[n]$ which are degenerated in the vertex $\Phi_k^{II}[n]$ (35):

$$C_{k+1} = C_k \backslash \left\{ \mathbf{x}_j[n]: (\mathbf{w}_k^{II}[n])^T \mathbf{x}_j[n] = 1 \right\} \quad (41)$$

This procedure allows to identify the family of the vertices $\mathbf{v}_k[r_k]$ (40) of the rank r_k with a relatively high degree of degeneration $d_k = m_k - r_k$ ($k = 1, \ldots, K$). Each vertex $\mathbf{v}_k[r_k]$ (40) is linked to the vertexical subspace $F_k[r_k]$ and to the vertexical plane $P_k(\mathbf{y}_{j(1)}[r_k],\ldots,\mathbf{y}_{j(rk)}[r_k])$ (11) equal to the hyperplane $H(\mathbf{v}_k[r_k], 1)$ (10) in $F_k[r_k]$.

7 Decision Rules Based on Collinearity Models

The vertexical plane $P_k(\mathbf{y}_{j(1)}[r_k], \ldots, \mathbf{y}_{j(rk)}[r_k])$ (11) based on the vertex $\mathbf{v}_k[r_k]$ (40) of the rank r_k can be treated as the model of collinearity between r_k features x_i contained in the vertexical subspace $F_k[r_k]$. Such K models of colinearity can be used among others in the *K-lines* or the *K-planes* algorithms [3, 4]. The k-th vertexical plane $P_k(\mathbf{y}_{j(1)}[r_k],\ldots,\mathbf{y}_{j(rk)}[r_k])$ (11) is equal to the hyperplane $H_k(\mathbf{v}_k[r_k], 1)$ (10) in the k-th feature subspace $F_k[r_k]$. The set C of m feature vectors $\mathbf{x}_j[n]$ (1) can be divided into K clusters C_k on the basis of the shortest distance $\delta_k(\mathbf{x}_j[n],$

$H_k(\mathbf{v}_k[r_k], 1))$ between the vector $\mathbf{x}_j[n]$ and the hyperplane $H_k(\mathbf{v}_k[r_k], 1)$ in the k-th feature subspace $F_k[r_k]$ [3]:

$$\begin{aligned} &\textit{if } \{(\forall k \in \{1, \ldots, K\})\delta_k(\mathbf{x}_j[n], H_k(\mathbf{v}_k[r_k], 1)) \geq \delta_{k'}(\mathbf{x}_j[n], H_{k'}(\mathbf{v}_{k'}[r_{k'}], 1))\}, \\ &\qquad\qquad\qquad\qquad\qquad\qquad\qquad\qquad \textit{then } x_j[n] \in C_{k'} \end{aligned} \tag{42}$$

where

$$\delta_k(\mathbf{x}_j[n], H_k(\mathbf{v}_k[r_k], 1)) = \left| \left(\mathbf{v}_k[r_k] / \mathbf{v}_k[r_k]^{\mathrm{T}} \mathbf{v}_k[r_k] \right)^{\mathrm{T}} \mathbf{y}_j[r_k] - 1 \right| \tag{43}$$

New vectors $\mathbf{x}[n]$ ($\mathbf{x}[n] \in F[n]$) can be also assigned to the clusters C_k by using the decision rule (42).

8 Concluding Remarks

The vertexical plane $P_k(\mathbf{y}_{j(1)}[r_k], \ldots, \mathbf{y}_{j(rk)}[r_k])$ (11) which is spanned by r_k reduced vectors $\mathbf{y}_j[r_k]$ (37) can be treated as a model of collinearity between r_k features x_i contained in the reduced feature space $F_k[r_k]$ ($k = 1, \ldots, K$). A large number m_k of such feature vectors $\mathbf{x}_j[n]$ (1) which are situated *on* (or *near*) the vertexical plane $P_k(\mathbf{y}_{j(1)}[r_k], \ldots, \mathbf{y}_{j(rk)}[r_k])$ (11) in the feature subspace $F_k[r_k]$ form the *flat pattern* $\mathbf{F}_k$ [4].

The flat pattern $\mathbf{F}_k$ is the set of such feature vectors $\mathbf{x}_j[n]$ (1) which are related by the linear relation (15) in the reduced feature subspace $F_k[r_k]$. We can remark, that discovering of the flat patterns is linked not only to selection of the subset (cluster) $\mathbf{F}_k$ of feature vectors $\mathbf{x}_j[n]$ ($\mathbf{F}_k \subset C$ (1)) but also to selection of the feature subspace $F_k[r_k]$ ($F_k[r_k] \subset F[n]$).

Acknowledgments This work was supported by the project S/WI/2/2016 from the Białystok University of Technology, Poland.

References

1. Hand, D., Smyth, P., Mannila, H.: Principles of Data Mining. MIT Press, Cambridge (2001)
2. Duda, O.R., Hart, P.E., Stork, D.G.: Pattern Classification. J. Wiley, New York (2001)
3. Bobrowski, L.: K-lines clustering with convex and piecewise linear. CPL) functions, MATHMOD, Vienna (2012)
4. Bobrowski, L.: Discovering main vertexical planes in a multivariate data space by using *CPL* functions. In: Perner, P. (ed.) ICDM 2014. Springer, Berlin (2014)
5. Duda, O.R., Hart, P.E.: Use of the hough transformation to detect lines and curves in pictures. Commun. Assoc. Comput. Mach. **15**(1), 11–15 (1972)
6. Ballard, D.H.: Generalizing the hough transform to detect arbitrary shapes. Pattern Recogn. **13** (2), 111–122 (1981)

7. Bobrowski, L.: Data Mining Based on Convex and Piecewise Linear Criterion Functions (in Polish). Technical University Białystok (2005)
8. Bobrowski, L.: Design of piecewise linear classifiers from formal neurons by some basis exchange technique. Pattern Recogn. **24**(9), 863–870 (1991)
9. Simonnard, M.: Linear Programming. Prentice-Hall (1966)

PLA Based Strategy for Solving MRCPSP by a Team of Agents

Piotr Jędrzejowicz and Ewa Ratajczak-Ropel

Abstract In this paper the dynamic interaction strategy based on the Population Learning Algorithm (PLA) for the A-Team solving the Multi-mode Resource-Constrained Project Scheduling Problem (MRCPSP) is proposed and experimentally validated. The MRCPSP belongs to the NP-hard problem class. To solve this problem a team of asynchronous agents (A-Team) has been implemented using multiagent system. An A-Team is the set of objects including multiple agents and the common memory which through interactions produce solutions of optimization problems. These interactions are usually managed by some static strategy. In this paper the dynamic learning strategy based on PLA is suggested. The proposed strategy supervises interactions between optimization agents and the common memory. To validate the proposed approach computational experiment has been carried out.

Keywords Multi-mode resource-constrained project scheduling · MRCPSP · Optimization · Agent · A-team · Population learning algorithm

1 Introduction

The Multi-mode Resource Constrained Project Scheduling Problem (MRCPSP) has attracted a lot of attention and many exact, heuristic and metaheuristic solution methods have been proposed in the literature in recent years [10, 15, 16]. The current approaches to solve these problems produce either approximate solutions or can only be applied for solving instances of the limited size. Hence, searching for more effective algorithms and solutions to the problems is still a lively field of research. One of the promising directions of such research is to take advantage of the parallel and

P. Jędrzejowicz · E. Ratajczak-Ropel (✉)
Chair of Information Systems, Gdynia Maritime University,
Morska 83, 81-225 Gdynia, Poland
e-mail: ewra@am.gdynia.pl

P. Jędrzejowicz
e-mail: pj@am.gdynia.pl

I. Czarnowski et al. (eds.), *Intelligent Decision Technologies 2016*,
Smart Innovation, Systems and Technologies 56,
DOI 10.1007/978-3-319-39630-9_25

distributed computation solutions, which are the common feature of the contemporary multiagent systems.

Modern multiagent system architectures are an important and intensively expanding area of research and development. There is a number of multiple-agent approaches proposed to solve different types of optimization problems. One of them is the concept of an A-Team [14]. The idea of the A-Team was used to develop the software environment for solving a variety of computationally hard optimization problems called JABAT [1]. JADE based A-Team (JABAT) system supports the construction of the dedicated A-Team architectures. Agents used in JABAT assure decentralization of computation across multiple hardware platforms. Parallel processing results in more effective use of the available resources and ultimately, a reduction of the computation time.

Population Learning Algorithm (PLA) proposed in [5] is a population-based method inspired by analogies to the social education systems in which a diminishing number of individuals enter more and more advanced learning and improvement stages. PLA divides the process of solving the problem into stages, in which the considered optimization problem is solved using a set of independent learning/improvement procedures.

A-Team is a system composed of the set of objects including multiple agents and the common memory which through interactions produce solutions of optimization problems. Several strategies controlling the interactions between agents and memories have been recently proposed and experimentally validated. The influence of such interaction strategy on the A-Team performance was investigated in [2]. In [7, 8] the dynamic interaction strategy based on Reinforcement Learning for A-Team solving the RCPSP and MRCPSP has been proposed. The similar topics were also considered for different multi-agent systems, e.g. [4, 12].

In this paper the PLA based dynamic interaction strategy for the A-Team solving the MRCPSP is proposed and experimentally validated. PLA strategy is used to control the parameters and manage the process of searching for solutions to the MRCPSP instances by a team of agents. In this approach the parameters depend on the current state of the environment and the stage of learning. Both are being changed dynamically during the computation.

The proposed A-Team produces solutions to the MRCPSP instances using four kinds of the optimization agents. They include local search, tabu search, crossover search and path relinking procedures.

The paper is constructed as follows: Sect. 2 contains the MRCPSP formulation. Section 3 provides details of the proposed PLA dynamic interaction strategy and its implementation in JABAT. Section 4 describes settings of the computational experiment carried-out with a view to validate the proposed approach and contains a discussion of the computational experiment results. Finally, Sect. 5 contains conclusions and suggestions for future research.

2 Problem Formulation

A single-mode Resource-Constrained Project Scheduling Problem (RCPSP) consists of a set of n activities, where each activity has to be processed without interruption to complete the project. The dummy activities 1 and n represent the beginning and the end of the project. The duration of an activity j, $j = 1, \ldots, n$ is denoted by d_j where $d_1 = d_n = 0$. There are r renewable resource types. The availability of each resource type k in each time period is r_k units, $k = 1, \ldots, r$. Each activity j requires r_{jk} units of resource k during each period of its duration, where $r_{1k} = r_{nk} = 0$, $k = 1, ..., r$. All parameters are non-negative integers. There are precedence relations of the finish-start type with a zero parameter value (i.e. $FS = 0$) defined between the activities. In other words activity i precedes activity j if j cannot start until i has been completed. The structure of a project can be represented by an activity-on-node network. SS_j (SP_j) is the set of successors (predecessors) of activity j, $j = 1, \ldots, n$. It is further assumed that $1 \in SP_j$, $j = 2, \ldots, n$, and $n \in SS_j$, $j = 1, \ldots, n-1$. The objective is to find a schedule S of activities starting times $[s_1, \ldots, s_n]$, where $s_1 = 0$ and precedence relations and resource constraints are satisfied, such that the schedule duration $T(S) = s_n$ is minimized.

In case of the MRCPSP each activity j, $j = 1, \ldots, n$ may be executed in one out of M_j modes. The activities may not be preempted and a mode once selected may not change, i.e., a job j once started in mode m has to be completed in mode m without interruption. Performing job j in mode m takes d_{jm} periods and is supported by a set R of renewable and a set N of non-renewable resources. Considering the time horizon, that is, an upper bound T on the projects makespan, one has the available amount of renewable (doubly constrained) resource as well as certain overall capacity of the non-renewable (doubly constrained) resource. The objective is to find a makespan minimal schedule that meets the constraints imposed by the precedence relations and the limited resource availability.

The MRCPSP is NP-hard as a generalization of the RCPSP [3]. Moreover, if there is more than one nonrenewable resource, the problem of finding a feasible solution for the MRCPSP is NP-complete [9].

3 A-Team with the Dynamic Interaction Strategy Controlled by PLA

JABAT was successfully used by the authors for solving RCPSP and MRCPSP [6] where static interaction strategies have been used. In [7, 8] the dynamic interaction strategies based on Reinforcement Learning have been proposed and successfully used. In this approach the dynamic interaction strategy based on PLA for solving MRCPSP is proposed.

To adapt JABAT to solving MRCPSP the sets of classes and agents were implemented. The first set includes classes describing the problem. They are responsible for reading and preprocessing of the data and generating random instances of the problem.

The second set includes classes describing the optimization agents. Each of them includes the implementation of an optimization algorithm used to solve the MRCPSP. All of them are inheriting from the OptiAgent class. Optimization agents are implementations of specialized algorithms introduced in [7]: CA, PRA, LSAm, LSAe, TSAm and TSAe. The prefix Opti is assigned to each agent with its embedded algorithm:

OptiCA implementing the Crossover Algorithm (CA),
OptiPRA implementing the Path Relinking Algorithm (PRA),
OptiLSAm implementing the Local Search Algorithm (LSAm),
OptiLSAe implementing the Local Search Algorithm (LSAe),
OptiTSAm implementing the Tabu Search Algorithm (TSAm),
OptiTSAe implementing the Tabu Search Algorithm (TSAe),

In the PRA, LSAm and TSAm the new solutions are obtained by moving the activities to new positions in the schedule and simultaneously changing their modes, while in the LSAe and TSAe by exchanging pairs of activities and simultaneously changing their modes.

In our earlier approaches to manage the interaction between agents and common memory different static strategies were used, including Blocking2. In Blocking2 strategy randomly selected worse solution from the population is replaced by the one sent by the optimization agent. Moreover, one new solution is generated randomly every fixed number of iterations, which replaces the worst solution from the population.

The dynamic strategies considered in our previous approaches are based on the Reinforcement Learning (RL) [8]. In this paper the strategy based on the Blocking2 where the PLA idea together with the Reinforcement Learning features have been combined. The basic features of the PLA based dynamic interaction strategy are as follows:

- All the individuals in the initial population of solutions are generated randomly and stored in the common memory.
- The individuals for improvement are selected from the common memory randomly and blocked which means that once selected individual (or individuals) cannot be selected again until the OptiAgent to which they have been sent returns the solution or the learning stage is finished.
- The returning individual, which represent the feasible solution, replaces its original version before the attempted improvement. It means that the solutions are blocked for the particular OptiAgent and the returning solution replaces the blocked one or the worst from the blocked one. If none of the solutions is worse, the random one is replaced. All solutions blocked for the considered OptiAgent are released and returned to the common memory.

- The new feasible solution is generated with fixed probability P_{mg} and replaces another one.
- For each level of learning the environment state is remembered. This state includes: the best individual, population average diversity, weights and probabilities. The state is calculated every fixed number of iterations *nITns*. To reduce the computation time, average diversity of the population is evaluated by comparison with the best solution only. Diversity of two solutions for the MRCPSP problem is evaluated as the sum of differences between activities starting times in a project.
- Computations are divided into learning stages. In each stage different parameter settings, different set of optimization agents and different stopping criteria are used.

To describe the PLA-based strategy the following notation will be used:

P	population of individuals;
avgdiv(P)	current average diversity of the population P;
nITwi	number of iterations without an improvement of the goal function which is used to stop their computation;
nITns	number of iterations after which a new environment state is calculated;
cTime	computation time (minutes);
nLS	number of learning stages in the PLA.

Additionally, two probability measures have been used:

p_{mg}	probability of selecting the method *mg* for selection of a new individual;
p_{mr}	probability of selecting the method *mr* for replacing an individual in the population;

There are four possible methods of generating a new individual:

mgr	randomly;
mgrc	using one point crossover operator for two randomly chosen individuals;
mgb	random changes of the best individual in the population;
mgbc	using crossover operator for two randomly chosen individuals from the five best individuals from the population.

The weight w_{mg} for each method is calculated, where $mg \in Mg$, $Mg = \{mgr, mgrc, mgb, mgbc\}$. The w_{mgr} and w_{mgrc} are increased where the population average diversity decreases and they are decreased in the opposite case. The w_{mgb} and w_{mgbc} are decreased where the population average diversity increases and they are increased in the opposite case. The probability of selecting the method *mg* is calculated as

$$p_{mg} = \frac{w_{mg}}{\sum_{mg \in Mg} w_{mg}}.$$

There are three methods of replacing an individual from the population by a new one:

mrr new solution replaces the random one in the population;
mrw new solution replaces the random worse one in the population;
mrt new solution replaces the worst solution in the population.

Experiments show that replacing the worse and worst solution is beneficial to intensify exploitation while replacing the random one intensifies exploration. The weight w_{mr} for each method is calculated, where $mr \in Mr, Mr = \{mrr, mrw, mrt\}$. The w_{mrr} is increased where the population average diversity decreases and it is decreased in the opposite case. The w_{mrw} and w_{mrt} are decreased where the population average diversity decreases and they are increased in the opposite case. The probability of selecting the method *mr* is calculated as

$$p_{mr} = \frac{w_{mr}}{\sum_{mr \in Mr} w_{mr}}.$$

The current environment state parameters are updated after any significant change: generating a new solution, receiving the solution from OptiAgent and replacing solution in the population. The update includes:

- set $w_{mgr}, w_{mgrc}, w_{mgb}, w_{mgbc}, w_{mrr}, w_{mrw}, w_{mrt}$;
- remember the best solution;
- calculate the avgdiv(P).

The general schema of the proposed PLA strategy for the A-Team is presented in Fig. 1. It is worth noticing that computations performed inside while loop are carried out independently and possibly in parallel, within the agent environment used.

Fig. 1 General schema of the PLA based strategy

```
PLA_strategy{
  generate the initial population P
  calculate environment state
  for(i=0; i < nLS; i = i + 1){
    while(none of the stopping criteria is met){
      use OptiAgents to improve solutions
      generate a new solution S_new with p_mg
      replace the individual in P by S_new with p_mr
      calculate environment state}}}
```

Table 1 PLA1S

Stage 1	Initial weights	$w_{mgr} = 25$, $w_{mgrc} = 25$, $w_{mgb} = 25$, $w_{mgbc} = 25$
		$w_{mrr} = 34$, $w_{mrw} = 33$, $w_{mrt} = 33$
	Optimization agents	OptCA, OptPRA, OptLSAm(10), OptLSAe(10), OptTSAm(20), OptTSAe(20)
	Stopping criteria	avgdiv(P) > 0.05 and $nSGS < 5000$

Table 2 PLA2S

Stage 1	Initial weights	$w_{mgr} = 50$, $w_{mgrc} = 50$, $w_{mgb} = 0$, $w_{mgbc} = 0$
		$w_{mrr} = 100$, $w_{mrw} = 0$, $w_{mrt} = 0$
	Optimization agents	OptCA, OptPRA, OptLSAm(10), OptLSAm(10)
	Stopping criteria	avgdiv(P) > 0.1 or $nSGS < 2500$
Stage 2	Initial weights	$w_{mgr} = 25$, $w_{mgrc} = 25$, $w_{mgb} = 25$, $w_{mgbc} = 25$ $w_{mrr} = 34$, $w_{mrw} = 33$, $w_{mrt} = 33$
	Optimization agents	OptLSAe(20), OptTSAm(20), OptTSAe(20)
	Stopping criteria	avgdiv(P) > 0.05 and $nSGS < 2500$

4 Computational Experiment

4.1 Settings

To evaluate the effectiveness of the proposed approach the computational experiment has been carried out using benchmark instances of MRCPSP from PSPLIB.[1] The test sets include: mm10 (multi mode, 10 activities), mm12, mm14, mm16, mm18, mm20. Each set includes 640 problem instances.

In the experiment the following global parameters have been used: $|P| = 30$, $nITns = 5$ and $nLS = 1, 2$ or 3.

To enable comparisons with other algorithms known from the literature, the number of schedules generated during computation is calculated. In the presented experiments the number of schedules $nSGS$ is limited to 5000.

To validate the approach three PLA based strategies are checked: PLA1S with one learning stage, PLA2S with two stages and PLA3S with three stages. In each stage a different set of parameters and OptiAgents are used. In the PLA more advanced stages are entered by a diminishing number of individuals from the initial population. Hence, at higher stages more advanced and complex learning techniques are used. The proposed sets of settings are shown in the Tables 1, 2 and 3.

[1] See PSPLIB at http://www.om-db.wi.tum.de/psplib/.

Table 3 PLA3S

Stage 1	Initial weights	$w_{mgr} = 50$, $w_{mgrc} = 50$, $w_{mgb} = 0$, $w_{mgbc} = 0$
		$w_{mrr} = 100$, $w_{mrw} = 0$, $w_{mrt} = 0$
	Optimization agents	OptCA, OptPRA
	Stopping criteria	avgdiv(P) > 0.1 and $nSGS$ < 2000
Stage 2	Initial weights	$w_{mgr} = 25$, $w_{mgrc} = 25$, $w_{mgb} = 25$, $w_{mgbc} = 25$ $w_{mrr} = 34$, $w_{mrw} = 33$, $w_{mrt} = 33$
	Optimization agents	OptLSAm(10), OptLSAe(10)
	Stopping criteria	avgdiv(P) < 0.5 and $nSGS$ < 1500
Stage 3	Initial weights	$w_{mgr} = 0$, $w_{mgrc} = 0$, $w_{mgb} = 50$, $w_{mgbc} = 50$ $w_{mrr} = 0$, $w_{mrw} = 20$, $w_{mrt} = 80$
	Optimization agents	OptTSAm(20), OptTSAe(20)
	Stopping criteria	avgdiv(P) > 0.05 and $nSGS$ < 1500

Table 4 MRE from the optimal solution for benchmark test sets mm10–mm20

Strategy	mm10 (%)	mm12 (%)	mm14 (%)	mm16 (%)	mm18 (%)	mm20 (%)
Blocking2	0.41	0.47	0.69	0.81	0.95	1.80
PLA1S	0.36	0.47	0.56	0.77	0.95	1.63
PLA2S	0.29	0.43	0.57	0.62	0.89	1.48
PLA3S	0.25	0.32	0.55	0.43	0.72	1.24

Table 5 MCT [s] for benchmark test sets mm10–mm20

Strategy	mm10	mm12	mm14	mm16	mm18	mm20
Blocking2	2.23	2.23	3.31	3.52	4.13	5.01
PLA1S	1.30	1.26	3.21	3.38	4.13	4.29
PLA2S	1.27	1.28	3.25	3.38	4.22	4.35
PLA3S	1.31	1.34	3.27	3.49	4.15	4.41

Table 6 MTCT [s] for benchmark test set mm10–mm20

Strategy	mm10	mm12	mm14	mm16	mm18	mm20
Blocking2	35.12	25.12	27.02	29.30	28.56	36.09
PLA1S	12.41	21.94	21.48	21.13	25.17	31.51
PLA2S	12.46	22.19	20.39	20.65	26.24	32.19
PLA3S	12.97	22.15	20.44	21.87	25.41	32.29

To calculate weights an effective approach based on reinforcement learning proposed in [7] have been used. The parameters values have been chosen experimentally based on earlier experiments for the RCPSP in JABAT and the preliminary experiments for the PLA strategy conducted using data set mm12.

Table 7 Literature reported results [11, 13, 15]

Set	Algorithm	Authors	MRE (%)	MCT (s)	Computer (GHz)
mm10	Distribution algorithm	Wang and Fang	0.01	1	2.2
	Genetic algorithm	Van Peteghem and Vanhoucke	0.01	0.12	2.8
	Hybrid genetic algorithm	Lova et al.	0.04	0.1	3
	Our approach		*0.25*	*1.31*	*1.4*
mm12	Distribution algorithm	Wang and Fang	0.02	1.8	2.2
	Genetic algorithm	Van Peteghem and Vanhoucke	0.09	–	–
	Hybrid genetic algorithm	Lova et al.	0.17	–	–
	Our approach		*0.32*	*1.34*	*1.4*
	Hybrid scatter search	Ranjbar et al.	0.65	10	3
mm14	Distribution algorithm	Wang and Fang	0.03	1	2.2
	Genetic algorithm	Van Peteghem and Vanhoucke	0.22	0.14	2.8
	Hybrid genetic algorithm	Lova et al.	0.32	0.11	3
	Our approach		*0.55*	*3.27*	*1.4*
	Hybrid scatter search	Ranjbar et al.	0.89	10	3
mm16	Distribution algorithm	Wang and Fang	0.17	1	2.2
	Genetic algorithm	Van Peteghem and Vanhoucke	0.32	0.15	2.8
	Our approach		*0.43*	*3.49*	*1.4*
	Hybrid genetic algorithm	Lova et al.	0.44	0.12	3
	Hybrid scatter search	Ranjbar et al.	0.95	10	3
mm18	Distribution algorithm	Wang and Fang	0.19	1	2.2
	Genetic algorithm	Van Peteghem and Vanhoucke	0.42	0.16	2.8
	Hybrid genetic algorithm	Lova et al.	0.63	0.13	3
	Our approach		*0.72*	*4.15*	*1.4*
	Hybrid scatter search	Ranjbar et al.	1.21	10	3
mm20	Distribution algorithm	Wang and Fang	0.32	1	2.2
	Genetic algorithm	Van Peteghem and Vanhoucke	0.57	0.17	2.8
	Hybrid genetic algorithm	Lova et al.	0.87	0.15	3
	Our approach		*1.24*	*4.41*	*1.4*
	Hybrid scatter search	Ranjbar et al.	1.64	10	3

The experiment has been carried out using nodes of the cluster Holk of the Tricity Academic Computer Network built of 256 Intel Itanium 2 Dual Core 1.4 GHz with 12 MB L3 cache processors and with Mellanox InfiniBand interconnections with 10Gb/s bandwidth. During the computation one node per four optimization agents was used.

4.2 Results

During the experiment the following characteristics of the computational results have been calculated and recorded: Mean Relative Error (MRE) calculated as the deviation from the optimal solution, Mean Computation Time (MCT) which has been needed to find the best solution and Mean Total Computation Time (MTCT) which has been needed to stop all optimization agents and the whole system. Each instance has been solved five times and the results have been averaged over these solutions. The number of schedules generated by SGS heuristics is limited to 5000 for all optimization agents used during search for the solution for each problem instance.

The results generated by the proposed approach, presented in Tables 4, 5 and 6, are good and promising. In each case all solutions were feasible. The mean relative errors for PLA3S below 1 % in case of 10–18 activities and below 1.3 % in case of 20 activities have been obtained (Table 7).

5 Conclusions

The computational experiment results show that the proposed dedicated A-Team architecture supported by the Population Learning Algorithm and Reinforcement Learning to control the interaction strategy is an effective and competitive tool for solving instances of the MRCPSP. Presented results are comparable with solutions known from the literature and in some cases outperform them. It can be also noted that they have been obtained in a comparable time. However, in this case time comparison may be misleading since the algorithms are run using different environments, operating systems, numbers and kinds of processors. In case of the agent-based environments the significant part of the time is used for agent communication which has an influence on both—computation time and quality of the results.

The presented experiment could be extended to examine different and additional parameters of the environment state and solutions as well as iteration numbers, probabilities and weights. On the other hand the additional or other learning stages should be examined. The kind and number of optimization agents (OptiAgents) used in each stage should be interesting to investigate. Additionally, an effective method for tuning optimization agents parameters including a number of iterations needed should be developed.

References

1. Barbucha, D., Czarnowski, I., Jędrzejowicz, P., Ratajczak-Ropel, E., Wierzbowska, I.: E-JABAT—An Implementation of the Web-Based A-Team. In: Nguyen, N.T., Jain, L.C. (eds.), Intelligent Agents in the Evolution of Web and Applications pp. 57–86, Springer, Heilderberg (2009)
2. Barbucha, D., Czarnowski, I., Jędrzejowicz, P., Ratajczak-Ropel, E., Wierzbowska, I.: Influence of the Working Strategy on A-Team Performance, Smart Information and Knowledge Management. In Szczerbicki, E., Nguyen, N.T. (eds.) Studies in Computational Intelligence, vol. 260, pp. 83–102 (2010)
3. Błażewicz, J., Lenstra, J., Rinnooy, A.: Scheduling subject to resource constraints: classification and complexity. Discrete Appl. Math. **5**, 11–24 (1983)
4. Cadenas, J.M., Garrido, M.C., Muñoz, E.: Using machine learning in a cooperative hybrid parallel strategy of metaheuristics. Inform. Sci. **179**(19), 3255–3267 (2009)
5. Jędrzejowicz, P.: Social learning algorithm as a tool for solving some difficult scheduling problems. Found. Comput. Decis. Sci. **24**(2), 51–66 (1999)
6. Jędrzejowicz, P., Ratajczak-Ropel, E.: New Generation A-Team for Solving the Resource Constrained Project Scheduling. In: Proceedings of the Eleventh International Workshop on Project Management and Scheduling, pp. 156–159. Istanbul (2008)
7. Jędrzejowicz, P., Ratajczak-Ropel, E.: Reinforcement learning strategies for A-team solving the resource-constrained project scheduling problem. Neurocomputing **146**, 301–307 (2014)
8. Jędrzejowicz, P., Ratajczak-Ropel, E.: Reinforcement Learning Strategy for Solving the MRCPSP by a Team of Agents. In: Neves-Silva, R., Jain, L.C., Howlett, R.J. (eds.) 7th KES International Conference on Intelligent Decision Technologies, KES-IDT 2015. Intelligent Decision Technologies, vol. 39, pp. 537–548. Springer International Publishing, Switzerland (2015)
9. Kolisch, R.: Project scheduling under resource constraints - Efficient heuristics for several problem classes. Ph.D. thesis. Physica, Heidelberg (1995)
10. Liu, S., Chen, D., Wang, Y.: Memetic algorithm for multi-mode resource-constrained project scheduling problems. J. Syst. Eng. Electron. **25**(4), 609–617 (2014)
11. Lova, A., Tormos, P., Cervantes, M., Barber, F.: An efficient hybrid genetic algorithm for scheduling projects with resource constraints and multiple execution modes. Int. J. Prod. Econ. **117**(2), 302–316 (2009)
12. Pelta, D., Cruz, C., Sancho-Royo, A., Verdegay, J.L.: Using memory and fuzzy rules in a cooperative multi-thread strategy for optimization. Inform. Sci. **176**(13), 1849–1868 (2006)
13. Ranjbar, M., Reyck, B., De Kianfar, F.: A hybrid scatter search for the discrete time/resource trade-off problem in project scheduling. Eur. J. Oper. Res. **193**(1), 35–48 (2009)
14. Talukdar, S., Baerentzen, L., Gove, A., De Souza, P.: Asynchronous Teams: Co-operation Schemes for Autonomous, Computer-Based Agents. Technical Report EDRC 18-59-96, Carnegie Mellon University, Pittsburgh (1996)
15. Van Peteghem, V., Vanhoucke, M.: A genetic algorithm for the preemptive and non-preemptive multi-mode resource-constrained project scheduling problems. Eur. J. Oper. Res. **201**(2), 409–418 (2010)
16. Węglarz, J., Józefowska, J., Mika, M., Waligora, G.: Project scheduling with finite or infinite number of activity processing modes—a survey. Eur. J. Oper. Res. **208**, 177–205 (2011)

Apache Spark Implementation of the Distance-Based Kernel-Based Fuzzy C-Means Clustering Classifier

Joanna Jędrzejowicz, Piotr Jędrzejowicz and Izabela Wierzbowska

Abstract The paper presents an implementation of a classification algorithm based on Kernel-based fuzzy C-means clustering. The algorithm is implemented in Apache Spark environment, and it is based on Resilient Distributed Datasets (RDDs) and RDD actions and transformations. The choice allows for parallel data manipulation.

Keywords Clustering · Classification · Fuzzy C-means

1 Introduction

Distance-based classifiers, also known as instance-based classifiers, belong to a family of learning algorithms that induce their decisions through comparing problem instances for which the class label is unknown, with instances with known class labels seen in training, which have been stored in memory. Among best known distance—based algorithms one can mention nearest neighbor classifiers, kernel-based classifiers and RBF networks (see, for example [3, 4, 9]). The discussed class of learning machines seems especially suited for constructing online classifiers due their ability to adopt to changes in the environment generating the data. Such an adaptation is usually based on updating the training set by incorporating newly arrived instances. Several variants of the online distance-based classifiers have been recently proposed by the authors [5–7]. In this paper we use the approach proposed in our earlier works, this time however, with different focus. Considering possible future

J. Jędrzejowicz
Institute of Informatics, Gdańsk University, Wita Stwosza 57, 80-952 Gdańsk, Poland
e-mail: jj@inf.ug.edu.pl

P. Jędrzejowicz · I. Wierzbowska (✉)
Department of Information Systems, Gdynia Maritime University,
Morska 81-87, 81-225 Gdynia, Poland
e-mail: iza@am.gdynia.pl

P. Jędrzejowicz
e-mail: pj@am.gdynia.pl

I. Czarnowski et al. (eds.), *Intelligent Decision Technologies 2016*,
Smart Innovation, Systems and Technologies 56,
DOI 10.1007/978-3-319-39630-9_26

application of the distance-based classifiers to mining big data we have decided to implement the proposed classifier in the Apache Spark environment using Resilient Distributed Datasets (RDDs) and RDD actions and transformations. Our secondary goal is to investigate how the distance and kernel based fuzzy C-means classifier would perform in the batch environment.

The paper is organized as follows. In Sect. 2 we present classifier based on fuzzy C-means clustering. In Sect. 3 some details of the Apache Spark environment implementation are given. Section 4 shows results of the computational experiment carried-out. Finally, Sect. 5 contains conclusions.

2 Classifier Based on Fuzzy C-Means Clustering

Let *trainingData* be a finite collection of data with known class labels:

$$trainingData = \sum_{c \in C} X^c.$$

Let $r \notin trainingData$.

The classifier is a modification of algorithms described in [8]. The approach proposed in [8] was oriented towards online problems. This paper deals with the batch classifier.

The algorithm proceeds in:

- **estimating** nc^c—the number of clusters for the data (X^c) in each class $c \in C$
- **clustering**—data in each class (X^c) is divided into nc^c clusters. The clusters are characterised by centroids.
- **classification**—the clusters obtained in the previous step help to classify new data: for each new vector r the class c is chosen for which the average distance from the x nearest centroids of X^c is minimal.

2.1 Estimating the Number of Clusters

The Algorithm 1 shows the steps for calculating the number of clusters in the data D [8, 11]. The algorithm is applied to the data in each class of the *trainingData* (X^c). As the result for each class $c \in C$ we obtain the number of clusters in this class: nc^c.

In [8] the number of clusters have been set as identical for all classes of the data. Here, the number of clusters is calculated for each class separately and it may differ between classes.

Algorithm 1 Number of clusters estimation

Input: data $D = \{x_1, \ldots, x_M\} \subset R^N$ with M feature vectors, kernel function K, threshold σ
Output: number of clusters nc

1: let $K_{ij} = K(x_i, x_j)$ be the quadratic matrix of size $M \times M$
2: calculate the eigenvalues of matrix (K_{ij})
3: $nc \leftarrow$ number of eigenvalues exceeding σ
4: return nc

2.2 Clustering

For clustering, the Kernel-based fuzzy C-means clustering algorithm is used, with centroids in kernel space and gaussian kernel function K:

$$K(\mathbf{x}, \mathbf{y}) = \exp \frac{-(\mathbf{x} - \mathbf{y})^2}{\sigma^2} \text{ for } \sigma^2 > 0. \tag{1}$$

Let $D = \{x_1, \ldots, x_M\} \subset R^N$ be a finite collection of data (feature or attribute vectors) to be clustered into nc clusters. Clusters are characterised by centroids c_i and fuzzy partition matrix $U = (u_{ik})$ of size $nc \times M$, where u_{ik} is the degree of membership of x_k in cluster i. The standard constraints for fuzzy clustering are assumed:

$$0 \le u_{ik} \le 1, \ \sum_{i=1}^{nc} u_{ik} = 1 \ \text{ for each } k, \ 0 < \sum_{k=1}^{M} u_{ik} < M \ \text{ for each } i \tag{2}$$

For each feature vectore x_k the maximal value of membership degree indicates the cluster to which x_k belongs: $clust(x_k) = \arg\max_{1 \le j \le nc} u_{jk}$.

Following [8], U may be calculated as follows:

$$u_{ik} = \frac{1}{\sum_{j=1}^{nc} \left(\frac{dist(\Phi(x_k), v_i)}{dist(\Phi(x_k), v_j)}\right)^{\frac{2}{m-1}}} \tag{3}$$

where

$$dist^2(\Phi(x_k), v_i) = K(x_k, x_k) - 2\frac{\sum_{j=1}^{M} u_{ij}^m K(x_k, x_j)}{\sum_{j=1}^{M} u_{ij}^m} + \frac{\sum_{j=1}^{M} \sum_{l=1}^{M} u_{ij}^m u_{il}^m K(x_j, x_l)}{(\sum_{j=1}^{M} i_{ij}^m)^2} \tag{4}$$

Like in [8] centroids are approximated using the following:

$$c_i = \frac{\sum_{k=1}^{M} u_{ik}^m \cdot K(x_k, c_i) \cdot x_k}{\sum_{k=1}^{M} u_{ik}^m \cdot K(x_k, c_i)} \tag{5}$$

The Algorithm 2 presents steps for calculating U and centroids for a given set D. The algorithm runs for each class of the *trainingData*. As the result for each class $c \in C$ a set of centroids is obtained: Cnt^c, consisting of nc^c centroids.

Algorithm 2 Kernel-based fuzzy C-means clustering

Input: data D, kernel function K, number of clusters nc
Output: fuzzy partition U, centroids $c_1, \ldots, c_n$

1: initialize U to random fuzzy partitions satysfying (2)
2: **repeat**
3: update U according to (3) and (4)
4: **until** stopping criterion satisfied or maximum number iterations reached
5: choose random vectors as centroids
6: update centroids according to (5)

2.3 *Classification*

To classify an unseen vector r the distances from r to each centroid in Cnt^c of every class $c \in C$ are calculated. Then, in each class c, the distances are sorted in non-decreasing order: $d_1, d_2, \ldots, d_{nc^c}$, where d_i is the distance to the i-th nearest centroid in class c. The coefficient S_c^x measures the mean distance from x nearest centroids:

$$S_c^x = \frac{\sum_{i=1}^{x} d_i}{x} \tag{6}$$

Row r is classified as class c, for which the value S_c^x (6) is minimal.

The distance between the vector r and a centroid may be measured with the use of any of the most popular metrics, for example chosen depending on whether data is continuous or non-continuous. In [8] several metrics were used, the same metrics have been used in this paper, and they include Eucliden, Manhattan, Chebychev, Camberra, Bray Curtis, cosine, chessboard and discrete metrics.

The Algorithm 3 [8] presents steps of the classification process for the unseen row r.

Algorithm 3 Classification based on fuzzy clustering partition

Input: partition of learning data $X = \bigcup_{c \in C} \bigcup_{j \le nc^c} X_j^c$, data row $r \notin X$, number of neighbours x
Output: class cl for row r
1: calculate S_{cj}^x for all $c \in C, j \le nc^c$ according to 6
2: $cl = \arg\min_{c \in C_{j,j \le nc^c}} S_{cj}^x$
3: return class cl

3 Implementation

The main contribution of this paper is implementation of the above algorithms in Apache Spark and Scala. Spark [1] is an open source cluster computing framework, and Scala is one of the languages supported by Spark.

The main concept introduced by Spark is Resilient Distributed Datasets (RDD)—immutable, fault-tolerant, distributed collection, that can be operated on in parallel. There are two types of operations that can be run on RDDs: *transformations and* actions. Transformations change every element of RDD, by applying a function or operation. Actions may aggregate all elements, returning for example the number or the sum of all elements.

The best known, and most often used pair of transformation/action is map and reduce (or, to be more precise reduceByKey). For example the number of instances of each class in the data may be counted as:

```
val counts = data.map(vector => (class(vector), 1))
                 .reduceByKey(_ + _)
```

In the example each vector is transformed into pair (class, count = 1) (*map*) and for each class all counts are added to receive the total number of each class (*reduceByKey*).

The implementation of the algorithms described in this paper is based on RDDs, and actions and transformations on RDDs. The choice allows for parallel manipulation on data.

The Algorithm 1 has been implemented in Spark as shown in Algorithm 4. This implementation does not fully reflect the steps of Algorithm 1. It makes use of function *computeSVD* from Spark's machine learning library (MLlib [10]). The function computes the singular values of singular value decomposition of a matrix. The number of the singular values given by *computeSVD* is limited, while in Algorithm 1 it is equal to the number of vectors in the data set—M.

So far, only smaller datasets has been processed with algorithms implemented for this paper. In the future, for bigger datasets, Algorithm 1 may be applied to a subset of the original data in order not to compute all M singular values.

Algorithm 4 Number of clusters estimation in Spark

```
//dataC = subset of data corresponding to particular class
var dataC = data.filter(a=>class(a)==classC).zipWithIndex
var dataCNo = dataC.count.toInt;

//change of data structures
val dataRows = dataC.cartesian( dataC)
.map(a=>(a._1._2, a._2._2,kernelFunc(a._1._1, a._2._1)))
.map(a=>MatrixEntry(a._1,a._2,a._3))
val mat: CoordinateMatrix = new CoordinateMatrix( dataRows)
val indexedRowMatrix = mat.toIndexedRowMatrix()

//singular value decomposition
val svdS = indexedRowMatrix.computeSVD(dataCNo, false).s

//calculating number of clusters
var nc = svdS.toArray.map( a=>if( a>threshold) 1 else 0).sum
```

4 Preliminary Results

The algorithm has been run on several datasets from [2]. The original data was divided into training and test data in proportions 0.9 and 0.1, and the results of the algorithm were compared with the original classification of the data in the test set. The test was repeated 100 times. Also, different metrics were tried. The Table 1 shows preliminary results of the tests, the accuracies present the percent of correctly classified instances. It has to be noted, that the algorithm was run on a standalone PC, and has yet to be run in parallel environment. Also, it has yet to be optimised.

Table 1 Classification results

Dataset	Instances	Attributes	Classes	Accuracy	Standard deviation (%)
Breast [2]	263	10	2	96.01	2.30
Chess [2]	503	17	2	80.85	0.05
Diabetes [2]	768	9	2	70.92	0.06
Heart [2]	303	14	2	82.17	6.51
Hepatitis [2]	155	20	2	76.53	13.87
Ionosphere [2]	352	35	2	84.55	7.78
Luxemburg [12]	1901	32	2	83.51	6.92
Sonar [2]	208	61	2	75.27	11.96
WBCdata [2]	630	11	2	91.46	3.09

Table 2 Mean computation times needed to classify a single instance using the described distance and kernel based fuzzy C-means classifier (seconds)

	Apache spark	Eclipse java
Breast	0.7814	0.8916
Chess	0.1728	0.3124
Diabetes	0.3562	2.5541
Heart	0.1373	0.7623
Hepatitis	0.1249	0.8823
Ionosphere	0.3022	1.8761
Luxemburg	0.6794	0.2667
Sonar	0.1361	3.6652
WBCdata	0.1885	1.4378

To compare efficiency of the distance-based classifier described in Sect. 2 and implemented with Apache Spark we have coded the classifier using traditional tool, which in this case was Eclipse Java. Next we have compared computation times of both versions. Table 2 shows mean computation times in seconds needed to classify a single instance for both discussed implementations. Both versions have been run on a standard PC with Intel Core i7 processor.

The above data are also shown in a graphical form in Fig. 1.

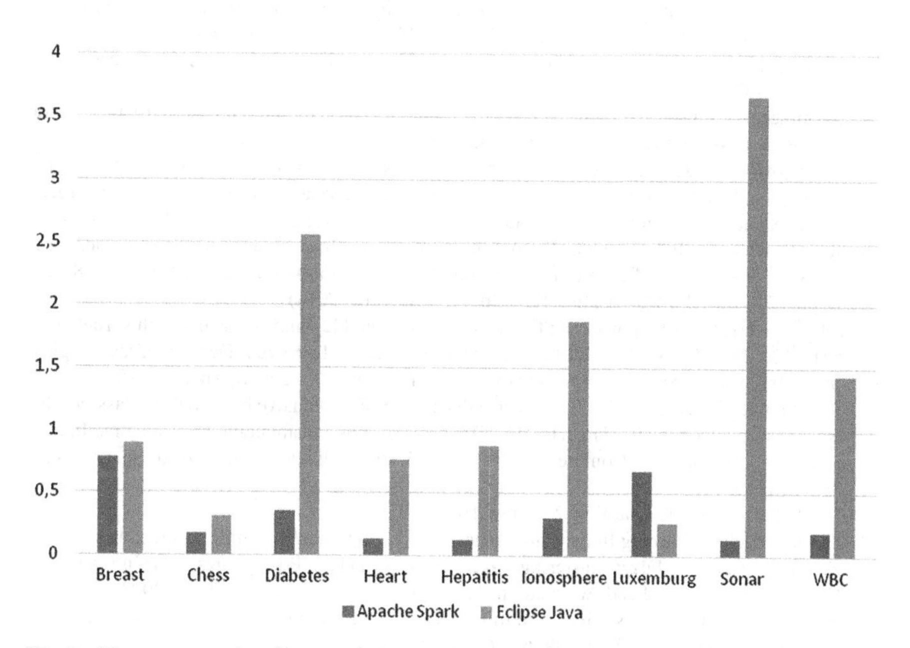

Fig. 1 Mean computation times needed to classify a single instance using both compared classifiers

It is easy to observe that the Apache Spark implementation outperforms traditional approach in terms of computation times while classification errors do not statistically differ. The above observation holds true even in case of the non-parallel environment.

5 Conclusion

The paper contributes through implementing distance and kernel based fuzzy C-means classifier proposed in [8] in the Apache Spark environment. Preliminary experiment carried-out using a standard PC showed that Apache Spark implementation has a promising potential to be used in a parallel environment dealing with mining big datasets. As the next step we will transfer our software to a computer cluster expecting further increase of the classifier efficiency. It is also worth noting that the approach assures a reasonable quality of classification. Further gains can be also expected from optimizing the Scala code, which is another planned task.

References

1. Apache Spark website. http://spark.apache.org/
2. Asuncion, A., Newman, D.J.: UCI Machine Learning Repository. University of California, School of Information and Computer Science (2007). http://archive.ics.uci.edu/ml/
3. Cover, T., Hart, P.: Nearest neighbor pattern classification. IEEE Trans. Inform. Theory IT-13, 21–27 (1967)
4. Crammer, K., Singer, Y.: On the algorithmic implementation of multiclass kernel-based vector machines. J. Mach. Learn. Res. **2**, 265–292 (2001)
5. Jędrzejowicz, J., Jędrzejowicz, P.: Online classifiers based on fuzzy c-means clustering. In: Badica, C., Nguyen, N.T., Brezovan, M. (eds.) Computational Collective Intelligence. Technologies and Applications, LNAI 8083, pp. 427–436. Springer, Berlin, Heidelberg (2013)
6. Jędrzejowicz, J., Jędrzejowicz, P.: A family of the on-line distance-based classifiers. In: Nguyen, N.T. et al. (eds.) Intelligent Information and Data-base Systems, LNAI 8398 Part II, pp. 177–186. Springer, Cham, Heidelberg. New York (2014)
7. Jędrzejowicz, J., Jędrzejowicz, P.: Distance-based ensemble online classifier with kernel clustering. In: Neves-Silva, R., Jain L.C., Howlett, R.J. (eds.), Intelligent Decision Technologies. Smart Innovation, Systems and Technologies, vol. 39, pp. 279–290. Springer (2015)
8. Jędrzejowicz J., Jędrzejowicz P.: A hybrid distance-based and naive bayes online classifier. In: Nnez, M., Nguyen, N.T., Camacho, D., Trawiski, B. (eds.) Computational Collective Intelligence: 7th International Conference, ICCCI 2015, Proceedings, pt. II. Madrid, Spain, 21–23 Sept 2015
9. Mitchell T.: Machine Learning. McGraw-Hill (1997)
10. Sparks' machine learning library. http://spark.apache.org/docs/latest/mllib-guide.html
11. Zhang, D., Chen, S.: Fuzzy clustering using kernel method. In: Proceedings of the International Conference on Control and Automation ICCA, pp. 162–163. Xiamen, China (2003)
12. Ẑliobaite, I.: Combining similarity in time and space for training set formation under concept drift. Intell. Data Anal. **15**(4), 589–611 (2011)

Ant Clustering Algorithm with Information Theoretic Learning

Urszula Boryczka and Mariusz Boryczka

Abstract In this paper a novel ant clustering algorithm with information theoretic learning consolidation is presented. Derivation of the information potential and its force from Renyi's entropy have been used to create an interesting model of ant's movement during clusterization process. In this approach each object is treated as a single agent-ant. What is more, in a local environment each agent-ant moves in accordance to information forces influence. The outcome of all information forces determines the direction and range of agent-ants' movement. Stopping criterion used in this approach indirectly emerges from Renyi entropy. This modified algorithm has been tested on different data sets and comparative study shows the effectiveness of the proposed clustering algorithm.

Keywords Ant clustering algorithm · Data mining · Information theoretic learning

1 Introduction

Clustering is a method of dividing objects into groups of certain resemblance. It means that we maximize the similarity measure between clusters as well as minimize the similarity between objects belonging to the same clusters. This method is widely examined by data mining scientists [17]. From the optimization problems' point of view it is easy to explain the evaluation of clustering output and more precisely define the goal of clustering. The definition of global "optimal" clusters without any subjective criterion is still an open problem and needs further analysis and experiments.

U. Boryczka (✉) · M. Boryczka
Institute of Computer Science, University of Silesia, Sosnowiec, Poland
e-mail: urszula.boryczka@us.edu.pl

M. Boryczka
e-mail: mariusz.boryczka@us.edu.pl

I. Czarnowski et al. (eds.), *Intelligent Decision Technologies 2016*,
Smart Innovation, Systems and Technologies 56,
DOI 10.1007/978-3-319-39630-9_27

Classical approaches in clusterization include methods based on partition or division methodology (k-means, k-medoids) [5], building hierarchy of dependences into attributes' values (BIRCH) and finally, methods based on density, grids and specific models (e.g. SOM or Neural Network) respectively [12–14].

Recently main focus of researchers was channeled into clusterization with swarm intelligence metaphor. In our work we want to concentrate on the ant colony behavior, especially a brood sorting and a cemetery organization as examples of sorting abilities of insects. Another interesting form of ants' adaptation to clustering is a foraging behavior, where pheromone plays the most revelant role. Key elements of this kind of algorithm are: the pheromone matrix updating rules, a probability of transition function and heuristic function [7].

The analyzed group of approaches, known as "piling" methods, concentrate on imitating ants' behavior concerning gathering of corpses and forming cemeteries or brood sorting. Some species of ants collect randomly dispersed around the environment dead bodies and then stack them into piles. The bigger the pile, the more ants are encouraged to put another corpse on the pile. The essence of this phenomenon is its positive feedback. In this field of work, the pioneering researcher was Deneubourg, who created the basic model (BM) of sorting objects [6]. Lumer and Fieta improved this model by incorporation dissimilarity/similarity measures (LF) [1, 14]. In analysis proposed by Lumer and Fieta, an object with n attributes can be examined as a point in n-dimensional space. This point can be projected into two-dimensional space and then an agent can analyze scattered objects and create piles of similar objects. To state whether the object should be picked up or dropped, the probability between a given object and other object in local vicinity is calculated. Handl et al. modified the LF algorithm by using non homogenical population of ants and in document retrieval [10, 11]. They also compared the ant-based clustering algorithm, providing several strategies to increase the reliability. Ramos et al. [8], using different speeds of ant movement, studied the performance of the modification in text document clustering. Boryczka proposed heterogenical populations of ants in ACA [2, 3] and the role of pheromone in learning process was examined [1, 2].

The hybridization of ant-based clustering with other methods of clusterization can be find in recent studies. The example can be an algorithm AntClass, which combines ant-based clustering and k-means approach. In initial phase, ants create preliminary clusters for k-means processing algorithm [15].

In this article we concentrate on the "piling method". Although this approach is constantly modified and improved, there are still two main problems. First one is the undirected randomness, which leads to low efficiency. The other issue rely on the fact that there are many coefficients or thresholds which values are difficult to adjust. Meanwhile, they have a great impact on the clusterization results. To overcome these difficulties, we propose a modification of ant clustering algorithm integrated with Renyi entropy. Various attempts to use information theory have been undertaken in the past. The usage of local entropy in clusterization was firstly described by Zhang [9]. However the approach grounded on this theory often impose unreal assumption concerning the various of coefficients in data factorization. Renyi entropy splendidly fits to the nonparametric estimation and eliminates difficulties involving the evaluation of the traditional entropy measure. In this work we derive the information potential and its force from Renyi entropy (based on the predecessor—Zhang [19]). This idea leads to establishing new model of ants' movement. This entropy is also used as a criteria of evaluation clusters created by ants. Furthermore, this criterion is responsible for stopping this process of clusterization.

This article is organized as follows. In section two the ant clustering algorithm is presented as well as its problems. Section three is devoted to the information theoretic learning and particularly Renyi entropy. Next part shows the description of proposed ant clustering algorithm with Renyi entropy. Section six contains experiments and comparative study. Future work and conclusions constitute the last part of this article.

2 Main Imperfections in Ant Clustering Algorithm (ACA)

In general, the process of clustering by ants may be presented in 5 steps:

1. Projection: all objects and agent-ants are randomly scattered on the grid file.
2. Probability counting's: the probability of dropping and picking up objects in each agent-ant's vicinity containing objects, which are the subject of estimations is calculated.
3. Picking up or dropping stage.
4. Agent-ants' movement.
5. Repetition of steps 2–4.

Some research works identified three impediments, which need to be overcome.

Effectiveness of the proposed method. Low effectiveness of the ACA is due to the omnipresent randomness. Randomly located objects on the grid file caused the small similarity measure observed in its vicinity. Objects are more frequently picked up than dropped into the grid file. The number of iterations is extremely high because of lack of significant improvement of the clustering algorithm effectiveness. This problem was analyzed by Chen et al. [4] and by Boryczka [2].

Ability of adjustment to the data with a specific structure. Approaches, which imitated ants' behavior had used distance measures such as: Euclidean or cosine. These measures are effective in case of elliptical or Gaussian structure. The worse performance could be noticed in nonlinear structure of data. This issue was pointed out in [18].

Parameters of the discussed approach. An abundance of the parameters' values caused some difficulties with appropriate adjustment. E.g. small value of parameter α has an influence on the lack of appropriately created groups of objects. In consequence, the more adaptive and simplified relations between some coefficients the better effectiveness in ACA can be observed [3].

3 Renyi Entropy and Information Potential

Hungarian mathematician A. Renyi proposed in 1960 a new information measure, which later has been called from his name—Renyi's entropy. For a stochastic variable X with the probability distribution f_x, Renyi's entropy is defined as follows:

$$H_R(X) = \frac{1}{1-\alpha} \log \int f_x^{\alpha} dx, \ \alpha \neq 1. \tag{1}$$

In particular case, when $\alpha = 2$ (known as a Renyi's quadratic entropy) is calculated as:

$$H_R(X) = -\log \int f_x^2 dx. \tag{2}$$

The most commonly used density function is the Gaussian function (normal distribution) and is presented as:

$$p(x) = \frac{1}{\sqrt{2\pi\sigma}} \exp\left(-\frac{(x-c)^2}{2\sigma^2}\right), \tag{3}$$

where c is the mean value, σ^2 is the variance and σ is the standard deviation.

For the probability distribution one of the approximation methods is a Parzen window. In this approach the value of the distribution function in a region R for samples located around a point x is determined. Let h be a length of hypercube's edge, N—number of dimensions and samples are located in the hypercube near the middle point x. The function below shows whether the point x_i is located in the middle of the square or not. The total number of k samples in region R can be calculated using the next formula:

$$k = \sum_{i=1}^{n} \phi\left(\frac{x_i - x}{h}\right). \tag{4}$$

The estimation of the Parzen window of the probability distribution for two dimensions can be expressed as:

$$p(x) = \frac{1}{N} \sum_{i=1}^{N} \frac{1}{h^2} \phi\left(\frac{x_i - x}{h}\right). \tag{5}$$

To assess the probability distribution Parzen window with Gaussian Kernel function can be utilized. Let $X = \{x_i : x_i R_m, i = 1, 2, ..., N\}$ be a set of samples for random variable X in M-dimensional space. The probability distribution estimation in accordance to data points is given by:

$$f_x = \frac{1}{N}\sum_{i=1}^{N} G(x - x_i, \sigma^2 I), \tag{6}$$

where N is a number of points.

The symmetric Gaussian kernel with covariance matrix is exploited here:

$$G(x, \sigma^2 I) = \frac{1}{(2\pi)^{\frac{M}{2}} \sigma^M)} \exp\left(\frac{x^T x}{2\sigma^2}\right). \tag{7}$$

Substituting (7) into Renyi's quadratic entropy (2) and exert the properties of Gaussian kernel, the entropy of X set is obtained in the following form:

$$H(\{X\}) = -\log V(\{X\}), \tag{8}$$

where: $$V(\{X\}) = \frac{1}{N^2}\sum_{i=1}^{N}\sum_{j=1}^{N} G(x_i - x_j, 2\sigma^2 I). \tag{9}$$

If X is a cluster, $H(\{X\})$ can be treated as an within-cluster entropy, because the entropy is calculated based on the points that belong to the same cluster. In general:

$$V(\{X\}) = H(\{X\}_1) + \cdots + H(\{X\}_k) = -\sum_{i=1}^{k} \log V(\{X\}_i). \tag{10}$$

In his science work [16] Principe defined the notion of the information potential (IP) as:

$$V(X) = \frac{1}{N^2}\sum_{i}\sum_{j} G(x_i - x_j, 2\sigma^2 I), \tag{11}$$

where $G(\cdot)$ has a positive value and inverse proportion to the distance between information particles. Authors have also defined the information force F_{ij} as a derivation of the information potential and being the strength with which the particle x_j effects the particle x_i:

$$F_{ij} = \frac{\partial}{\partial x_i} G(x_i - x_j, \sigma^2 I) = -G(x_i - x_j, 2\sigma^2 I)\frac{(x_i - x_j)}{2\sigma^2}. \tag{12}$$

Zhang et al. [19] modified the information force by dividing this value by d_{ij} (they obtained so called information field strength):

$$S_{ij} = \frac{F_{ij}}{d_{ij}} = -\frac{G(x_i - x_j, 2\sigma^2 I)}{2\sigma^2} = -\frac{1}{2^{M+1}\Pi^{\frac{M}{2}}\sigma^{M+2}} \exp\left(-\frac{d_{ij}^2}{(2\sigma)^2}\right)$$

$$= -J(M,\sigma)\exp\left(-\frac{d_{ij}^2}{(2\sigma)^2}\right). \quad (13)$$

4 New Ant Clustering with Information Strength Influence

Assignment of objects to the appropriate group or cluster may be explained as follows. If an object x is wrongly assigned to the cluster C_1 then the entropy of C_1 cluster is increased more than the entropy of cluster C_2 (when the object x was assigned to the cluster C_2, suitably). When we have k clusters, the assignment of object x into cluster C is observed, then

$$H(C_i + x) - H(C_i) < H(C_k + x) - H(C_k), \quad (14)$$

for $k = 1, \dots, K, k \neq i$, where $H(C_k)$ determines the entropy for k-th cluster. This method is called a differential entropy clusterization.

In ant clustering approach each ant is identified as an object. Each ant/object is located in the point (x_i, y_i) in k-th iteration. Its vicinity is denoted as Neigh(o_i). We assume that o_i is treated as a set of all objects in Neigh(o_i) before the object o_i is joined with the set. Consequently, this connection is denoted as $C_i + o_i$.

In this situation the differential entropy for clustering process is reflected by:

$$\Delta H(C_i) = H(C_i + o_i) - H(C_i). \quad (15)$$

$$\text{Let} \quad \delta C_i = \left|\frac{\Delta H(C_i)}{H(C_i)}\right|. \quad (16)$$

If the object o_i is very similar to others in Neigh(o_i) then the entropy is not changing or very small changes are observed. In this situation objects o_i should be sustainable in Neigh(o_i). Then $\Delta H(C_i) < 0$ or $\delta(C_i) < \delta_0$, where δ_0 ought to have a small positive value.

Parameters of the acquainted algorithm are as follows:

- the size of the grid file is $N \times N$, where N is the number of objects. The parameter N determines the region where agent-ants analyze perceived information. This parameter influences the speed of creating clusters in initial steps of our approach,
- the parameter r determines the range of agent-ant's activity; this value is changeable during the time. The smallest value of r is equal to $r_{min} = \left\lfloor \frac{N}{10} \right\rfloor$, and the greatest value of discussed parameter is $r_{max} = \left\lfloor \frac{N}{2} \right\rfloor$,

- the parameter σ is an important parameter, because the window width is settled by this value, which can be determined by the researcher,
- the parameter δ_0 is a threshold, which determines the H entropy's changes after agent-ant's movement,
- the parameter t_{max} is the maximum number of algorithm's iterations.

Modifications employed in our approach are strongly connected with the σ parameter, which is the key parameter because of its role in correctly partitioning the data. We decided to use the self-adaptive version of our approach for each data set, individually.

Let d_{mean} is a mean value of the distances between all objects in analyzed data set:

$$d_{mean} = \frac{1}{N^2} \sum_{i=1}^{N} \sum_{j=1}^{N} Nd_{ij}, \tag{17}$$

where N is a number of all objects. We set the lower and upper values of parameter σ in which it has been examined:

$$0 < \sigma < -\frac{d_{mean}}{2\sqrt{\ln 2}}. \tag{18}$$

```
1  Algorithm: ACA-ITL
2  // Random location of objects in the grid file
3  // Initialization of parameters' values: r, t_max, σ_0
4  for t = 1 to t_max do
5  |   for i = 1 to N do
6  |   |   find all objects in vicinity of o_i
7  |   |   calculate H(C_i + o_i) and H(C_i)
8  |   |   if ΔH(C_i) < 0) or (δ(C_i) < δ_0) then
9  |   |   |   object/ant is still
10 |   |   else
11 |   |   |   object/agent o_i moves to the point (x, y) in Neigh(o_i)
12 |   |   end
13 |   end
14 |   arrange r, δ_0
15 |   t = t + 1
16 end
17 Indicate the objects' location
```

Algorithm 1: Pseudo-code of ACA-ITL algorithm

Because of dynamic character of σ (the kernel size)—in self adaptation scheme its value should be updated in accordance to the equation:

$$\sigma(i+1) = \begin{cases} 0.95\sigma(i) & \text{when } (i \bmod 10) = 0 \\ \sigma(i) & \text{otherwise.} \end{cases} \tag{19}$$

5 Experiments and Results

The benchmarking experiments show that two ant-based clustering approaches perform very well compared to each other for synthetic data sets (see Table 1). The ACA-ITL algorithm always achieves results that are slightly better than the results of ACA clustering algorithm (for the presented evaluation metrics). In the second part of our work results concerning four real data sets: Iris, Ionosphere, Parkinsons and Wines are reported. The key parameter σ is examined in the 0.01–0.2 range. In classical version of our approach the step of sampling was established to 0.05. Each of the data sets was tested in accordance to the four distance measures: cosine, Euclidean, Gower and Minkowski. The number of iterations is a significant parameter and for each data set it is established separately, with condition of the best performance. In Table 2 most promising results of analyzing quality measures were chosen. We designated for which values of σ parameter the best results were obtained. It can be observed that in most cases, the lowest values of parameter σ (for cosine measure) guarantee good results, while for Euclidean measure the same situation can be pointed out when parameter $\sigma = 0.20$. Outcomes for Manhattan, Gower and Minkowski measures were not included in Table 2 because of article size limitation.

Table 1 Synthetic data sets—quality measures of clustering

Data set	Quality measure	ACA	ACA-ITL
Square1	Rand index	0.9784	0.9246
	Error rate	0.0258	**0.0005**
Square2	Rand index	0.9728	0.9645
	Error rate	0.0327	**0.0002**
Square3	Rand index	0.9812	0.9593
	Error rate	0.0205	**0.0002**
Square4	Rand index	0.9121	0.8052
	Error rate	0.1025	**0.0035**
Square5	Rand index	0.7197	**0.8621**
	Error rate	0.1031	**0.0042**
Halfrings	Rand index	0.8163	**0.8391**
	Error rate	0.1285	**0.0003**

Table 2 Quality measures of ACA-ITL approach for different distance measures

Data set	Distance measure	σ	Rand measure	F measure	Error rate
Iris	Cosine	0.01	0.8462	0.7031	0.0087
	Euclidean	0.15	0.8075	0.6380	0.0105
Ionosphere	Cosine	0.01	0.7274	0.7046	0.0028
	Euclidean	0.20	0.7173	0.6801	0.0027
Parkinsons	Cosine	0.05	0.7755	0.7688	0.0049
	Euclidean	0.20	0.7639	0.7711	0.0049
Wine	Cosine	0.05	0.7680	0.7429	0.0096
	Euclidean	0.01	0.7560	0.6745	0.0109

The problem concerning the best value of parameter σ establishment is very difficult, therefore in the next step of our research the algorithm ACA-ITL will be improved by adaptation of parameter σ.

Another direction of this study was the comparison of ACA-ITL with its modification and well-known clustering algorithms: *k*-means, *k*-medoids and DBSCAN. In case of Iris data set, the work was repeated 30 times. The outcome of this process is comprised in Table 3—mean values for Rand index, F-measure, and rate error are presented there. This experiment was conducted for 5 distance measures, mentioned above.

Table 3 Comparative study with classical approaches

Distance measure	Quality measure	ACA-ITL	mACA-ITL	*K*-means	DBSCAN
Cosine	Rand index	**0.7900**	0.7546	0.5590	0.6552
	F index	**0.6564**	0.6560	0.3373	0.5949
	Error rate	**0.0107**	0.0136	0.0125	0.0130
Euclidean	Rand index	0.7800	0.7593	0.5621	**0.8277**
	F index	0.6095	0.6512	0.3402	**0.7952**
	Error rate	0.0118	0.0123	0.0118	**0.0058**
Gower	Rand index	**0.7545**	0.7540	0.5527	0.6644
	F index	0.5640	**0.6731**	0.3264	0.6205
	Error rate	0.0131	0.0132	0.0126	**0.0100**
Manhattan	Rand index	0.7692	0.7541	0.5578	**0.8800**
	F index	0.6000	0.6746	0.3320	**0.8678**
	Error rate	**0.0125**	0.0131	0.0131	0.0203
Minkowski	Rand index	**0.7820**	0.7561	0.5578	0.6552
	F index	0.6527	**0.6736**	0.3325	0.5949
	Error rate	**0.0111**	0.0132	0.0123	0.0130

Table 4 The significance of modification of ACA-ITL

Data set	Distance measure	Quality measure	ACA-ITL	mACA-ITL
Iris	Cosine	Rand index	**0.7900** (0.0344)	0.7546 (**0.0008**)
		F index	0.6564 (0.0404)	0.6560 (**0.0137**)
		Error rate	**0.0107** (0.0012)	0.0136 (**0.0004**)
Iris	Euclidean	Rand index	**0.7800** (0.0223)	0.7593 (**0.0047**)
		F index	0.6095 (0.0341)	**0.6512** (**0.0125**)
		Error rate	**0.0118** (0.0015)	0.0123 (**0.0014**)
Wine	Cosine	Rand index	**0.7629** (0.0128)	0.7521 (**0.0095**)
		F index	**0.7216** (0.0344)	0.6607 (**0.0180**)
		Error rate	**0.0098** (0.0007)	0.0114 (**0.0005**)
Wine	Euclidean	Rand index	**0.7540** (**0.0120**)	0.7504 (0.0137)
		F index	0.6637 (**0.0200**)	**0.6662** (0.0221)
		Error rate	**0.0111** (**0.0006**)	0.0116 (0.0006)

For cosine measure the best results have been obtained for ACA-ITL, in each quality measures checked. For Euclidean measure the best outcome was observed for DBSCAN approach. For the Gower measure two algorithms won: ACA-ITL and mACA-ITL ex aequo. For Manhattan measure again the best turned out the DBSCAN algorithm. The lowest error rate in analyzed measure come off for ACA-ITL. In case of Minkowski measure once more the best results were obtained for ACA-ITL as well as for its modification.

Two data sets: Iris and Wine have been tested in order to point out the superiority of either ACA-ITL or mACA-ITL. In comparison average results of repetitions for Rand index, F measure and error rate were examined (Table 4). Smaller standard deviations were obtained for mACA-ITL in every case. This indicates that the mACA-ITL is stable and easier adapts to the analyzed data.

6 Conclusions

The results of this research illustrate that algorithms that mimic the nature (ants' colony behavior) combined with the information theoretic learning performs well in data clusterization. For distance measures: cosine, Euclidean, Gower, Manhattan and Minkowski average results are better than in case of DBSCAN approach. In each analyzed case of distance measure ACA-ITL is better than k-means approach. BDSCAN is more suitable only in case of Euclidean and Manhattan measures. The key problem was to establish the parameter σ value, which was improved by self-adaptation incorporated in modified version of our proposition. In this research we did not analyze the time consuming issue, which is worse in our approach than it was in classical approaches. Further studies should focus on the speed of the performance of the ACA-ITL approach as well as an optimization of memory usage. Another significant parameter which will be tested in the future is a stopping criterion.

References

1. Bonabeau, E., Dorigo, M., Theraulaz, G.: Swarm Intelligence. From Natural to Artificial Systems. Oxford University Press, New York (1999)
2. Boryczka, U.: Finding groups in data: cluster analysis with ants. Appl. Soft Comput. **9**(1), 61–70 (2009)
3. Boryczka, U.: Ant colony metaphor in a new clustering algorithm. Control Cybern. **39**(2), 343–358 (2010)
4. Chen, L., Tu, L., Chen, H.: A novel ant clustering algorithm with digraph. In: Wang, L., Chen, K., Ong, J. (eds.) Advances in Natural Computation: First International Conference, ICNC 2005, Changsha, China. pp. 1218–1228. Volume 3611 of the series Lecture Notes in Computer Science, Springer (2005)
5. Chen, Q., Mo, J.: Optimizing the ant clustering model based on k–means algorithm. In: Proceeding of the 2009 WRI World Congerss on Computer Science and Information Engineering. vol. 3, pp. 699–702 (2009)
6. Deneubourg, J.L., Goss, S., Franks, N., Sendova-Franks, A., Detrain, C., Chretien, L.: The dynamics of collective sorting: Robot–like ant and ant–like robot. In: Meyer, J.A., Wilson, S.W. (eds.) First Conference on Simulation of Adaptive Behavior. From Animals to Animats. pp. 356–365 (1991)
7. Dorigo, M., Stützle, T.: Ant Colony Optimization. MIT Press, Cambridge (2004)
8. Fernandes, C., Merelo, J.J., Mora, A.M., Ramos, V., Laredo, J.L.J.: Kohonants: A self–organizing algorithm for clustering and pattern classification. Artificial Life pp. 428–435 (XI 2008)
9. Gutowitz, H.: Complexity—seeking ants (1993), unpublished report
10. Handl, J., Meyer, B.: Improved ant–based clustering and sorting in a document retrieval interface. In: Verlag, S. (ed.) PPSN — VII. Seventh international Conference on Parallel Problem Solving from Nature. pp. 913–923. Berlin (2002)
11. Handl, J., Knowles, J., Dorigo, M.: Ant–based clustering: a comparative study of its relative performance with respect to k–means, average link and id–som. Technical Report 24, IRIDIA, Universite Libre de Bruxelles, Belgium (2003)
12. Jain, A., Murty, M., Flynn, P.: Data clustering: a review. ACM Comput. Surv. **31**(3), 264–323 (1999)
13. Kaufman, L., Russeeuw, P.: Finding groups in data: an introduction to cluster analysis. Wiley (1990)
14. Lumer, E., Faieta, B.: Diversity and adaptation in populations of clustering ants. In: Third international Conference on simulation of Adaptive Behavior: From animals to Animats 3. pp. 489–508. MIT Press, Cambridge (1994)
15. Monmarché, N., Slimane, M., Venturini, G.: Antclass: discovery of clusters in numeric data by an hybridization of an ant colony with the k-means algorithm. Technical report, Laboratoire d'Informatique, Écolé d'Ingénieurs en Informatique pour l'Industrie, Université de Tours (1999), internal Report no. 213
16. Principe, J.: Information Theoretic Learning. Springer, Renyi's Entropy and Kernel Perspectives (2010)
17. Runkler, T.: Ant colony optimization of clustering model. Int. J. Intell. Syst. **20**(12), 1233–1261 (2005)
18. Vizine, A., de Castro, L., Hruschka, E.: Towards improving clustering ants: adaptive clustering algorithm. Inform. J. **29**(2), 143–154 (2005)
19. Zhang, L., Cao, Q., Lee, J.: A novel ant-based clustering algorithm using Renyi entropy. Appl. Soft Comput. **13**(5), 2643–2657 (2013)

Kernel-Based Fuzzy C-Means Clustering Algorithm for RBF Network Initialization

Ireneusz Czarnowski and Piotr Jędrzejowicz

Abstract Designing an effective structure of the RBF network is the task carried-out at the network initialization phase. Usual approach to deal with the problem is to decide on the number of hidden units and to apply a clustering algorithm to calculate cluster centroids. Clustering techniques have a strong influence on the performance of the RBF networks. The paper focuses on the radial basis function neural network initialization problem and the implementation of the kernel-based fuzzy C-means clustering algorithm, as an alternative method for the RBF networks initialization. Performance of the RBFNs initialized using the kernel-based fuzzy clustering algorithm is compared with several other clustering techniques, including k-means, fuzzy C-means and X-means.

Keywords Classification • Neural networks • Radial basis function • Clustering • Kernel-based clustering

1 Introduction

The paper focuses on a radial basis function network (RBFN) initialization problem. RBFNs, introduced by Broomhead and Low [3], can be placed among best-known neural networks types and are a popular type of the feedforward networks. RBF networks are used as universal approximation tools, similarly to the multilayer perceptrons (MLPs). However, radial basis function networks usually achieve faster convergence since only one layer of weights is required [9].

RBFNs have been successfully used to solve many different problems, including classification, regression, pattern recognition, image recognition, time series

I. Czarnowski (✉) · P. Jędrzejowicz
Department of Information Systems, Gdynia Maritime University, Morska 83, 81-225 Gdynia, Poland
e-mail: irek@am.gdynia.pl

P. Jędrzejowicz
e-mail: pj@am.gdynia.pl

I. Czarnowski et al. (eds.), *Intelligent Decision Technologies 2016*,
Smart Innovation, Systems and Technologies 56,
DOI 10.1007/978-3-319-39630-9_28

analysis, etc., with applications in science, engineering, economy, medicine and many other fields. Advantage of the RBFNs over traditional methods results from their ability to deal with a non-linearity in the data [3].

The RBF network is constructed from a three-layer architecture with a feedback. The input layer consisting of a set of source units connecting the network to the environment. The hidden layer consists of hidden neurons with radial basis functions [10]. Each hidden unit in the RBFN represents a particular point in space of the input instances. The output of the hidden unit depends on the distance between the processed instances and the particular point in the input space of instances (the point is called an initial seed point, prototype, centroid or kernel of the basis function). The distance is calculated as a value of the activation function. Next, the distance is transformed into a similarity measure that is produced through a nonlinear transformation carried-out by the output function of the hidden unit called the radial basis function. RBFNs can use different functions in each hidden unit. The output of the RBF network is a linear combination of the outputs of the hidden units.

The performance of RBF networks depends on numerous factors. Especially, the performance of the RBF network will be influenced by the number of the radial basis functions, the number of centroids and their locations that is associated with the RBFN initialization phase. Furthermore, connection weights used to form a combination of structure. Making choices with respect to the structure of the RBF network involves the outputs of the hidden units, have also a direct influence on the RBFN performance. These respective weights are determined during the RBF network training phase.

In a traditional approach to the RBF network design, the network structure is usually fixed through some compromise setting or is based on the designer experience, and only the output layer weights are updated with time. However, this might not be sufficient to track non-stationary systems, and the RBFN design is not straightforward [4].

The traditional two-stage design schema is based on initializing the hidden layer using specialized strategy for identifying or calculating centroids. It is a very important stage from the point of view of achieving a good approximation by the RBF network under development. RBF networks design process can be also seen as a hybrid process, which combines unsupervised and supervised learning [20].

Usual approach to deal with the initialization problem is to decide on the number of hidden units and to apply the k-means algorithm, as the unsupervised technique, to calculate cluster centroids [5, 20]. Such approach is probably inspired by the simplicity of the k-means algorithm. However, k-means clustering algorithm is sensitive to the initial centroids selection and searching for the optimal centroid locations may result in a poor local minima [12]. Generally, major problem with the k-means-based algorithms is likely convergence of the clustering process to a non-optimal configuration, hence resulting in a non-optimal RBFN structure [19].

Other disadvantage of the k-means based approach is that RBFNs trained in a conventional way may also not be accurate since the number of clusters at the initialization stage of the RBFN design must be set up a priori. Hence, several alternative approaches to cluster initialization for the RBF networks have been proposed [21]. However, none of the approaches proposed so far can be considered as being superior, guaranteeing optimal results in terms of the learning error reduction or increased efficiency of the learning process. Hence, searching for a robust and efficient approaches to RBFN design is still a lively field of research [8].

The paper focuses on the radial basis function neural initialization problem. The implementation of the kernel-based fuzzy C-means clustering algorithm (KFCM), as an alternative method for the RBF networks initialization, is discussed. In this paper the RBFNs performance with the kernel-based fuzzy clustering algorithm used as the initialization tool, is analyzed and compared with other approaches. A comparative study of four cluster-based approaches to RBFNs initialization, including k-means, fuzzy C-means, X-means and discussed KFCM, is carried out using benchmark artificial and real datasets. The KFCM, as a tool for designing of the RBF network, has been suggested in earlier author's paper [7]. In [7] the KFCM was used for cluster initialization, and next, at the second stage, centroids from thus obtained clusters have been selected by applying the population-based metaheuristic. The metaheuristic has been also applied for determination of connection weights. However, in this paper the KFCM is used for RBF network initialization in a traditional way. KFCM initialized clusters and their centroids are next used for calculating values of the activation functions applying the agent-based population learning algorithm (PLA) metaheuristics. For comparison, in each considered case, the RBF network is also trained using the back-propagation algorithm.

The paper is organized as follows. Section 2 provides a short description of the RBFNs processing. In the next section the initialization problem is outlined, including a short review of several selected clustering algorithms. Section 3 also depicts the kernel-based fuzzy C-means clustering algorithm. A detailed description of the computational experiment setup and the discussion of the experiment results are given in Sect. 4. Finally, the last section contains conclusions and suggestions for future research.

2 RBF Neural Network Processing

The output of the RBF network is a linear combination of outputs of the hidden units, i.e. a linear combination of the nonlinear radial basis function generating approximation of the unknown function. In general, the RBF neural network output function has the following form:

$$f(x) = \sum_{i=1}^{M} w_i G_i((\|x - c_i\|), p_i), \tag{1}$$

where M defines the number of hidden neurons, G_i denotes a radial basis function associated with i-th hidden neuron, x is an n-dimensional input instance ($x \in \mathfrak{R}^n$), p_i is a vector of parameters (like, for example locations of centroids),

$\{c_i \in \Re^n: i = 1, \ldots, M\}$ are cluster's centers (centroids or prototypes), $\|\cdot\|$ is the Euclidean norm and $\{w_i \in \Re^n: i = 1, \ldots, M\}$ represents the output layer weights.

One of most popular radial basis functions is the Gaussian function. In such case the output function takes the following form:

$$G(r, b) = e^{-\left(\frac{\|x-c\|}{b}\right)^2}, \tag{2}$$

where b defines a function dispersion.

3 RBF Neural Network Initialization

3.1 *The Initialization Problem*

The RBF network initialization is a process, where the set of parameters of the radial basis functions needs to be calculated or induced. Gaussian centroids belongs to such a set of parameters.

Several different approaches to RBFN initialization have been proposed in the literature (see for example [14, 23, 26]). Among classical methods used to RBFN initialization one should mention clustering techniques. In [17] it has been shown that there is a strong correlation between good clustering and the performance of the RBF network.

Besides clustering methods, the support vector machine or the orthogonal forward selection methods are used. In [15] the following strategies for RBFNs initialization were reviewed: random selection, sequential growing, systematic seeding, editing techniques and clustering. A short review of the RBFN initialization approaches can be found in [6].

In general, the clustering problem can be solved through applying some unsupervised techniques. The aim of the clustering is to find a natural data groups in a non-classified data set [19]. Clustering is a process for partitioning a set of N instances into t subsets, named clusters.

As it has been mentioned, k-means algorithm is an usual approach to deal with the RBFN initialization problem. It is known to converge to a local minimum and is sensitive to the initial centroids selection. Unfortunately, it is also known to be rather slow. Another weakness of the k-means is its inability to determine the number of clusters, which has to be set by the user.

X-means is k-means extended by adding special procedures dedicated to improve structure of the clusters and evaluation of their quality. The modifications led to improve the convergence of the clustering process and to easier estimation of the number of clusters [22].

Alternative to the k-means algorithm is the fuzzy C-means algorithm (FCM), introduced by Ruspini [24] and extended by Bezdek [2]. FCM considers each cluster as a fuzzy set with membership function measuring the possibility for each data to belong to a given cluster. The advantage of this approach, in comparison to k-means algorithm, is that by fixing the so-called threshold, all instances having memberships' degrees which exceed this threshold are assigned to the respective clusters. From the implementation point of view, using the FCM algorithm means that the number of clusters can be determined automatically [17]. Although the FCM has considerable advantages, in comparison to k-means algorithm, its main limitation is sensitivity to noises [18].

In this paper the kernel-based fuzzy clustering algorithm is considered as an alternative approach to RBFN initialization.

3.2 Kernel-Based Fuzzy Clustering

Kernel-based fuzzy C-means (KFCM) was introduced to overcome noise and outliers sensitivity in FCM [27] by transforming input data X into a higher dimensional kernel space via a non-linear mapping Φ which increases the possibility of linear scalability of the instances in the kernel space and allows for fuzzy C-means clustering in the feature space [16].

The kernel method uses the fact that product in the kernel space can be expressed by a Mercer kernel K as follows [27]:

$$K(x, y) \equiv \Phi(x)^T \Phi(y). \tag{3}$$

This allows to replace the computation of distance in the kernel space by a Mercer kernel function, which is known as a "kernel trick" [27]. One of most common kernel function is the Gaussian function.

To describe the proposed approach in a formal manner, the following notation is introduced. Let x denote a training instance, N—the number of instances in the original training set D and, n—the number of attributes with the total length of each instance (i.e. training example) equal to $n + 1$. When the training set includes data of the classification problem, the element numbered $n + 1$ contains a class label that can take any value from the finite set of class labels $C = \{c^l : l = 1, \ldots, k\}$. Also, let $X = \{x_{ij} : j = 1, \ldots, n + 1;\ i = 1, \ldots, N\}$ denote the matrix of $n + 1$ columns and N rows containing values of all instances from the training dataset.

From mathematical point of view, let clusters be characterized by prototypes c_i and fuzzy partition matrix $U = (u_{ij})$ of size t x N, where u_{ij} is the degree of membership of x_j in cluster i. Then the standard constrains for fuzzy clustering are assumed, that can be expressed as:

$$0 \leq u_{ij} \leq 1, \quad \underset{j}{\forall} \sum_{i=1}^{t} u_{ij} = 1, \quad \underset{i}{\forall} 0 < \sum_{j=1}^{N} u_{ij} < N \tag{4}$$

and the partition of data D into clusters is performed in accordance with the maximal value of membership degree:

$$clust(x_j) = \arg\max_{1 \le i \le t} u_{ij}. \tag{5}$$

For KFCM-K clustering it is assumed that prototypes v_i are located in the kernel space and centroids further need to be approximated by an inverse mapping $c_i = \Phi^{-1}(v_i)$ to the feature space [13].

The KFCM-K algorithm minimizes the following objective function [11, 27]:

$$J = \sum_{i=1}^{t} \sum_{j=1}^{N} u_{ij}^{m} dist^2(\Phi(x_j), v_i), \tag{6}$$

where $m \in \mathbf{N}$ and controls the fuzziness of the memberships and is fixed natural, most often equal 2. Optimizing J with respect to v_i gives [27]:

$$v_i = \frac{\sum_{j=1}^{N} u_{ij}^{m} \Phi(x_j)}{\sum_{j=1}^{N} u_{ij}^{m}}, \tag{7}$$

which leads to the formula for partition matrix

$$u_{ij} = \frac{1}{\sum_{i=1}^{t} \left(\frac{dist(\Phi(x_j), v_i)}{dist(\Phi(x_j), v_i)} \right)^{\frac{2}{m-1}}}, \tag{8}$$

where:

$$dist^2(\Phi(x_j), v_i) = K(x_j, x_j) - 2\frac{\sum_{k=1}^{N} u_{ik}^{m} K(x_j, x_k)}{\sum_{k=1}^{N} u_{ik}^{m}} + \frac{\sum_{k=1}^{N} \sum_{l=1}^{N} u_{ik}^{m} u_{il}^{m} K(x_k, x_l)}{\left(\sum_{k=1}^{N} u_{ik}^{m}\right)^2} \tag{9}$$

Centroids c_i are approximated in the feature space using the minimization of

$$V = \sum_{i=1}^{t} dist(\Phi(c_i), v_i), \tag{10}$$

which in case of the Gaussian kernel gives [27]:

$$c_i = \frac{\sum_{j=1}^{N} u_{ij}^{m} \cdot K(x_j, c_i) \cdot x_j}{\sum_{j=1}^{N} u_{ij}^{m} \cdot K(x_j, c_i)}. \tag{11}$$

Advantage of using kernel functions is the possibility to determine the number of clusters based on significant eigenvalues of a matrix determined by the kernel function applied to feature vectors.

The algorithm of kernel based fuzzy C-means clustering is shown as Algorithm 1.

ALGORITHM 1. Kernel-based fuzzy C-means clustering

Input: D - dataset; K - kernel function; δ - threshold for determining the number of clusters.
Output: clusters and their centroids.

For l=1 **to** k **do** // *for instances from X, belonging to the class* c^l
  Let $K_{ij}^{l} = K(x_i^l, x_j^l)$ be a quadratic matrix of size N^l x N^l.
  Calculate the eigenvalues of matrix (K_{ij}^{l}).
  t^l := number of eigenvalues exceeding δ.
  Initialize U^l to random fuzzy partition satisfying the conditions (4) and with respect to t^l.
  Repeat
    Update U^l according to (8), (9).
  Until stopping criteria satisfied or maximum number iterations reached.
  Repeat
    Update centroids according to (11).
  Until maximum number iteration reached.
End for

Map input vectors from D according to U^l (l=1,...,k) into t clusters, where $t = \sum_{l=1}^{k} t^l$.

Let $C_1,...,C_t$ denote the obtained clusters such that $D = \bigcup_{i=1}^{t} C_i$, $\forall_{i \neq j:i,j=1,...,t}\ C_i \cap C_j = \varnothing$.

Return $C_1,...,C_t$ and their centroids.

The KFCM-K produces clusters of prototypes and their centroids. Hence, it can be in a simple way used for the RBF initialization.

4 Computational Experiment

This section contains results of the computational experiment carried out with a view to evaluate the performance of the RBFNs initialized using KFCM-K. In particular, the reported experiments aimed at comparing quality of the clustering, using KFCM-K and k-means, X-means and FCM, and influence of the clustering algorithms on the performance of the RBF network. In the reported experiment, all clustering approaches have been used for RBF initialization, i.e. for cluster initialization. Next, from thus obtained clusters the centroids have been selected using the agent-based population learning algorithm. In such way initialized RBF networks have been trained using the back-propagation algorithm.

Evaluation of the clustering approaches and performance comparisons are based on the classification problems. The individual RBF classifiers have been applied to solve respective problems using several benchmark datasets obtained from the UCI Machine Learning Repository [1]. Basic characteristics of these datasets are shown in Table 1.

Each benchmark problem has been solved 50 times, and the experiment plan involved 10 repetitions of the 10-cross-validation scheme. The reported values of the goal function have been averaged over all runs. The goal function was the correct classification ratio—accuracy (in %).

The evaluation has been carried out by estimating classification accuracy of the RBFN, which is initialized using prototypes and trained using the backpropagation algorithm. The number of epoch for backpropagation cycles has been set to 1000. Values of these parameters have been set arbitrarily.

During the experiment population size for each PLA was set to 60. The process of searching for the best set of centroids has been stopped either after 100 iterations or after there has been no improved of the current best solution for 1 min of computation. Probability of mutation for the implemented mutation procedures has been set to 20 %.

For algorithms based on k-means the number of prototypes has been set in a way assuring comparability with respect to the number of prototypes produced by the other procedures.

Among other settings, the fuzziness of the memberships has been set to 2. The Gaussian function has been used as a kernel function of the KFCM-K. The Gaussian function has been also used for computations within RBFNs hidden units. The dispersion of the Radial function has been calculated as a double value of the minimum distance between basic functions [25].

Table 2 shows performance comparison involving the clustering algorithms. From the results it can be observed that the KFCM-K assures comparable results in comparison to other algorithms. The KFCM-K proved to be competitive in six cases. Two best results have been obtained using the FCM. Comparing the

Table 1 Datasets used in the reported experiment

Dataset	Number of instances	Number of attributes	Number of classes
WBC	699	9	2
ACredit	690	15	2
GCredit	999	21	2
Sonar	208	60	2
Satellite	6435	36	6
Diabetes	768	9	2
Customer	24000	36	2
Heart	303	13	2
Ionosphere	351	34	2
Mushroom	8124	22	2

Table 2 RBFN performance (in %) obtained for different clustering algorithms applied to the task of the RBF's initialization

	KFCM-K	FCM	k-means	X-means
Sonar	82.41	**82.95**	81.15	82.45
Heart	**83.65**	81.7	80.5	81.75
WBC	93.12	92.47	**95.83**	93.84
ACredit	83.71	**84.79**	84.16	80.78
GCredit	72.46	**73.45**	70.07	70.56
Diabetes	**73.12**	72.84	73.69	73.04
Satellite	**84.04**	83.97	83.57	83.74
Customer	**71.38**	70.74	70.8	71.2
Ionosphere	**95.43**	94.17	91.24	92.6
Mushroom	**99.42**	99.3	99.01	99.15

KFCM-K and the FCM, better results have been obtained by the KFCM-K. Furthermore, it can be observed that results obtained by the X-means outperform results of k-means.

5 Conclusions

The paper investigates influence of the kernel-based fuzzy clustering algorithm on the quality and performance of the RBF networks. The algorithm is used for the RBF initialization, i.e. as the initialization tool. From induced clusters prototypes are selected using the agent-based population learning algorithm, which next are used for calculation of value of RBFs activation functions in each hidden unit.

It can be observed that the KFCM-K is better suited to overcome problems caused by noise and outliers as compared to the FCM. The above property is statistically valid as confirmed by the computational experiment, where the performance of RBF networks, initialized using the KFCM, was investigated. The computational experiment results show also that the KFCM-K is more effective than k-means and its extension, i.e. X-means.

Future research will focus on improving quality of the KFCM-based clustering, where the role of the population learning algorithm will be to find optimal allocation of instances to clusters.

References

1. Asuncion, A., Newman, D.J.: UCI Machine Learning Repository. http://www.ics.uci.edu/~mlearn/MLRepository.html. University of California, School of Information and Computer Science, Irvine (2007)
2. Bezdek, J.C.: Pattern Recognition with Fuzzy Objective Function Algorithms. Springer Publishers, New York (1981)

3. Broomhead, D.S., Lowe, D.: Multivariable Functional Interpolation and Adaptive Networks. Complex Syst. **2**, 321–355 (1988)
4. Chen, H., Gong, Y., Hong, X., Chen. S.: A Fast Adaptive Tunable RBF Network for Non-stationary Systems. IEEE Trans. Syst. Man Cybern. Part B (99), 1–10 (2015). doi:10.1109/TCYB.2015.2484378
5. Chen, S., Billings, S.A., Grant, P.M.: Recursive hybrid algorithm for non-linear system identification using radial basis function networks. Int. J. Control **55**(5), 1051–1070 (1992). doi:10.1080/00207179208934272
6. Czarnowski, I., Jędrzejowicz, P.: Agent-based approach to the design of RBF networks. Cybern. Syst. **44**(2–3), 155–172 (2013)
7. Czarnowski, I., Jędrzejowicz, J., Jędrzejowicz, P.: Designing RBFNs with similarity-based and kernel-based fuzzy C-means clustering algorithm. ACM Trans. Knowl. Discov. Data 2016 (manuscript submitted for publication)
8. De Corvalho, A., Brizzotti, M.M.: Combining RBF networks trained by different clustering techniques. Neural Process. Lett. **14**, 227–240 (2001)
9. Du, K.-L.: Clustering: A neural network approach. J. Neural Netw. **23**(1), 89–107 (2010). doi:10.1016/j.neunet.2009.08.007
10. Gao, H., Feng, B., Zhu, L.: Training RBF neural network with hybrid particle swarm optimisation. In: Weng, J., et al. (eds.) ISNN 2006, LNCS 3971, pp. 577–583. Springer, Berlin Heidelberg (2006)
11. Garg, S., Patra, K., Khetrapal, V., Pal, S.K., Chakraborty, D.: Genetically evolved radial basis function network based prediction of drill flank wear. Eng. Appl. Artif. Intell. **23**, 1112–1120 (2010)
12. Graves, D., Pedrycz, W.: Kernel-based fuzzy clustering and fuzzy clustering: a comparative experimental study. Fuzzy Sets Syst. **161**, 522–543 (2010). doi:10.1016/j.fss.2009.10.021
13. Grover, N.: A study of various fuzzy clustering algorithms. Int. J. Eng. Res. **3**(3), 177–181 (2014)
14. Havens, T.C., Bezdek, J.C., Palaniswami, M.: Cluster validity for kernel fuzzy clustering. In: Proceedings of 2012 IEEE International Conference on Fuzzy Systems (FUZZ-IEEE). IEEE. Brisbane, QLD, pp. 1–8 (2012). doi:10.1109/FUZZ-IEEE.2012.6250820
15. Huang, G.-B., Saratchandra, P., Sundararajan, N.: A generalized growing and pruning RBF (GGAP-RBF) neural network for function approximation. IEEE Trans. Neural Netw. **16**(1), 57–67 (2005). doi:10.1109/TNN.2004.836241
16. Krishnaiah, P.R., Kanal, L.N.: Handbook of Statistics 2: Classification, Pattern Recognition and Reduction of Dimensionality. North Holland, Amsterdam (1982)
17. Li, Z., Tang, S., Xue, J., Jiang, J.: Modified FCM clustering based on kernel mapping. In: Proceedings of the International Conference on Society for Optical Engineering, vol. 4554, pp. 241–245 (2001). doi:10.1117/12.441658
18. Mashor, M.Y.: Hybrid training algorithm for RBF network. Int. J. Comput. Internet Manag. **8** (2), 50–65 (2000)
19. Mekhmoukh, A., Mokrani, K., Cheriet, M.: A modified Kernelized Fuzzy C-Means algorithm for noisy images segmentation: application to MRI images. IJCSI Int. J. Comput. Sci. Issues **9** (1), 1172–1176 (2012)
20. Moody, J., Darken, C.J.: Fast learning in networks of locally-tuned processing units. Neural Comput. **1**(2), 281–294 (1989)
21. Niros, A.D., Tsekouras, G.E.: A novel training algorithm for RBF neural network using a hybrid fuzzy clustering approach. J. Fuzzy Sets Syst. **193**, 62–84 (2012). doi:10.1016/j.fss.2011.08.011
22. Pelleg, D., Moore, A.: X-means: Extending K-means with efficient estimation of the number of clusters. In: Proceedings of the Seventeenth International Conference on Machine Learning, pp. 727–734 (2000)
23. Platt, J.C.: A resource-allocating network for function interpolation. Neural Comput. **3**(2), 213–225 (1991)
24. Ruspini, E.H.: Numerical methods for fuzzy clustering. J. Inf. Sci. **2**(3), 319–350 (1970)

25. Sánchez, A.V.D.: Searching for a solution to the automatic RBF network design problem. Neurocomputing **42**(1–4), 147–170 (2002)
26. Wong, Y.W., Seng, K.P., Ang, L.-M.: Radial basis function neural network with incremental learning for face recognition. IEEE Trans. Syst. Man Cybern. Part B—Cybern. **41**(4), 1–16 (2011). doi:10.1109/TSMCB.2010.2101591
27. Zhou, S., Gan, J.Q.: Mercel kernel fuzzy c-means algorithm and prototypes of clusters, In: Proceedings of the International Conference on Data Engineering and Automated Learning. Lecture Notes in Computer Science, vol. 3177, pp. 613–618 (2004). doi:10.1007/978-3-540-28651-6_90

Properties of the Island-Based and Single Population Differential Evolution Algorithms Applied to Discrete-Continuous Scheduling

Piotr Jędrzejowicz and Aleksander Skakovski

Abstract In this paper we have studied by experiment the properties of two models of the DE search: a model based on a single population, and a model based on multiple populations, known as the island model (IBDEA). We consider two versions of the island model: with migration of individuals between islands and without migration. We investigated how the effectiveness of models depends on such parameters as the size of a single population, and in the case of the island model, also the number of islands and migration rate between them. The general conclusion is that both models can be equally effective when used with proper parameter settings, which have been determined by the experiment. In addition, conditions for higher effectiveness of the IBDEA were discussed. The discrete-continuous scheduling with continuous resource discretisation was used as the test problem.

Keywords Island model • Differential evolution • Population size • Number of islands • Migration rate • Effectiveness

1 Introduction

The performance of evolutionary algorithm (EA) to a great extent depends on its ability to cope with the obstacles that prevent or hinder the progress of the search. One of these obstacles is premature convergence, which results in getting stuck in a local optimum. Another one—too large search area, which is a natural obstacle on

---

P. Jędrzejowicz
Chair of Information Systems, Gdynia Maritime University, Gdynia, Poland
e-mail: pj@am.gdynia.pl

A. Skakovski (✉)
Department of Navigation, Gdynia Maritime University, Gdynia, Poland
e-mail: askakow@am.gdynia.pl

I. Czarnowski et al. (eds.), *Intelligent Decision Technologies 2016*,
Smart Innovation, Systems and Technologies 56,
DOI 10.1007/978-3-319-39630-9_29

the way to finding a global optimum. One way to cope with these difficulties is to diversify the population, which is being evolved by EA. In the literature, we find a variety of approaches and techniques to maintain an appropriate level of the population diversification, see [15]. One of the simplest ways to diversify the population, is to determine such size of the evolving population, that ensures the highest efficiency of EA. Alternatively, one can initiate and perform search in various distant separated from each other regions of the search area. In such case, the algorithm operates on multiple and independent to a certain degree subpopulations, thus extending the search space, which may contribute to faster convergence towards the global optimum. Such method of performing the search was the subject of interest of many researchers e.g. [1, 3–5, 7–9]. It is also known in the literature as an island-based model and is often reported as more effective, than a search performed on a single population, e.g. [14, 21]. The island model is well suited for processing in parallel, using distributed, and agent systems which may provide greater effectiveness of search and a reduction of the response time. It is known from the literature that most important parameters of island model are: the size of population on an island and the number of islands [5, 22], migration size [4, 19] and migration rate [13, 19], migration policy [4], migration topology [7, 17], and the heterogeneity of the island models [18].

In this paper we study a special method of the evolutionary search, namely, differential evolution method. Differential evolution is a stochastic direct search and global optimization method proposed in [20].

In our work we examine the properties of two models of the DE search: the model based on a single population (the implementation of which we denote as DEA), and the model based on multiple populations, known as the island model. We considered two versions of the island model: with migration of individuals between islands (IBDEA^{m}) and without migration ($\text{IBDEA}^{-\text{m}}$). In the IBDEA^{m}, the islands periodically exchange among themselves their best solutions with a migration rate *ex*. We investigated how the effectiveness of the models under consideration depends on such parameters as the size of a single population x_P, and in the case of the island model, also the number of islands K and the migration rate *ex*.

The main goal of the research was to determine the setting of the parameters which would provide the highest effectiveness of the considered models. The secondary objective was to contribute to the knowledge on the behavior of a single and multiple population algorithms with respect to the parameters under concern.

As a test problem, we used the discrete-continuous scheduling problem with continuous resource discretisation (DCSPwCRD), which stems from DCSP, described in [10] and dealt with in [11, 12]. The brief definition of DCSPwCRD is given in Sect. 2, whereas detailed description might be found in [9, 16]. In order to conduct our tests, we adapted the differential evolution algorithm (DEA) proposed in [6] for solving the DCSPwCRD. A brief description of the DEA and the IBDEA for solving the DCSPwCRD is given in Sect. 3 and the extended description might be found in [9]. The description of the assumptions of the experiment, as well as discussion on the results and illustrating them charts are given in Sect. 4. Section 5 includes conclusions and idea for future research.

2 Problem Formulation

We define a problem Θ_Z in the same way as in [16]. Namely, let $\boldsymbol{J} = \{J_1, J_2, \ldots, J_n\}$ be a set of nonpreemptable tasks, with no precedence relations and ready times $r_i = 0$, $i = 1, 2, \ldots, n$, and $\boldsymbol{P} = \{P_1, P_2, \ldots, P_m\}$ be a set of parallel and identical machines, and there is one additional renewable discrete resource in amount $U = 1$ available. A task J_i can be processed in one of the modes $l_i = 1, 2, \ldots, D_i$ (D_i—the number of processing modes of task J_i), for which J_i requires a machine from $\boldsymbol{P}$ and amount of the additional resource known in advance. The processing mode of J_i cannot change during the processing. For each task two vectors are defined: a processing times vector $\tau_i = [\tau_i^1, \tau_i^2, \ldots, \tau_i^{D_i}]$, where $\tau_i^{l_i}$ is the processing time of task J_i in mode $l_i = 1, 2, \ldots, D_i$ and a vector of additional resource quantities allocated in each processing mode $u_i = [u_i^1, u_i^2, \ldots, u_i^{D_i}]$. The problem is to find processing modes for tasks from $\boldsymbol{J}$ and their sequence on machines from $\boldsymbol{P}$ such that schedule length $Q = \max\{C_i\}$, $i = 1, \ldots, n$ is minimized.

The formulated problem is a particular case of more general Multi-Mode Resource Constrained Project Scheduling Problem (MMRCPSP), which is known to be NP-hard [2].

3 DEA and IBDEA

In this section we give only a brief description of the DEA and the IBDEA. Since the IBDEA was first proposed in [9], the reader might find more extended description of it there. All individuals (solutions) used in the IBDEA can be characterized in the following manner:

- an individual (a solution) is represented by an n-element vector $S = [c_i| \; 1 \leq i \leq n]$,
- all processing modes of all tasks are numbered consecutively. Processing mode l_b of task J_b has the number $c_b = \sum_{i=1}^{b-1} D_i + l_b$. Thus c_b represents task i and the mode in which task i is processed.

The general description of the IBDEA is given below.

```
IBDEA procedure
Begin

Set K ≥ 2 - the number of the islands. Assign DEA procedure to
all islands.
```

Assume the population of individuals P_k on an island *k* consists of two halves P_k^1 and P_k^2, i.e. $P_k = P_k^1 + P_k^2$, and $|P_k| = x_P = 2 \cdot x_{DE}$, $|P_k^1| = x_{DE}$, $|P_k^2| = x_{DE}$,
Generate an initial population of individuals P_k^1 of the size x_{DE} on every island *k*, $k = 1, 2, \ldots, K$.
Improve individuals on all islands with the DEA procedure, cyclically exchanging best individuals among randomly chosen pairs of islands.
Stop after n_{ev} number of fitness function evaluations on the archipelago have been carried out.
Output the best solution to the problem.

End.

The random interconnection topology among the islands was chosen as the most efficient according to [7]. The DEA procedure used in the IBDEA is described in the following pseudo code.

For the purpose of our research we adapted the differential evolution algorithm for solving the multi-mode resource-constrained project scheduling problem proposed in [6] for solving the discrete-continuous scheduling problem with continuous resource discretisation (DCSPwCRD). In this section we give only a general idea of the DEA for solving DCSPwCRD, and the extended description of the algorithm the reader might find in [9].

The DEA is an evolutionary algorithm that evolves population P_k^1 of *target vectors* S_{tg} in cycles *k*. In every cycle *k*, for every target vector S_{tg} in P_k^1 *a trial vector* *T* in P_k^2 is created. This way P_k^2 is filled with trial vectors *T*. The pseudo code of the DEA is given below.

The DEA:
Begin

For every target vector S_{tg} in the current population P_k^1 do:

Create a mutant vector *M* from three vectors S_0, S_1, S_2 randomly chosen from P_k^1, using formula: $M = S_0 + A^* r^* (S_1 - S_2)$, where $A > 0$ - is a scale factor, that controls the evolution rate of the population and $r \in [0, 1]$.
Create a trial vector *T* in P_k^2 applying crossover operator to each element of mutant vector *M* and corresponding element of target vector S_{tg} according to the rule:

if the random number $r \leq C_r, C_r \in [0, 1]$), then the trial element is inherited from mutant vector *M*, otherwise from target vector S_{tg}.

```
Create a new population P^1_{k+1} selecting the best vectors from P^1_k
and P^2_k.
Repeat evolution cycles until the stop criterion is met.
End.
```

The crossover constant C_r controls in the DEA probability, that trial vector T will inherit the element either from target vector S_{tg}, or mutant vector M.

4 Computational Experiments

4.1 Parameter Set up

In our experiments values of the parameters of the DEA were assumed to be the same as in [6], namely the scale factor A which controls the evolution rate of the population was set to $A = 1.5$ and values of the variable $rand \in [0, 1]$. The crossover constants Cr_p and Cr_l which control the probability that the trial individual will receive the actual individual's tasks or modes were set $Cr_p = 0.2$ and $Cr_l = 0.1$, where p and l in the notations Cr_p and Cr_l stand for tasks positions and modes respectively. In the experiment, we considered the following population sizes: 20, 50, 100, 200, 1000, and the numbers of the fitness function evaluations $\#ev$ necessary for the DEA and the IBDEA to yield one solution was set to $\#ev = 10^6$. An initial population of feasible individuals in the DEA was generated using the uniform distribution equal $1/n$ for the tasks, and $1/D$ for the task's modes. Our assumptions concerning the test problem are as follows. We considered three combinations of $n \times m$: 10×2, 10×3, and 20×2, where n is the number of tasks and m is the number of machines, and three levels of the continuous resource discretisation D: 10, 20, 50. This way we considered nine sizes n x m x D of the problem Θ_Z: $10 \times 2 \times 10$, $10 \times 2 \times 20$, $10 \times 2 \times 50$, $10 \times 3 \times 10$, ..., $20 \times 2 \times 50$. For each of the sizes, we considered 6 instances of the problem Θ_Z, which makes 54 instances of the problem in total. These 54 instances were used for testing one value of the parameter under consideration, where the considered parameters were: the declining period T^d, population size x_P, and the number of fitness function evaluations $\#ev$. The schedule lengths determined by the algorithm for these 54 instances were summed up and the obtained result was used for evaluation and comparison purposes. To make the tests credible, the tested algorithm was always run with the same seed of the random number generator. Thus, the only factor that could cause change in the results was the value of the parameter that was tested.

In our tests, we considered different sizes of population x_P on an island and numbers of islands K. The values of x_P, and the ranges of K and max x_A used in the tests are given in Table 1.

Table 1 The values of x_P and the ranges of K and x_A used in the tests

x_P	min K	max K	min x_A	max x_A
5	2	2150	10	10750
10	2	1062	20	10620
17	2	565	34	9605
27	2	237	54	6399
44	2	237	88	10428
71	2	126	142	8946
115	2	92	230	10580
186	2	49	372	9114
301	2	36	602	10836
487	2	19	974	9253
788	2	14	1576	11032
1275	2	10	2550	12750
2063	2	5	4126	10315
3338	2	5	6676	16690
5401	2	3	10802	16203
8739	2	3	17478	26217
14140	2	2	28280	28280

All tests were carried out on a PC under 64-bit operating system Windows 7 Enterprise with Intel(R) Core(TM) i5-2300 CPU @ 2.80 GHz 3.00 GHz, RAM 4 GB compiled with aid of Borland Turbo Delphi for Win32. When the number of fitness function evaluations was set to 720000, mean time required by the DEA to find a solution for the problem sizes 10×2 and 10×3 for all discretisation levels was approximately 2–3 s and for the problem size 20×2 for all discretisation levels approximately 5–6 s. The total time taken by the DEA to process all 54 instances was approximately 206 s.

4.2 Test Results

The aim of the experiment was to compare the effectiveness of two models of DE search. We compared DE based on a single population, implemented as a differential evolution algorithm (DEA) and DE based on multiple populations, implemented as an island-based differential evolution algorithm (IBDEA). The island model was implemented in two versions: with and without solution migration among the islands, denoted in the paper as IBDEA^{m} and $\text{IBDEA}^{-\text{m}}$ respectively. In the IBDEA^{m}, the islands exchanged their best solutions with different rates. In our tests, the best solutions exchange was carried out after 1, 2, 3, 4, 5, 7, 9, 11, 13, 15, 17 populations had been generated on each island. In the article, we will be using a notation *ex##* for denoting the migration rate, where ## denotes the number of populations after which the migration had been carried out.

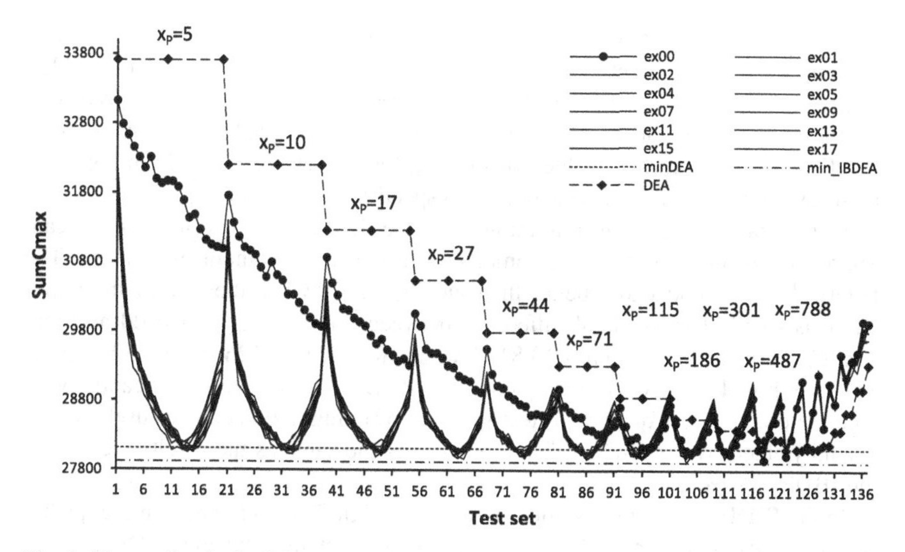

Fig. 1 The results obtained for the test sets 1–137

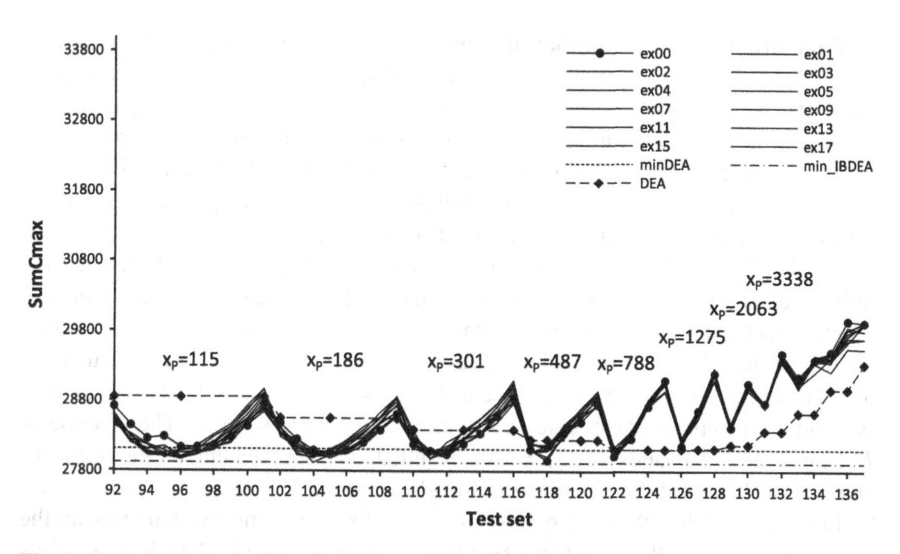

Fig. 2 The results obtained for the test sets 92–137

Figures 1 and 2 present the results of our tests. Figure 2 illustrates in more detail the results obtained for test sets 92–137 shown in Fig. 1. Our observations on the results of the tests are as follows.

DEA versus IBDEA. Both IBDEAs perform much better than the DEA when populations are small, namely, $x_P \in [5, 71]$ (test sets 1–91), see Fig. 1. The value of $sumC_{max}$ yielded by the DEA for $x_P = 5$ is greater than the minimum value of $sumC_{max}$ obtained by the IBDEA^{-m} by about 8,8 % (test set 20, $K = 2150$,

$x_A = 10750$) and about 20 % greater than the minimum value of $sumC_{max}$ obtained by the IBDEA^m (test sets 13–15, $K = 237$, 325, 445, $x_A = 1185$, 1625, 2225 respectively). However, these differences decrease with increase of x_P. When x_P increases up to 788 and 1275, the DEA performs practically the same as both IBDEAs, see the minimum values of $sumC_{max}$ for test sets 122 and 126 in Fig. 2. In these two test sets $x_A = 1576$ and 2250 respectively.

Migration. Introducing migration into the IBDEA results in a significant improvement in performance, compared to the case without migration. The IBDEA^m works noticeably better than the IBDEA^{-m} when the size of the population is small. The maximal difference between min($sumC_{max}$) of $ex00$ and min ($sumC_{max}$) of $ex01$–17 reaches 13.81 % when $x_P = 5$, $K = 126$, $x_A = 630$ (test set 11), see Fig. 1. However, the results of IBDEA^{-m} improve significantly with increase of x_P. When $x_P \geq 115$ (test sets 92–137), migration becomes useless. The results yielded by the IBDEA^{-m} are practically the same as the results yielded by the IBDEA^m, see Fig. 2.

In the IBDEA^m, the rate of migration has no significant influence on the quality of the results. The curves representing $sumC_{max}$ obtained by the IBDEA^m with $ex01$–$ex17$ differ from each other in the area of the minima at most 0.23 %, see Fig. 1.

Population Size and Number of Islands. In the case of the IBDEA^m, the values of minima of $sumC_{max}$ for all $x_P < 2500$ are almost the same, they differ at most about 0.6 %, see Figs. 1 and 2. These minima are not dependent on any particular number of islands K, but the number of individuals throughout the archipelago x_A, which can be defined approximately as an interval [1200, 2500]. Hence, in order to ensure maximal performance of the IBDEA^m, the values of x_P and K should be selected such, that x_A would take values from this range.

For the IBDEA^{-m}, the number of individuals in the population x_P determines the trend and has a major influence on the quality of the results, that are improved together with the increase of the population size. The increase of x_P from 5 to 487 resulted in improvement of the best value of $sumC_{max}$ by 10.83 %. The results can also be improved by increasing K, e.g. for $x_P = 5$ up to 6.89 %. However, this possibility of improvement decreases along with the increase of x_P. The increase of K from 2 to 10 islands for $x_P = 301$ resulted in improvement of the value of $sumC_{max}$ only by 0.5 %, see test sets 110–112 in Fig. 2.

In the case of the DEA, the quality of the results is also improved along with the increase of x_P. The DEA yields the best results when $x_P \approx [800, 2000]$, see test sets 122-130 in Fig. 2. The minimum value of the $sumC_{max}$ yielded by the DEA in this range of x_P differs from the minimum $sumC_{max}$ yielded by both IBDEAs only by 0.69 %, see Fig. 2.

Finally, it must be said that the exaggerated increase of x_P or K causes deterioration of the quality of results. The increase of x_P over 2500 worsens the results yielded by the DEA and both IBDEAs, see test sets 131–137 in Fig. 2. The increase of K, leading to the increase of x_A over 2500, also worsens the quality of results yielded by the IBDEA^m for all x_P. However, it is only partially true in the case of the IBDEA^{-m}. The increase of K even to its maximum value when $x_P \in [5,71]$

improves the quality of results. In these cases x_A might take values up to about 10000. When $x_P > 71$, the IBDEA^{-m} starts to behave in the same way as the IBDEA^{m}.

5 Conclusions

In the paper, we examined properties of the two models of evolutionary search: the model based on a single population (DEA) and the model based on multiple populations, known as the island model. We considered two versions of the island model: with migration of individuals between islands (IBDEA^{m}) and without migration (IBDEA^{-m}). In the IBDEA^{m}, the islands periodically exchanged among themselves their best solutions. We investigated how the effectiveness of the models under consideration depends on such parameters as the size of a single population x_P, and in the case of the island model, also the number of islands K and the migration rate ex. The main goal of this work was to determine the setting of parameters which would provide the highest efficacy of the considered models. Conclusions of our research are as follows.

The general conclusion is that both models can be equally effective when used with proper parameter settings. Therefore none of them is better than the other. However, it follows from the experiment, that when the population size is small the quality of the results yielded by the IBDEA^{m} and IBDEA^{-m} might be better than the quality of the results yielded by the DEA up to 20 and 8.8 % respectively. Such advantage of the island model is caused by a possibility to increase the total number of individuals on the archipelago x_A by increasing the number of the islands K and additionally by the migration of individuals among the islands as it is in the IBDEA^{m}. This advantage deteriorates along with the increase of x_P. It should also be mentioned, that the IBDEA^{m} showed practically the same efficacy for all considered migration rates, so it is hard to chose any of them as the most preferable. The approximate values of the parameters for which the algorithms show their best efficacy are as follows. The DEA is most efficacious when $x_P \approx [800, 2000]$, the IBDEA^{m} and the IBDEA^{-m}—when $x_A \approx [1200, 2500]$, however in the case of the IBDEA^{-m}, it is true when $x_P > 100$. It should be added, that there is no reason of using solution migration mechanism in the island model for the indicated values of x_A. Thus, it would be more practical to use the DEA with $x_P \approx [800, 2000]$, as the simpler implementation of DE.

Finally, we are obliged to emphasize, that the results and conclusions following from them, are true for the number of fitness function evaluations $\#ev$, which in our experiments has been set to $\#ev = 10^6$. Properties of the algorithms may differ from those observed in our experiment, when the search process will be limited by smaller values of $\#ev$. Thus, our proposal for future work would involve determining the properties of the examined models of DE for smaller values of $\#ev$, and verifying whether the established in our work properties of the models are true for other types of evolutionary search.

References

1. Alba, E., Troya, J.: Analysis of synchronous and asynchronous parallel distributed genetic algorithms with structured and panmictic Islands. In: Rolim. J. et al., (eds.) In: Proceedings of the 10th Symposium on Parallel and Distributed Processing, pp. 248–256. San Juan, Puerto Rico, USA, 12–16 April 1999
2. Bartusch, M., Rolf, H.M., Radermacher, F.J.: Scheduling project networks with resource constraints and time windows. Ann. Oper. Res. **16**, 201–240 (1988)
3. Belding, T.C.: The distributed genetic algorithm revisited. In: Eshelman, L.J. (ed.) Proceedings of the Sixth International Conference on Genetic Algorithms, pp. 114–121. Morgan Kaufmann, San Francisco, CA (1995)
4. Cantu-Paz, E.: Migration policies, selection pressure, and parallel evolutionary algorithms. J. Heuristics **7**(4), 31–334 (2001)
5. Cantu-Paz, E., Goldberg, D.E.: Are multiple runs of genetic algorithms better than one? In: Proceedings of the Genetic and Evolutionary Computation Conference (2003)
6. Damak, N., Jarboui, B., Siarry, P., Loukil, T.: Differential evolution for solving multi-mode resource-constrained project scheduling problems. Comput. Oper. Res. **36**(9), 2653–2659 (2009)
7. Jędrzejowicz, P., Skakovski, A.: Structure vs. efficiency of the cross-entropy based population learning algorithm for discrete-continuous scheduling with continuous resource discretisation. In: Czarnowski, I., Jędrzejowicz, P., Kacprzyk, J. (eds.) Studies in Computational Intelligence. Agent-Based Optimization, vol. 456, pp. 77–102. Springer (2013)
8. Jędrzejowicz, P., Skakovski, A.: Population learning with differential evolution for the discrete-continuous scheduling with continuous resource discretisation. In: IEEE International Conference on Cybernetics (CYBCONF), pp. 92–97. Lausanne Switzerland, 13–15 June 2013
9. Jędrzejowicz, P., Skakovski, A.: Island-based differential evolution algorithm for the discrete-continuous scheduling with continuous resource discretisation. Procedia Comput. Sci. **35**, 111–117 (2014)
10. Józefowska, J., Węglarz, J.: On a methodology for discrete-continuous scheduling. Europ J Oper Res. **107–2**, 338–353 (1998)
12. Józefowska, J., Różycki, R., Waligóra, G., Węglarz, J.: Local search metaheuristics for some discrete-continuous scheduling problems. Europ J Oper Res **107–2**, 354–370 (1998)
11. Józefowska, J., Mika, M., Różycki, R., Waligóra, G., Węglarz, J.: Solving discrete-continuous scheduling problems by Tabu Search. In: 4th Metaheuristics International Conference MIC'2001, Porto, Portugal, pp. 667–671, 16–20 July 2001
13. Krink, T., Mayoh, B.H., Michalewicz, Z.: A PACHWORK model for evolutionary algorithms with structured and variable size populations. In: Banzhaf, W., Daida, J., Eiben, A.E., Garzon, M.H., Honavar, V., Jakiela, M., Smith, R.E. (eds.) Proceedings of the Genetic and Evolutionary Computation Conference, vol. 2, pp. 1321–1328. Orlando, Florida, USA, Morgan Kaufmann, (1999)
14. Muhlenbein, H.: Evolution in time and space: the parallel genetic algorithm. In: Rawlins, G. (ed.) FOGA-1. pp. 316–337. Morgan Kaufman (1991)
15. Pandey, H.M., Chaudharyb, A., Mehrotra, D.: A comparative review of approaches to prevent premature convergence in GA. Appl. Soft Comput. **24**, 1047–1077 (2014)
16. Różycki, R.: Zastosowanie algorytmu genetycznego do rozwiązywania dyskretno-ciągłych problemów szeregowania. PhD diss, Poznań University of Technology, Poland (2000)
17. Sekaj, I.: Robust parallel genetic algorithms with re-initialisation. In: Proceedings of Parallel Problem Solving from Nature—PPSN VIII. 8th International Conference, LNCS, vol. 3242, pp. 411–419. Springer, Birmingham, UK, 18–22 Sept 2004
18. Skolicki, Z.: An analysis of Island models in evolutionary computation. In: Proceedings of GECCO'05, pp. 386–389. Washington, DC, USA, 25–29 June 2005
19. Skolicki, Z., Kenneth, D.J.: The influence of migration sizes and intervals on Island models. In: Proceedings of GECCO'05, pp. 1295–1302. Washington, DC, USA, 25–29 June 2005

20. Storn, R., Price, K.: Differential evolution—a simple and efficient heuristic for global optimization over continuous spaces. J. Global Opt. **11**, 341–359 (1997)
22. Whitley, D., Starkweather, T.: GENITOR II: a distributed genetic algorithm. J. Exper. Theor. Artif. Intel. **2**, 33–47 (1990)
21. Whitley, D., Rana, S., Heckendorn, R.B.: The island model genetic algorithm: on separability, population size and convergence. J. Comp. and Infor. Tech. **7–1**, 33–47 (1999)

An Improved Agent-Based Approach to the Dynamic Vehicle Routing Problem

Dariusz Barbucha

Abstract The paper proposes a multi-agent approach to the Dynamic Vehicle Routing Problem, where the process of solving instances of the problem is performed by a set of software agents with different abilities. The agents are responsible for generating new requests, managing a set of requests, allocating them to the available vehicles, and monitoring the behavior of the system. The main steps of the algorithm implemented in the system include dispatching the static and dynamic requests to the available vehicles. In order to increase the efficiency of these processes, a request buffering strategy has been implemented. Computational experiment confirmed its positive impact on the results obtained by the proposed approach.

Keywords Dynamic vehicle routing problem · Multi-agent systems · Buffering strategy

1 Introduction

Several real-world Vehicle Routing Problems (VRP) arise in practice of contemporary transport companies. Having a set of vehicles, they aim at serving a set of customer requests in order to minimize the cost of transport, and satisfying several customers and vehicles constraints. Most representative examples of transport of goods arise in dynamic fleet management and courier services. Others, focused on personal transportation services, include dial-a-ride problems and taxi cab services.

As Psaraftis [13] suggests, such real-world routing applications often include two important dimensions: *evolution* and *quality of information*. Evolution of information is related to the fact that in some problems the information available to the planner may change during the execution of the routing, for instance with the arrival of new requests or changes in the traveling times. Quality of information reflects

D. Barbucha (✉)
Department of Information Systems, Gdynia Maritime University,
Morska 83, 81-225 Gdynia, Poland
e-mail: d.barbucha@wpit.am.gdynia.pl

I. Czarnowski et al. (eds.), *Intelligent Decision Technologies 2016*,
Smart Innovation, Systems and Technologies 56,
DOI 10.1007/978-3-319-39630-9_30

possible uncertainty on the available data, for example when the demand of a client is only known as an estimation of its real demand. Taking into account these dimensions, four categories of problems can be identified [12]:

- *Static* and *deterministic*—input data are known beforehand and vehicle routes do not change once they are in execution.
- *Static* and *stochastic*—problems are characterized by input partially known as random variables, which realizations are only revealed during the execution of the routes. Uncertainty may affect any of the input data; the three most studied cases include: stochastic customers (a customer needs to be serviced with a given probability), stochastic times (either service or travel times are modeled by random variables), and stochastic demands.
- *Dynamic* and *deterministic*—part or all of the input is unknown and revealed dynamically during the design or execution of the routes. For these problems, vehicle routes are redefined in an ongoing fashion.
- *Dynamic* and *stochastic*—they have part or all of their input unknown and revealed dynamically during the execution of the routes. But in contrast with the latter category, exploitable stochastic knowledge is available on the dynamically revealed information.

The Dynamic Vehicle Routing Problems (DVRP) cover a broad class including a number of variants having similar structure. Besides the classical DVRP considered in the paper, the class also includes: Dynamic Vehicle Routing Problem with Time Windows (DVRPTW), where the customers have to be visited during a specific time interval, Dynamic Pickup and Delivery Problem (DPDP), where goods have to be either picked-up or delivered in specific amounts in each of the vertices, and its variation with time windows (DPDPTW), etc.

The goal of the paper is to propose a multi-agent approach with request buffering to the DVRP, where the dynamism is represented by a set of input requests which are revealed dynamically and unpredictably during the execution of already arrived requests. Because of the fact, that new customer requests can continuously arrive over time, at any moment of time, there may exist customers already under servicing and new customers which need to be serviced. As a consequence, each newly arriving dynamic request, needs to be incorporated into the existing vehicles tours and the current solution may need to be reconfigured to minimize the goal functions. The proposed approach is based on the agent paradigm and extends the multi-agent platform for solving dynamic VRP and VRPTW proposed in [1, 2], respectively.

The remainder of the paper consists of four sections. Section 2 includes formulation of the DVRP and presents different measures of dynamism of VRP. Section 3 presents main assumptions and elements of the proposed agent-based approach with request buffering to the DVRP. Main assumptions, goal and results of the computational experiment are presented in Sect. 4. Finally, Sect. 5 concludes the paper and suggests the directions of the future research.

2 Dynamic Vehicle Routing Problem

The *static* VRP can be formulated as the problem of determining optimal routes passing through a given set of locations (customers). It is represented as an undirected graph $G = (V, E)$, where $V = \{0, 1, \dots, N\}$ is the set of nodes and $E = \{(i,j)|i,j \in V\}$ is the set of edges. Node 0 is a central depot with NV identical vehicles of capacity W. Each other node $i \in V \backslash \{0\}$ denotes the customer characterized by a non-negative demand d_i, and a service time s_i. Each link $(i,j) \in E$ denotes the shortest path from customer i to j and is described by the cost c_{ij} of travel from i to j by shortest path $(i,j \in V)$. It is assumed that $c_{ij} = c_{ji} (i,j \in V)$. It is also often assumed that c_{ij} corresponds to the travel time t_{ij}.

The goal is to minimize the total distance needed to pass by vehicles in order to supply all customers satisfying the following constraints: each route starts and ends at the depot, each customer $i \in V \backslash \{0\}$ is serviced exactly once by a single vehicle, the total load on any vehicle associated with a given route does not exceed vehicle capacity.

The *dynamic* VRP considered in the paper is an extension of the VRP defined above, where a certain number of customers' requests are available in advance and the remaining requests arrive while the system is already running. Let the planning horizon starts at time 0 and ends at time T. Let $t_i \in [0, T]$ $(i = 1, \dots, N)$ denotes the time when the ith customer request is submitted. Let N_s denotes the number of *static* (i.e. submitted in advance) requests available in $t_i = 0$ $(i = 1, \dots, N_s)$ and N_d—the number of *dynamic* requests arriving within the $(0, T]$ interval $(N_s + N_d = N)$.

A few measures of dynamism have been defined for vehicle routing problems. The natural and the simplest one is the *Degree of Dynamism* (*DoD*) proposed by Lund et al. [8] and Larsen [7] for classical VRP. It is defined as a proportion of the number of dynamic requests to the number of all requests ($DoD = N_d/N$).

Larsen [7] generalized the above definition and introduced a new measure of dynamism—an *Effective Degree of Dynamism* (*eDoD*) which takes into account the time in which the dynamic requests occur ($eDoD = \sum_{i=1}^{N} (\frac{t_i}{T})/N$). According to it, the problem has a higher degree of dynamism if dynamic requests occur at the end of the operational interval $[0, T]$.

For problems with time windows, Larsen [7] also extended the effective degree of dynamism in order to reflect the level of urgency of the requests. He defined an *Effective Degree of Dynamism with Time Windows* (*eDoDTW*) by defining the reaction time r_i as the difference between the arrival date t_i and the end of the corresponding time window l_i $(r_i = l_i - t_i)$, and thus highlighting that greater reaction times mean more flexibility to insert the request into the current routes ($eDoDTW = \sum_{i=1}^{N} (1 - \frac{r_i}{T})/N$).

It is worth noting that all three measures defined above take values belonging to the interval [0, 1] and increase with the level of dynamism of a problem. The level equal to 0 or 1 denotes fully static or dynamic problem, respecively.

Whereas many important advances have been proposed during the recent years in the field of static versions of different variants of VRP, definitely much less has been done with respect to solving their dynamic versions. Different variants of DVRP, methods of solving them and examples of their practical applications can be found for example in [12]. Among approaches dedicated to DVRP one can list a few of them: an adaptive granular local search heuristic [3], a parallel tabu search heuristic with an adaptive memory mechanism [5], a genetic algorithm [6], and ant colony system [11].

3 Multi-agent Approach to DVRP

The proposed approach assumes that the process of simulating and solving DVRP is performed by four types of agents with different abilities and responsibilities. They include: `GlobalManager` (an agent which runs first and initializes all others agents), `RequestGenerator` (it generates or reads new requests), `RequestManager` (it manages the list of static and dynamic requests), and `Vehicle` agents (they represent vehicles and are responsible for serving the customers' requests). More detailed description of the agents the reader can find in [2].

Generally, the process of solving the DVRP includes the main three tasks: initialization, the process of dispatching the static requests to the available vehicles, and the process dispatching the newly arrived requests (dynamic requests) to the available vehicles. The proposed approach assumes that in order to dispatch the requests (static or dynamic) to the available vehicles, the series of partial static instances of VRP have to be solved. Each partial instance includes all requests available at the particular moment of time. The details of the approach are presented in Algorithm 1.

The algorithm starts with initialization of all sets of requests, variables (states of requests, vehicles, etc.) and all agents which will be engaged in the process of solving the DVRP. At first, the `GlobalManager` agent is created, which next initializes all other agents (`RequestGenerator`, `RequestManager`, and `Vehicle`). After receiving answers from all agents about their readiness to act, the system is ready to develop a routing plan, which will allow to dispatch all static requests to the available vehicles. To perform this, the `RequestManager` agent solves the static problem P_{DVRP} using a dedicated constructive procedure proposed in [1].

Let s be a solution of P_{DVRP}, and let $R_s = [R_s^1, R_s^2, \dots, R_s^{NV}]$ be a global plan including a list of routes constituting the solution of the static problem with N_s requests, where each R_s^i is a route associated with vehicle $v(i)$ ($i = 1, \dots, NV$). Basing on this plan, the `RequestManager` agent gradually incorporates the requests belonging to each route to the vehicles (i.e. each vehicle $v(i)$, $i = 1, \dots, NV$ receives requests defined by R_s^i in solution of the static problem). After receiving the requests, all vehicles begin realization of their plans, and the system is waiting for the next events.

Three kinds of events are generated and observed in the system: *system events* (generated by the system and being sent to `RequestManager`), *request events* (generated by `RequestGenerator` and being sent to `RequestManager`), and *vehicle events* (generated by `Vehicle` agents and being sent to `RequestManager`, in which the vehicles report about their states during serving the requests assigned to

Algorithm 1 MAS-DVRP

1: Initialize agents: `GlobalManager`, `RequestGenerator`, `RequestManager`, and `Vehicle`
2: Let *StaticR* be a set of N_s static requests.
3: Let *BufferR* be a set of buffered requests (i.e. requests which have arrived but have not been sent yet to vehicles).
4: Let $BufferR = \emptyset$.
5: Let P_{DVRP} be a DVRP for requests belonging to the *StaticR* set.
6: Solve P_{DVRP}
7: Let s be a solution of P_{DVRP}, and let $R_s = [R_s^1, R_s^2, \ldots, R_s^{NV}]$ be a global plan including a list of routes constituting the solution of the P_{DVRP}, where each R_s^i is a route associated with vehicle $v(i)$.
8: **for** $i = 1, \ldots, NV$ **do**
9:   **if** $(R_s^i \neq \emptyset)$ **then**
10:     Send R_s^i to vehicle $v(i)$
11:   **end if**
12: **end for**
13: **while** (`systemEvent::endOfRequests` has not been received) **do**
14:   The system is waiting for events:
15:   **if** (event = `requestEvent::newRequest`) **then**
16:     {message with a new request o_j received by `RequestManager` from `RequestGenerator` agent}
17:     $BufferR = BufferR \cup \{o_j\}$
18:     Resolve the problem P_{DVRP} taking into account the requests belonging to *BufferR* set.
19:     Update the routing plan R_s, where s is a new solution of P_{DVRP}.
20:   **end if**
21:   **if** (event = `vehicleEvent::vehicleStopAtLocation`) **then**
22:     {message received by `RequestManager` from `Vehicle` agent $v(i)$ informing about reaching current location}
23:     Get next request o_j from R_s^i (i.e. assigned to $v(i)$) and send it to $v(i)$
24:     $BufferR = BufferR \setminus \{o_j\}$
25:   **end if**
26:   Update states of requests, vehicles, and other system variables.
27: **end while**
28: All vehicles waiting at their last locations are driving to the depot.

them). In order to increase the readability of the Algorithm 1, only the most important messages belonging to these groups are presented in it and described below.

`requestEvent::newRequest`

Receiving the message `requestEvent::newRequest` means that a new request has been registered and it has to be allocated to one of the available vehicles. Following the results presented in one of the last author's paper [2], the Centralized Dispatching Strategy (CDS) has been adapted in the process of dispatching dynamic requests to the available vehicles. In general, CDS assumes that the list of all requests is maintained by the `RequestManager` agent, and it is responsible for planning each vehicle route. On the other hand, `Vehicle` agent maintains a short-term list of requests including only the location of the next request.

Assume that $R^i = [r^i_1, r^i_2, \ldots, r^i_k]$ be a current route assigned to the vehicle $v(i)$ ($i = 1, \ldots, NV$), and r^i_k is a customer (location), the vehicle $v(i)$ is currently driving to. Thus, the part of route $[r^i_1, r^i_2, \ldots, r^i_k]$ is fixed, and the process of assigning a new request to the existing route is possible only on position $k + 1$ of the route of $v(i)$. It is easy to see that if a particular request is arriving close to the end of planning horizon (highly dynamic request), possibility of re-optimization of the route assigned to the vehicle $v(i)$ is limited.

The above mentioned dispatching strategy applied in the proposed approach has been additionally extended here by adapting a *request buffering strategy* which has been proposed and described by Pureza and Laporte [14] and by Mitrovic-Minic et al. [10] for dynamic Pickup and Delivery Problems with Time Windows. The main idea of this strategy is to postpone a request assignment decision. It means that allocation of each new request to one of the available vehicles and route adjustment are not performed whenever new request arrives. After arriving, the request is stored in the buffer, and it is considered to allocate at later stages. The proposed approach assumes that the decisions about storing in a buffer and assignment are based on the request and on the current and future routes characteristics. It is expected that by buffering the requests, better routing decisions are more likely to be achieved due to the larger number of accumulated requests available [14].

`vehicleEvent::vehicleStopAtLocation`

After reaching a location of the current customer r^i_k and finishing its service, the `Vehicle` agent $v(i)$ sends the message `vehicleEvent::vehicleStopAtLocation` to the `RequestManager` informing it about readiness for serving next requests. According to the current global route plan, the `RequestManager` sends the request of the next customer to the `Vehicle` agent $v(i)$.

`systemEvent::endOfRequests`

Receiving the message `systemEvent::endOfRequests` means that the set of all requests has been exhausted. All vehicles finish their mission and return to the depot. The process of solving the DVRP stops.

4 Computational Experiment

A computational experiment has been carried out to validate the proposed multi-agent approach to the DVRP. Its main goal was to evaluate the influence of the request buffering strategy on the performance of the algorithm. Quality of the results have been measured as the percentage increase in cost of allocating all requests of dynamic instance as compared to the best known solution of the static instance.

The proposed agent-based approach was tested on seven VRP instances of Christofides et al. [4] (vrpnc01-vrpnc05, vrpnc11-vrpnc12) containing 50-199 customers.

The instances tested in the experiment reflected weakly, moderately and strongly dynamic problems, where the proportion of the number of dynamic requests to all requests (N_d/N) was equal to 20 %, 50 %, and 80 %, respectively. Moreover, arrivals of the dynamic requests have been generated using the Poisson distribution with λ parameter denoting the mean number of requests occurring in the unit of time (1 h in the experiment). For the purpose of the experiment λ was set to 5, 10, and 20. It has been assumed that the vehicle speed was set at 60 km/h.

Two cases have been tested and compared in the experiment, where buffering has been applied to the dynamic requests only or to all requests:

- D-BUF—it assumes that only dynamic requests are buffered (Algorithm 1, line 4: $BufferR = \emptyset$),
- SD-BUF—all requests (static and dynamic) are buffered (Algorithm 1, line 4: $BufferR = StaticR$).

In order to discover the influence of the proposed buffering strategy on the obtained results, the above cases have been also compared with the case, where no buffering has been applied (N-BUF). It means that a newly arrived request is assigned immediately to one of the available vehicles.

Table 1 Results obtained by the proposed approach (weakly dynamic instances)

Instance	Best static	λ	N-BUF	%	D-BUF	%	SD-BUF	%
vrpnc01	524.61	5	696.86	32.8	671.09	27.9	635.29	21.1
		10	657.10	25.3	631.59	20.4	604.48	15.2
		20	630.74	20.2	601.51	14.7	568.81	8.4
vrpnc02	835.26	5	1108.71	32.7	1067.66	27.8	1027.03	23.0
		10	1045.79	25.2	1004.26	20.2	962.44	15.2
		20	998.66	19.6	962.03	15.2	915.17	9.6
vrpnc03	826.14	5	1106.38	33.9	1060.58	28.4	1018.88	23.3
		10	1039.23	25.8	995.04	20.4	922.70	11.7
		20	997.47	20.7	948.62	14.8	903.23	9.3
vrpnc04	1028.42	5	1363.55	32.6	1319.86	28.3	1239.09	20.5
		10	1282.57	24.7	1237.40	20.3	1184.55	15.2
		20	1242.64	20.8	1171.27	13.9	1116.37	8.6
vrpnc05	1291.29	5	1725.59	33.6	1646.34	27.5	1558.31	20.7
		10	1617.92	25.3	1546.99	19.8	1467.69	13.7
		20	1554.60	20.4	1483.61	14.9	1410.92	9.3
vrpnc11	1042.11	5	1376.90	32.1	1332.62	27.9	1257.29	20.6
		10	1301.79	24.9	1245.02	19.5	1176.69	12.9
		20	1251.07	20.1	1192.59	14.4	1132.83	8.7
vrpnc12	819.56	5	1087.54	32.7	1045.54	27.6	999.09	21.9
		10	1023.09	24.8	980.48	19.6	925.96	13.0
		20	978.79	19.4	942.74	15.0	883.02	7.7

Table 2 Results obtained by the proposed approach (moderately dynamic instances)

Instance	Best static	λ	N-BUF	%	D-BUF	%	SD-BUF	%
vrpnc01	524.61	5	814.28	55.2	780.74	48.8	760.06	44.9
		10	771.50	47.1	761.92	45.2	704.46	34.3
		20	742.14	41.5	718.97	37.0	679.60	29.5
vrpnc02	835.26	5	1293.19	54.8	1277.57	53.0	1204.00	44.1
		10	1242.32	48.7	1192.09	42.7	1117.73	33.8
		20	1189.44	42.4	1148.45	37.5	1076.15	28.8
vrpnc03	826.14	5	1286.59	55.7	1260.16	52.5	1182.97	43.2
		10	1216.29	47.2	1187.73	43.8	1114.45	34.9
		20	1176.98	42.5	1140.58	38.1	1043.75	26.3
vrpnc04	1028.42	5	1591.28	54.7	1571.63	52.8	1481.88	44.1
		10	1514.44	47.3	1481.57	44.1	1390.30	35.2
		20	1468.24	42.8	1436.38	39.7	1337.66	30.1
vrpnc05	1291.29	5	2024.82	56.8	1955.59	51.4	1822.40	41.1
		10	1910.16	47.9	1835.61	42.2	1758.69	36.2
		20	1833.03	42.0	1778.83	37.8	1679.95	30.1
vrpnc11	1042.11	5	1622.92	55.7	1590.64	52.6	1507.92	44.7
		10	1524.11	46.3	1495.78	43.5	1426.90	36.9
		20	1486.81	42.7	1440.01	38.2	1365.50	31.0
vrpnc12	819.56	5	1280.38	56.2	1241.17	51.4	1170.62	42.8
		10	1211.28	47.8	1185.02	44.6	1112.53	35.7
		20	1161.82	41.8	1130.35	37.9	1082.93	32.1

For each case, each instance was repeatedly solved five times and mean results from these runs were recorded. All simulations have been carried out on PC Intel Core i5-2540M CPU 2.60 GHz with 8 GB RAM running under MS Windows 7.

The experiment results for weakly, moderately, and strongly dynamic instances of the DVRPTW are presented in Tables 1, 2, and 3, respectively. They include: the name of the instance set, the best known solution for static case of this instance, the value of λ parameter, and for each tested case, the average cost and the percentage increase in cost of allocating all requests of dynamic instance as compared to the best known solution of the static instance.

Results presented in the tables provide a few interesting observations. The first one is that comparison of the results obtained for dynamic instances with their static counterparts proves that dynamization of the VRP always results in deterioration of the results. Such deterioration often depends on level of dynamism of the instance: when a low ratio of request arrivals ($\lambda = 5, 10$) holds, the results are worse in comparison with the case where customers arrive at early stage of computation ($\lambda = 20$). But deterioration also depends on the degree of dynamism of the instance: for weakly dynamic instances the averaged results are closer to the best known results of static instances than for moderately or strongly dynamic ones.

Table 3 Results obtained by the proposed approach (strongly dynamic instances)

Instance	Best static	λ	N-BUF	%	D-BUF	%	SD-BUF	%
vrpnc01	524.61	5	1020.25	94.5	979.69	86.7	924.27	76.2
		10	1000.03	90.6	936.12	78.4	877.80	67.3
		20	947.56	80.6	892.07	70.0	846.72	61.4
vrpnc02	835.26	5	1677.58	100.8	1538.66	84.2	1464.44	75.3
		10	1598.36	91.4	1461.92	75.0	1405.42	68.3
		20	1512.37	81.1	1410.27	68.8	1358.60	62.7
vrpnc03	826.14	5	1627.22	97.0	1527.87	84.9	1453.18	75.9
		10	1531.68	85.4	1468.40	77.7	1386.42	67.8
		20	1521.25	84.1	1423.83	72.3	1342.84	62.5
vrpnc04	1028.42	5	1979.54	92.5	1903.89	85.1	1809.14	75.9
		10	1927.24	87.4	1837.06	78.6	1733.24	68.5
		20	1904.35	85.2	1772.56	72.4	1677.94	63.2
vrpnc05	1291.29	5	2590.60	100.6	2407.26	86.4	2275.65	76.2
		10	2396.62	85.6	2256.12	74.7	2163.70	67.6
		20	2309.68	78.9	2181.90	69.0	2090.76	61.9
vrpnc11	1042.11	5	2058.47	97.5	1944.59	86.6	1832.02	75.8
		10	1960.29	88.1	1867.41	79.2	1750.82	68.0
		20	1955.05	87.6	1795.16	72.3	1693.89	62.5
vrpnc12	819.56	5	1605.75	95.9	1502.84	83.4	1425.89	74.0
		10	1568.27	91.4	1440.91	75.8	1370.54	67.2
		20	1593.22	94.4	1427.55	74.2	1329.72	62.2

The second interesting conclusion refers to the benefits of applying buffering strategies as a part of the proposed approach. Buffering strategy applied either only for dynamic requests (D-BUF) or for all requests (SD-BUF) gives better results when compare to the case without request buffering (N-BUF) for all tested instances. By focusing observation on cases where requests buffering (D-BUF or SD-BUF) have been applied, the outperformance of strategy where all requests are buffered over strategy buffering only dynamic requests has been observed.

5 Conclusions

This paper proposes an approach based on multi-agent paradigm to the Dynamic Vehicle Routing Problem. The process of solving instances of the problem is performed by the set of agents which pose different abilities and responsibilities. One of the important tasks performed by the agents is to dispatch the sets of static and dynamic requests to the available vehicles. In order to increase the efficiency of these processes, the request buffering strategy has been implemented in the approach.

Computational experiment confirmed the influence of the request buffering strategy on the results obtained by the proposed approach. Although the deterioration of the results were observed in all cases, its strength was different depending on the level of applying the buffering strategy.

Future research may aim at extending the proposed approach with buffering strategy to other problems like for example DVRPTW or DPDP. Another direction is to incorporate other strategies which may have positive impact on the results. One of them is waiting strategy [9] which aims at deciding whether a vehicle should wait after servicing a request, before heading toward the next customer or planning a waiting period on a strategic location. Its positive impact on the results has been confirmed by the authors of it.

References

1. Barbucha, D., Jędrzejowicz, P.: Multi-agent platform for solving the dynamic vehicle routing problem. In: Proceedings of the 11th International IEEE Conference on Intelligent Transportation Systems (ITSC 2008), pp. 517–522. IEEE Press, New York (2008)
2. Barbucha, D.: A multi-agent approach to the dynamic vehicle routing problem with time windows. In: Badica, C., Nguyen, N.T., Brezovan, M. (eds.) Computational Collective Intelligence. Technologies and Applications—5th International Conference, ICCCI 2013. LNCS, vol. 8083, pp. 467–476. Springer. Berlin (2013)
3. Branchini, R.M., Armentano, A.V., Lokketangen, A.: Adaptive granular local search heuristic for a dynamic vehicle routing problem. Comput. Oper. Res. **36**(11), 2955–2968 (2009)
4. Christofides, N., Mingozzi, A., Toth, P., Sandi, C. (eds.): Combinatorial Optimization. John Wiley, Chichester (1979)
5. Gendreau, M., Guertin, F., Potvin, J.-Y., Taillard, E.: Parallel Tabu search for real-time vehicle routing and dispatching. Transp. Sci. **33**(4), 381–390 (1999)
6. Hanshar, F.T., Ombuki-Berman, B.M.: Dynamic vehicle routing using genetic algorithms. Appl. Intell. **27**, 89–99 (2007)
7. Larsen, A.: The on-line vehicle routing problem. Ph.D. Thesis, Institute of Mathematical Modelling, Technical University of Denmark (2001)
8. Lund, K., Madsen, O.B.G., Rygaard, J.M.: Vehicle routing problems with varying degrees of dynamism. Technical report, Institute of Mathematical Modelling, Technical University of Denmark (1996)
9. Mitrovic-Minic, S., Laporte, G.: Waiting strategies for the dynamic pickup and delivery problem with time windows. Transp. Res. Part B **38**, 635–655 (2004)
10. Mitrovic-Minic, S., Krishnamurti, R., Laporte, G.: Double-horizon based heuristics for the dynamic pickup and delivery problem with time windows. Transp. Res. Part B **38**, 669-685 (2004)
11. Montemanni, R., Gambardella, L.M., Rizzoli, A.E., Donati, A.V.: A new algorithm for a dynamic vehicle routing problem based on ant colony system. J. Comb. Optim. **10**, 327–343 (2005)
12. Pillac, V., Gendreau, M.: Guret, C., Medaglia, A.L.: A review of dynamic vehicle routing problems. Eur. J. Oper. Res. **225**, 1–11 (2013)
13. Psaraftis, H.: A dynamic-programming solution to the single vehicle many-tomany immediate request dial-a-ride problem. Transp. Sci. **14**(2), 130–154 (1980)
14. Pureza, V., Laporte, G.: Waiting and buffering strategies for the dynamic pickup and delivery problem with time windows. INFOR **46**(3), 165–175 (2008)

Pattern Recognition for Decision Making Systems

Predictive Strength of Bayesian Networks for Diagnosis of Depressive Disorders

Blessing Ojeme and Audrey Mbogho

Abstract Increasing cases of misdiagnosis of mental disorders in Nigeria despite the use of the international standards provided in the Diagnostic and Statistical Manual of Mental Disorders (DSM-V) and International Classification of Diseases (ICD-10) calls for an approach that takes cognizance of the socio-economic difficulties on the ground. While a growing recognition of the potential of artificial intelligence (AI) techniques in modeling clinical procedures has led to the design of various systems to assist clinicians in decision-making tasks in physical diseases, little attention has been paid to exploring the same techniques in the mental health domain. This paper reports the preliminary findings of a study to investigate the predictive strength of Bayesian networks for depressive disorders diagnosis. An automatic Bayesian model was constructed and tested with a real-hospital dataset of 580 depression patients of different categories and 23 attributes. The model predicted depression and its severity with high efficiency.

Keywords Artificial intelligence · Bayesian networks · Mental health · Depression disorders · Psychiatric diagnosis

1 Introduction

Developed in 1763 by Thomas Bayes [1], an 18th-century British mathematician and theologian, the primary objectives of Bayesian networks were to extract informative knowledge from real-world data even in the presence of uncertain and missing entries, to allow one to learn about causal relationships, and to facilitate the

B. Ojeme (✉)
Department of Computer Science, University of Cape Town,
Rondebosch 7701, South Africa
e-mail: bojeme@cs.uct.ac.za

A. Mbogho
Department of Mathematics and Physics, Pwani University, Kilifi, Kenya
e-mail: a.mbogho@pu.ac.ke

I. Czarnowski et al. (eds.), *Intelligent Decision Technologies 2016*,
Smart Innovation, Systems and Technologies 56,
DOI 10.1007/978-3-319-39630-9_31

combination of domain knowledge and data for modeling tasks [2]. To achieve these objectives, Bayesian networks must first be constructed, either by hand (manually) or by software (from data). As a real-world problem-solving tool, Bayesian networks have been used to address problems in different areas of medicine [3].

Efforts have been consistently made to build Bayesian networks to assist clinicians in the tasks of screening, diagnosis, prediction and treatment of physical diseases [4, 5]. Though diagnostic activities in physical and psychiatric disorder domains are generally seen as the same, Gangwar et al. [6] have shown that each domain possesses different conditions, with psychiatric diagnosis being more complex and controversial due to the conflicting, overlapping and confusing nature of the symptoms. While Ganasen et al. [7] noted that the much higher risks in the area of psychiatry in the developing world is as a result of the acute shortage of mental health facilities, more studies have identified the relatively little research on the potential of Bayesian networks in the psychiatric domain, attributing it to lack of availability of combined domain expertise in both psychiatry and computer science [8], difficulties of interdisciplinary collaboration, and lack of access to mental health data [9]. In Nigeria, for instance, a significant number of depression sufferers resort to self-medication because they are either undiagnosed, misdiagnosed or wrongly treated [10].

This paper is a report on the preliminary findings of a study investigating the predictive strength of Bayesian networks and other machine learning techniques in clinical decision-support systems, with an application in depressive disorders. Depression is an important problem because of its high prevalence [11], high comorbidity with other mental and/or physical illnesses [12, 13] and rampant misdiagnosis by primary care physicians [13, 14]. In addition to looking at literature from current conference papers, journal publications, books and technical reports in the field of AI, medicine and psychiatry, a Bayesian network was constructed and tested with real hospital data.

1.1 Overview of Depression and Challenges of Diagnosis

Depression is a chronic mood disorder characterised by sadness, loss of interest or pleasure, social withdrawal, decreased energy, feelings of guilt or low self-worth, disturbed sleep or appetite, and poor concentration [15]. Depression is the second most common psychiatric disorder seen by primary care physicians, affecting over 14 % of individuals worldwide [14]. Depression affects society as a whole, regardless of sex, age, colour, religion, culture, race or status.

In most developing countries like Nigeria, with scarce mental health services [7], traditional diagnostic practice in depression services typically involves clinician-to-patient interview where judgments are made from the patient's appearance and behaviour, subjective self-reported symptoms, depression history, and current life circumstances [16]. The views of relatives or other third parties may be taken into account. A physical examination to check for ill health, the effects of medications or other drugs may be conducted. This intuitive model, though still in use today, is slow and leaves diagnostic decision-making entirely to the subjective clinical skills and opinion of the clinicians [8].

Depression is difficult to detect by clinicians because it shares symptoms with other physical and/or mental disorders [8]. A robust model is therefore needed for detection and diagnosis. A robust depression diagnostic model must be able to make accurate decisions regardless of the presence of missing data and classify the presented depression cases correctly as being, for example, 'no depression', 'mild depression', 'moderate depression' or 'severe depression'. The results should also be presented in an easily interpretable format [2]. A poorly constructed model makes detection and diagnosis difficult and then renders treatment ineffective. The challenge facing researchers and practitioners, therefore, is how to construct a robust depression diagnostic model to achieve optimal results. Given the complexity of depression, an intuitive model would prove to be poorly constructed or weak [8].

2 Related Work

The concept of Bayesian networks as a probabilistic reasoning tool for management of imprecision of data and uncertainty of knowledge was pioneered by Bayes [1] who considered mapping the dependencies between a set of random variables to a directed acyclic graph (DAG), both increasing human readability and simplifying the representation of the joint probability distribution of the set of variables.

Sumathi [17] proposed a Bayesian networks framework as the main knowledge representation technique in diagnosing depression level and suggested treatments according to the patient details provided to the framework. Though not implemented, the framework systematizes domain expert knowledge and observed datasets of patients and helps to map cause-effect relationship between variables. The limitation of this framework is that it was unable to identify key factors of depression.

Curiac et al. [18] presented a Bayesian network-based analysis of four major psychiatric diseases: schizophrenia (simple and paranoid), mixed dementia (Alzheimer disease included), depressive disorder and maniac depressive psychosis.

Chang [19] built a prototype that utilizes ontologies and Bayesian network techniques for inferring the possibility of depression.

3 Methodology

This study seeks to investigate the predictive strength of Bayesian networks to diagnose depressive disorders in Nigeria. The steps taken to achieve the objectives are as follows:

(a) Collect depression data from the mental health unit of University of Benin Teaching Hospital (UBTH) and primary care centres in Nigeria.
(b) Extract the features (symptoms of depression).
(c) Build a Bayesian model using Weka (Waikato Environment for Knowledge Analysis), a popular, free machine learning tool [20].
(d) Test the performance of the built model on a set of real-world depression cases.

3.1 *Data Gathering*

In order to train the Bayesian networks, data were collected from one primary care centre and the mental health unit of the UBTH. Given that the accuracy of a diagnostic Bayesian network is dependent on the reliability of the clinical information of patients entered into the system, the researcher started by presenting a seminar on the concepts and applications of Bayesian networks in medical diagnosis to the domain experts (a group of psychiatrists and psychologists) at the UBTH. This was followed by a series of interviews by the researcher to obtain information on the clinical procedures for depression diagnosis. The responses and feedback from the domain experts at the interviews and seminar provided the expert knowledge for the Bayesian networks. The seminar and the interview sessions were necessary to establish a relationship for smooth communication with the domain experts and also to reduce all the biases in extracting knowledge from experts, as identified by Tversky et al. [21]. The second phase was the extraction of 580 records of previously diagnosed patients from the hospital and primary care centres to construct the Bayesian network structure and to estimate the conditional probability tables. There were 254 male and 326 female patients from 12 to 92 years old (with a mean age of 41.8 and standard deviation of 16.3). The features shown in Table 1 were identified as relevant for the screening and diagnosis of depression.

3.2 *Model Building and Testing*

Witten et al. [22] identified two techniques for constructing Bayesian networks: manual construction and automatic construction (data-driven construction). This study adopted the data-driven Bayesian network because of its high accuracy and

Table 1 Features of depression extracted from the dataset

S/N	Features	Code	Data type
1	Age	ag	integer
2	Sex	se	Integer
3	Marital status	ms	Integer
4	Sad mood	sm	Integer
5	Suicidal	su	Integer
6	Loss of pleasure	lp	Integer
7	Insomnia	in	Integer
8	Hypersomnia	hy	Integer
9	Loss of appetite	la	Integer
10	Psychomotor agitation	pa	Integer
11	Psychomotor retardation	pa	Integer
12	Loss of energy	le	Integer
13	Feeling of worthlessness	fw	Integer
14	Lack of thinking	Lt	Integer
15	Indecisiveness	id	Integer
16	Recurrent thoughts of death	rt	Integer
17	Impaired function	if	Integer
18	Weight gain	wg	Integer
19	Weight loss	wl	Integer
20	Stressful life events	sl	Integer
21	Financial pressure	fp	Integer
22	Depression in family	df	Integer
23	Employment status	es	Integer
24	Depression diagnosis		Nominal
25	Comorbidity		Nominal
26	Treatment		Nominal

simplicity. Software tools available for the data-driven Bayesian network construction include Weka, BigML, Orange, Netica, Rapidminer, GeNIe, KNIME and Hugin [23]. The data-driven Bayesian network is less dependent on human experts for knowledge acquisition, thereby helping to eliminate the biases associated with manual construction, as identified by Oteniya [24]. The major drawbacks of the data-driven Bayesian network are its dependence on a large volume of data and exponential growth of the model with an increase in the number of variables [24]. These shortcomings were overcome by using a small amount of data along with a stratified ten-fold cross-validation technique [22] to split the data into three equal folds, in which two-thirds (387) of the dataset was used for training the model while the remaining one-third (193) was used for testing. This procedure was repeated three times to ensure an even representation in training and test sets, and outputs the averaged results across all folds. Weka provided the platform for the data analysis, preparation, model testing and result evaluation shown in Fig. 1.

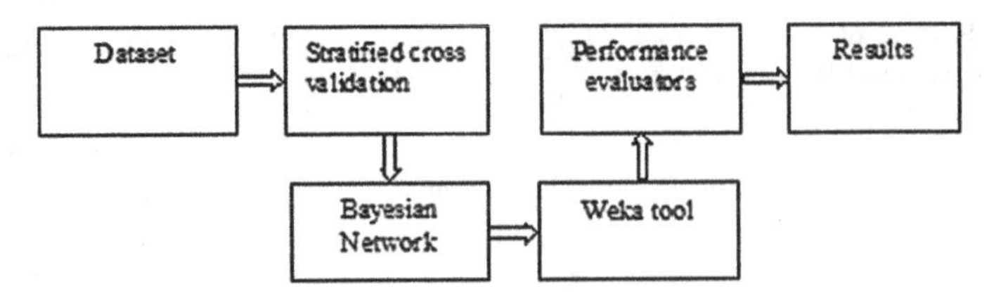

Fig. 1 Data preparation, analysis and model testing

3.3 Mathematical Formalism of Bayesian Networks

Bayes theorem, derived from the product rule of probability, is stated mathematically as follows:

$$P(B/A) = \frac{P(A/B) * P(B)}{P(A)} = \frac{\text{likelihood} * \text{prior}}{evidence} \quad (1)$$

P(B/A) = Posterior or conditional probability; P(B) = prior probability of B;

P(A/B) = probability that B being true will result in A; P(A) = prior probability of A

Extending Bayes' theorem in (1) for reasoning under uncertainty in medical diagnosis: variables (A and B) are replaced with disease and symptoms (D and S) respectively [23]:

$$P(D/S) = \frac{P(S/D) * P(D)}{P(S)} \approx \frac{P(S/D) * P(D)}{(P(S/D) * P(D) + P(S/\neg D) * P(\neg D))} \quad (2)$$

$P(\neg D)$ = prior probability of disease being false

$P(S/\neg D)$ = probability of finding symptom S even when disease D is false

To overcome the difficulty of diagnosis where P(S) is unknown, we used the following formalism:

$$P(D/S) = \frac{P(S\neg D) * P(\neg D)}{P(S)} \quad (3)$$

Dividing Eq. (2) by Eq. (3) gives

$$\frac{P(S/D) * P(D)}{P(S/\neg D) * P(\neg D)} \quad (4)$$

Expressing the conditional probability in Eq. (4) for multiple diseases D_1, D_2, …, D_m and multiple symptoms S_1, S_2, …, S_n looks like this:

$$P(D_i/S_1S_2..S_n) = \frac{P(S_1/D_i) * P(S_2D_i)*\cdots*P(S_n/D_i) * P(D_i)}{\sum_{k=1}^{m} P(S_1/D_k) * P(S_2/D_k)*\cdots*P(S_n/D_k) * P(D_k)} \quad (5)$$

4 Analysis of Results

Table 2 shows the different evaluation metrics used in this study include: True positive rate (TPR), False Positive Rate (FPR), Precision, Recall, F-Measure, Matthew's Correlation Co-efficient (MCC), Receiver Operating Characteristics (ROC) and Precision-Recall (PRC) area. The metrics perform differently under balanced and imbalanced datasets [25]. For instance, MCC was used because it is calculated on all four values (True Positive, False Negative, False Positive and True Negative) of the confusion matrix. The True Positive and True Negative represented the number of persons diagnosed by the diagnosis system correctly either as having or not having depression. The False Positive represented the number of persons diagnosed by the system as depressed whereas they were actually not depressed while False Negative represented the number of persons classified as not depressed whereas they were actually depressed. The ROC is a metric used to check the quality of classifiers. ROC provide the area under the curve (AUC) of the plot of the true positive rate (y-axis) against the false positive rate (x-axis). An excellent classifier will have ROC area values between 0.9 and 1.0 while a poor classifier will have ROC area values between 0.6 and 0.7 [25]. Similar to ROC, precision-recall (PRC) provided a very powerful way of evaluating the performance of the Bayesian networks classifier given the imbalanced dataset used for the study. A precision of 0.876 achieved by the model is interpreted as 87.6 % correct predictions among the positive predictions.

From the results of Tables 2 and 3, Bayesian networks demonstrates an outstanding performance. There were 271 severe depression cases, 23 mild depression cases, 272 moderate depression cases and 14 not depressed cases, as diagnosed by mental health professionals, out of which the Bayesian networks algorithm classified 252 as being severe, 0 as being mild, 264 as being moderate and 7 as being not depressed. This represented a total of 523 correctly classified cases and 57 wrongly classified cases. Again the ROC area of 0.975 shows the Bayesian networks as an excellent classifier.

Table 2 Results from the Bayesian network

TPR	FPR	Precision	Recall	F-Measure	MCC	ROC area	PRC area	Class
0.930	0.019	0.977	0.930	0.953	0.914	0.997	0.997	Severe
0.000	0.004	0.000	0.000	0.000	−0.012	0.869	0.193	Mild
0.971	0.159	0.843	0.971	0.903	0.812	0.963	0.945	Moderate
0.500	0.000	1.000	0.500	0.667	0.703	0.972	0.665	Not depressed
0.902	0.084	0.876	0.902	0.885	0.825	0.975	0.932	Weighted avg

Table 3 Confusion matrix from the Bayesian network

a	b	c	d	Classified as
252	0	19	0	A = severe
0	0	23	0	B = mild
6	2	264	0	C = moderate
0	0	7	7	D = not depressed

5 Conclusions and Future Work

In this paper, we reviewed a few available papers on Bayesian networks for the detection and diagnosis of depression risk in patients. We constructed a Bayesian network from labelled patient data and tested it with real-hospital data, consisting of 580 data instances and 23 attributes (symptoms). Though the research is still in the very early stage, the preliminary results indicate that the Bayesian network diagnostic tool, as compared with the diagnosis of mental health professionals, was reasonably efficient to support clinicians in the task of screening and diagnosis of depression in underserved areas such as Nigeria where there are severe shortages of medical professionals, a situation not expected to significantly improve in the near term. From the interviews and dataset collected, depression has high comorbidity with other physical and/or mental disorders. The discussed technique in the study is limited to depression and its severity. Future work would construct Bayesian networks and other types of machine learning models to separate patients having other diseases in addition to depression.

References

1. Bayes, T.: An essay towards solving a problem in the doctrine of chances. By the Late Rev. Mr. Bayes, F. R. S. communicated by Mr. Price, in a letter to John Canton, A. M. F. R. S. Philos. Trans. R. Soc. London **53**(0), 370–418 (1763)
2. Heckerman, D.: Bayesian networks for data mining. Data Min. Knowl. Discov. **119**(1), 79–119 (1997)

3. Nikovski, D.: Constructing Bayesian networks for medical diagnosis from incomplete and partially correct statistics. IEEE Trans. Knowl. Data Eng. **12**(4), 1–18 (2000)
4. Tylman, W., Waszyrowski, T., Napieralski, A., Kamiński, M., Trafidło, T., Kulesza, Z., Kotas, R., Marciniak, P., Tomala, R., Wenerski, M.: Real-time prediction of acute cardiovascular events using hardware-implemented Bayesian networks. Comput. Biol. Med. (2015)
5. Su, C., Andrew, A., Karagas, M.R., Borsuk, M.E.: Using Bayesian networks to discover relations between genes, environment, and disease. BioData Min. **6**(1), 6 (2013)
6. Gangwar, M., Mishra, R.B., Yadav, R.S.: Classical and intelligent computing methods in psychiatry and neuropsychitry : an overview. Int. J. Adv. Res. IT Eng. **3**(12) (2014)
7. Ganasen, K.A., Parker, S., Hugo, C.J., Stein, D.J., Emsley, R.A., Seedat, S.: Mental health literacy: focus on developing countries. Afr. J. Psychiatry **11**(1), 23–28 (2008)
8. Chattopadhyay, S.: A neuro-fuzzy approach for the diagnosis of depression. Elsevier Appl. Comput. Inform. in Press, 19 (2014)
9. Doherty, G., Coyle, D., Matthews, M.: Design and evaluation guidelines for mental health technologies. Interact. Comput. **22**(4), 243–252 (2010)
10. James, B., Jenkins, R., Lawani, A., Omoaregba, J.: Depression in primary care: the knowledge, attitudes and practice of general practitioners in Benin City, Nigeria. S Afr Fam Pr. **54**(1), 55–60 (2012)
11. Ferrari, A.J., Charlson, F.J., Norman, R.E., Patten, S.B., Freedman, G., Murray, C.J.L., Vos, T., Whiteford, H.: Burden of depressive disorders by country, sex, age, and year: findings from the global burden of disease study 2010. PLoS Med. **10**(11) (2013)
12. Salihu, A.S.: Impact of somatic symptoms on identification of depression among general outpatients by family physicians. Open J. Psychiatry **5**, 278–284 (2015)
13. Ahmed, K., Bhugra, D.: Depression across ethnic minority cultures: diagnostic issues. World Cult. Psychiatry Res. Rev. **2**(3), 51 (2007)
14. Huang, S.H., LePendu, P., Iyer, S.V., Tai-Seale, M., Carrell, D., Shah, N.H.: Toward personalizing treatment for depression: predicting diagnosis and severity. J. Am. Med. Inform. Assoc. 1–7 (2014)
15. WHO: Depression: a global public health concern. WHO Dep. Ment. Heal. Subst. Abus. 1–8 (2012). http://www.who.int/mental_health/management/depression/who_paper_depression_wfmh_2012.pdf
16. Baasher, T.A., Carstairs, G.M., Giel, R., Hassler, F.R.: Mental health services in developing countries. WHO Seminar on the Organisation of Mental Health Services. World Health Organisation, Geneva (1975)
17. Sumathi, M.R., Poorna, B.: A bayesian framework for diagnosing depression level of adolescents. Int. Conf. Comput. Intell. Syst. **4**(March), 1350–1354 (2015)
18. Curiac, D.-I., Vasile, G., Banias, O., Volosencu, C., Albu, A.: Bayesian network model for diagnosis of psychiatric diseases. In: Proceedings of ITI 2009 31st International Conference on Information Technology Interfaces, pp. 61–66 (2009_
19. Chang, Y.-S., Fan, C.-T., Lo, W.-T., Hung, W.-C., Yuan, S.-M.: Mobile cloud-based depression diagnosis using an ontology and a Bayesian network. Futur. Gener. Comput. Syst. **43–44**, 87–98 (2015)
20. Bouckaert, R.R., Frank, E., Hall, M., Kirkby, R., Reutemann, P., Seewald, A., Scuse, D.: WEKA Manual for Version 3-7-12. University of Waikato, Hamilton, New Zealand (2014)
21. Tversky, A., Kahneman, D.: Judgment under Uncertainty: Heuristics and Biases. American Association for the Advancement of Science, New Series, pp. 1124–1131 (1974)
22. Witten, I., Frank, E., Hall, M.: Data Mining: Practical Machine Learning Tools and Techniques. Morgan Kaufmann Publishers, Third Edit (2011)
23. Shojaei Estabragh, Z., Riahi Kashani, M.M., Jeddi Moghaddam, F., Sari, S., Taherifar, Z., Moradi Moosavy, S., Sadeghi Oskooyee, K.: Bayesian network modeling for diagnosis of social anxiety using some cognitive-behavioral factors. Netw. Model. Anal. Heal. Inform. Bioinforma. **2**(4), 257–265 (2013)

24. Oteniya, L.: Bayesian Belief Networks for Dementia Diagnosis and Other Applications: A Comparison of Hand-Crafting and Construction using A Novel Data Driven Technique. A PhD Thesis, Department of Computing Science and Mathematics, University of Stirling, 2008
25. Saito, T., Rehmsmeier, M.: The precision-recall plot is more informative than the ROC plot when evaluating binary classifiers on imbalanced datasets. PLoS ONE **10**(3), e0118432 (2015)

Automatic Human Activity Segmentation and Labeling in RGBD Videos

David Jardim, Luís Nunes and Miguel Sales Dias

Abstract Human activity recognition has become one of the most active research topics in image processing and pattern recognition. Manual analysis of video is labour intensive, fatiguing, and error prone. Solving the problem of recognizing human activities from video can lead to improvements in several application fields like surveillance systems, human computer interfaces, sports video analysis, digital shopping assistants, video retrieval, gaming and health-care. This paper aims to recognize an action performed in a sequence of continuous actions recorded with a Kinect sensor based on the information about the position of the main skeleton joints. The typical approach is to use manually labeled data to perform supervised training. In this paper we propose a method to perform automatic temporal segmentation in order to separate the sequence in a set of actions. By measuring the amount of movement that occurs in each joint of the skeleton we are able to find temporal segments that represent the singular actions. We also proposed an automatic labeling method of human actions using a clustering algorithm on a subset of the available features.

Keywords Human motion analysis · Motion-based recognition · Action recognition · Temporal segmentation · Clustering · K-means · Labeling · Kinect · Joints · Video sequences

D. Jardim (✉) · M.S. Dias
MLDC, Lisbon, Portugal
e-mail: t_dajard@microsoft.com

M.S. Dias
e-mail: miguel.dias@microsoft.com

D. Jardim · L. Nunes
Instituto de Telecomunicações, Lisbon, Portugal
e-mail: luis.nunes@iscte.pt

D. Jardim · L. Nunes · M.S. Dias
University Institute of Lisbon (ISCTE-IUL), Lisbon, Portugal

D. Jardim · M.S. Dias
ISTAR-IUL, Lisbon, Portugal

I. Czarnowski et al. (eds.), *Intelligent Decision Technologies 2016*,
Smart Innovation, Systems and Technologies 56,
DOI 10.1007/978-3-319-39630-9_32

1 Introduction

Human activity recognition is a classification problem in which events performed by humans are automatically recognized. Detecting specific activities in a live feed or searching in video archives still relies almost completely on human resources. Detecting multiple activities in real-time video feeds is currently performed by assigning multiple analysts to simultaneously watch the same video stream. Manual analysis of video is labour intensive, fatiguing, and error prone. Solving the problem of recognizing human activities from video can lead to improvements in several application fields like surveillance systems, human computer interfaces, sports video analysis, digital shopping assistants, video retrieval, gaming and health-care [8, 9, 12, 15]. Ultimately, we are interested in recognizing high-level human activities and interactions between humans and objects. The main sub-tasks of this recognition are Usually achieved using manually labeled data to train classifiers to recognize a set of human activities. An interesting question is how far can we take the automatic labeling of human actions using unsupervised learning? From our experiments we have found that this labeling is possible, but still with a large margin for improvement.

2 Related Work

Human activity recognition is a classification problem in which events performed by humans are automatically recognized by a computer program. Some of the earliest work on extracting useful information through video analysis was performed by O'Rourke and Badler [13] in which images were fitted to an explicit constraint model of human motion, with constraints on human joint motion, and constraints based on the imaging process. Also Rashid [16] did some work on understanding the motion of 2D points in which he was able to infer 3D position. Driven by application demands, this field has seen a relevant growth in the past decade. This research has been applied in surveillance systems, human computer interfaces, video retrieval, gaming and quality-of-life devices for the elderly. Initially the main focus was recognizing simple human actions such as walking and running [4]. Now that that problem is well explored, researchers are moving towards recognition of complex realistic human activities involving multiple persons and objects. In a recent review written by [1] an approach-based taxonomy was chosen to categorize the activity recognition methodologies which were divided into two categories. Single-layered approaches [2, 18, 20] typically represent and recognize human activities directly based on sequences of images and are suited for the recognition of gestures and actions with sequential characteristics. Hierarchical approaches represent high-level human activities that are composed of other simpler activities [1]. Hierarchical approaches can be seen as statistical, syntactic and description-based [3, 6, 8, 14, 17, 21].

The previous approaches all used computer vision (CV) techniques to extract meaningful features from the data. Motion capture data (MOCAP) has also been used in this field, a relevant approach found was [22] where they pose the problem of learning motion primitives (actions) as a temporal clustering one, and derive an unsupervised hierarchical bottom-up framework called hierarchical aligned cluster analysis (HACA). HACA finds a partition of a given multidimensional time series into m disjoint segments such that each segment belongs to one of k clusters representing an action. They were able to achieve competitive detection performances (77 %) for human actions in a completely unsupervised fashion. Using MOCAP data has several advantages mainly the accuracy of the extracted features but the cost of the sensor and the required setup to obtain the data is often prohibitive.

With the cost in mind Microsoft released a sensor called Kinect which captures RGB-D data and is also capable of providing joint level information in a non-invasive way allowing the developers to abstract away from CV techniques. Using Kinect [10] the authors consider the problem of extracting a descriptive labeling of the sequence of sub-activities being performed by a human, and more importantly, of their interactions with the objects in the form of associated affordances. Given a RGB-D video, they jointly model the human activities and object affordances as a Markov random field where the nodes represent objects and sub-activities, and the edges represent the relationships between object affordances, their relations with sub-activities, and their evolution over time. The learning problem is formulated using a structural support vector machine (SSVM) approach, where labelings over various alternate temporal segmentations are considered as latent variables. The method was tested on a dataset comprising 120 activity videos collected from 4 subjects, and obtained an accuracy of 79.4 % for affordance, 63.4 % for sub-activity and 75.0 % for high-level activity labeling.

In [7] the covariance matrix for skeleton joint locations over time is used as a discriminative descriptor for a sequence of actions. To encode the relationship between joint movement and time, multiple covariance matrices are deployed over sub-sequences in a hierarchical fashion. The descriptor has a fixed length that is independent from the length of the described sequence. Their experiments show that using the covariance descriptor with an off-the-shelf classification algorithm one can obtain an accuracy of 90.53 % in action recognition on multiple datasets.

In a parallel work [5] authors propose a descriptor for 2D trajectories: Histogram of Oriented Displacements (HOD). Each displacement in the trajectory votes with its length in a histogram of orientation angles. 3D trajectories are described by the HOD of their three projections. HOD is used to describe the 3D trajectories of body joints to recognize human actions. The descriptor is fixed-length, scale-invariant and speed-invariant. Experiments on several datasets show that this approach can achieve a classification accuracy of 91.26 %.

Recently [11] developed a system called Kintense which is a real-time system for detecting aggressive actions from streaming 3D skeleton joint coordinates obtained from Kinect sensors. Kintense uses a combination of: (1) an array of supervised learners to recognize a predefined set of aggressive actions, (2) an unsupervised learner to discover new aggressive actions or refine existing actions, and (3) human feedback to reduce false alarms and to label potential aggressive actions. The system is 11–16 % more accurate and 10–54 % more robust to changes in distance, body orientation, speed, and subject, when compared to standard techniques such as dynamic time warping (DTW) and posture based gesture recognizers. In two multi-person households it achieves up to 90 % accuracy in action detection.

3 Temporal Segmentation

This research is framed in the context of a doctoral program where the final objective is to predict the next most likely action that will occur in a sequence of actions. In order to solve this problem we divided it in two parts, recognition and prediction. This paper will only refer to the recognition problem. Human activity can be categorized into four different levels: gestures, actions, interactions and group activities. We are interested in the actions and interactions category.

An initial research was conducted to analyze several datasets from different sources like LIRIS (Laboratoire d'InfoRmatique en Image et Systèmes d'information) dataset [19], CMU (Carnegie Mellon University) MoCap dataset,[1] MSR-Action3D and MSRDailyActivity3D dataset [12] and verify it's suitability to our problem. All these datasets contain only isolated actions, and for our task we require sequences of actions. We saw this as an opportunity to create a new dataset that contains sequences of actions.

We used Kinect to record the dataset which contains 8 aggressive actions like punching and kicking, 6 distinct sequences (each sequence contains 5 actions). Recorded 12 subjects, each subject performed 6 sequences. Total of 72 sequences, 360 actions. An example of a recorded sequence is illustrated on Fig. 1. Kinect captures data at 30 frames per second. The data is recorded in .xed files which contains RBG, depth and skeleton information, and also a light version in .csv format containing only the skeleton data. We expect to make the dataset available to public in a near future on a dedicated website.

Since our dataset contains sequences of actions our very first task was to automatically decompose the sequence in temporal segments where each segment represents an isolated action. We went for a very simple approach. From visual observation we've noticed that during each action there were joints that moved more than others. If we could measure that movement and compare it with other joints we could be able to tell which joint is predominant in a certain action and then assign a temporal segment to a joint. Figure 2 shows a timeline which represents the movement of the right ankle. It is perfectly visible that there are two regions where that joint has a significant higher absolute speed. These two regions represent moments in time where an action was performed that involved mainly the right leg.

Our first step was to create these regions which we called regions of interest. This was achieved by selecting frames in which the absolute speed value was above the

[1] http://mocap.cs.cmu.edu/.

Fig. 1 Example of a sequence of actions in the dataset

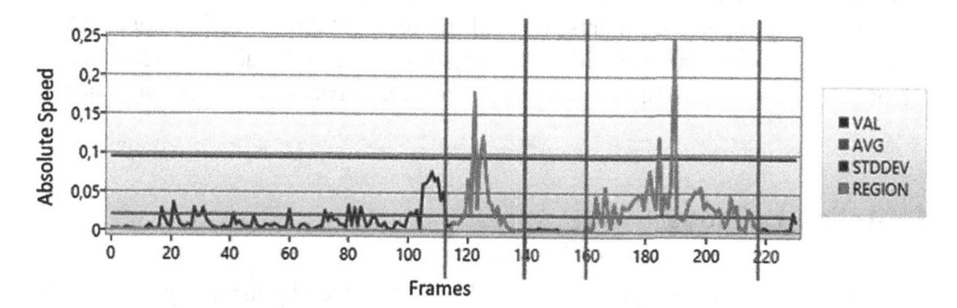

Fig. 2 Absolute speed of the right ankle while performing actions and regions of interest found

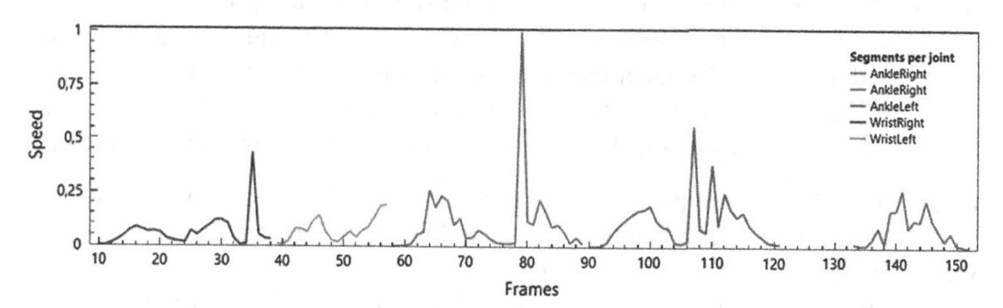

Fig. 3 Visual representation of our action segmentation method

standard deviation multiplied by a factor of two. Then we selected all the neighboring frames that were above the average value with a tolerance of 3 frames below of the average. This data was collected for four different joints: right and left ankle, right and left wrist. Then we searched for overlapping regions. While the user performs a kick the rest of his body moves, specially the hands to maintain the body's balance. Overlapping regions were removed by considering only the joint moving at a higher average speed in each frame. Figure 3 illustrates an example result of our automatic segmentation method. Each color of the plot represents a temporal segment to which we assigned a joint as being the dominant joint for that action. We obtained 5 temporal segments which successfully correspond to the number of actions that the sequence contains, in this case: right-punch; left-punch; front-right-kick; front-left-kick; side-right-kick.

Table 1 Automatic temporal segmentation per sequence

Sequence	1 (%)	2 (%)	3 (%)	4 (%)	5 (%)	6 (%)	Average (%)
Segmentation accuracy	94.23	89.66	73.55	89.44	76.31	75.03	83.04

Table 1 shows that the automatic segmentation can be improved. These results reflect the measurements between the frames of the annotated data and the frames of our automatic temporal segments. Overall the segmentation is satisfactory and we believe that the segments have the most important part of the actions. This method might be revisited in the future to improve the overall performance of our system.

4 Action Labeling

In most cases, as seen in [11], action labeling is achieved by manually labeling the segments obtained by the segmentation algorithm and then use that data to train one classifier per action. Those classifiers would then be used in an application capable of recognizing actions in real-time. Instead we thought that it would be more interesting if we could automatically label equal actions performed by different subjects. For example a right-punch performed by subject 1 should be very similar to a right-punch performed by subject 2. This process is composed of the following stages:

1 Automatically find temporal segments that represent the actions of the sequence
2 Sample the dataset based on the previously found temporal segments
3 Extract meaningful features for each segment
4 Use clustering to automatically group similar actions and thus label them.

4.1 Sampling

To sample the data for the clustering algorithm the program automatically selects the automatically found temporal segments which ideally should be 5 per sequence, which corresponds to the number of actions that compose the sequence. The most active joint is assigned to that segment. Based on the window-frame of the segment found for a specific joint we create new temporal segments for the remaining joints on the same exact window-frame. This can be portrayed has stacking the joints time-line one on top of another and making vertical slices to extract samples of data that correspond to temporal segments where an action has occurred.

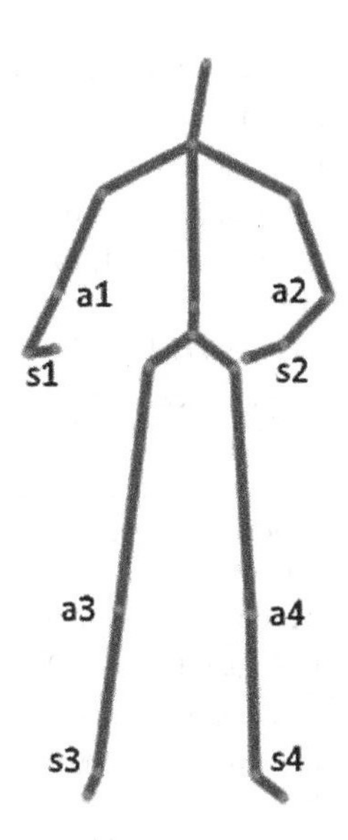

Fig. 4 Visual representation of body relative features used

4.2 Feature Extraction

An action can be seen as a sequence of poses over time. Each pose respects certain relative positions and orientation of joints of the skeleton. Based on the positions and orientations of the joints we extracted several features that will be used to model the movements performed by the subjects. We have experimented with several features (speed; absolute speed; speed per axis; joint flexion in angles; bone orientation). After a comparison of these different approaches (to be published) we selected the angles of the elbows and the knees *a1, a2, a3, a4* and the relative position of the wrists and ankles *s1, s2, s3, s4* (Fig. 4) and used these to calculate other features like relative speed of each joint. Different subsets of these features combined will constitute the feature vectors that will be used by the clustering algorithm.

4.3 Clustering Experiments

As previously mentioned, the objective is to cluster similar actions performed by different subjects (or by the same subject in different recordings). For that purpose we will use k-means which is one of the simplest unsupervised learning algorithms. We made several experiments with different combinations of features.

Our initial experiments using simply the average speed of each joint over the whole segment as a feature. Results of this experiment are shown in Table 2. Clustering all the segments of the same sequence of actions being performed by different subjects brought interesting results. All the actions were correctly labeled except for the side-right-kick. As shown in the table this action was classified as

Table 2 Clustering results for Sequence 1 using average speed as a feature

Action	Cluster 1 (%)	Cluster 2 (%)	Cluster 3 (%)	Cluster 4 (%)	Cluster 5 (%)
Right punch	8.33	91.67	0.0	0.0	0.0
Left punch	0.0	0.0	0.0	100.0	0.0
Front right kick	91.67	0.0	8.33	0.0	0.0
Side right kick	33.33	0.0	66.67	0.0	0.0
Side left kick	8.33	8.33	0.0	0.0	83.33

Table 3 Clustering results for Sequence 1 using average speed and angles of the knees and elbows as a feature

Action	Cluster 1 (%)	Cluster 2 (%)	Cluster 3 (%)	Cluster 4 (%)	Cluster 5 (%)
Right punch	0.0	0.0	0.0	100.0	0.0
Left punch	0.0	0.0	0.0	100.0	0.0
Front right kick	100.0	0.0	0.0	0.0	0.0
Side right kick	100.0	0.0	0.0	0.0	0.0
Side left kick	0.0	16.67	50.0	0.0	33.33

a front-right-kick 33.33 % of the time. These two actions are similar and originate from the same body part. These results lead us to believe that maybe more features could help distinguish these movements more clearly. Table 3 shows the results of clustering using also the angles of the knees and the elbows. Surprisingly the results are worst. The right-punch and the left-punch even when they are from different arms are labeled with the same cluster-label, the same happened to the front-right-kick and the side-right-kick.

This can be explained by the angle features becoming more relevant than the speed features. Given that angles are less discriminative of these movements, this results in more miss-classifications. When considering the amplitude of the movements of the lower body members the differences between a right-punch and a left-punch become minor. To prove this a simple experiment was performed using only the temporal segments originated by an action from the upper part of the body. Table 4 shows that

Table 4 Clustering results for all the sequences using only upper body actions

Action	Cluster 1 (%)	Cluster 2 (%)
Right punch	2.08	97.92
Left punch	91.67	8.33
Back fist	2.08	97.92
Elbow strike	16.67	83.33

Table 5 Clustering results for Sequence 1 using angles as a feature

Action	Cluster 1 (%)	Cluster 2 (%)	Cluster 3 (%)	Cluster 4 (%)	Cluster 5 (%)
Right punch	0.0	0.0	0.0	75.0	25.0
Left punch	0.0	0.0	0.0	60.0	100.0
Front right kick	50.0	0.0	25.0	0.0	25.0
Side right kick	58.33	0.0	41.67	0.0	0.0
Side left kick	0.0	25.0	50.0	0.0	25.0

using only the upper body k-means is perfectly capable of distinguishing the actions of the right arm from the actions of the left arm using the same features as in Table 3.

Table 5 shows the results using only the angles of the knees and elbows as features. In this case the kicking actions are diluted amongst several clusters. So using only the angles as features has proven insufficient to correctly label the actions.

Since a single value (average) is used to represent a temporal segment a loss in the granularity of information might be a problem. In the following experiments temporal segments were divided in equal parts to increase the feature vector, but the results were 30–40 % lower. This can be explained in Fig. 5 where we show a comparison between the same movement performed by two different subjects. The curves are similar but they start at different frames, which if we divide the temporal segment in four parts, for subject 10 the first two parts will have a higher value and for subject 1 the last two will have a higher value. For this reason these two actions would probably be assigned different clusters. Overall the results are very similar to all the other sequences that we have in our dataset (total of 8). Due to space limitations we were unable to include the clustering results for each sequence. Our final experiment (Table 6) was to see how well k-means coped with all the sequences at the same time using only the average speed as a feature since it was the feature that proved to have the best results. Again there is a clear separation from actions from the right and left side of the body. As for actions that are from the same part of the body there is room for improvement.

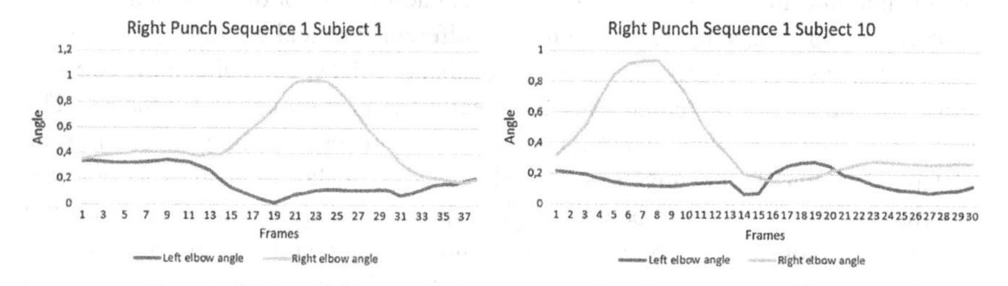

Fig. 5 Temporal segment of a right punch performed by subject 1 and subject 10

Table 6 Clustering results for all the sequences using speed as a feature

Action	Cluster 1 (%)	Cluster 2 (%)	Cluster 3 (%)	Cluster 4 (%)	Cluster 5 (%)	Cluster 6 (%)	Cluster 7 (%)	Cluster 8 (%)
Right punch	2.08	4.17	0.0	2.08	0.0	2.08	75.0	14.58
Left punch	0.0	0.0	2.78	0.0	88.89	2.78	2.78	2.78
Front right kick	0.0	0.0	93.75	0.0	0.0	4.17	0.0	2.08
Side right kick	5.00	0.0	50.00	0.0	0.0	45.00	0.0	0.0
Front left kick	5.56	0.0	2.78	86.11	0.0	0.0	0.0	5.56
Side left kick	36.11	0.0	0.0	38.89	1.39	2.78	2.78	18.06
Back fist	0.0	52.08	0.0	0.0	2.08	0.0	45.83	0.0
Elbow strike	8.33	16.67	0.0	0.0	25.0	0.0	33.33	16.67

5 Conclusion

In this paper, we described a new dataset of sequences of actions recorded with Kinect, which is, to the best of our knowledge, the first to contain whole sequences. We proposed a method to achieve automatic temporal segmentation of a sequence of actions trough a simple filtering approach. We also proposed and evaluated an automatic labeling method of human actions using a clustering algorithm. In summary, our results show that, for the type of actions used, k-means is capable of grouping identical actions performed by different users. This is evident when the clustering is performed with all of the subjects performing the same sequence of actions. When all the sequences are used the accuracy decreases. This might be explained by the effect that the neighboring actions have on the current action. So for different neighboring actions, the same current action will have a different start and ending.

By using several features (absolute speed, absolute 3D speed, joint angle) we also show that the choice of features affects greatly the performance of k-means. The poor results achieved when using the angles of the knees and elbows, appear to be related to how the flexion angles are calculated using the law of cosines. In our next experiment Euler angles will be used which represent a sequence of three elemental rotations (rotation about X, Y, Z). We also think that we could improve the results if we applied dynamic time warping to the temporal segments. This technique is often used to cope with the different speed with which the subjects perform the actions.

Our study showed how clustering and filtering techniques can be combined to achieve unsupervised labeling of human actions recorded by a camera with a depth sensor which tracks skeleton key-points.

References

1. Aggarwal, J.K., Ryoo, M.S.: Human activity analysis: a review. ACM Comput. Surv. **43**(3), 1–43 (2011)
2. Bobick, A.F., Wilson, A.D.: A state-based approach to the representation and recognition of gesture. IEEE Trans. Pattern Anal. Mach. Intell. **19**(12), 1325–1337 (1997)
3. Damen, D., Hogg, D.: Recognizing linked events: searching the space of feasible explanations. In: 2009 IEEE Conference on Computer Vision and Pattern Recognition, pp. 927–934 (2009). http://ieeexplore.ieee.org/lpdocs/epic03/wrapper.htm?arnumber=5206636
4. Gavrila, D.: The visual analysis of human movement: a survey. Comput. Vis. Image Underst. **73**(1), 82–98 (1999). http://www.sciencedirect.com/science/article/pii/S1077314298907160
5. Gowayyed, M.A., Torki, M., Hussein, M.E., El-Saban, M.: Histogram of oriented displacements (HOD): describing trajectories of human joints for action recognition. In: International Joint Conference on Artificial Intelligence, vol. 25, pp. 1351–1357 (2013)
6. Gupta, A., Srinivasan, P., Shi, J., Davis, L.S.: Understanding videos, constructing plots learning a visually grounded storyline model from annotated videos. In: IEEE Conference on Computer Vision and Pattern Recognition, 2009. CVPR 2009, pp. 2012–2019 (2009)
7. Hussein, M.E., Torki, M., Gowayyed, M.a., El-Saban, M.: Human action recognition using a temporal hierarchy of covariance descriptors on 3D joint locations. In: International Joint Conference on Artificial Intelligence pp. 2466–2472 (2013)
8. Intille, S.S., Bobick, A.F.: A framework for recognizing multi-agent action from visual evidence. In: Proceedings of the Sixteenth National Conference on Artificial Intelligence and the Eleventh Innovative Applications of Artificial Intelligence Conference Innovative Applications of Artificial Intelligence, vol. 489, pp. 518–525 (1999). http://dl.acm.org/citation.cfm?id=315149.315381
9. Keller, C.G., Dang, T., Fritz, H., Joos, A., Rabe, C., Gavrila, D.M.: Active pedestrian safety by automatic braking and evasive steering. IEEE Trans. Intell. Transp. Syst. **12**(4), 1292–1304 (2011). http://ieeexplore.ieee.org/lpdocs/epic03/wrapper.htm?arnumber=5936735
10. Koppula, H., Gupta, R., Saxena, A.: Learning human activities and object affordances from RGB-D videos. Int. J. Robot. Res. **32**(8), 951–970 (2013). arxiv:1210.1207v2
11. Nirjon, S., Greenwood, C., Torres, C., Zhou, S., Stankovic, J.a., Yoon, H.J., Ra, H.K., Basaran, C., Park, T., Son, S.H.: Kintense: A robust, accurate, real-time and evolving system for detecting aggressive actions from streaming 3D skeleton data. In: 2014 IEEE International Conference on Pervasive Computing and Communications, PerCom 2014 pp. 2–10 (2014). http://ieeexplore.ieee.org/lpdocs/epic03/wrapper.htm?arnumber=6813937
12. Niu, W., Long, J., Han, D., Wang, Y.F.: Human activity detection and recognition for video surveillance. In: 2004 IEEE International Conference on Multimedia and Exp (ICME), vols. 1-3. pp. 719–722 (2004)
13. O'Rourke, J., Badler, N.: Model-based image analysis of human motion using constraint propagation. IEEE Trans. Pattern Anal. Mach. Intell. **6**(6), 522–536 (1980). http://ieeexplore.ieee.org/lpdocs/epic03/wrapper.htm?arnumber=6447699
14. Pinhanez, C.S., Bobick, A.F.: Human action detection using pnf propagation of temporal constraints. In: IEEE Computer Society Conference on Computer Vision and Pattern Recognition, pp. 898–904. IEEE (1998)
15. Popa, M., Kemal Koc, A., Rothkrantz, L.J.M., Shan, C., Wiggers, P.: Kinect sensing of shopping related actions. Commun. Comput. Inf. Sci. 277 CCIS, 91–100 (2012)

16. Rashid, R.F.: Towards a system for the interpretation of moving light displays. IEEE Trans. Pattern Anal. Mach. Intell. **2**(6), 574–581 (1980). http://scholar.google.com/scholarhl=enŹbtnG=SearchŹq=intitle:Towards+a+System+for+the+Interpretation+of+Moving+Light+Displays#0
17. Ryoo, M.S., Aggarwal, J.K.: Semantic representation and recognition of continued and recursive human activities. Int. J. Comput. Vis. **82**(1), 1–24 (2009). http://link.springer.com/10.1007/s11263-008-0181-1
18. Starner, T., Weaver, J., Pentland, A.: Real-time american sign language recognition using desk and wearable computer based video. Trans. Pattern Anal. Mach. Intell. **20**(466), 1371–1375 (1998). http://ieeexplore.ieee.org/xpls/abs_all.jsp?arnumber=735811
19. Wolf, C., Mille, J., Lombardi, E., Celiktutan, O., Jiu, M., Dogan, E., Eren, G., Baccouche, M., Dellandrea, E., Bichot, C.E., Garcia, C., Sankur, B.: Evaluation of video activity localizations integrating quality and quantity measurements. Comput. Vis. Image Underst. **127**, 14–30 (2014). http://liris.cnrs.fr/Documents/Liris-5498.pdf
20. Yamato, J., Ohya, J., Ishii, K.: Recognizing human action in time-sequential images using hidden Markov model. Comput. Vis. Pattern Recognit. 379–385 (1992). http://ieeexplore.ieee.org/xpls/abs_all.jsp?arnumber=223161, http://ieeexplore.ieee.org/ielx2/418/5817/00223161.pdf?tp=Źarnumber=223161Źisnumber=5817, http://ieeexplore.ieee.org/xpl/articleDetails.jsp?tp=Źarnumber=223161ŹcontentType=Conference+Publication
21. Yu, E., Aggarwal, J.K.: Detection of fence climbing from monocular video. In: 18th international conference on pattern recognition, vol. 1, pp. 375–378 (2006). http://ieeexplore.ieee.org/lpdocs/epic03/wrapper.htm?arnumber=1698911
22. Zhou, F., Torre, F.D.L., Hodgins, J.: Hierarchical aligned cluster analysis (HACA) for temporal segmentation of human motion. IEEE Trans. Pattern Anal. Mach. Intell. **35**(3), 1–40 (2010). http://scholar.google.com/scholar?hl=enŹbtnG=SearchŹq=intitle:Hierarchical+Aligned+Cluster+Analysis+(+HACA+)+for+Temporal+Segmentation+of+Human+Motion#1

Smart Under-Sampling for the Detection of Rare Patterns in Unbalanced Datasets

Marco Vannucci and Valentina Colla

Abstract A novel resampling approach is presented which improves the performance of classifiers when coping with unbalanced datasets. The method selects the frequent samples, whose elimination from the training dataset is most beneficial, and automatically determines the optimal unbalance rate. The results achieved test datasets put into evidence the efficiency of the method, that allows a sensible increase of the rare patterns detection rate and an improvement of the classification performance.

1 Introduction

Many real world applications related to monitoring and control purposes cope with the automatic detection and classification of infrequent events. If machine learning approaches are adopted, datasets are required for their tuning phase and such datasets are named *unbalanced* (*or uneven*), as the infrequent events are scarcely represented. In this kind of applications, the rare patterns convey the most relevant information. For instance, in the medical field they represent diseases [1], in the industrial field they are related to faults [2] or defective products to be discarded or downgraded [3], in the commercial sector they can be associated to frauds in credit card transactions. Thus the correct detection of the rare patterns is of utmost importance and the misclassification of patterns related to frequent situations (the so–called *false alarms*) is strongly preferable than the missed detection of rare ones. In the context of binary classification problems, unbalanced datasets are characterized by the marked preponderance of samples belonging to the frequent class (C_F) with respect to the rare class (C_R). The predictive performance of standard classifiers when dealing with such problems is heavily affected by the class imbalance issues, as they

M. Vannucci (✉) · V. Colla
TeCIP Institute, Scuola Superiore Sant'Anna, Via G. Moruzzi, 1, 56124 Pisa, Italy
e-mail: mvannucci@sssup.it

V. Colla
e-mail: colla@sssup.it

I. Czarnowski et al. (eds.), *Intelligent Decision Technologies 2016*,
Smart Innovation, Systems and Technologies 56,
DOI 10.1007/978-3-319-39630-9_33

are conceived to maximize the overall classification performance [4]. In practice, very rarely there is a clear spatial distinction among patterns belonging C_R and C_F, thus class imbalance prevents the effective characterization and detection of the rare patterns belonging to C_R and makes the separation of the classes difficult for any classifier. In [5, 6] it was shown that the detrimental effect of class imbalance affects any kind of standard classifier. Despite this intrinsic criticality, due to the importance and generality of the issue, many methods have been developed to detect rare patterns in unbalanced datasets.

In this paper a novel approach to the binary classification of unbalanced dataset is proposed, based on a selective and optimized reduction of the frequent samples within training dataset.

The paper is organized as follows: in Sect. 2 the most widely adopted techniques for handling class imbalance within classification tasks are briefly analysed by paying particular attention to the methods based on the reduction of the unbalance rate. In Sect. 3 the proposed method is described. The experimental set-up and test results related to the evaluation of the methods are reported in Sect. 4. Finally in Sect. 5 conclusions are drawn and future perspectives of the proposed approach are discussed.

2 Review of Existing Approaches

Traditional classification methods fail when facing unbalanced datasets mainly because the misclassification of rare samples is for them as penalizing as the misclassification of frequent ones. As a consequence, the classifiers tend to focus on most probable events since neglecting the rare ones does not strongly affect the overall performance.

In literature it is possible to find a high number of different methods for contrasting this behaviour that can be grouped into two major categories according to *the way* they operate: *external* and *internal* approaches. The internal approaches are based on the development or improvement of existing classifiers in order to face uneven datasets; thus they exploit the *original* unbalanced dataset for training the classifier. In contrast, the external methods are based on a modification of the training dataset which reduces the unbalance rate in order to mitigate its detrimental effects: the *modified* dataset is then used to train standard classifiers.

The basic idea behind **internal methods** is the promotion of the correct detection of rare patterns despite the generation of some false positive. A widely used class of techniques that implements in a direct manner this idea is based on the different weighting of misclassification errors [7] and emphasize the cost of the missed detection of rare patterns during the learning phase of the classifier through a suitable cost matrix which compensates the class imbalance [8]. This approach can be coupled to standard classifiers such as Artificial Neural Networks (ANNs) or Decision Trees (DTs). For instance, the LASCUS method [9] combines the cost matrix approach to a self-organizing map and a fuzzy inference system obtaining very interesting results.

Internal methods have been developed as adaptations of many existing classifiers. For instance, Rectangular Basis Functions (RecBF) networks are particular kind of radial basis function networks proposed in [10], which are characterized by neurons with hyper-rectangular activation functions in the hidden layer allowing a higher precision in the detection of the boundary of the input space regions of each class.

Support Vector Machines (SVMs) have been widely used as a basis for designing classifiers suitable to uneven datasets. For instance in [11] an ensemble of SVMs (one for each class) is proposed, while a variation of SVM, the v-SVM introduced in [12], is used for recognizing patterns belonging to a specific class within a dataset, as information on a single class is used for its training. Due to its peculiarity, this latter method can be exploited to detect whether a pattern belongs to the rare ones.

ANN-based approaches have been adopted as well: the TANN (Thresholded ANN) method has achieved interesting results in fault detection and product quality assessment [13]. The TANN is based on a two-layers feed–forward ANN with one single output neuron coupled to a *threshold operator* that activates this latter neuron when its natural activation overcomes a threshold τ automatically determined so as to improve the sensitivity of the system to rare patterns.

2.1 Focus on External Methods

The basic idea of external methods consists in reducing the training dataset unbalance rate by operating a so–called *data resampling* in order to overcome the classification problems faced by standard methods.

The resampling can be achieved in a twofold manner:

- **under-sampling**, which rebalances the class distribution by removing samples belonging to the majority classes from the training dataset until the desired unbalance ratio is reached;
- **over-sampling**, operates in the opposite direction, by adding samples belonging to the minority class to the dataset.

Usually resampling techniques do not completely rebalance the classes by matching the number of their samples within the training dataset but rather increase the C_R frequency until a predetermined ratio. Unfortunately a fix unbalance ratio which optimizes the performance of any classifier does not exist: it depends on the problem and original dataset characteristics and, in practice, must be experimentally determined.

Under-sampling techniques operate by removing frequent samples from the training dataset. The basic approach of randomly removing C_F samples, although efficient in some cases, is risky since it may involve the removal of samples with high informative content. More sophisticated techniques have been developed for the focused removal of frequent samples by selecting only those located in the extreme regions of the input space, in order to reduce the size of input space areas which the classifier would assign to frequent patterns and to create a more specific concept of the corresponding class. These techniques are discussed and tested in [14] achieving sufficiently good results but suffering from the uncertainty related to the optimal unbalance rate.

Oversampling methods add samples of C_R to the training dataset. Also in this case, the straightforward approach that randomly duplicates existing samples can be risky: for instance, the creation of very compact *clusters* of minority class samples can even reduce the area of the input space associated to C_R. A smarter approach consists in the selective replication of samples lying on the boundary zones of the input space between C_F and C_R, in order to spread the regions that the classifier associates to the C_R and to limit eventual classification conflicts that standard classifiers would solve in favour of C_F [4]. Among the oversampling approaches it is worth to note the SMOTE algorithm, proposed in [15]. Its main element of novelty lies in the fact that the additional C_R samples are different from existing ones as they are synthetically created. In facts, an undesired effect of the replication of existing patters is the reduction and over-specification of the input space areas devoted to C_R. Replication does not add any further information to the dataset but just overbalances the decision metrics of the classifier. On the contrary, the creation of new infrequent samples broadens the zones of the input space associated to them. SMOTE synthetically creates minority samples where they *likely* could be; in particular, new minority samples are located along the lines connecting existing minority samples. Depending on the number of synthetic samples required to rebalance the dataset, different numbers of couples of rare patterns are selected for the generation of new samples.

Both undersampling and oversampling techniques are very used in practice and it is not possible to identify the best ones: their performances are strictly related to the characteristics of the problems, classifiers and datasets. A combination of the two approaches has been attempted in [16, 17] where focused undersampling and oversampling are performed.

3 Smart Under-Sampling for Unbalanced Datasets

A novel approach named *Smart Undersampling* (SU) belonging to the family of external techniques is proposed here. The method performs a selective undersampling of the training dataset and is specifically designed to handle unbalanced dataset where the detection of rare patterns is the key target whilst the generation of some false alarm is acceptable.

The basic idea behind this approach is the selection and elimination of those C_F samples whose presence within the training dataset mostly affects the correct characterization of C_R patterns. Multiple criteria are taken into consideration and an *optimal* resampling rate associated to each of them is determined by means of Genetic Algorithms (GAs). In the following paragraphs the operation of the proposed method is described in detail. The variables within the dataset are assumed to lie in the range [0; 1]. This latter assumption does not compromise the validity of the proposed method. Moreover, the original dataset DS is partitioned (keeping constant the unbalance rate) into a training D_{TR}, validation D_{VD} and a test D_{TS} datasets which are exploited as usual. The division quotes are discussed in Sect. 4 together with the experimental setup.

All the samples belonging to the frequent class C_F within the training dataset are evaluated according to a set of criteria that rate them through a score $s \in [0, 1]$ that assesses their impact on the classifier performance: the higher s, the more detrimental the sample. For each sample $x \in C_F$ the following values are calculated:

1. **Distance to Closest Rare sample (DCR)** is the distance between x and the closest rare pattern in the dataset and is computed as:

$$DCR(x) = 1 - \min_{p \in C_R} \|x - p\| \tag{1}$$

   where C_R represents the set of frequent patterns within the training dataset and the $\|\|$ operator represents the Euclidean distance. This value is higher (i.e. closer to 1) for those frequent patterns that are closest to rare ones thus their pruning contributes to the spreading of the input space associated to the interested rare pattern.
2. **Distance to Closest Frequent sample (DFR)** is the minimum distance between x and the closest frequent sample calculated as

$$DFR(x) = \min_{p \in C_F} \|x - p\| \tag{2}$$

   This value is lower for those frequent patterns that are close to other C_F samples. The pruning of one of them reduces the redundancy within the samples in C_F without loosing informative content and rebalances the dataset.
3. **Average Distance to Rare samples (ADR)** is the average distance of x from the rare samples within the dataset, calculated as

$$ADR(x) = 1 - \underset{p \in C_R}{\text{average}}(\|x - p\|) \tag{3}$$

   that is closer to 1 for those frequent samples whose neighbourhood includes a higher number of infrequent samples.

Once calculated for each sample in C_F, these indexes are sorted in a decreasing order and used to create three distinct rankings of the C_F samples. At the top of these rankings the samples whose removal is most fruitful are located. The proposed method determines three removal percentages R_{DCR}, R_{DFR} and R_{ADR} associated to the rankings related to DCR, DFR and ADR respectively. These percentages are used to select the frequent patterns to be pruned from the training dataset. More in detail, each one represents the rate of samples, picked from the associated ranking, to be removed from the dataset according to their rating (i.e. starting from the top of the ranking) and managing eventual intersections. For instance, in the case $R_{DCR} = 10\,\%$, the 10 % frequent samples characterized by the highest DCR values will be selected for pruning.

Optimal R_{DCR}, R_{DFR} and R_{ADR} are determined through a GAs–based optimization in order to maximize the resampling benefits for the classification of unbalanced datasets. Within this work the problem is set up as a minimization task. The fitness function of the GAs engine, given an arbitrary candidate solution $[R_{DCR}, R_{DFR}, R_{ADR}]$ works as follows:

1. the resampled training dataset $\widehat{D_{TR}}$ is built according to the candidate solution;
2. a DT classifier is trained by means of the C4.5 algorithm [18] exploiting $\widehat{D_{TR}}$;
3. the performance of the trained classifier is evaluated on both the not–resampled training dataset D_{TR} and the validation dataset D_{VD} according to the following expression designed in [13] to this aim:

$$E = 1 - \frac{\gamma TPR - FPR}{ACC + \mu FPR} \tag{4}$$

where *TPR* represents the True Positive Rate, *FPR* the False Positive Rate (commonly known as *false alarms*), *ACC* the overall classifications accuracy while γ and μ are two empirical parameters. The calculated value E varies in the range [0, 1] and is close to 0 for well performing classifiers. Equation (4) is used for the calculation of the performance E_{TR} on D_{TR} and E_{VD} on D_{VD}.

4 The final fitness for the candidate is calculated as

$$Fitness = E_{TR} + |E_{TR} - E_{VD}| \tag{5}$$

where the $|E_{TR} - E_{VD}|$ term is added to take into account the effect of overfitting, since it is a measure of the discrepancy between the performance of the classifier on training and validation data.

Real–coded GAs have been employed for the determination of optimal removal rates. Candidate solutions are coded as three elements vectors where each element varies in the range [0, 100 %]. The set of candidate solutions—whose cardinality is a parameter of the method—is initialised at random and evolved through the generations by means of the *crossover* and *mutation* operators. Crossover is implemented as the so–called *single point crossover* while mutation randomly varies in a range [−5, +5 %] one random element within selected chromosomes. Selection operator is based on the well–known *roulette wheel* technique while the terminal condition is satisfied when an arbitrary number of generations of GAs has been completed or a stall condition is reached. At the end of the optimization process the GAs returns the optimal values $[R^*_{DCR}, R^*_{DFR}, R^*_{ADR}]$ that are employed to resample the dataset D_{TR} and generate $\widehat{D_{TR}}$ that are actually used for the classifier training. Currently the method is based on the use of a DT as classifier but this choice does not affect the validity of the method neither prevents the selection of a different type of classifier.

4 Test of the Proposed Approach

The SU method has been tested on several unbalanced datasets coming from the UCI dataset repository [19] and from industrial frameworks. The main characteristics of the employed datasets are summarized in Table 1 in terms of unbalance rate, number of samples and input variables. A description of the dataset drawn from the UCI repository can be found in [19] while the industrial ones are briefly described in the following.

Occlusion Refers to the problem of the detection of the occlusions of casting nozzles during steel manufacturing from liquid steel characteristics and casting process measured parameters. Nozzles occlusion is highly detrimental and its detection is fundamental. Within the employed datasets 1.2 % of the observations belong to occlusion situations.

Metal sheet quality (MSQ1, MSQ2) Concerns the automatic grading of metal sheets quality on the basis of the information provided by a set of sensors that inspect sheets surface. The grading, according to the number and type and characteristics of the reported defects, determines whether a product can be put into market or not. The two datasets used for testing purpose in this work refer to two different processes.

As already mentioned, data are normalized and divided into three groups keeping the original unbalance rate: training (60 %), validation (20 %) and test (20 %) datasets. The number of candidate solutions evaluated within each generation of the GAs engine employed within SU is 100. The terminal condition allows a maximum of 50 generations if convergence is not reached. The results achieved by SU are compared to those obtained by other widely used resampling approaches: random oversampling and undersampling and SMOTE. The results achieved by the standard classifier when no resampling is performed are also presented. In all these cases a DT trained by means of the C4.5 algorithm has been used as classifier. Since standard resampling methods do not automatically determine the optimal unbalance rate, different rates (i.e. spanning from 5 to 50 %) have been tested for these approaches and the best results are reported. The results achieved are reported in Table 2: for each method the ACC, TPR, FPR are shown. The optimal unbalance rate of the training dataset is also reported: for SU it is automatically determined while in the other cases it is the value associated to the best classification performance.

Table 1 Main characteristics of the original datasets used in the tests

Source	Dataset	Unb. rate (%)	Samples	Vars
UCI rep.	CARDATA	3.8	1728	6
	NURSERY	2.5	1296	8
	SATELLITE	9.7	6435	36
Industrial	OCCLUSION	1.2	3756	6
	MSQ-1	24	1915	11
	MSQ-2	0.3	21784	9

Table 2 Results achieved on the datasets coming from both the UCI repository and some industrial cases study

Source	Database	Method	Unb. rate (%)	ACC (%)	TPR (%)	FPR (%)
UCI repository	CARDATA	No resampling	–	98	79	1
		Rand. Unders.	8	98	92	2
		SU	8	98	99	2
		Rand. Overs.	10	99	91	1
		SMOTE	25	99	81	1
	NURSERY	No resampling	–	99	83	1
		Rand. Unders.	10	97	97	3
		SU	4	97	99	3
		Rand. Overs.	25	99	93	1
		SMOTE	10	99	84	0
	SATELLITE	No resampling	–	91	55	5
		Rand. Unders.	50	82	80	18
		SU	14	87	77	12
		Rand. Overs.	25	91	54	5
		SMOTE	45	90	62	7
Industrial field	OCCLUSION	No resampling	–	98	5	2
		Rand. Unders.	20	90	52	10
		SU	8	88	67	11
		Rand. Overs.	10	99	11	1
		SMOTE	25	96	0	4
	MSQ-1	No resampling	–	86	65	7
		Rand. Unders.	30	83	68	12
		SU	26	84	73	13
		Rand. Overs.	25	85	67	9
		SMOTE	50	84	67	11
	MSQ-2	No resampling	–	99	0	0
		Rand. Unders.	2	99	4	1
		SU	6	90	46	10
		Rand. Overs.	20	99	13	1
		SMOTE	5	99	10	1

The analysis of the performance of the tested methods takes into account the issues related to unbalance datasets: the detection of rare patterns (i.e. defects, malfunctions) is privileged rather that the overall performance and the generation of *false positives* (i.e. false alarms) is tolerated. The performance of SU on UCI datasets is satisfactory, since in the CARDATA it drastically improves the TPR (+7 %) and keeps FPR constant, while for the SATELLITE dataset the TPR improvement is slighter less significant (+2 %) with respect to the random undersampling but much more marked with respect to other methods. In the case of SATELLITE dataset, SU reaches a different trade–off between TPR, FPR and ACC through the evaluation function of Eq. (4) for which TPR, although satisfactory (77 %), is slightly lower (−2 %) with respect to random undersampling. However the FPR is significantly reduced (−6 %). Also when processing the industrial datasets SU outperforms the other resampling methods. In the OCCLUSION dataset the TPR is improved by 15 % with respect to the best competitor, while the FPR is kept satisfactorily low. Further, in this particular industrial application, the detection of an occlusion is fundamental since it results in a significant time and resources saving whilst a false alarm just leads to the application of simple countermeasures. SU achieves similar results on the MSQ-1 dataset where the TPR improvement is sensible (+5 %) with a reasonable increase of FPR (+1 %). Also in this case the detection of defective products is fundamental from an industrial point of view. Finally the strongest improvement is achieved in the MSQ-2 dataset where, due to the extreme unbalance of the original dataset, other methods totally fail while SU is able to detect an appreciable rate of rare patterns.

Globally the performance of SU is very satisfactory since in most cases it improves the TPR while keeping FPR and ACC under control. The tests show that SU is mostly beneficial when handling datasets characterized by high unbalance rates (the case of MSQ-2 is emblematic). The optimal unbalance rate determined by means of SU is lower with respect to the one that performs better when coupled to the other methods. It is a consequence of the efficient selection carried out by SU on the samples to prune from the original data. In effect, only those samples whose elimination is most fruitful are deleted, by maximizing the performance even without purging a large amount of frequent samples and by consequence limiting the information content loss.

5 Conclusions

In this paper a novel resampling method for the classification of unbalanced dataset was presented. The method performs a smart undersampling by automatically determining the optimal unbalance rate of the training dataset and by selecting, among the frequent patterns, those ones whose elimination maximizes the classifier performance according to multiple criteria related to samples distribution. The method has been tested on several dataset coming both from literature and the industrial field and affected by problems related to unbalance. The performance of the proposed method on these datasets is encouraging since it sensibly improves classifiers performance by increasing the detection of rare patterns without affecting the performance in terms of false positives and overall accuracy. Moreover the automatic setting of the optimal unbalance rate results particularly advantageous. The interesting performance of the method and its flexibility encourage future developments: in particular future research will focus on the combination of SU with a suitable oversampling approach and on its integration with internal methods.

References

1. Stepenosky, N., Polikar, R., Kounios, J., Clark, C.: Ensemble techniques with weighted combination rules for early diagnosis of Alzheimer's disease. In: International Joint Conference on Neural Networks, IJCNN'06 (2006)
2. Shreekant, G., Bin Y., Meckl, P.: Fault detection for nonlinear systems in presence of input unmodeled dynamics. In: International Conference on Advanced Intelligent Mechatronics, pp. 1-5, IEEE/ASME (2007)
3. Borselli, A., Colla, V., Vannucci, M., Veroli, M: A fuzzy inference system applied to defect detection in flat steel production. In: 2010 World Congress on Computational Intelligence, Barcelona (Spain), 18–23 July 2010, pp. 148-153 (2010)
4. Estabrooks, A.: A combination scheme for inductive learning from imbalanced datasets. MSC Thesis. Faculty of Computer Science, Dalhouise University (2000)
5. Estabrooks, A., Japkowicz, N.: A multiple resampling method for learning from imbalanced datasets. Comput. Intell. **20**(1), 18–36 (2004)
6. Japkowicz, N.: The class imbalance problem: significance and strategies In: International Conference on Artificial Intelligence, Las Vegas, Nevada pp. 111–117 (2000)
7. Pazzani, M., Marz, C., Murphy, P., Ali, K., Hume, T., Brunk, C.: Reducing misclassification cost. In: 11th International Conference on Machine Learning, pp. 217–225 (1994)
8. Elkan, C.: The foundations of cost–sensitive learning. In: 17th International Joint Conference on Artificial Intelligence, pp. 973–978. Morgan Kaufmann Publishers Inc., San Francisco (2001)
9. Vannucci, M., Colla, V.: Novel classification method for sensitive problems and uneven datasets based on neural networks and fuzzy logic. Appl. Soft Comput. **11**(2), 2383–2390 (2011)
10. Soler, V., Prim, M.: Rectangular basis functions applied to imbalanced datasets. Lecture Notes in Computer Science, vol. 4668. pp. 511–519. Springer (2007)
11. Li, P., Chan, K.L., Fang, W.: Hybrid kernel machine ensemble for imbalanced data sets. In: 18th International Conference on Pattern Recognition. IEEE (2006)
12. Scholkopf, B., et al.: New support vector algorithms. Neural Comput. **12**, 1207–1245 (2000)
13. Vannucci, M., Colla, V., Sgarbi, M., Toscanelli, O.: Thresholded neural networks for sensitive industrial classification tasks. Lecture Notes in Computer Science vol. 5517 LNCS, pp. 1320-1327 (2009)
14. Chawla, N.V.: C4.5 and imbalanced data sets: investigating the effect of sampling method, probabilistic estimate, and decision tree structure. In: Workshop on Learning from Imbalanced Dataset II, ICML, Washington DC (2003)
15. Chawla, N.V., Bowyer, K.W., Hall, L.O., Kegelmeyer, W.P.: SMOTE: synthetic minority oversampling technique. J. Artif. Intell. Res. **16**, 321–357 (2002)
16. Ling, C., Li, C.: Data mining for direct marketing problems and solutions. In: Fourth International Conference on Knowledge Discovery and Data Mining, New York, vol. 2, pp. 73–78 (1998)
17. Cateni, S., Colla, V., Vannucci, M.: A method for resampling imbalanced datasets in binary classification tasks for real-world problems. Neurocomputing **135**, 32–41 (2014)
18. Quinlan, J.R.: C4.5: Programs for Machine Learning. Morgan Kaufmann, San Francisco (1993)
19. Lichman, M.: UCI ML Repository. University of California, School of Information and Computer Science, Irvine, CA (2013). http://archive.ics.uci.edu/ml

Personal Recommendation System for Improving Sleep Quality

Patrick Datko, Wilhelm Daniel Scherz, Oana Ramona Velicu, Ralf Seepold and Natividad Martínez Madrid

Abstract Sleep is an important aspect in life of every human being. The average sleep duration for an adult is approximately 7 h per day. Sleep is necessary to regenerate physical and psychological state of a human. A bad sleep quality has a major impact on the health status and can lead to different diseases. In this paper an approach will be presented, which uses a long-term monitoring of vital data gathered by a body sensor during the day and the night supported by mobile application connected to an analyzing system, to estimate sleep quality of its user as well as give recommendations to improve it in real-time. Actimetry and historical data will be used to improve the individual recommendations, based on common techniques used in the area of machine learning and big data analysis.

1 Introduction

Health care is becoming an upcoming topic in modern societies, with the basic goal to improve living quality by increasing personal state of health. By the use of modern mobile devices (e.g. Smartphones, Smart Watches, and sensor bracelets), companies

P. Datko · W.D. Scherz · O.R. Velicu · R. Seepold (✉)
HTWG Konstanz, Ubiquitous Computing Lab, 78462 Konstanz, Baden-Württemberg, Germany
e-mail: Ralf.Seepold@htwg-konstanz.de

P. Datko
e-mail: Patrick.Datko@htwg-konstanz.de

W.D. Scherz
e-mail: wscherz@htwg-konstanz.de

O.R. Velicu
e-mail: OanaRamona.Velicu@htwg-konstanz.de

N.M. Madrid
Reutlingen University, Internet of Things Lab, 72762 Reutlingen, Baden-Württemberg, Germany
e-mail: natividad.martinez@reutlingen-university.de

I. Czarnowski et al. (eds.), *Intelligent Decision Technologies 2016*,
Smart Innovation, Systems and Technologies 56,
DOI 10.1007/978-3-319-39630-9_34

such as Google, Apple, Jawbone, Polar and others track different kind of information such as users steps, movement, heart rate, etc., analyse them and notify the user about its current health state as well as provide recommendations on how to improve. Sleeping has a huge influence of personals health care. Poor sleep quality as well as short of sleep can lead to psychological and physical health disorders such as depressions, cardiovascular disease, diabetes and others. Therefore, it is necessary to improve personal sleeping quality with the help of modern technologies to provide a healthier living quality.

Recommendations and predictions of sleeping quality, with common machine learning techniques, are based on a huge amount of personal data, represented by concrete vital data such as heart rate as well as contextual data like the current mood, activity and so on. There it is necessary to build a modern software system based on a new kind of architectural paradigm, which combines aspects such as distributed big data analysis, real-time recommendations/feedback as well as users privacy and mobility.

In the following section, related works and approaches in the area of sleep detection and sleep quality analysis will be shown. In Sect. 3 the evaluation of different body sensors for measuring vital data will be presented. The basic system as well as the Information Processing Model for pre-processing data will be explained in Sect. 4. Concluding an outlook will be given on the future work according to a short conclusion.

2 State of the Art

Research as well as industry are very interested in the area of sleep analysis and detection. Different papers were published addressing this area. In [1] a method was introduced to estimate the quality of sleep based on the so called *Sleep Index*. The authors compared the seconds spent in each sleep stage of health and users with obstructive sleep apnea (OSA). As a result they showed, that OSA users spend more time in the beginning sleep stages (1 and 2) but miss almost one stage (4). But the results of the paper only represent the difference between normal healthy users and persons with an obstructive sleep apnea and is not focusing concretely on the sleep quality. Another approach is described in [4], where the authors try to estimate sleep quality based on acoustic signal processing. They use the ratio between the different sleep stages to estimate the sleep quality. Such as [1], this approach is useful for persons with sleep apnea, not for healthy users. In [3], the sleep quality calculation is based on the ratio between time spend in different sleep stages, estimated by the movement of the user during the sleep. The calculation is based on fixed rules, which are not adapted for individual's sleeping behavior. A similar approach is shown in [5], where the authors use a wristband to detect movements during the sleep. For the sleep quality estimation they use a similar approach, the ratio between deep sleep duration and total sleep duration. To achieve a more refined result, they additionally try to estimate the sleeping latency, the period of time where the user is in bed but not sleeping. The same principle is used by the authors in [6], instead of a movement body sensor they used video observation and passive infrared to detect movements. As an additionally parameter they used a heart rate monitor to estimate the different stages. As the previous result, the sleep quality is based on the ratio between deep sleep duration and total sleep duration. Based on the same calculation the authors in [7] use heart-rate variability for the detection of the sleep stage and in [2] ECG is used. Due the fact the estimation of sleep qualitity is based on the ratio between the different sleep stages, neither of the discussed approaches take care of additional influences. They are not using historical information, such as records of previous sleep analysis, as well as the actimetry during the day. Furthermore, all explained approaches are only giving a feedback about the sleep qualitiy and do not recommend, for example physical exercises, the user that might lead to an improvement.

3 Methodology

The basic concept of the proposed approach is based on two cyclical repetitive parts. Alternating the user will be observed with its current vital data during the day, as well as the sleeping behavior during the night. Collected information can be seen as the data input used for analyzing data, finding patterns, classify behaviors for the final evaluation and recommendation. Figure 1 represents the project's methodology. User's vital data, such as heart rate, ECG or activity level, will be monitored 24 h during the day, either by invasive or noninvasive sensors.

The basic goal is to compare a user's behavior represented by its vital information, with its sleep quality during the night. Daily activities have an impact on the sleeping behavior and therefore is the most important input parameter, which has to be observed, to increase the sleeping quality of a person. Because of the alternating behavior of this method, all observations will be used in the analyzing process to give a direct feedback to the user with a recommendation to change his behavior in some way (e.g. try to rest now for an half hour), and therefore, back in the observation phase, try to compare if the estimated recommendation has a positive influence on the current state of the user and corresponding on the sleep quality. Because all individuals are different, same recommendations, even for persons with a congruent vital classification (e.g. same age, sex, weight), can lead to complete interracial results in

Fig. 1 Method principle

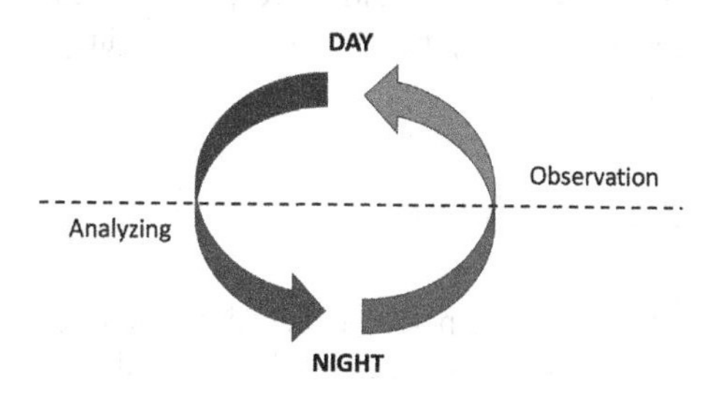

the resulting change of health state. This is related on multiple factor's influencing the individual, such as work conditions, private issues, finances, current/former diseases and others. Therefore the whole system, will be based on a self-learning system, adapting detailed recommendations to the individual's circumstances based on abstract recommendations provided by a rule-based engine targeting the classification domain of its user (e.g. class for male, 20–30 years, regular weight). The procedure of alternating monitoring and analyzing periods are basically inspired by the pulse-width modulation (PWM) with the goal to periodically change the input of a system to close as possible to fit the desired goal.

3.1 Observation

The observation phase is the starting point of the procedure circle. Here, all necessary vital information will be collected for the later analysis and recommendations to the user. The basic intention is to improve sleeping quality of the user. Therefore, gathered vital data are chosen, which have directly influence such as the body movement (accelerometer) respectively indirectly but representative influence, such as ECG for stress estimation, on the sleeping behavior. For the analyzing, user's heart rate, accelerometer, ECG and temperature will be collected as raw input data. Furthermore, aggregated information will be used such as Activity level (based on accelerometer) and stress level (based on ECG and temperature). During the day, all this data will be gathered by a body sensor, in this case a chest strap, monitoring vital data of its user and provide them in real time, thru on standardized wireless information exchange protocol. This solution is a trade-off between data gathering possibilities and minor limitation in everyday life of its user.

During the the sleep period, a chest strap could be uncomfortable and disturbing to wear for some users. Because of movements during sleep the sensor can slip resulting to inappropriate data. Therefore another solution will be designed, focused on the direct environment of the user. An own delevoped intelligent bed will be used for gathering all necessary information, as previously described. A sensor equipped slatted frame is used for measuring the movement of the user as well as the heart rate based on acceleration. Body's temperature is estimated by sensor stripes on the mattress. All gathered information are exchanged with the same protocols as for the body sensor. The main benefit solution of the intelligent bed, the user has a complete free moving space and no responsibilities of configuration or manually controlling the monitoring, the environment is managing itself on its own.

3.2 Analyzing

The analyzing period is based on the previously mentioned observation phase. All collected vital data either by the body sensor or the intelligent bed are used for a deeper analysis. Based on the analysis, later recommendations for the user will be determined to improve sleeping quality. For initial calibration user's context information will be assigned, such as age, gender, wheight, height and current athleticism level. This information will be used for the individual's personal classification. The analyzing phase is context-based, thus it has different functions. During the day, all gathered information will be used to determine the current vital status and provide recommendations to improve sleep quality. During the night, the analyzing phase is used to estimate the sleep quality. Though, collected vital data will be used to determine each sleep stage, before the sleep qualitiy is calculated based on a rule set. The estimated sleeping quality can be seen as a reference value for the observed activities during the day.

The whole method is based on a self-learning aspect, therefore the recommendations will lead to better results if it is used for a long period. A determined recommendation is based on three parts: history data, rule-based engine, individual recommender. Historical data are important as a base for future recommendations. This data can be seen as a "log", where all recommendations and vital data, as well as sleep quality estimations are stored. Pattern can be derived, which result in finding recommendations having a positive influence on the sleeping quality. Using only historical data, can be leading to useless over time, because it is possible that the user is adapting to some recommendations and they have no influence anymore because they often occurred. Therefore a rule-based engine and a individiual recommender are added. The rule-based engine is a library using common rules and patterns improving health state for a special grouped person for example a male user between 20–30 year and a higher athleticism level, needs more recommendations with sporting activities, than another male with the same age but a lower athleticism level. Recommendations only using history & rule-based information can also lead to inappropriate results, because they are not taking into account the current status of the individiual, e.g. a sudden occurred health issue such as a broken leg. Therefore an inidvidiual recommender is necessary, trying to monitor the current status, give small recommendations and try to estimate the current influence and therefore try to adjust further recommendations to approximate the best individiual's behavior to improve sleeping quality.

4 System Overview

The basic system is based on a client-server architecture. The server part is represented by a backbone accessible via internet, which handles the major data analysis with common machine learning techniques. The client part is composed of a body sensor, which gather vital data of its user and an Android-based smartphone, which requests collected information of the sensor via the standardized Bluetooth LE interface. The abstract architecture is shown in Fig. 2. The data exchange between client smartphone and backend is based on common mobile internet protocols such as Global System or Mobile Communications (GSM), Universal Mobile

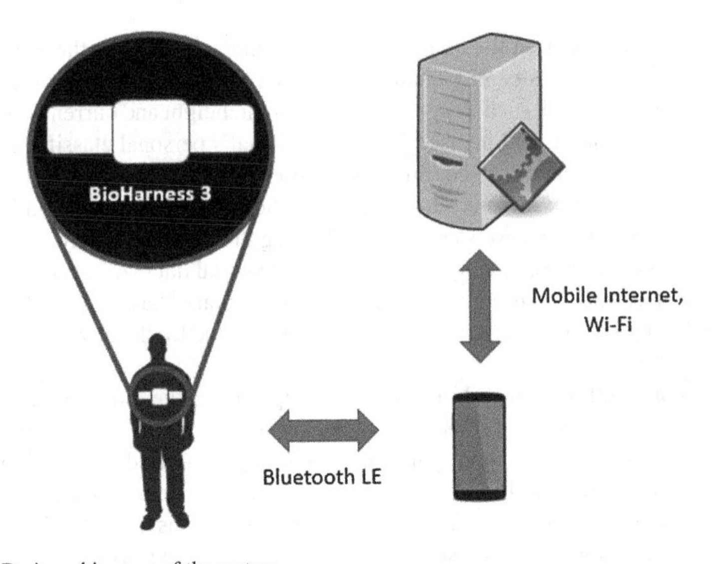

Fig. 2 Basic architecture of the system

Telecommunications System (UMTS) or Long Term Evolution (LTE) as well as on centralized internet possibilities such as Wi-Fi.

A major aspect of the system is the real-time functionality that calculates results by the backend (e.g. recommendations) and that should deliver as a live feedback. Therefore the connection between client smartphone and the backend is based on a socket communication, which provides a constant information exchange connection.

Besides the collection of sensor data, an Android smartphone will be used for pre-processing of received information on the one hand to reduce computing power on the backend side and on the other hand to filter information which are not relevant for further analysis or it can be computed directly on the smartphone. Furthermore, the client uses pre-processed and filtered information for basic computation directly on the device to provide feedback for the user even, if there is no mobile or Wi-Fi internet connection.

Information Processing Model

The Information Processing Model (IPM) is the basic core of the Android application on the client side. The main functionality of the IPM is the pre-processing and analyzing of received data from the sensor via Bluetooth LE, to achieve a basic feedback functionality, even without any internet connection as well as reduce computation resources on the backend side. Therefore the IPM defines a pipeline of steps of procedures to aggregate data for the given needs, which is shown in Fig. 3. The stages in the IPM, which will be processed successively, so data will be forwarded from one stage to another. At first, received raw data will be filtered based on common techniques (e.g. Kalman filter), to reduce noise in the data set. After that filtered information will be used for pre-processing and data compression as for example

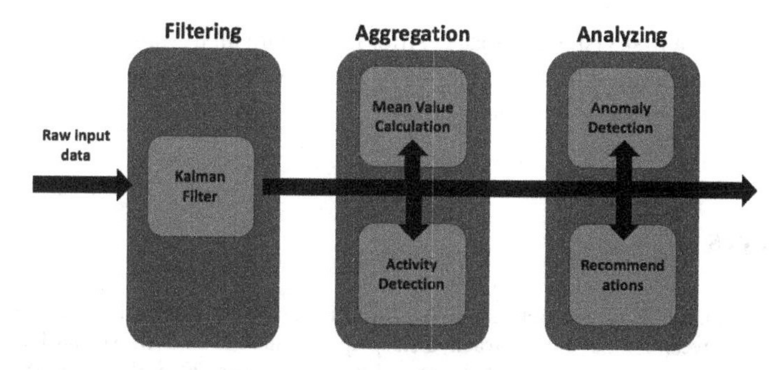

Fig. 3 Information processing model (IPM)

current speed will be used to estimate the mean speed per hour of its user. Finally, in the analyzing stage, pre-processed data will be used for real-time and stationary (client-side) calculations to give a direct feedback (e.g. current heart rate mean value) as well as basic recommendations (e.g. "You moved only 100 steps in the last hour. You should go for a walk.") to the user.

The functionality within the different processing stages are build with an abstract defined interface. Based on this technique, it is possible to reuse components and attach them among each other to estimate different combined results. For example a "Mean Value Calculation" component can be used on the one hand to calculate the mean heart rate in the last hour and on the other hand to calculate the mean steps made by the user. The outcome of both calculations can be combined in another component, which tries to detect a correleation between mean heart rate and mean steps. Thus, it is possible to define pipes between components to calculate and determine different results, depending on the needed use cases.

5 Conclusion

Sleep has an important influence of an individual's health status. A poor quality of sleep as well as sleep disorders can lead to diseases such as diabetes, cancer, overweight, high blood pressure and others. Therefore it is important improve the sleep quality based on modern and available technologies integrated in the everyday life of its users. In this paper, a system will be proposed that is an approach targeting the area of improving the quality of sleep by monitoring vital paramaters of its user during a whole day, show its current status as well as give recommendations. The basic requirement is to provide an advisor for everybody with a simple use, so that the user will disordered or handicapped. After a basic investigation in the domain of sleeping, an evaluation of professional body sensors in the market have been done. After the selection, the basic implementation of the connection between the body sensor and an Android application gathering provided information, is currently in progress. The next steps are defined as, visualize received information and calculated patterns to give a first visual feedback. After that a data analysing part, integrated in the mobile application will be developed to pre-process information and give basic real-time recommendations to the user.

References

1. Balakrishnan, G., Burli, D., Behbehani, K., Burk, J., Lucas, E.: Comparison of a sleep quality index between normal and obstructive sleep apnea patients. In: 27th Annual International Conference of the Engineering in Medicine and Biology Society, 2005. IEEE-EMBS 2005, pp. 1154–1157 (2005)
2. Bsoul, M., Minn, H., Nourani, M., Gupta, G., Tamil, L.: Real-time sleep quality assessment using single-lead ecg and multi-stage svm classifier. In: 2010 Annual International Conference of the IEEE Engineering in Medicine and Biology Society (EMBC), pp. 1178–1181 (2010)
3. Cheng, S.P., Mei, H.: A personalized sleep quality assessment mechanism based on sleep pattern analysis. In: 2012 Third International Conference on Innovations in Bio-Inspired Computing and Applications (IBICA), pp. 133–138 (Sept 2012)
4. Dafna, E., Tarasiuk, A., Zigel, Y.: Sleep-quality assessment from full night audio recordings of sleep apnea patients. In: 2012 Annual International Conference of the IEEE Engineering in Medicine and Biology Society (EMBC), pp. 3660–3663 (2012)
5. Miwa, H., Sasahara, S.i., Matsui, T.: Roll-over detection and sleep quality measurement using a wearable sensor. In: 29th Annual International Conference of the IEEE Engineering in Medicine and Biology Society, 2007. EMBS 2007, pp. 1507–1510 (2007)
6. Peng, Y.T., Lin, C.Y., Sun, M.T., Landis, C.: Multimodality sensor system for long-term sleep quality monitoring. IEEE Trans. Biomed. Circuits Syst. **1**(3), 217–227 (2007)
7. Wakuda, Y., Hasegawa, Y., Fukuda, T., Noda, A., Arai, F., Kawaguchi, M.: Estimation of sleep cycle and quality based on nonlinear analysis of heart rate variability. In: Proceedings of the 2004 International Symposium on Micro-Nanomechatronics and Human Science, 2004 and the Fourth Symposium Micro-Nanomechatronics for Information-Based Society, pp. 181–186 (2004)

Multivariate Direction Scoring for Dimensionality Reduction in Classification Problems

Giorgio Biagetti, Paolo Crippa, Laura Falaschetti, Simone Orcioni and Claudio Turchetti

Abstract Dimensionality reduction is the process of reducing the number of features in a data set. In a classification problem, the proposed formula allows to sort a set of directions to be used for data projection, according to a score that estimates their capability of discriminating the different data classes. A reduction in the number of features can be obtained by taking a subset of these directions and projecting data on this space. The projecting vectors can be derived from a spectral representation or other choices. If the vectors are eigenvectors of the data covariance matrix, the proposed score is aimed to take the place of the eigenvalues in eigenvector ordering.

1 Introduction

Dimensionality reduction is the process that transforms a high-dimensional data problem in an adequate representation with a reduced number of components. This can improve the performances of a classifier, but also it can be useful for data visualization. This problem was faced for decades and several techniques were developed to this end [4–6, 10].

Principal component analysis (PCA) [6] is one of them. It is based on the transformation of the initial features into an equal number of uncorrelated variables. These new variables are obtained by means of projections of the original features on new orthogonal vectors that represent the directions of maximum variance of the data. These directions can be obtained as the eigenvectors of the covariance matrix of the data. The dimensionality reduction can be achieved by choosing a significant number of eigenvectors to project the data onto, in order to obtain a reduced number of uncorrelated features.

G. Biagetti · P. Crippa (✉) · L. Falaschetti · S. Orcioni · C. Turchetti
DII—Dipartimento di Ingegneria dell'Informazione, Università Politecnica delle Marche, Via Brecce Bianche 12, 60131 Ancona, Italy
e-mail: p.crippa@univpm.it

I. Czarnowski et al. (eds.), *Intelligent Decision Technologies 2016*,
Smart Innovation, Systems and Technologies 56,
DOI 10.1007/978-3-319-39630-9_35

PCA is used in a vast field of applications, from classical face recognition and object recognition [7, 9, 11] to, more recently, ECG beat classification [2] or speaker identification [3].

Linear discriminant analysis (LDA) [4] is based on the Fisher's discriminant coefficient. It cannot provide a number of directions greater than the number of classes less one [10]. LDA is based on taking into account the variance "between the classes" and that "within the class"; the LDA discriminant directions are those who minimize a functional obtained as the ratio of the two types of variance. Since LDA deals directly with discrimination between the classes, intuitively it must be preferred over PCA in classification problems.

In the case of face classification problem this has been demonstrated unless the database possesses a small number of training samples [8]. In this case, it is possible that PCA outperforms LDA.

The proposed method, albeit somehow related to LDA, lets you order any number of directions, irrespectively if they come from LDA, PCA or any other method. Then this new order can be used to choose the best directions for the given classification problem.

2 Multivariate Direction Scoring

Let us consider a P-dimensional space, each point of which can belong to only one of C different classes. The training set is composed of N points x, and arranged as a matrix X, with dimensions equal to $N \times P$. The training set points x_c that belong to the class c are arranged in a matrix X_c, with $c = 1, \dots, C$. We want to evaluate a set of linear independent vectors ϕ_i, with $i = 1, \dots, P$, in order to choose a subset of these vectors, with respect to which we obtain the projections of the point x so as to reduce the number of features of the classification problem.

The score we propose for vector sorting is defined as

$$\mathrm{mds}_i = \frac{\phi_i^T X^T X \phi_i}{\sum_{c=1}^{C} \phi_i^T X_c^T X_c \phi_i} = \frac{\phi_i^T \Sigma \phi_i}{\sum_{c=1}^{C} \phi_i^T \Sigma_c^T \phi_i} \tag{1}$$

where $\Sigma = \frac{X^T X}{N-1}$ is the covariance matrix of x points, while $\Sigma_c = \frac{X_c^T X_c}{N_c-1}$ is the covariance matrix of the N_c points that belong to the c class.

The numerator in (1) represents the variance of the projections of all the points x in the direction ϕ_i, while the denominator is the sum of the variances of the projections of the points belonging to the different classes. The projections are performed along the same direction ϕ_i.

The functional defined by (1) is optimized in LDA in order to find the directions ϕ_i, with respect to which to project the points x. At this stage, the vectors ϕ_i are not necessarily coming from LDA, but from any method of choice.

Now we can sort the ϕ_i vectors according to the score got from (1), take a reduced number of them, ϕ_m, $m = 1, \dots, M$, with $M < P$, and create a reduced base $\Phi_M = [\phi_1, \dots, \phi_M]$ in order to project the training set

$$k_x = x\Phi_M^T \,. \tag{2}$$

As an application of this technique we want to improve the classification capabilities of PCA. So as directions ϕ_i, we use the eigenvectors of the data covariance matrix. They natively have an order given by their eigenvalues. We want to investigate if it is possible to outperform this ordering by using (1).

In this case, we have

$$X^T X = \Phi \Lambda \Phi^T \tag{3}$$

with $\Phi = [\phi_1, \dots, \phi_P]$ and $\Lambda = \text{diag}(\lambda_1, \dots, \lambda_P)$. For each class, we can also write the spectral representation of its data as

$$X_c^T X_c = \Phi_c \Lambda_s \Phi_c^T \,, \tag{4}$$

with $\Phi_s = [\phi_{1c}, \dots, \phi_{Pc}]$ and $\Lambda_c = \text{diag}(\lambda_{1c}, \dots, \lambda_{Pc})$.

The new score of the eigenvectors, directly from (1), can be written as

$$\text{mds}_i = \frac{\lambda_i}{\sum_{c=1}^{C} \phi_i^T \Phi_c \Lambda_c \Phi_c^T \phi_i} \tag{5}$$

where λ_i is the eigenvalue of ϕ_i.

In the application of PCA to a classification problems, it is not always true that a higher global variance is directly due to a separation “between” the classes and not a greater and common variance “within” the classes. An improvement will be reached if we choose to order the PCA vectors according to their ability to discriminate between classes.

3 Experimental Results

This section provides some experimental results obtained by the application of the proposed score to different classification problems. In order to achieve unbiased results datasets with real world data were used. They were obtained by taking all databases from the Standard Classification data set, of the KEEL dataset repository [1], after discarding those with variables that takes letteral (non-numeric) values.

A standard stratified cross-validation (SCV) method is chosen, so the 5-folds SCV data sets were used. For each database, the data were centered and z-score normalized.

Then a KNN-classifier was tested on the five couples of training and testing sets and the performances were then averaged. The number of nearest neighbors was chosen equal to three.

As concise indexes of performance, precision and sensitivity (recall) were used. The precision, or positive predicted value (PPV), is the ratio between the correctly predicted samples for that class and all the predicted samples for that class. It is thus the probability that a randomly selected sample among those assigned to a certain class is classified correctly:

$$\mathrm{PPV} = \frac{\mathrm{TP}}{\mathrm{TP} + \mathrm{FP}} \tag{6}$$

where TP are the true positive classified samples and FP the false positive classified samples.

Sensitivity, or true positive predicted value (TPR), is the fraction of correctly classified samples of a class with respect to all the samples belonging to that class. So it is the probability that a randomly selected sample of a certain class is classified

Table 1 Binary classification database performances

Dataset	Features	Samples	F1 diff.
Appendicitis	7	106	−5.96
Australian	14	690	2.78
Bands	19	365	−4.16
Bupa	6	345	4.74
Coil2000	85	9822	−0.36
Haberman	3	306	0.80
Heart	13	270	0.90
Hepatitis	19	80	−0.99
Ionosphere	33	351	0.61
Magic	10	19020	2.55
Mammographic	5	830	0.22
Monk-2	6	432	8.04
Phoneme	5	5404	−0.26
Pima	8	768	−1.34
Ring	20	7400	0.32
Sonar	60	208	0.01
Spambase	57	4597	−0.21
Spectfheart	44	267	3.73
Titanic	3	2201	0
Twonorm	20	7400	−0.01
Wdbc	30	569	−0.55
Wisconsin	9	683	0.18

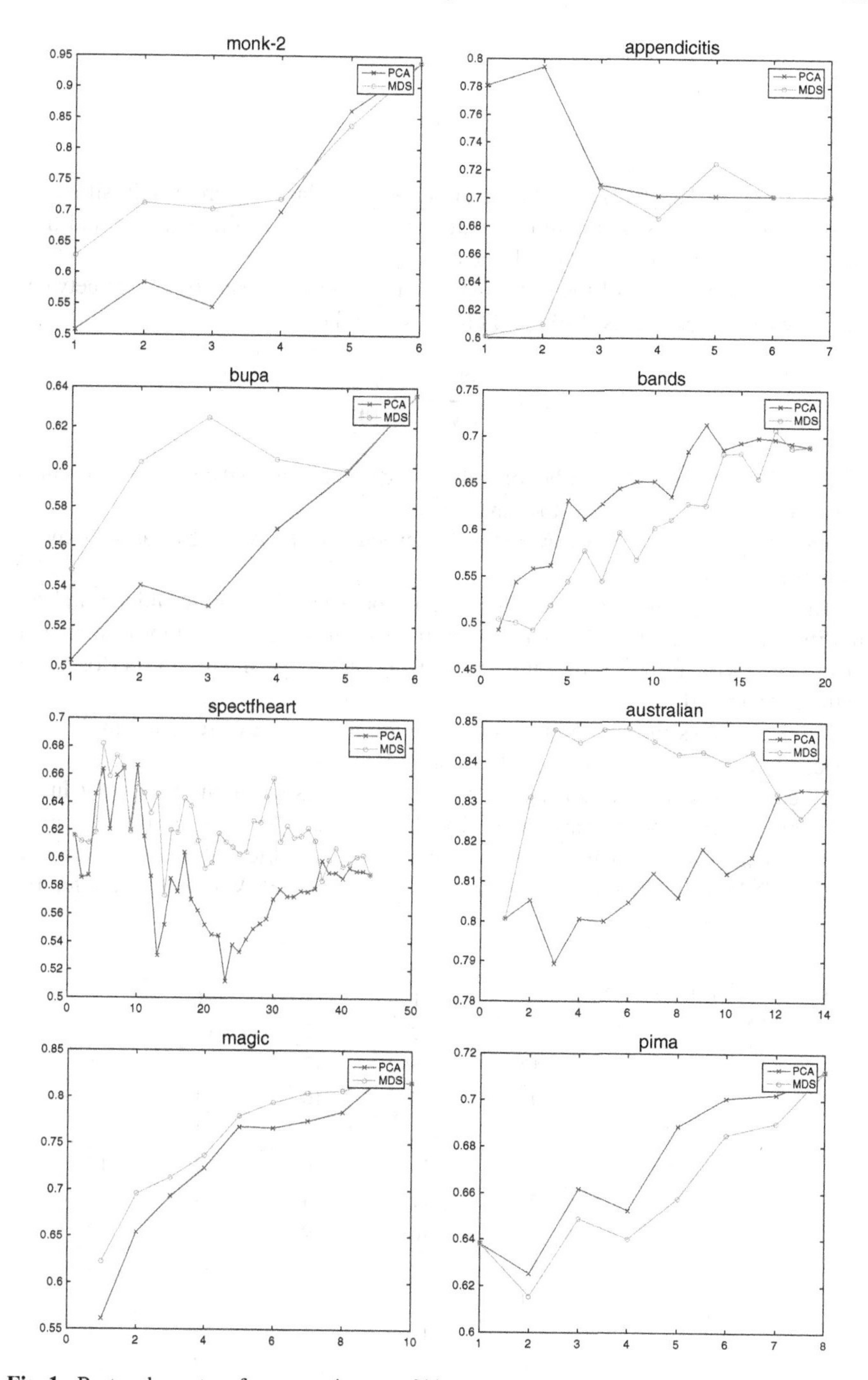

Fig. 1 Best and worst performances in case of binary classification problems

correctly. It can be written as

$$\mathrm{TPR} = \frac{\mathrm{TP}}{\mathrm{TP} + \mathrm{FN}} \tag{7}$$

where FN (false negative) are all the samples of that class that are misclassified.

Precision and sensitivity are calculated for each class of the database and then their values are averaged among all the classes.

As a single measure of the performance the harmonic mean (F1 score) between these two averaged scores ($\mathrm{PPV}_{\mathrm{av.}}$, $\mathrm{TPR}_{\mathrm{av.}}$) is calculated:

$$\mathrm{F1} = 2\frac{\mathrm{PPV}_{\mathrm{av.}}\mathrm{TPR}_{\mathrm{av.}}}{\mathrm{PPV}_{\mathrm{av.}} + \mathrm{TPR}_{\mathrm{av.}}} . \tag{8}$$

For each database, the number of features are reduced from the maximum number down to 1 and the F1 score is then averaged.

The directions to be sorted were the eigenvectors of PCA, so (5) was the expression of the used score.

Table 1 reports the difference of F1 score between the new technique and the traditional ordering of PCA eigenvectors. In most of the cases an improvement can be obtained with the new score, although in 9 out of 22 experiments, the performance actually degraded.

Figure 1 shows the first eight results, among the databases reported in Table 1, sorted by the absolute value of the F1 score difference.

As can be seen, also by sorting the databases with respect to the absolute value of performance variation, we have a majority of cases with a performance improvement.

To get an insight on how the new score works, Table 2 shows the permutations of PCA directions (eigenvectors) for all the five SCV training sets for the

Table 2 Australian database permutations

1	1	1	1	1
12	12	12	12	12
13	6	13	6	5
9	8	10	13	8
5	2	8	10	10
6	13	6	2	13
2	11	4	9	9
7	10	2	3	7
8	7	9	5	2
11	9	7	8	6
3	14	11	14	11
10	3	5	7	14
14	5	14	11	3
4	4	3	4	4

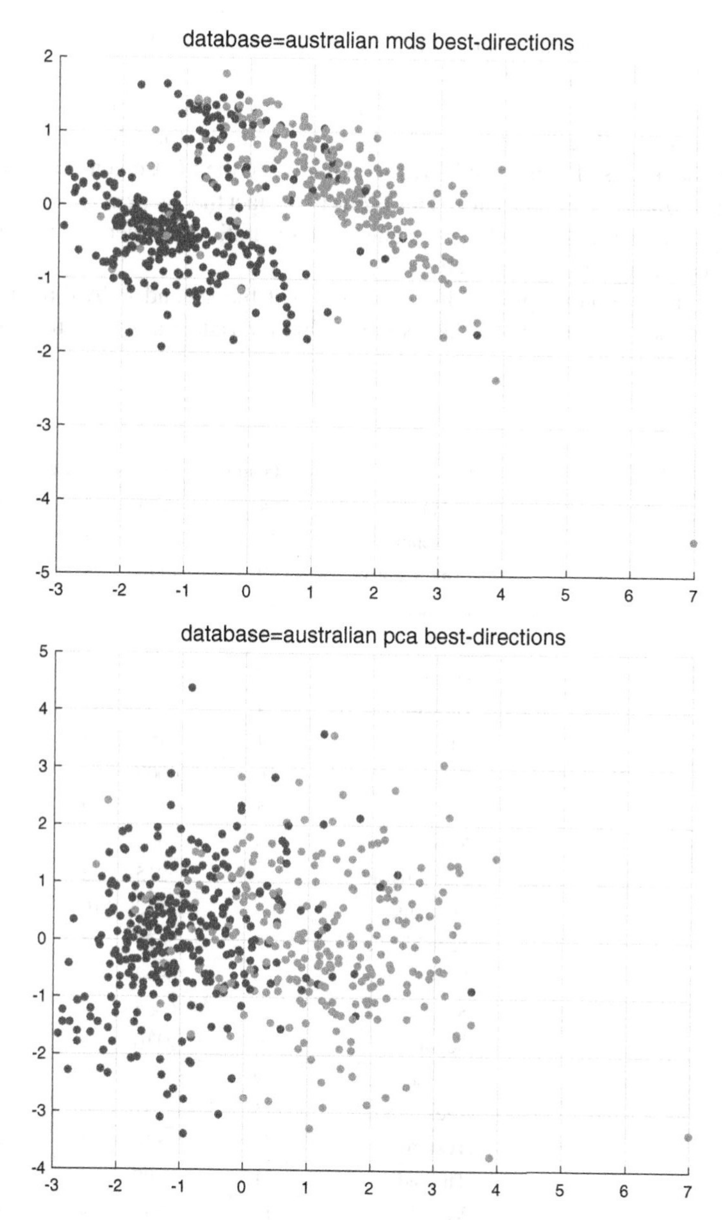

Fig. 2 Projections of training set of Australian database on the two best directions chosen by MDS or PCA

"australian" database. As can be seen, a PCA direction with a very low eigenvalue (the twelfth) always got promoted to the second position in the new ranking, achieving an improvement in performance.

Figure 2 shows the projections of the fifth SCV partition of the “Australian” database. The x-axis is common to the two graphs, while the y-axis is relative to the projection on the twelfth PCA direction in the top plot (which ranked second with our new score), and on the second PCA direction on the bottom plot.

From the results of Table 1, it is worth noticing that the PCA eigenvalue ordering can achieve good performance too. This is due to the fact that, generally, if a direction has a higher variance this can likely be due to the separation between the classes. But this, in general, can not be ensured.

In fact in the bottom plot of Fig. 2 we see that the second PCA direction has a high variance, because both classes have a high variance although they remain superimposed.

Table 3 Multi-class classification database performances

Dataset	Features	Samples	Classes	F1 diff.
Balance	4	625	3	16.91
Cleveland	13	297	5	−0.96
Contraceptive	9	1473	3	1.71
Dermatology	34	358	6	−0.04
Ecoli	7	336	8	−2.36
Glass	9	214	7	−1.31
Hayes-Roth	4	160	3	7.97
Iris	4	150	3	0
Letter	16	20000	26	4.17
Marketing	15	6876	9	−0.02
Movement libras	90	360	15	0.37
New thyroid	5	215	3	−0.11
Optdigits	64	5620	10	0.30
Page blocks	10	5473	5	2.93
Penbased	16	10992	10	−0.08
Satimage	36	6435	7	−2.67
Segment	19	2310	7	1.98
Shuttle	9	58000	7	−15.50
Tae	5	151	3	7.60
Texture	40	5500	11	−0.01
Thyroid	21	7200	3	−5.00
Vehicle	18	846	4	−0.68
Vowel	13	990	11	−1.72
Wine	13	178	3	−0.18
Yeast	8	1484	10	0.34
Zoo	17	101	7	4.56

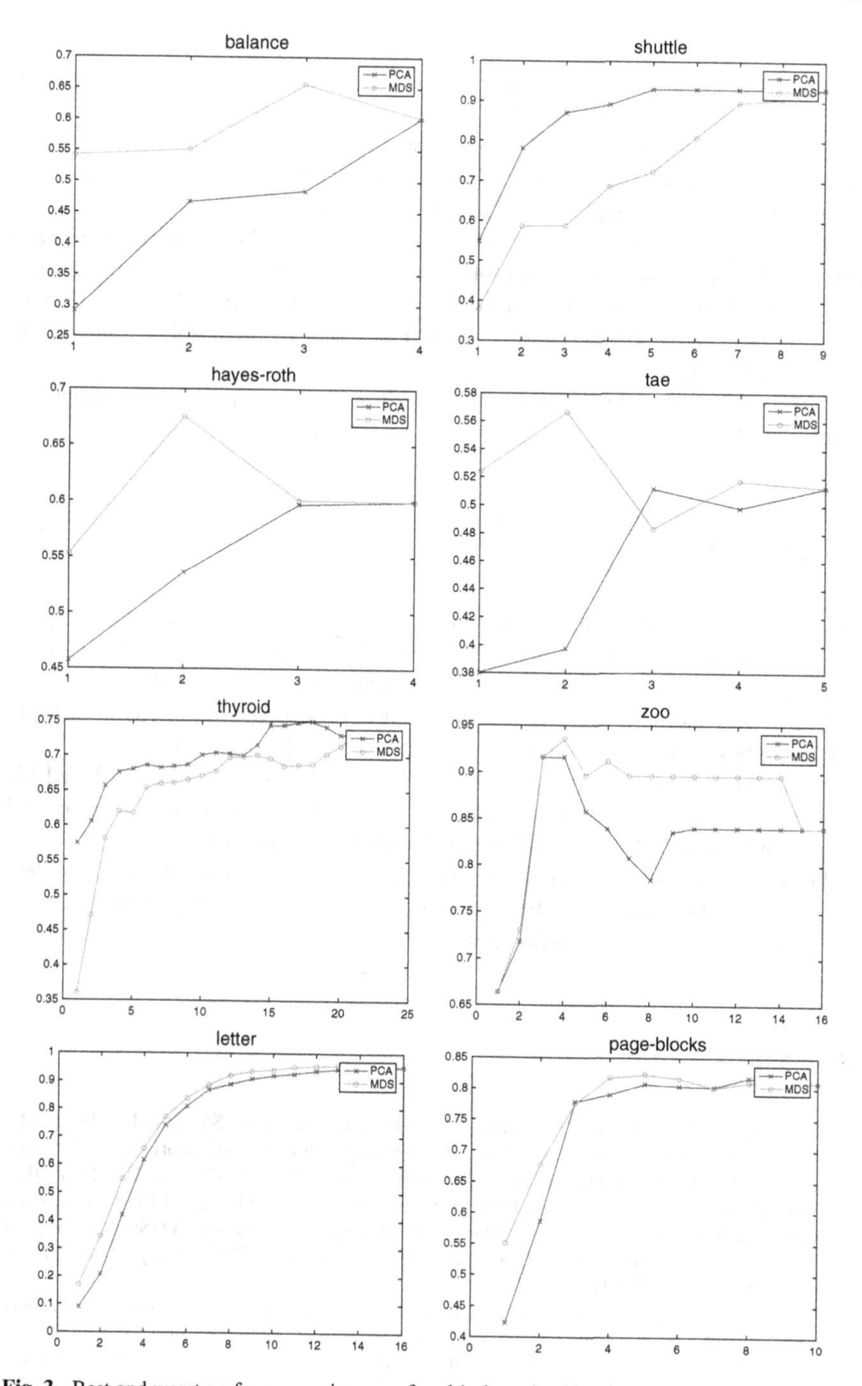

Fig. 3 Best and worst performances in case of multi-class classification problems

The alternative direction, presented in the above plot, possesses a lower global variance, but it better separates among the two classes, although a partial superposition between the classes is present.

Notwithstanding this, data from real world not always follow normal distributions, so a technique based on second moments only, as that proposed, is of course not guaranteed to always give better results.

Major improvement in absolute performances can also be achieved with multi-class databases, as can be seen in Table 3, although the opposite case is also true.

Figure 3 shows, for the multi-class databases, the performance of the first eight of them, among the databases reported in Table 3, sorted by the absolute value of the F1 score difference.

Also in this case, the proposed technique allows to earn a performance improvement over PCA.

4 Conclusion

In dimensionality reduction data are projected over a subspace of the original space in order to obtain a reduced number of features. This is achieved by selecting a subset from some set of independent directions that spans the entire data space.

The aim of the proposed score is to sort any set of given directions in order to obtain the best separation among the projected data of the different classes.

Extensive experimental results were obtained by using databases from the KEEL dataset repository and a standard stratified cross-validation method. The proposed score proved to be able to outperform PCA over the majority of cases.

For the other cases, it can be stated that the departure of real data from an ideal normal distribution appears to be the main cause of trouble, as expected for a technique based only on second-order moments.

References

1. Alcalá-Fdez, J., Fernández, A., Luengo, J., Derrac, J., García, S., Sánchez, L., Herrera, F.: KEEL data-mining software tool: Data set repository, integration of algorithms and experimental analysis framework. J. Multiple-Valued Logic Soft Comput. **17**(2–3), 255–287 (2011)
2. Biagetti, G., Crippa, P., Curzi, A., Orcioni, S., Turchetti, C.: A multi-class ECG beat classifier based on the truncated KLT representation. In: Proceedings of UKSim-AMSS 8th European Modelling Symposium on Computer Modelling and Simulation, EMS 2014, pp. 93–98. IEEE Computer Society (2014)
3. Biagetti, G., Crippa, P., Curzi, A., Orcioni, S., Turchetti, C.: Speaker identification with short sequences of speech frames. In: Proceedings of the 4th International Conference on Pattern Recognition Applications and Methods (ICPRAM 2015), pp. 178–185. SCITEPRESS (2015)
4. Fisher, R.A.: The use of multiple measurements in taxonomic problems. Ann. Eugenics **7**(2), 179–188 (1936)
5. Frisch, R.: Correlation and scatter in statistical variables. Nordic Stat. J. 36–102 (1929)

6. Hotelling, H.: Analysis of a complex of statistical variables into principal components. J. Educ. Psychol. **24**(6), 417–441 (1933)
7. Kirby, M., Sirovich, L.: Application of the Karhunen-Loève procedure for the characterization of human faces. IEEE Trans. Pattern Anal. Mach. Intell. **12**(1), 103–108 (1990)
8. Martínez, A.M., Kak, A.C.: PCA versus LDA. IEEE Trans. Pattern Anal. Mach. Intell. **23**(2) (2001)
9. Moghaddam, B., Pentland, A.: Probabilistic visual learning for object representation. IEEE Trans. Pattern Anal. Mach. Intell. **19**(7), 696–710 (1997)
10. Rao, C.R.: The utilization of multiple measurements in problems of biological classification. J. Royal Stat. Soc. Ser. B (Methodological) **10**(2), 159–203 (1948)
11. Turk, M., Pentland, A.: Eigenfaces for recognition. J. Cognit. Neurosci. **3**(1), 71–86 (1991)

An Efficient Technique for Real-Time Human Activity Classification Using Accelerometer Data

Giorgio Biagetti, Paolo Crippa, Laura Falaschetti, Simone Orcioni and Claudio Turchetti

Abstract Accurate estimation of biometric parameters recorded from subjects' wrist or waist, when the subjects are performing various physical exercises, is often a challenging problem due to the presence of motion artifacts. In order to reduce the motion artifacts, data derived from a triaxial accelerometer have been proven to be very useful. Unfortunately, wearable devices such as smartphones and smartwatches are in general differently oriented during real life activities, so the data derived from the three axes are mixed up. This paper proposes an efficient technique for real-time recognition of human activities by using accelerometer data that is based on singular value decomposition (SVD) and truncated Karhunen-Loève transform (KLT) for feature extraction and reduction, and Bayesian classification for class recognition, that is independent of the orientation of the sensor. This is particularly suitable for implementation in wearable devices. In order to demonstrate the validity of this technique, it has been successfully applied to a database of accelerometer data derived from static postures, dynamic activities, and postural transitions occurring between the static postures.

Keywords Activity detection · Accelerometer · Real-time · Smartphone · Health · Fitness · Bayesian classification · Singular value decomposition · SVD · Expectation maximization · EM · Feature extraction

---

G. Biagetti · P. Crippa (✉) · L. Falaschetti · S. Orcioni · C. Turchetti
DII – Dipartimento di Ingegneria dell'Informazione,
Università Politecnica delle Marche, Via Brecce Bianche 12, I-60131 Ancona, Italy
e-mail: p.crippa@univpm.it

G. Biagetti
e-mail: g.biagetti@univpm.it

L. Falaschetti
e-mail: l.falaschetti@univpm.it

S. Orcioni
e-mail: s.orcioni@univpm.it

C. Turchetti
e-mail: c.turchetti@univpm.it

I. Czarnowski et al. (eds.), *Intelligent Decision Technologies 2016*,
Smart Innovation, Systems and Technologies 56,
DOI 10.1007/978-3-319-39630-9_36

1 Introduction

Human activity recognition is one of the most interesting research topics for several areas such as pervasive and mobile computing, ambient assisted living, surveillance-based security, sport and fitness activities, healthcare.

Over the recent years sensor technologies especially high-capacity, low-power, low-cost, miniaturized and lightweight sensors, wired, wireless and hybrid communication network protocols as well as signal processing theory have greatly progressed.

Wearable sensors, i.e. sensors that are positioned directly or indirectly on the human body, generate signals (accelerometric, PPG, ECG, sEMG, …) when the user performs activities. Therefore they can monitor features that are descriptive of the person's physiological state or movement.

These sensors can be embedded into clothes, shoes, belts, sunglasses, smartwatches and smartphones, or positioned directly on the body and can be used to collect information such as body position and movement, heart rate, muscle fatigue and skin temperature [3, 4].

Among wearable sensors, accelerometers are probably the most frequently used for activity monitoring. In particular, they are effective in monitoring actions that involve repetitive body motions, such as walking, running, cycling, sitting, standing, and climbing stairs.

On the one hand, activity classification using accelerometers can be obtained using one or more sensors on the body [7, 12]. However single sensor systems are more practical and in this case common location choices are the waist, upper arm, wrist, and ankle [13, 15, 16]. The waist location has been used extensively in physical activity measurements because it captures major body motions, but algorithms using waist data can underestimate overall expenditure on activities such as bicycling or arm ergometry, where the waist movement is uncorrelated to the movement of the limbs. Therefore several recent studies addressed the problem of detecting human activities from smartphones and smartwatches [1, 8, 11].

On the other hand, accurate estimation of biometric parameters recorded from subjects' wrist or waist, when the subjects are performing various physical exercises, is often a challenging problem due to the presence of motion artifacts. In order to reduce the motion artifacts, data derived from a triaxial accelerometer have been proven to be very useful [6].

Wrist-worn sensor devices can be comfortably used during activities of daily living, including sleep, and can remain active during changing of clothes and do not require special belts or clips, thus increasing the wear time. Therefore smartwatches for human activity monitoring are becoming very important tools in personal health monitoring. In particular, exercise routines and repetitions can be counted in order to track a workout routine as well as determine the energy expenditure of individual movements. Indeed, mobile fitness coaching has involved topics ranging from quality of performing such sports actions to detection of the specific sports activity [5].

However, wearable devices such as smartphones and smartwatches are in general differently oriented during human activities, so the data derived from the three axes are mixed up.

This paper proposes an efficient technique for real-time recognition of human activities, and transitions between them, by using accelerometer data. The proposed technique is based on singular value decomposition (SVD) and truncated Karhunen-Loève transform (KLT) for feature extraction and reduction, and Bayesian classification for class recognition. The algorithm is independent of the orientation of the sensor making it particularly suitable for implementation in wearable devices such as smartphones where the orientation of the sensor can be unknown or its placement could be not always correct. In order to demonstrate the validity of this technique, it has been successfully applied to a database of accelerometer data derived from static postures, dynamic activities, and postural transitions occurring between the static postures.

The paper is organized as follows. Section 2 provides a brief overview of the human activity classification algorithm. Section 3 presents the experimental results carried out on a public domain data set in order to show the effectiveness of the proposed approach. Finally Sect. 4 summarizes the conclusions of the present work.

2 Recognition Algorithm

This section presents a description of the overall algorithm, based on the dimensionality-reduced singular value spectrum of the data Hankel matrix and on the Bayesian classifier, that is able to identify human activity classes from accelerometer data.

A schematic diagram of the activity detection algorithm, is shown in Fig. 1.

2.1 *Data Preprocessing and Feature Extraction*

Let x, y, z be the accelerometer signals. After a preprocessing stage where the raw data have been windowed into windows $N+L-1$ samples long, the resulting accelerometer signals have been manipulated as follows.

Let $x_t = [x(t) \ldots x(t+N-1)]^T$, $x_t^{(i)} = x_{t+i-1}$, and $X_t = [x_t^{(1)} \ldots x_t^{(L)}]$. Analogously, let $Y_t = [y_t^{(1)} \ldots y_t^{(L)}]$ and $Z_t = [z_t^{(1)} \ldots z_t^{(L)}]$. The matrices X_t, Y_t, and Z_t so built are the Hankel data matrices of the three accelerometer signals, where $x_t^{(i)}$, $y_t^{(i)}$, $z_t^{(i)}$, $i = 1, \ldots, L$, represent the observations achieved from the three-axes accelerometer, each shifted in time by i samples.

The complete matrix of sample signals

$$H_t = [X_t\, Y_t\, Z_t] \in \mathbb{R}^{N \times 3L} \tag{1}$$

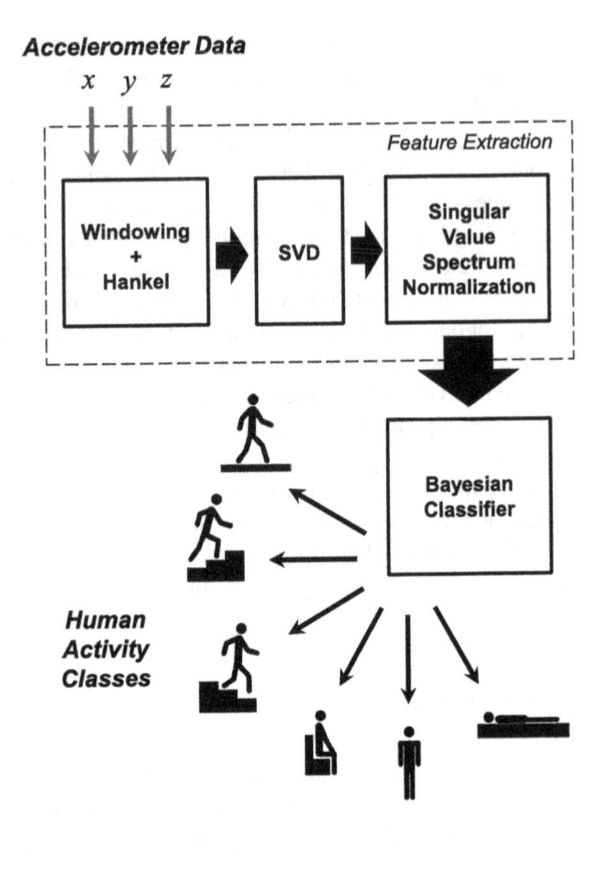

Fig. 1 Flow chart of the proposed framework for human activity classification (x, y, z are the 3-axial accelerometer signals)

can be represented by the singular value decomposition (SVD) as

$$H_t = S_t \Lambda_t R_t^T = \sum_{i=1}^{N} \lambda_i s_i r_i^T, \tag{2}$$

where, if $N < 3L$, $S_t = [s_1 \dots s_N]$, $R_t = [r_1 \dots r_N]$, with s_i, r_i being the corresponding left and right singular vectors, and λ_i are the singular values in decreasing order $\lambda_1 \geq \lambda_2 \geq \cdots \geq \lambda_N$.

By denoting with $H_t \in \mathbb{R}^{N\times 3L}$ the data matrix of the accelerometer signals at each time instant t, in order to apply the human activity classification algorithm, a feature vector ξ_t has to be derived from this matrix.

We noticed that different types of activities lead to different distributions of the energy of the accelerometer signals among its eigenvectors. Thus, a suitable candidate for identifying the type of activity is the normalized spectrum of singular values $\Lambda_t = [\lambda_1 \dots \lambda_N]$, so as to avoid dependence on the intensity of the activity. Therefore we choose $\xi_t = \Lambda_t / ||\Lambda_t||$ where $|| \cdot ||$ represents the norm of a vector.

In order to face the problem of dimensionality, the usual choice [10] is to reduce the vector ξ_t to a vector k_{tM} of lower dimension by a linear non-invertible transform Ψ (a rectangular matrix) such that

$$k_{tM} = \Psi \, \xi_t \, , \tag{3}$$

where $\xi_t \in \mathbb{R}^N$, $k_{tM} \in \mathbb{R}^M$, $\Psi \in \mathbb{R}^{M \times N}$, and $M \ll N$.

It is well known that, among the allowable linear transforms $\Psi : \mathbb{R}^N \rightarrow \mathbb{R}^M$, the Karhunen-Loève transform truncated to $M < N$ orthonormal basis functions, is the one that ensures the minimum mean square error.

This normalized singular value spectrum can easily be computed immediately after having performed the SVD on the accelerometer signals, and used as input to the Bayesian classifier after a KLT-based dimensionality reduction from $N = 96$ to $M = 10$.

The algorithm developed in this section follows the approach reported in [5] as it was successfully adopted in the physical exercise identification for photoplethysmography artifact reduction.

2.2 Bayesian Classification

Let us refer to a frame $k_{tM}[n]$, $n = 0, \dots, M-1$, containing features extracted from the accelerometer signals.

We assume that the observations for all human activities that need to be identified, are acquired and divided in two sets, $\mathcal{W}$ for training and $\mathcal{Z}$ for testing.

For Bayesian classification, a group of Γ activities is represented by the probability density functions (pdfs) $p_\gamma(k_{tM}) = p(k_{tM} \mid \theta_\gamma)$, $\gamma = 1, 2, \dots, \Gamma$, where θ_γ are the parameters to be estimated during training. Thus we can define the vector $p = [p_1(k_{tM}), \dots, p_\Gamma(k_{tM})]^T$.

The objective of classification is to find the model θ_γ corresponding to the activity γ which has the maximum a posteriori probability for a given frame $k_{tM} \in \mathcal{Z}$. Formally:

$$\widehat{\gamma}(k_{tM}) = \underset{1 \le \gamma \le \Gamma}{\operatorname{argmax}} \left\{ p(\theta_\gamma \mid k_{tM}) \right\} = \underset{1 \le \gamma \le \Gamma}{\operatorname{argmax}} \left\{ \frac{p(k_{tM} \mid \theta_\gamma) p(\theta_\gamma)}{p(k_{tM})} \right\} . \tag{4}$$

Assuming equally likely activities (i.e. $p(\theta_\gamma) = 1/\Gamma$) and noting that $p(k_{tM})$ is the same for all activity models, the Bayesian classification is equivalent to

$$\widehat{\gamma}(k_{tM}) = \underset{1 \le \gamma \le \Gamma}{\operatorname{argmax}} \left\{ p_\gamma(k_{tM}) \right\} . \tag{5}$$

Thus Bayesian identification reduces to solving the problem stated by (5).

The most generic statistical model one can adopt for $p(k_{tM} \mid \theta_\gamma)$ is the Gaussian mixture model (GMM) [14]. The GMM for the single exercise is a weighted sum of F components densities and given by the equation

$$p(k_{tM} \mid \theta_\gamma) = \sum_{i=1}^{F} \alpha_i \mathcal{N}(k_{tM} \mid \mu_i, C_i) \tag{6}$$

where α_i, $i = 1, \ldots, F$ are the mixing weights, and $\mathcal{N}(k_{tM} \mid \mu_i, C_i)$ represents the density of a Gaussian distribution with mean μ_i and covariance matrix C_i. It is worth noting that α_i must satisfy $0 \leq \alpha_i \leq 1$ and $\sum_{i=1}^{F} \alpha_i = 1$ and θ_γ is the set of parameters needed to specify the Gaussian mixture, defined as $\theta_\gamma = \{\alpha_1, \mu_1, C_1, \ldots, \alpha_F, \mu_F, C_F\}$.

The choice for obtaining an estimate of the mixture parameters is an unsupervised algorithm for learning a finite mixture model from multivariate data, that overcomes the main lacks of the standard expectation maximization (EM) algorithm, i.e. sensitiveness to initialization and selection of number F of components [9]. This algorithm integrates both model estimation and component selection, i.e. the ability of choosing the best number of mixture components F according to a predefined minimization criterion, in a single framework.

3 Experimental Results

We used the "Human Activity Recognition Using Smartphones" data set [2]. This data set includes data recorded from experiments made by a group of 30 volunteers, each of which performed six different activities, three belonging to the "static" class, i.e., standing, sitting, and lying, and three belonging to the "cyclic" class, i.e., walking, climbing stairs, descending stairs. Data recorded during the transitions occurring between static postures were labeled accordingly as "transitions". 3-axial linear acceleration was recorded at a 50 Hz sampling rate and the experiments were video-recorded to allow accurate manual data labeling.

The signals so gathered were split in 2.56 s long windows with 50 % overlap. Windows containing unlabeled portions of signal had been discarded, as were windows containing more than 25 % of signal with inconsistent labeling (i.e., a label that differs from that of the majority of the data points within the window). This yielded a total of 10991 windows to process, 7808 of which were used for training and the remaining 3183 for testing. Separation was done so that data recorded by any given person never occurred both in the training and testing sets.

Data was pre-processed by using these windows to build three $N \times L$ Hankel matrices, one for each acceleration direction, with $N = 96$ and $L = 33$. These are then fed together to the SVD, so as to remove the effect of sensor orientation, and all the ensuing normalized singular values used, after dimensionality reduction to $M = 10$ principal components, as the feature vectors for the classifier.

Table 1 Confusion matrix of the exercise type identifier evaluated on the whole testing set

	Standing	Sitting	Lying	Walking	W. down-stairs	W. upstairs	Stand-to-sit	Sit-to-stand	Sit-to-lie	Lie-to-sit	Stand-to-lie	Lie-to-stand
Standing	420	27	26	0	0	0	1	0	0	0	1	0
Sitting	98	273	64	0	0	0	1	0	2	0	12	0
Lying	66	63	266	0	0	0	0	0	0	0	0	0
Walking	0	0	0	292	125	102	2	3	0	0	0	0
W. downstairs	0	0	0	228	178	151	5	1	0	0	0	0
W. upstairs	0	0	0	269	128	158	1	3	0	1	0	0
Stand-to-sit	0	0	0	2	2	0	9	6	4	3	5	3
Sit-to-stand	0	0	0	0	2	0	5	4	0	1	3	1
Sit-to-lie	0	0	0	0	0	0	5	4	12	5	11	4
Lie-to-sit	0	0	0	0	0	0	2	3	1	10	7	12
Stand-to-lie	0	0	0	2	0	0	6	4	3	6	28	10
Lie-to-stand	0	1	0	0	0	0	2	2	3	6	7	10

As a first experiment, the performance of the classifier in recognizing the exact activity being performed was assessed. Of course, this method was never intended to be able to discriminate between all these activities, as by mixing the axes of the accelerometer output is clearly nigh impossible to discriminate, e.g., between the static postures. This is clearly shown if Table 1, where the confusion matrix of this classification experiment is reported. The resulting performance is reported in Table 2, yielding an overall accuracy of 52.15 % with an F1-score of 43.22 %.

Table 2 Performance (sensitivity, precision, and F1-score) of the exercise type identifier evaluated on the whole testing set

Activities	Sensitivity (%)	Precision (%)	F1-score (%)
Standing	88.42	71.92	79.32
Sitting	60.67	75.00	67.08
Lying	67.34	74.72	70.84
Walking	55.73	36.82	44.34
Walking downstairs	31.62	40.92	35.67
Walking upstairs	28.21	38.44	32.54
Stand-to-sit	26.47	23.08	24.66
Sit-to-stand	25.00	13.33	17.39
Sit-to-lie	29.27	48.00	36.36
Lie-to-sit	28.57	31.25	29.85
Stand-to-lie	47.46	37.84	42.11
Lie-to-stand	32.26	25.00	28.17

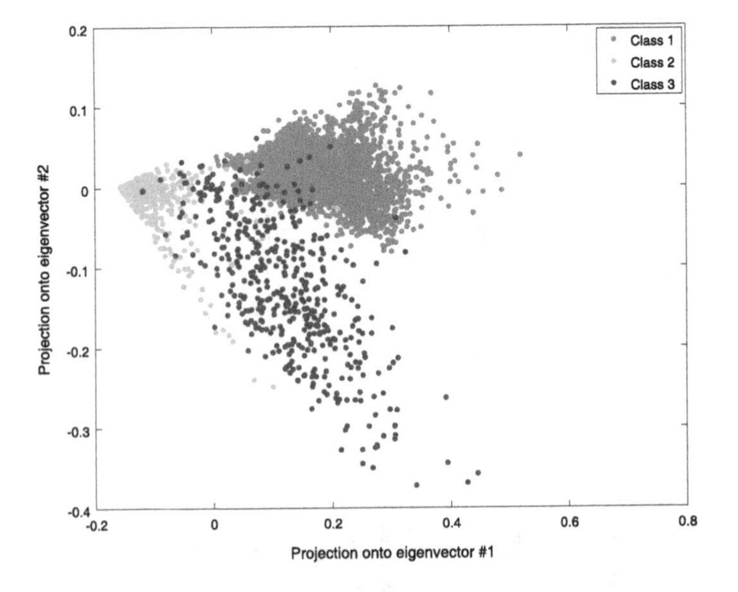

Fig. 2 Eigenvector projections

Table 3 Confusion matrix of the human activity class identifier evaluated on the whole testing set

	Static	Cyclic	Transitional
Static	1303	0	17
Cyclic	0	1629	18
Transitional	1	7	208

Table 4 Performance (sensitivity, precision, and F1-score) of the activity class identifier evaluated on the whole testing set

Activities	Sensitivity (%)	Precision (%)	F1-score [%]
Static	98.71	99.92	99.31
Cyclic	98.91	99.57	99.24
Transitional	96.30	85.60	90.63

Despite these results, it is quite clear from observation of Table 1 that the recognizer very seldom makes mistakes between the three classes of exercises (static postures, cyclic movements, transitions). A scatter plot of the first two features, shown in Fig. 2, also confirms this idea, as in just two dimensions the three classes are very well separated.

This is better assessed by the second experiment, where only these three classes were considered for training the models. The confusion matrix is shown in Table 3, and the performance in Table 4. The overall accuracy in this context is 98.65 % and the F1-score is 96.48 %, denoting the high reliability of the chosen features to discriminate between these classes. It should also be recalled that a number of windows, 5.8 % to be exact, contain up to 25 % of "noise", i.e., data belonging to different classes, so some classification error is to be expected.

4 Conclusion

In this paper we present a feature set, based on the dimensionality-reduced singular value spectrum of the data Hankel matrix, that is suitable to identify human activity classes from accelerometer data. Since the singular value spectrum is inherently invariant with respect to rotation matrices, classification accuracy is independent of the orientation of the sensor, freeing the user from having to worry about its correct placement.

Experimental results conducted on a public domain data set show the effectiveness of the proposed approach.

References

1. Anguita, D., Ghio, A., Oneto, L., Parra, X., Reyes-Ortiz, J.L.: Energy efficient smartphone-based activity recognition using fixed-point arithmetic. J. Univ. Comput. Sci. **19**(9), 1295–1314 (2013)
2. Anguita, D., Ghio, A., Oneto, L., Parra, X., Reyes-Ortiz, J.L.: A public domain dataset for human activity recognition using smartphones. In: European Symposium on Artificial Neural Networks, Computational Intelligence and Machine Learning (ESANN 2013), pp. 437–442 (2013)
3. Bacà, A., Biagetti, G., Camilletti, M., Crippa, P., Falaschetti, L., Orcioni, S., Rossini, L., Tonelli, D., Turchetti, C.: CARMA: a robust motion artifact reduction algorithm for heart rate monitoring from PPG signals. In: 23rd European Signal Processing Conference (EUSIPCO 2015), pp. 2696–2700 (2015)
4. Biagetti, G., Crippa, P., Curzi, A., Orcioni, S., Turchetti, C.: Analysis of the EMG signal during cyclic movements using multicomponent AM-FM decomposition. IEEE J. Biomed. Health Inf. **19**(5), 1672–1681 (2015)
5. Biagetti, G., Crippa, P., Falaschetti, L., Orcioni, S., Turchetti, C.: A rule based framework for smart training using sEMG signal. In: Intelligent Decision Technologies: Proceedings of the 7th KES International Conference on Intelligent Decision Technologies (KES-IDT 2015), pp. 89–99. Sorrento, Italy (2015)
6. Biagetti, G., Crippa, P., Falaschetti, L., Orcioni, S., Turchetti, C.: Artifact reduction in photoplethysmography using Bayesian classification for physical exercise identification. In: Proceedings of the 5th International Conference on Pattern Recognition Applications and Methods (ICPRAM 2016), pp. 467–474. Rome, Italy (2016)
7. Catal, C., Tufekci, S., Pirmit, E., Kocabag, G.: On the use of ensemble of classifiers for accelerometer-based activity recognition. Appl. Soft Comput. **37**, 1018–1022 (2015)
8. Dernbach, S., Das, B., Krishnan, N.C., Thomas, B.L., Cook, D.J.: Simple and complex activity recognition through smart phones. In: 8th International Conference on Intelligent Environments (IE), pp. 214–221 (2012)
9. Figueiredo, M.A.F., Jain, A.K.: Unsupervised learning of finite mixture models. IEEE Trans. Pattern Anal. Mach. Intell. **24**(3), 381–396 (2002)
10. Jain, A.K., Duin, R.P.W., Mao, J.: Statistical pattern recognition: a review. IEEE Trans. Pattern Anal. Mach. Intell. **22**(1), 4–37 (2000)
11. Khan, A., Lee, Y.K., Lee, S., Kim, T.S.: Human activity recognition via an accelerometer-enabled-smartphone using kernel discriminant analysis. In: 2010 5th International Conference on Future Information Technology (FutureTech), pp. 1–6 (2010)
12. Mannini, A., Sabatini, A.M.: Machine learning methods for classifying human physical activity from on-body accelerometers. Sensors **10**(2), 1154–1175 (2010)
13. Mannini, A., Intille, S.S., Rosenberger, M., Sabatini, A.M., Haskell, W.: Activity recognition using a single accelerometer placed at the wrist or ankle. Med. Sci. Sports Exerc. **45**(11), 2193–2203 (2013)
14. Reynolds, D.A., Rose, R.C.: Robust text-independent speaker identification using Gaussian mixture speaker models. IEEE Trans. Speech Audio Process. **3**(1), 72–83 (1995)
15. Rodriguez-Martin, D.: Samà, A., Perez-Lopez, C., Català, A., Cabestany, J., Rodriguez-Molinero, A.: SVM-based posture identification with a single waist-located triaxial accelerometer. Expert Syst. Appl. **40**(18), 7203–7211 (2013)
16. Torres-Huitzil, C., Nuno-Maganda, M.: Robust smartphone-based human activity recognition using a tri-axial accelerometer. In: 2015 IEEE 6th Latin American Symposium on Circuits Systems (LASCAS), pp. 1–4 (2015)

New Advances of Soft Computing in Industrial and Management Engineering

A Double Layer Neural Network Based on Artificial Bee Colony Algorithm for Solving Quadratic Bi-Level Programming Problem

Junzo Watada and Haochen Ding

Abstract In this study, we formulate a double layer neural network based hybrid method to solve the quadratic bi-level programming problem. Our proposed algorithm comprises an improved artificial bee colony algorithm, a Hopfield network, and a Boltzmann machine in order to effectively and efficiently solve such problems. The improved artificial bee colony algorithm is developed for dealing with the upper level problem. The experiment results indicate that compared with other methods, the proposed double layer neural network based hybrid method is capable of achieving better optimal solutions for the quadratic bi-level programming problem.

Keywords Bi-level programming problem · Double layer neural networks · Artificial bee colony algorithm

1 Introduction

In this section, we will introduce the background of this research, including an introduction of bi-level programming problem, a short survey of existing methods for solving this problem and real life based applications developed from bi-level programming problems. First, we will elaborate the mathematical formulation of the bi-level programming problems, its relationship between game theory, important definitions around this problem and discuss its computational complexity. Second, we will introduce some existing methods, such as extreme point methods, branch and bound methods, descent methods and recently developed meta-heuristic based methods. Those methods, though not perfect, have shown some good properties in

J. Watada (✉) · H. Ding
Graduate School of Information, Production and Systems, Waseda University,
2-7 Hibikino of Wakamatsu-ku, Kitakyushu-shi, 808-0135 Fukuoka, Japan
e-mail: junzow@osb.att.ne.jp

H. Ding
e-mail: haochen.ding@ruri.waseda.jp

I. Czarnowski et al. (eds.), *Intelligent Decision Technologies 2016*,
Smart Innovation, Systems and Technologies 56,
DOI 10.1007/978-3-319-39630-9_37

dealing with bi-level programming problems and provide us good enlightenments for developing a new algorithm.

1.1 Bi-Level Programming Problem

The Bi-Level Programming Problem (BLPP) is an optimization problem which involves two decision makers (or two decision aspects) in a hierarchical structure, where the upper level decision maker takes priority to optimize his objective, while the lower level decision maker, which is regarded as constraints of the upper level, will optimize his objective accordingly by following the choices of the upper level decision maker's. In this mathematical programming model, the two decision makers will usually have conflict in constraint resources, and both want to reach his own optimization objective. Thus, they need to take their counterpart's choice into consideration. An important feature of the bi-level programming problem is the essential hierarchical relationship between two autonomous, and possibly conflict decision makers. In the sense of hierarchy and conflict decision making, it is very related to Stackelberg (leader–follower) game in the economics.

The general form of bi-level programming problem can be mathematically stated as

$$\begin{aligned} &\min_x F(x,y) \\ &s.t.\ \ G(x,y) \leq 0 \\ &\qquad \min_y f(x,y) \\ &\qquad s.t.\ \ g(x,y) \leq 0 \end{aligned} \tag{1}$$

where $x \in R^{n_1}$ and $y \in R^{n_2}$ are respectively the upper-level variables and lower level variables. Similarly, the objective function $F : R^{n_1} \times R^{n_2} \to R$ is the upper level objective function and $f. : R^{n_1} \times R^{n_2} \to R$ is the lower level objective function, while the vector valued objective functions $G : R^{n_1} \times R^{n_2}. \times .R^{m_1}$ and $g : R^{n_1} \times R^{n_2}. \times .R^{m_2}$ are respectively called the upper level and lower level constraints.

A bi-level programming problem is called the Nonlinear Bi-Level Programming Problem, when at least one of the objective functions, $F(x,y)$ and $f(x,y)$, is nonlinear.

In this paper, we mainly focus on studying the case both $F(x,y)$ and $f(x,y)$ are quadratic. which can be more specifically formulated as:

$$\begin{aligned} &\min_x F(x,y) = [x^T, y^T]P \begin{bmatrix} x \\ y \end{bmatrix} + [a^T, b^T] \begin{bmatrix} x \\ y \end{bmatrix} \\ &s.t.\ \ Ax + By \leq r_1 \\ &\qquad \min_y f(x,y) = [x^T, y^T]Q \begin{bmatrix} x \\ y \end{bmatrix} \\ &\qquad s.t.\ \ Cx + Dy \leq r_2 \end{aligned} \tag{2}$$

where $a, c \in R^{n_1}$ and $b, d \in R^{n_2}$, $A \in R^{m_1 \times n_1}$, $B \in R^{m_1 \times n_2}$, $C \in R^{m_2 \times n_1}$, $D \in R^{m_2 \times n_2}$, $r_1 \in R^{m_1}$, $r_2 \in R^{m_2}$. P and Q are $n_1 + n_2$ dimension symmetrical matrices.

The Linear Bi-Level Programming Program is a strong NP-hard problem.

Checking strict local optimality in Linear Bi-level Programming Problem is NP-hard.

Based on the settings in the previous section, Sherali [18] mentioned that if there is only one leader, the Stackelberg problems has a hierarchical structure similar to that of BLPP, and these two kinds of problems are equivalent with each other, although in the Stackelberg game, lower level problem is an equilibrium rather than an optimization problem. Based on the above discussion, we introduce some further definitions followed with the bi-level programming problem [11, 19].

The bi-level programming model arises in many useful engineering and economic application fields such as transportation and network design [24], power system management [10], government policy making [2], financial decision making. For further study, Bard [3] and Dempe [6] are good general references on introducing the bi-level programming problem. Besides them, Dempe [7] provides a survey which covers applications as well as major theoretical development, and the latest overview for this topic is [4]. But despite the bi-level programming problem would appropriately model the decision process in a wide range of real life situation, the truth of the matter is that the usage of BLPP model are quite scarce. Hansen et al. [9] has proved that the bi-level programming problem is a strong NP hard problem by using reduction from KERNEL. By using reduction from 3-SAT, Vicente and Calamai [20] has further proved that merely checking strict or local optimality of a bi-level programming problem is also NP Hard. Here, we give a brief introduction on the complexity of the bi-level programming problem based on [15].

Algorithms based on this idea were proposed by Savard and Gauvin [17], and Fortuny-Amat and McCarl for solving linear bi-level programming problems [4, 5, 8]. The approach was adapted by Bard and Moore to linear-quadratic problems to the quadratic case [1]. Combining branch-and-bound, monotonicity principles and penalties similar to those used in mixed-integer programming, Hansen et al. [9] have developed a code capable of solving medium-sized linear bi-level programming programs [4].

For which the authors developed a method to get the optimal solution. Alternatively, Vicente and Calamai [20] proposed a descent method for convex quadratic bi-level programming programs, i.e., problems where both objectives are quadratic, and where constraints are linear [4]. They extend the work of Savard and Gauvin [17] by solving problem (1.29) using the sequential LCP method, and propose a way to compute exact step sizes. Motivated by the fact that checking local optimality in the sequential LCP approach is very difficult, Vicente and Calamai [20] have designed a hybrid algorithm using both the above-mentioned features and a pivot step strategy that enforces the complementarity constraints.

2 Proposed Method for Solving Quadratic Bi-Level Programming Problem

In this section, we will introduce an algorithm developed by us for solving quadratic bi-level programming problem, a problem harder to solve than linear BLPP. As mentioned before, although researchers have published a certain amount of methods for solving BLPP problem, the fact is that the usage of meta-heuristic methodology just started for about a decade and the neural network based methods are quite scarce. The proposed algorithm is a hybrid method which combines double layer neural network and artificial bee colony algorithm. The double layered neural network was proposed by Watada and Oda [22], Yaakob et al. [23]. Li et al. [11] also consider such problem in [12]. Such combination can provides many advantages. At first, the double layer neural network is essentially a parallel computing model, which provides a capability of fast computing. Second, the neural network part in this algorithm has an exponential convergence rate to the solution of the lower level problem of the quadratic BLPP, which plays an important role in finding the final optimal solution fast and accurately. Thirdly, the artificial bee colony algorithm provides a fast global search ability which can avoid situations that the searching process sticks into local optimal solution. In the first Sect. 2.1, we will firstly give an overview of our proposed method.

2.1 *Overview of the Proposed Method*

The overview structure of the proposed method is shown in Fig. 1. This hybrid method mainly consists two parts, the first part is the artificial bee colony algorithm for dealing with the upper level problem, then for each fixed upper level variable x, which is selected by the artificial bee colony algorithm by the upper level, the second part is the double layer neural network for dealing with the lower level problem.

Now we will take a closer look at the equation of the quadratic bi-level programming problem (1.2), since the parameter Q is a symmetrical matrix, we let $Q = \begin{bmatrix} Q_2 & Q_1^T \\ Q_1 & Q_0 \end{bmatrix}$, where $Q_0 \in R^{n_2} \times R^{n_2}$, $Q_1 \in R^{n_2} \times R^{n_1}$, $Q_2 \in R^{n_1} \times R^{n_1}$, thus we have:

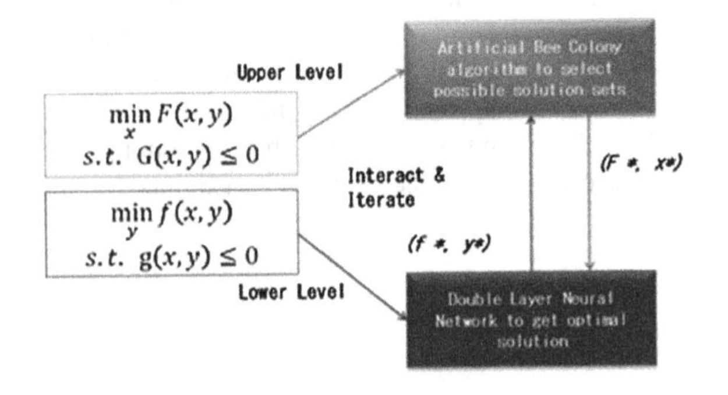

Fig. 1 Overview of the proposed method for solving quadratic bi-level programming problem

Theorem 1 *The quadratic bi-level programming problem of (1.2) is equivalent to the problem:*

$$\begin{aligned} &\min_x F(x,y) = \frac{1}{2}\left[x^T y^T\right] 2P \begin{bmatrix} x \\ y \end{bmatrix} + \left[a^T b^T\right] \begin{bmatrix} x \\ y \end{bmatrix} \\ &s.t. \quad Ax + By \le r_1 \\ &\qquad \min_y f(x,y) = y^T Q_0 y + (d + 2Q_1 x)^T y \\ &\qquad s.t. \quad Cx + Dy \le r_2 \end{aligned} \tag{3}$$

Proof With the definition of Q, the lower level problem is transformed into

$$f(x,y) = y^T Q_0 y + (d + 2Q_1 x)^T y + c^T x + x^T Q_2 x \tag{4}$$

For each fixed upper level value x_0 in the constraint region Ω, the term $c^T x_0 + x_0^T Q_2 x_0$ is a constant to the lower level problem $f(x_0, y)$, i.e., the lower level rational reaction set $R(x_0)$ will not be effected by the value of $c^T x_0 + x_0^T Q_2 x_0$. Hence, the lower level problem can be simplified as $f(x_0, y) = y^T Q_0 y + (d + 2Q_1 x_0)^T y$ by letting $c = 0$ and $Q_2 = 0$ for each fixed x_0.

With Theorem 1, we can accelerate the algorithm when calculating the lower level problem without calculating the value of $c^T x_0 + x_0^T Q_2 x_0$.

The artificial bee colony algorithm (ABC) is inspired by the foraging behavior of bee, it was proposed by KaraBoga. ABC is like other swarm intelligence algorithm, it also has the characteristics of distributed control, global information transmission and self-organization. Thanks to the characteristic of distributed control, the system will become more stable, if one bee can't work well, it will not have much impact on the group. Another feature is the system is very easy to achieve because each bee is very simple. The function of the whole system is realized through the complex interaction of the simple individual.

3 Experiments and Discussions

In this section, we will discuss the experiments about the neural network based hybrid method for solving quadratic bi-level programming problem. In order to test the efficiency and the accuracy in terms of finding global optimal solution for the proposed method, we execute the algorithm based on 4 examples as below:

Example 1 [14]

$$\begin{aligned} &\min_x F = x^2 - 4x + y_1^2 + y_2^2 \\ &s.t. \quad 0 \le x \le 2 \\ &\qquad \min_y f = y_1^2 + 0.5y_2^2 + y_1 y_2 + (1 - 3x)y_1 + (1 + x)y_2 \\ &\qquad s.t. \quad y_1 \ge 0, \qquad y_2 \ge 0 \\ &\qquad\qquad 2y_1 + y_2 - 2x \le 1 \end{aligned} \tag{5}$$

Example 2 [16]

$$\begin{aligned} &\min_x F = -7x_1 + 4x_2 + y_1^2 + y_3^2 - y_1y_3 - 4y_2 \\ &s.t. \quad x_1, x_2 \geq 0 \\ &\quad\quad x_1 + x_2 \leq 1 \\ &\quad\quad \min_y f = (1 - 3x_1)y_1 + (1 + x_2)y_2 + y_1^2 + 0.5y_2^2 + 0.5y_3^2 + y_1y_2 \\ &\quad\quad s.t. \quad y_1, y_2, y_3 \geq 0 \\ &\quad\quad\quad\quad x_1 - 2x_2 + 2y_1 + y_2 - y_3 + 2 \leq 0 \end{aligned} \tag{6}$$

Example 3 [13]

$$\begin{aligned} &\min_x F = x^2 + 2y_1^2 - 4x + 1 \\ &s.t. \quad x \geq 0 \\ &\quad\quad \min_y f = 4y_1^2 + 4y_2^2 - 16y_1 - 4y_2 + xy_1 + 17 \\ &\quad\quad s.t. \quad y_1 \geq 0, \qquad y \geq_2 \geq 0 \\ &\quad\quad\quad\quad 4x + 5y_1 + 4y_2 \leq 12 \\ &\quad\quad\quad\quad -4x - 5y_1 + 4y_2 \leq -4 \\ &\quad\quad\quad\quad 4x - 4y_1 + 5y_2 \leq 4 \\ &\quad\quad\quad\quad -4x + 4y_1 + 5y_2 \leq 4 \end{aligned} \tag{7}$$

Example 4 [11]

$$\begin{aligned} &\min_x F = x_1^2 + x_2^2 - 60x_1 - 40x_2 - 20y_1 + 20y_2 + 1300 \\ &s.t. \quad 11x_1 + x_2 - 25 \leq 0 \\ &\quad\quad -x_1 - 2x_2 + 30 \leq 0 \\ &\quad\quad x_2 \leq 15 \\ &\quad\quad \min_y f = x_1^2 + x_2^2 + y_1^2 + y_2^2 - 2x_1y_1 - 2x_2y_2 \\ &\quad\quad s.t. \quad 0 \leq y_1 \leq 10 \\ &\quad\quad\quad\quad 0 \leq y_2 \leq 10 \end{aligned} \tag{8}$$

For a better understanding of our proposed method, we will first elaborate the mechanism of the proposed method with Example 2. For example, after the GA operator and Scout-bee phase, the improved ABC algorithm part will possibly generate 4 feasible selection for the upper level variable x as: $(0.15, 0.15)$, $(0.2, 0.8)$, $(0.74, 0.22)$ and $(0.61, 0.39)$. For those selections, the formulation of the lower level objective function remains unchanged as: $f = (1 - 3x_1)y_1 + (1 + x_2)y_2 + y_1^2 + 0.5y_2^2 + 0.5y_3^2 + y_1y_2$. But since the value of x has been fixed by those selections, the f becomes a quadratic function that only related to y. At the same time, the lower level constraints are also becomes $y_1 \geq 0$, $y_2 \geq 0$, $y_3 \geq 0$ and $2y_1 + y_2 - y_3 \leq 2x_2 - x_1 - 2$ with constant value of x. Then, during the execution of the proposed method, a double layer neural network depicted will be formed for the lower level problem. The important parameters for Example 2 are as follows:

$$P = \begin{bmatrix} 0\,0 & 0 & 0 & 0 \\ 0\,0 & 0 & 0 & 0 \\ 0\,0 & 1 & 0 & -0.5 \\ 0\,0 & 0 & 0 & 0 \\ 0\,0 & -0.5 & 0 & 1 \end{bmatrix}, \quad A = \begin{bmatrix} -1 & 0 \\ 0 & -1 \\ 1 & 1 \end{bmatrix}, \quad B = 0 \tag{9}$$

$$a = \begin{bmatrix} -7 \\ 4 \end{bmatrix}, \quad b = \begin{bmatrix} 0 \\ -4 \\ 0 \end{bmatrix}, \quad c = 0, \quad d = \begin{bmatrix} 1 \\ 1 \\ 0 \end{bmatrix}, \quad r_1 = \begin{bmatrix} 0 \\ 0 \\ 1 \end{bmatrix} \tag{10}$$

$$Q_0 = \begin{bmatrix} 1 & 0.5x & 0 \\ 0.5 & 0.5 & 0 \\ 0 & 0 & 0.5 \end{bmatrix}, \quad Q_1 = \begin{bmatrix} -1.5 & 0 \\ 0 & 0.5 \\ 0 & 0 \end{bmatrix}, \quad Q_2 = 0 \tag{11}$$

$$C = \begin{bmatrix} 0 & 0 \\ 0 & 0 \\ 0 & 0 \\ 1 & -2 \end{bmatrix}, \quad D = \begin{bmatrix} -1 & 0 & 0 \\ 0 & -1 & 0 \\ 0 & 0 & -1 \\ 2 & 1 & -1 \end{bmatrix}, \quad r_2 = \begin{bmatrix} 0 \\ 0 \\ 0 \\ -2 \end{bmatrix} \tag{12}$$

For the selection $(0.15, 0.15)$, we can see from the picture, the neural network finally converges, we get an optimal y as $(0, 0, 1.8499)$. The corresponding upper level and lower level value are $F = 2.9721$ and $f = 1.7111$.

Finally, for these 4 selections, the algorithm get 4 result pairs of (F, f, x, y), they are listed in the following Table 1. The result pairs will then be sorted in terms of the value of F with ascending order. Then the algorithm selects k optimal for the next iteration. Here, k equals to the population size popsize of the improved ABC algorithm. Besides, with above figures showing the transition behaviors, we can see that each time the double layer neural network converges to the optimal solution of the lower level problem very fast.

Now we end elaborations for the algorithm's mechanism. Next, we perform experiments for these 4 examples. The experiments are performed with MATLAB 7.14, based on a Windows 8 PC with 8GB RAM and Intel Core i5-4300U 1.9GHz CPU. Although this Intel Core i5 is a multi-processor CPU, we only use single CPU core during our experiments on this PC (Fig. 2).

Table 1 Four result pairs of Example 2 during the iteration of the proposed algorithm

F	f	x	y
0.6385	1.6743	(0.6100, 0.3900)	(0, 0, 1.8299)
2.1600	0.1800	(0.2000, 0.8000)	(0, 0, 0.6000)
2.9721	1.7111	(0.1500, 0.1500)	(0, 0, 1.8499)
4.6994	4.4997	(0.7400, 0.2200)	(0, 0, 2.9999)

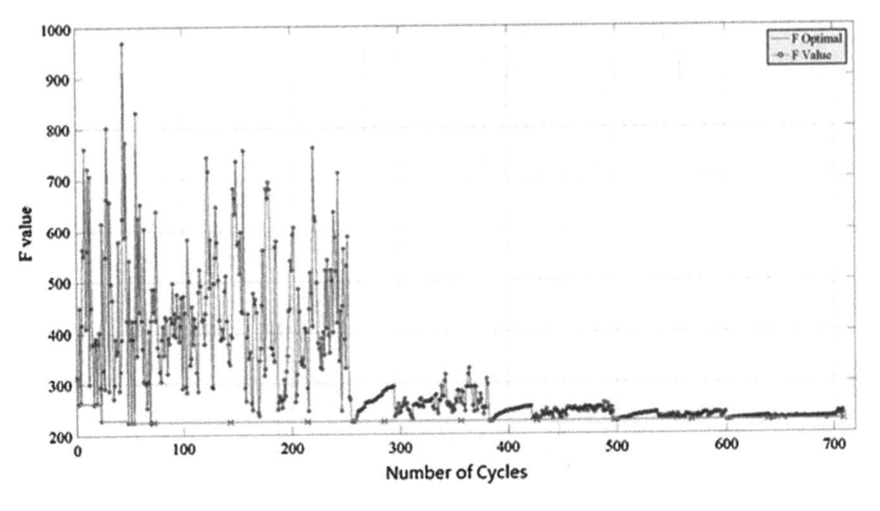

Fig. 2 Transition behavior of the proposed method for finding upper level optimal value in Example 4

In these figures, the blue lines and circles show the F value of each feasible selection. It can be easily seen that the proposed algorithm has a wide range searching ability when trying to find the optimal value. The curve with red lines and green crosses show the F optimal value, i.e. the smallest F value in these four examples during the whole execution process of the algorithm. In Examples 1 and 4, the algorithm stays stable to a smallest F value very quickly while in Examples 2 and 3, the algorithm goes stable to a smallest F value relatively slower. In total, the algorithm deals with these examples in a short time, all examples are executed within 3 min. The optimal solutions achieved by the proposed are listed in the Table 2. F^* corresponds to the optimal value of the upper=level objective function.

f^* corresponds to the optimal value of the lower level objective function. x^* and y^* respectively correspond to the upper level variables and lower level variables when the F^* is achieved. The column t stands for the average execution time achieved by 50 separate runs of the algorithm for each example. Since currently there is no mathematical method for proof of the optimal solution in the research of the bi-level programming problem, we will show our optimal solution compared with

Table 2 Optimal solutions of four examples achieved by the proposed algorithm

	F^*	f^*	x^*	y^*	t (s)
Example 1	−2.0769	−0.6017	(0.8505, 0.7757)	(0)	80.0
Example 2	0.6385	1.6743	(0.6100, 0.3800)	(0, 0, 1.8299)	66.1
Example 3	−1.4073	7.6170	(1.8889)	(0.8889, 0)	143.8
Example 4	225	100	(20, 5)	(10, 5)	41.3

Table 3 Optimal solutions of four examples achieved by other scholars' methods

	F^*	f^*	x^*	y^*	
Example 1	[30]	−2.0	2.5	(1.0, 0)	(1.0)
Example 2	[99]	0.6426	1.6708	(0.6090, 0.3910)	(0, 0, 1.8280)
Example 3	[28]	−1.3985	7.5853	(1.883)	(0.891, 0.003)
Example 4	[45]	225	100	(20, 5)	(10, 5)

other scholars' in the literature as Table 3 listed. In comparison with other scholars' methods, the proposed neural network based hybrid method has found equal or better optimal solution for the four examples of the quadratic bi-level programming problem.

4 Conclusions

In this research, we have developed a double layer neural network based hybrid method for solving the quadratic bi-level programming problems. This proposed algorithm combines an improved ABC algorithm and a Double-Layered Boltzmann machine together in order to solve such problems efficiently and accurately. The improved ABC algorithm is developed for dealing with the upper level problem.

After implemented this double layer neural network based hybrid method, we performed experiments for the proposed methods based on four examples of quadratic bi-level programming problem. In order to show the performance and accuracy of the proposed method, we have divided the experiments into two parts. In the first part, we give an elaboration to demonstrate how the improved ABC algorithm and double layer neural network work together. Then, we executed the algorithm with the four examples and compared the optimal solutions with other scholars' method.

References

1. Bard, J.F., Moore, J.T.: A branch and bound algorithm for the bilevel programming problem. SIAM J. Sci. Stat. Comp. **11**(2), 281–292 (1990)
2. Bard, J.F., Plummer, J., Sourie, J.C.: A bilevel programming approach to determining tax credits for biofuel production. Eur. J. Oper. Res. **120**(1), 30–46 (2000)
3. Bard, J.F.: Practical Bilevel Optimization: Algorithms and Applications, vol. 30, 476 pages. Springer (1998)
4. Colson, B., Marcotte, P., Savard, G.: An overview of bilevel optimization. Ann. Oper. Res. **153**(1), 235–256 (2007)
5. Colson, B., Marcotte, P., Savard, G.: Bilevel programming: a survey. OR **43**(2), 87–107 (2005)
6. Dempe, S.: Foundations of Bilevel Programming. Springer (2002)
7. Dempe, S.: Annotated bibliography on bilevel programming and mathematical programs with equilibrium constraints. Optim.: J. Math. Progr. Oper. Res. **52**(3), 333–359 (2003)

8. Fortuny-Amat, J., McCarl, B.: A representation and economic interpretation of a two-level programming problem. J. Oper. Res. Soc. **32**(9), 783–792 (1981)
9. Hansen, P., Jaumard, B., Savard, G.: New branch-and-bound rules for linear bilevel programming. SIAM J. Sci. Stat. Comput. **13**(5), 1194–1217 (1992)
10. Hobbs, B.F., Carolyn, B.M., Pang, J.S.: Strategic gaming analysis for electric power systems: an MPEC approach. IEEE Trans. Power Syst. **15**(2), 638–645 (2000)
11. Li, H., Jiao, Y., Zhang, L.: Orthogonal genetic algorithm for solving quadratic bilevel programming problems. J. Syst. Eng. Electr. **21**(5), 763–770 (2010)
12. Li, J., Watada, J., Yaakob, S.B.: A genetic algorithm based double layer neural network for solving quadratic bilevel programming problem. In: 2014 International Joint Conference on Neural Networks (IJCNN), pp. 382–389. IEEE (2014)
13. Lv, Y., Tiesong, H., Wang, G., Wan, Z.: A neural network approach for solving nonlinear bilevel programming problem. Comput. Math. Appl. **55**(12), 2823–2829 (2008)
14. Lv, Y., Chen, Z., Wan, Z.: A neural network for solving a convex quadratic bilevel programming problem. J. Comput. Appl. Math. **234**(2), 505–511 (2010)
15. Marcotte, P., Gilles, S.: Bilevel programming: a combinatorial perspective. In: Graph Theory and Combinatorial Optimization, pp. 191–217. Springer US (2005)
16. Muu, L.D., Van Quy, N.: Global optimization method for solving convex quadratic bilevel programming problems. J. Global Optim. **26**(2), 199–219 (2003)
17. Savard, G., Gauvin, J.: The steepest descent direction for the nonlinear bilevel programming problem. Oper. Res. Lett. **15**(5), 265–272 (1994)
18. Sherali, H.D., Soyster, A.L., Murphy, F.H.: Stackelberg-Nash-Cournot equilibria: characterizations and computations. Oper. Res. **31**(2), 253–276 (1983)
19. Shih, H.-S., Wen, U.-P., Lee, S., Lan, K.-M., Hsiao, H.-C.: A neural network approach to multiobjective and multilevel programming problems. Comput. Math. Appl. **48**(1), 95–108 (2004)
20. Vicente, L.N., Calamai, P.H.: Bilevel and multilevel programming: a bibliography review. J. Global Optim. **5**(3), 291–306 (1994)
21. Vicente, L., Savard, G., Judice, J.: Discrete linear bilevel programming problem. J. Optim. Theory Appl. **89**(3), 597–614 (1996)
22. Watada, J., Oda, K.: Formulation of a two-layered Boltzmann machine for portfolio selection. Int. J. Fuzzy Syst. **2**(1), 39–44 (2001)
23. Yaakob, S.B., Watada, J., Fulcher, J.: Structural learning of the Boltzmann machine and its Application to life cycle management.Neurocomputing **74**(12–13), 2193–2200 (2011)
24. Yang, Hai, Bell, M.G.H.: Models and algorithms for road network design: a review and some new developments. Transp. Rev. **18**(3), 257–278 (1998)

A Memetic Fuzzy ARTMAP by a Grammatical Evolution Approach

Shing Chiang Tan, Chee Peng Lim and Junzo Watada

Abstract This paper presents a memetic fuzzy ARTMAP (mFAM) model constructed using a grammatical evolution approach. mFAM performs adaptation through a global search with particle swarm optimization (PSO) as well as a local search with the FAM training algorithm. The search and adaptation processes of mFAM are governed by a set of grammatical rules. In the memetic framework, mFAM is constructed and it evolves with a combination of PSO and FAM learning in an arbitrary sequence. A benchmark study is carried out to evaluate and compare the classification performance between mFAM and other state-of-art methods. The results show the effectiveness of mFAM in providing more accurate prediction outcomes.

Keywords Grammatical evolution · Fuzzy ARTMAP · Particle swarm optimization · Data classification

1 Introduction

Artificial neural network (ANN) represents a computing paradigm that imitates the biological nervous system of the brain to process information. ANNs learn information from data through a training session. An ANN model can be constituted by a supervised or unsupervised method. Examples of unsupervised ANNs are growing neural gas [1] and self-organizing map [2], while examples of supervised

S.C. Tan (✉)
Multimedia University, Cyberjaya, Malaysia
e-mail: sctan@mmu.edu.my

C.P. Lim
Deakin University, Waurn Ponds, Australia
e-mail: chee.lim@deakin.edu.au

J. Watada
Waseda University, Tokyo, Japan
e-mail: junzow@osb.att.ne.jp

I. Czarnowski et al. (eds.), *Intelligent Decision Technologies 2016*,
Smart Innovation, Systems and Technologies 56,
DOI 10.1007/978-3-319-39630-9_38

ANNs are radial basis function [3], multilayer layer perceptron (MLP) [4] and adaptive resonance theory (ART) [5] ANNs. An ANN can also be trained in a semi-supervised manner (e.g. growing self-organizing map [6]). Notably, ANN learning is not optimized in nature and, therefore, its generalization performance is likely to be sub-optimal.

To improve the generalization capability of an ANN, evolutionary-based ANNs (EANNs) have been introduced, e.g. [7, 8]. EANNs adopt the heuristic evolutionary algorithm as an additional mode of adaptation on top of conventional ANN learning. ANNs have also been integrated with other heuristic methods inspired from the nature, such as particle swarm optimization (PSO) [9], gravitational search algorithm [10], and ant colony algorithm [11] to improve its generalization capability.

Another increasingly popular way to devise a computing method is to complement an evolutionary algorithm with a local search technique, i.e., the *memetic algorithm* [12]. According to [13], *memes* represents a social and cultural entity equivalent to a gene. *Memes* is a unit of knowledge that can be available in multiple copies in the human brain. These memes are modifiable, and they can be combined with another memes to produce a new memes. In this paper, a memetic ANN is composed through grammatical evolution (GE) [14]. Specifically, a supervised Adaptive Resonance Theory (ART) network, i.e., Fuzzy ARTMAP (FAM) [15], is applied as the local search algorithm to elicit information from a given training data set as well as to fine tune the solution space. FAM performs incremental learning by absorbing new information continuously without forgetting previously learned information. On the other hand, a particle swarm optimization (PSO) [16] algorithm is deployed as the global search algorithm to explore the basin of the optimal solution space. Both PSO and FAM constitute the basic components of the memetic FAM model for integration with GE. During the course of GE, output languages are generated from a set of rules through a grammatical inference mechanism. These output languages consist of a combination of artifacts/programs for evolving the memetic FAM model (mFAM) with local learning and global search. In other words, each FAM in the population-based mFAM is deemed as a memes; it is modified and combined with another memes with these artifacts/programs.

2 Method

2.1 Grammatical Evolution

In GE [14], grammatical rules are formulated using an evolutionary approach. The grammar of a language is defined in the Backus-Naus Form (BNF) [17]. BNF specifies production rules generated from the grammars to solve the problem under scrutiny.

The genotype of GE is a group of variable-length binary strings, in which each codon (of an 8-bit size) represents an integer. These integers are mapped through a function to determine a set of production rules based on the BNF definition. The search for all constituents of a production rule is accomplished by using an evolutionary algorithm.

2.2 *Fuzzy ARTMAP*

FAM [15] is a constructive ANN for which its structure is built from scratch through an incremental learning process. FAM is made up of a pair of fuzzy ART models that is linked by a map field, F^{ab}. One of the fuzzy ART model is used as the input module (i.e. ART_a) that receives the input vectors. Another fuzzy ART model is used s the output module (i.e. ART_b) that produces the predicted output class of a training sample. Each fuzzy ART model consists of three layers of nodes. F_0^a of ART_a is the normalization layer for implementing complement-coding [15], in which an M-dimensional input vector, $\boldsymbol{a}$, to normalised a *2 M-dimensional* vector, $\boldsymbol{A} = (\boldsymbol{a}, 1 - \boldsymbol{a})$. F_1^a is the input layer for receiving $\boldsymbol{A}$; while F_2^a is the recognition layer for keeping a group of prototypes of clustered information learned from data samples. With an incremental learning property, new recognition nodes can be inserted into F_2^a to learn new information wherever it is necessary.

The dynamics of ART_a is described, as follows. Upon complement-coding an input vector as $\boldsymbol{A}$, it is propagated to F_2^a. The activation of each prototype node in response to the input vector is computed using a choice function [15],

$$T_j = \frac{\left|\boldsymbol{A} \wedge \boldsymbol{w}_j^a\right|}{\alpha + \left|\boldsymbol{w}_j^a\right|} \tag{1}$$

where α is the choice parameter that assumes a small positive value close to 0 [15]; $\boldsymbol{w}_j^a$ is the weight of the j-th prototype node. The winning node, denoted as node J, is determined by selecting the highest activation score. A vigilance test is then conducted to compare the similarity measure between $\boldsymbol{w}_J^a$ and $\boldsymbol{A}$ against a vigilance parameter [15] $\rho_a \in [0,1]$.

$$\frac{\left|\boldsymbol{A} \wedge \boldsymbol{w}_J^a\right|}{|\boldsymbol{A}|} \geq \rho_a \tag{2}$$

A new search cycle for another winning node is initiated if the vigilance test is violated. The process of searching for a new winning node is continued until it is able to pass the vigilance test. If none of the existing prototype nodes can satisfy the vigilance test, a new node is created in F_2^a to learn the input sample. The output module, ART_b, also performs the same pattern-matching procedure as in ART_a.

In this regards, ART_b receives the target vector, and a winning node that represents the target class is identified.

After the winning node has been identified in both ART_a and ART_b, a map-field vigilance test is executed to determine accuracy of the prediction [15].

$$\frac{|\boldsymbol{y}^b \wedge \boldsymbol{w}_J^{ab}|}{|\boldsymbol{y}^b|} \geq \rho_{ab} \quad (3)$$

where $\boldsymbol{y}^b$ is the output vector of ART_b; $\boldsymbol{w}_J^{ab}$ is the weight of the winning node from F_2^a to F^{ab}; and $\rho_{ab} \in [0,1]$ is the map-field vigilance parameter. If the map-field vigilance test is not successful, this means an incorrect prediction of the target class is made. This triggers the match-tracking process [15], in which ρ_a is increased from its baseline setting of $\bar{\rho}_a$ to

$$\rho_a = \frac{|\boldsymbol{A} \wedge \boldsymbol{w}_J^a|}{|\boldsymbol{A}|} + \delta \quad (4)$$

where δ is a small positive value. The ART_a vigilance test becomes unsuccessful as a result of the increase in ρ_a. Subsequently, a new search cycle is initiated in ART_a with a higher setting of ρ_a. This search process is continued until a correct prediction is made in the map field vigilance test. As such, learning ensues. During learning, $\boldsymbol{w}_J^a$ is updated [15] to

$$\boldsymbol{w}_J^{a(new)} = \beta_a\left(\boldsymbol{A} \wedge \boldsymbol{w}_J^{a(old)}\right) + (1 - \beta_a)\boldsymbol{w}_J^{a(old)} \quad (5)$$

where $\beta_a \in [0,1]$ is the learning parameter of ART_a. ART_b undergoes the same aforementioned mechanism in ART_a by replacing a in Eqs. (1)–(5) with b.

2.3 *Particle Swarm Optimization*

PSO [16] is a population-based search method that finds optimal solutions in an iterative manner. Initially, a population of potential solutions (particles) is randomly generated at different positions in the solution space. The initial velocity of each particle is also randomly set. The particles are flown through the solution space from one session of search to another. In any search session, each particle records the individual best position (solution) that it has found thus far (namely, ***pBest***). The PSO algorithm also keeps track of the overall best position found so far from the entire population (namely, ***gBest***). Assume a standard PSO algorithm is applied to search for a likely particle j in a D-dimensional solution space, where $d = 1, 2, \ldots, D$. Then, the velocity, $\boldsymbol{V}_j$, and position, $\boldsymbol{X}_j$, are updated as follows.

$$V_{j,d}(t) = wV_{j,d}(t-1) + c_1 r_{j,d}(pBest_{j,d}(t-1) - X_{j,d}(t-1)) + c_2 r'_{j,d}(gBest_{j,d}(t-1) - X_{j,d}(t-1)) \quad (6)$$

$$X_{j,d} = X_{j,d}(t-1) + V_{j,d}(t) \quad (7)$$

where t denotes the index of iteration; c_1, c_2 are the acceleration coefficients for adjusting the importance of the acceleration terms associated with $\boldsymbol{pBest}_j$ and $\boldsymbol{gBest}_j$; w is the inertia weight; and, $r_{j,d}$, $r'_{j,d}$ are randomly generated numbers in [0,1]. The updated velocity is intended to make a better movement for a particle to reach a new position.

In this study, FAM is evolved with PSO. A population of FAM is initialized before evolution with PSO. In this case, all connection weights of F_2^a nodes are grouped to constitute a particle.

$$\boldsymbol{X}_0 = [\boldsymbol{w}_1^a \, \boldsymbol{w}_2^a \, \boldsymbol{w}_3^a \ldots \boldsymbol{w}_S^a] \quad (8)$$

where S is the number of F_2^a nodes. A population of particles is created from a perturbation process [18] as follows.

$$\boldsymbol{X}_k = \boldsymbol{X}_0 + \text{RMF} \times \text{rand}(1, n) \quad (9)$$

where $\boldsymbol{X}_k$ is the k-th (for $k = 1, 2, \ldots, K-1$) replicated particle of $\boldsymbol{X}_0$ ($K = 10$ in this study); RMF $\in [0, 1]$ is the range multiplication factor (set to 0.30 in this study); rand(1, n) is vector of n elements generated from a uniform distribution; n is the length of $\boldsymbol{X}_0$. The goodness of a particle is evaluated by considering the number of training samples that are correctly classified. Each FAM-PSO adapts and arrives at a new position according to Eqs. (6)–(7) from one iteration to another. The process of search and adaptation is continued until a pre-specified number of iteration is met.

2.4 Memetic FAM

A context free grammar (Fig. 1) is defined in BNF to evolve a population of mFAM models. A grammar is expressed as a tuple of {N, T, P, S}, where N indicates a set of non-terminals, i.e. N = {meme, S} by referring to Fig. 1; T indicates a set of terminals, i.e. T = {FAM, FAM-PSO, +} where the symbol "+" means combination; P denotes a set of production rules that associate the elements of N with T; S denotes a start symbol. An element of N can be identified by selecting one of the production rules, where each option is separated by a "|" symbol.

Each variable-length binary string of the GE genotype is used to map S to T by converting a codon of 8 bits to an integer from which a production rule is chosen using this mapping function.

Predecessor	::=	Productions	Index
<S>	::=	<meme>	0
		| <meme> + <meme>	1
		| <meme> + <meme> + <meme>	2
<meme>	::=	FAM	0
		| FAM-PSO	1

Fig. 1 The proposed grammar to develop a population of mFAM models

$$\text{Rule} = (\text{codon integer value})\ \text{MOD}\ (\text{number of rules for the current } N \text{ element}) \quad (10)$$

where MOD denotes the modulus operator. Notably, GE does not conduct an evolutionary process directly on mFAM. Instead, a variable-length genetic algorithm is employed to evolve the binary strings through the single-point crossover, mutation and selection operations. Whenever another non-terminal element exists, the process of mapping it to a production rule is carried out again by reading another codon. The output language for a variable-length binary string is a set of elements of T in an arbitrary combination.

In the proposed population-based mFAM models, a number of q units of the trained FAM (particles) models are first identified. Each FAM is trained separately with different sequences of training samples. Then, the entire FAM models proceed to evolution with GE. A group of q variable-length binary strings are defined, and then used to find different combinations of FAM and FAM-PSO for a number of generations. In this study, $q = 4$. Each binary string finds a combination set of FAM and FAM-PSO by referring to the grammar as in Fig. 1 to train an mFAM model. The most accurate mFAM model is selected for evaluation using unseen data samples from the test set.

3 Experiment

Three benchmark data sets were used to evaluate the efficacy of mFAM. They are the syntactic Ripley data set [19] and two UCI Machine Learning data sets (Iris flowers and steel plate faults) from [20]. Each mFAM model was trained in a single-epoch with $\beta = 1$ and $\bar{\rho}_a = 0.0$. Unless otherwise mentioned, the experiment was conducted for 100 runs with different orders of training samples. The details are as follows.

Table 1 Results of the Ripley data set (the 95 % confidence intervals are in square brackets)

Algorithm	Classification correct rate (%)	Network size	Training time (s)
mFAM	86.17 [85.79 86.54]	16.6 [15.9 17.3]	34.53 [30.20 38.80]
FAM (single-epoch)	81.56 [80.83 82.22]	10.2 [9.8 10.8]	0.45 [0.44 0.46]
FAM (multi-epoch)	83.40 [82.64 84.09]	23.4 [20.7 27.0]	9.27 [9.12 9.41]

3.1 The Ripley Data Set

The Ripley data set [18] is a two-class nonlinear classification problem. The data samples are artificially created from two Gaussian mixture distributions with equal covariance. The training and test sets comprised 250 and 1000 data samples, respectively. The performance of mFAM was compared with those of original FAM trained with both single- and multi-epoch methods. The bootstrapping method [21] was utilized to estimate the average classification results at the 95 % confidence interval.

Table 1 presents the results. mFAM could achieve the highest average accuracy rate (86.17 %) as compared with 81.56 and 83.32 % of single-epoch and multi-epoch FAM. The accuracy mean of mFAM fell within a range higher than those of FAM. These results show that mFAM is statistically more accurate than FAM in classification of the Ripley data. This finding indicates the benefits of deploying a global search process with PSO and a local adaptation with FAM learning within a memetic framework for improving the generalization capability of mFAM. An interesting observation is the network size of mFAM is between those of FAM trained with the single- and multi-epoch methods. In other words, mFAM is able to perform with a more accurate classification rate at the expense of using a moderate number of prototype nodes, especially when compared with an FAM model with multi-epoch training. The training session of mFAM is the longest. This is within expectation because mFAM adapts its connection weights and structure by a mixture of adaptive learning, as well as stochastic search and adaptation with PSO.

3.2 The Iris Data Set

The performance between mFAM and other evolutionary ANNs [22] was compared using the Iris flower data set. The evolutionary ANNs comprised three RBF variants trained with a genetic algorithm, namely, GA-RBF-1, GA-RBF-2, and GA-RBF-3. The data set consists of four attributes pertaining to the petal width and length of three categories of Iris flowers. In this experiment, the number of the data samples

Table 2 Results of the Iris data set [22]

Algorithm	Accuracy rate (%)	Number of nodes	Execution time (s)
mFAM	95.62	6.3	11.58
GA-RBF-2	95.04	32	1.81 + 0.07
GA-RBF-3	93.42	32	1.81 + 0.49
RBF-2	91.80	50	0.75
RBF-1	90.91	50	0.12
GA-RBF-1	86.71	32	1.81

in the training and test sets was set to a ratio of 7:3. Table 2 lists the classification results. The network size of mFAM is smaller than all RBF-based networks [22]. Overall, mFAM forms a compact and accurate model as compared to the RBF networks at the expense of a longer duration for training.

3.3 *The Steel Plate Data Set*

The steel plate data set [20] comprised a total number of 1941 samples of faulty conditions of a stainless steel leaf. Each data sample had 27 input attributes of a steel-plate image, consisting of geometric shape, luminosity, orientation, edges, and contour information. The output class was one of seven faults, i.e., pastry, Z-scratch, K-scratch, stains, dirtiness, bumps, and others.

mFAM was evaluated using a stratified 2-fold cross validation strategy [23], with 1457 samples for training set and 484 samples for test. To have a fair comparison, we followed the same experimental setup in [24] with 30 runs. In [24], four types of feature selection technique were proposed to combine with a two-stage EANN to build TSEAFS, in which its connection weights and structure were optimized in two separate sessions of evolution. The performance of TSEAFS was compared with other state-of-art classification methods using the original data and data sets with a portion of input features. In this study, we compared with the results from the original data set taken from [24].

Table 3 lists the average classification results of mFAM and other machine learning methods reported in [24], i.e., an EANN (TSEAFS), two ANNs (MLP and RBF), two rule-based classifiers (C4.5 and PART), a statistical classifier (SVM), and an instance-based classifier (1-NN). The results in Table 3 show that mFAM outperforms all other classifiers. This finding indicates the effectiveness of mFAM in improving the detection rate of faulty conditions of steel plates.

Table 3 Results of the steel plate data set [22]

Algorithm	Accuracy rate (%)
mFAM	66.44
FAM	64.84
RBF	59.94
SVM	57.02
MLP	53.50
TSEAFS	51.46
PART	46.69
1-NN	49.17
C4.5	39.05

4 Summary

In this paper, a memetic mFAM model has been introduced. mFAM undergoes a population-based search and adaptation process by a mixture of supervised ART learning and PSO algorithm in arbitrary sequences. The empirical results from three benchmark data sets positively show the effectiveness of mFAM in data classification. The outcomes show that mFAM is an evolving ANN capable of providing more accurate prediction rates than other state-of-art machine-learning classifiers.

For future work, the current version of mFAM can be extended to an ensemble model for dealing with either single- or multi-objective classification problems. The potentials of mFAM can be further explored by incorporating other computing methods within the memetic framework. Besides that, additional benchmark and real-world data sets from different application areas can be used to further demonstrate the usefulness of mFAM and its extended variants.

References

1. Shen, F., Ouyang, Q., Kasai, W., Hasegawa, O.: A general associative memory based on self-organizing incremental neural network. Neurocomputing **104**, 57–71 (2013)
2. Fišer, D., Faigl, J., Kulich, M.: Growing neural gas efficiently. Neurocomputing **104**, 72–82 (2013)
3. Ploj, B., Harb, R., Zorman, M.: Border Pairs method—constructive MLP learning classification algorithm. Neurocomputing **126**, 180–187 (2014)
4. Reiner, P., Wilamowski, B.M.: Efficient Incremental construction of RBF networks using quasi-gradient method. Neurocomputing **150**, 349–356 (2015)
5. Zhang, Y., Ji, H., Zhang, W.: TPPFAM: use of threshold and posterior probability for category reduction in fuzzy ARTMAP. Neurocomputing **124**, 63–71 (2014)
6. Allahyar, A., Yazdi, H.S., Harati, A.: Constrained semi-supervised growing self-organizing map. Neurocomputing **147**, 456–471 (2015)
7. Yao, X., Liu, Y.: A new evolutionary system for evolving artificial neural networks. IEEE Trans. Neural Networks **8**, 694–713 (1997)

8. Dragoni, M., Azzini, A., Tettamanzi, A.G.B.: SimBa: a novel similarity-based crossover for neuro-evolution. Neurocomputing **130**, 108–122 (2014)
9. Karami, A., Guerrero-Zapata, M.: A hybrid multiobjective RBF-PSO method for mitigating Dos attacks in named data networking. Neurocomputing **151**, 1261–1282 (2015)
10. Tan, S.C., Lim, C.P.: Evolving an adaptive artificial neural network with a gravitational search algorithm. In: Neves-Silva R. et al. (eds) Intelligent Decision Technologies, Smart Innovation, Systems and Technologies, vol. 39, pp. 599–609 (2015)
11. Mavrovouniotis, M., Yang, S.: Training Neural Networks with Ant Colony Optimization Algorithms for Pattern Classification. Soft. Comput. **19**, 1511–1522 (2015)
12. Moscato P.: Memetic Algorithms: a short introduction. In: Corne, D., Dorigo, M., Glover, F. (eds.) New Ideas in Optimization, McGraw-Hill, pp. 219–234 (1999)
13. Dawkins, R.: The Selfish Gene. Clarendon Press, Oxford (1976)
14. O'Neil, M., Ryan, C.: Grammatical evolution. IEEE Trans. Evol. Comput. **5**, 349–358 (2001)
15. Carpenter, G.A., Grossberg, S., Markuzon, N., Reynolds, J.H., Rosen, D.B.: Fuzzy Artmap: a neural network architecture for incremental supervised learning of analog multidimensional maps. IEEE Trans. Neural Networks **3**, 698–713 (1992)
16. Kennedy, J., Eberhart, R.C.: Particle swarm optimization. In: Proceeding of the IEEE International Conference on Neural Networks, pp. 1942–1948. Perth, Australia (1995)
17. Naur, P.: Revised report on the algorithmic language ALGOL 60. Commun. ACM **6**, 1–17 (1963)
18. Baskar, S., Subraraj, P., Rao, M.V.C.: Performance of hybrid real coded genetic algorithms. Int. J. Comput. Eng. Sci. **2**, 583–602 (2001)
19. Ripley, B.D.: Neural networks and related methods for classification. J. Roy. Stat. Soc.: Ser. B (Methodol.) **56**, 409–456 (1994)
20. Asuncion, A., Newman, D.J.: UCI machine learning repository. University of California, School of Information and Computer Science, Irvine, CA (2007). http://www.ics.uci.edu/~mlearn/MLRepository.html
21. Efron, B.: Bootstrap methods: another look at the jackknife. Ann. Stat. **7**, 1–26 (1979)
22. Ding, S., Xu, L., Su, C., Jin, F.: An optimizing method of rbf neural network based on genetic algorithm. Neural Comput. Appl. **21**, 333–336 (2012)
23. Kohavi, R.: A study of cross validation and bootstrap for accuracy estimation and model selection. In: Proceedings of the 14th International Joint Conference Artificial Intelligence (IJCAI), pp. 1137–1145. Morgan Kaufmann (1995)
24. Tallón-Ballesteros, A.J., Hervás-Martínez, C., Riquelme, J.C., Ruiz, R.: Feature selection to enhance a two-stage evolutionary algorithm in product unit neural networks for complex classification problems. Neurocomputing **114**, 107–117 (2013)

Particle Swarm Optimization Based Support Vector Machine for Human Tracking

Zhenyuan Xu, Chao Xu and Junzo Watada

Abstract Human tracking is one of the most important researches in computer vision. It is quite useful for many applications, such as surveillance systems and smart vehicle systems. It is also an important basic step for content analysis for behavior recognition and target detection. Due to the variations in human positions, complicated backgrounds and environmental conditions, human tracking remains challenging work. In particular, difficulties caused by environment and background such as occlusion and noises should be solved. Also, real-time human tracking now seems a critical step in intelligent video surveillance systems because of its huge computational workload. In this paper we propose a Particle Swarm Optimization based Support Vector Machine (PSO-SVM) to overcome these problems. First, we finish the preliminary human tracking step in several frames based on some filters such as particle filter and kalman filter. Second, for each newly come frame need to be processed, we use the proposed PSO-SVM to process the previous frames as a regression frame work, based on this regression frame work, an estimated location of the target will be calculated out. Third, we process the newly come frame based on the particle filter and calculate out the target location as the basic target location. Finally, based on comparison analysis between basic target location and estimated target location, we can get the tracked target location. Experiment results on several videos will show the effectiveness and robustness of the proposed method.

Keywords Human tracking · Occlusion · Real-time · Particle filter · PSO-SVM

Z. Xu (✉) · C. Xu
Nanjing Audit University, West Yushan Road 86, Pukou, Nanjing, China
e-mail: xzyzx05@nau.edu.cn

C. Xu
e-mail: xuchao@nau.edu.cn

J. Watada
Graduate School of Information, Production and Systems,
Waseda University, 2-7 Hibikino, Wakamatsu-ku, Kitakyushu, Fukuoka, Japan
e-mail: junzow@osb.att.ne.jp

I. Czarnowski et al. (eds.), *Intelligent Decision Technologies 2016*,
Smart Innovation, Systems and Technologies 56,
DOI 10.1007/978-3-319-39630-9_39

1 Introduction

Since Video Motion Detection (VMD) first used image processing to perform security work in the early 1980s, various methods have been developed for human tracking and detection systems, which are widely used in many fields including security, robots and surveillance systems.

To support security, surveillance and many other systems, tracking systems today require increasingly high tracking accuracy. Researchers have focused on finding method to track objects increasingly accurately [1]. For this purpose, human tracking systems use various methods such as Kalman filter, particle swarm optimization (PSO) and particle filter. The particle filter is also called the Sequential Monte Carlo method (SMC) [2].

Based on a sequence of noisy measurements in a security system, estimates are made to pursue various changing states of a system over time [3]. As a technique that estimates the next view by considering the sequences of the front ground, human tracking also requires a dynamic estimation method to realize highly accurate measurement in the presence of noise.

In any tracking system, such analysis and inference lead to focusing on the following two points: First, we must take into consideration system noise, that is, we must estimate the system by using the front sequence under noise in the system. Second, we must overcome difficulties caused by environment such as occlusion and multi object problems. In video sequences, as the position of the camera and the target object are in the same line, the target will be occluded by obstacles or other objects in the environment, the system can not get the effective information from the video sequence.

The particle filter provides highly-accurate results by expressing the distribution of the random state particles from the posterior probability. However, the resampling stage will cause the sample effectiveness, diversity loss and sample impoverishment, which may fail to solve complicated environment human tracking problems. Thus, we use the PSO-SVM as regression framework to assist the conventional tracking methods. After finishing the preliminary human tracking step in several frames based on particle filter, we use the proposed PSO-SVM to process the previous frames as a regression framework, based on this regression framework, an estimated location of the target will be calculated out, then we process the newly come frame based on the particle filter and calculate out the target location as the basic target location. Finally, based on comparison analysis between basic target location and estimated target location,

The objective of this paper is to show how to solve the human tracking step under occlusion in the real-time by the proposed PSO-SVM.

This paper consists of the following: Sect. 2 explains details on the PSO-SVM. In Sect. 3 we give some experiments and compare results to show the effectiveness of our proposed method. We finally summarize our conclusions in Sect. 4.

2 Method

2.1 Human Tracking Part

The tracking part generally consists of three main modules: (1) data collection, (2) target detection and (3) target tracking. Figure 1 shows the framework of the tracking step.

2.2 Preparation

We will finish the preliminary human tracking step in several frames based on particle filter in the preparation step.

In statistics, particle filters, also known as Sequential Monte Carlo (SMC) methods [4], are sophisticated model estimation techniques based on simulation. Particle filters have important applications in econometrics. An SMC performs important functions in many fields. This method is also known as bootstrap filtering [5], the condensation algorithm [6], particle filtering [7], interacting particle approximations [8] and survival of the fittest [2]. In tracking problems, particle filtering is an important tool, especially when methods require nonlinear and non-Gaussian environments [9]. This method can effectively track moving objects robustly in a cluttered environment [9].

In human tracking applications, many researchers apply particle filters. McKenna and Nait-Charif, for example, successfully tracked human motion with iterated likelihood weighting using this method [9], and Tung and Matsuyama proposed color-based particle filters driven by optical flow to track human motion. Hierarchical genetic particle filters are a new variation for performing articulated human tracking. This technique has been proposed by Ye et al. [7].

Sampling importance resampling (SIR), the original particle filtering algorithm [10], is a very commonly used particle filtering algorithm, which approximates the filtering distribution $p(x_k|y_0 \cdots y_k)$ by a weighted set of P particles $\{(w_k^{(L)}, x_k^{(L)}) : L \in \{1 \cdots P\}\}$.

Importance weights $w_k^{(L)}$ are approximations to the relative posterior probabilities (or densities) of particles such that $\sum_{L=1}^{P} w_k^{(L)} = 1$. SIR is a sequential (i.e., recursive) version of importance sampling. As in importance sampling, the expectation of a function $f(\cdot)$ can be approximated as a weighted average

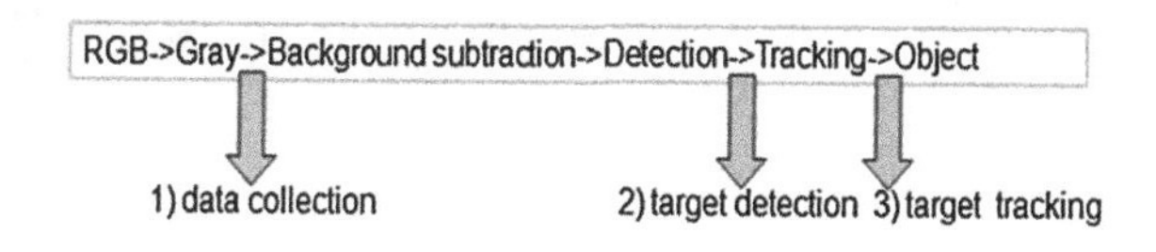

Fig. 1 Framework of tracking part

$$\int f(x_k)p(x_k|y_0,\ldots,y_k)dx_k \approx \sum_{L=1}^{P} w^{(L)} f(x_k^{(L)}) \tag{1}$$

For a finite set of particles, the algorithm performance is dependent on the choice of the proposal distribution

$$\pi(x_k|x_{0:k-1},y_{0:k}). \tag{2}$$

The optimal proposal distribution is given as the target distribution

$$\pi(x_k|x_{0:k-1},y_{0:k}) = p(x_k|x_{k-1},y). \tag{3}$$

However, the transition prior is often used as importance function, since using the transition prior makes it easier to draw particles (or samples) and perform subsequent importance weight calculations:

$$\pi(x_k|x_{0:k-1},y_{0:k}) = p(x_k|x_{k-1}). \tag{4}$$

Sampling Importance Resampling (SIR) filters with the transition prior as an importance function are commonly known as bootstrap filters and condensation algorithms.

Resampling is used to avoid the problem of degeneracy of the algorithm [11], that is, avoiding a situation in which all but one of the importance weights are close to zero. Algorithm performance can also be affected by the proper choice of resampling method. The stratified resampling proposed by [11] is optimal in terms of variance.

A single step of sequential importance resampling is as follows:

(1) For $L = 1,\ldots,P$, draw samples from the proposal distribution

$$x_k^{(L)} \; \pi(x_k|x_{0:k-1}^{(L)},y_{0:k}) \tag{5}$$

(2) For $L = 1,\ldots,P$, update importance weights up to a normalizing constance:

$$w_k^{(L)} = w_{k-1}^{(L)} \frac{p(y_k|x_k^{(L)})p(x_k^{(L)}|x_{k-1}^{(L)})}{\pi(x_k^{(L)}|x_{0:k-1}^{(L)},y_{0:k})} \tag{6}$$

Note that this simplifies to the following:

$$w_k^{(L)} = w_{k-1}^{(L)} p(y_k|x_k^{(L)}) \tag{7}$$

When we use:

$$\pi(x_k^{(L)}|x_{0:k-1}^{(L)},y_{0:k}) = p(x_k^{(L)}|x_{k-1}^{(L)}) \tag{8}$$

(3) For $L = 1, \ldots, P$, compute the normalized importance weights:

$$w_k^{(L)} = \frac{w_k^{(L)}}{\sum_{j=1}^{P} w_k^{(j)}} \tag{9}$$

(4) Compute an estimate of the effective number of particles as

$$\mathbb{N}_{eff} = \frac{1}{\sum_{L=1}^{P} (w_k^{(L)})^2} \tag{10}$$

(5) If the effective number of particles is less than a given threshold $\mathbb{N}_{eff} < N_{thr}$ then perform resampling:

Draw P particles from the current particle set with probabilities proportional to their weights. Replace the current particle set with this new one.

(b) For $L = 1, \ldots, P$, set $w_k^{(L)} = \frac{1}{P}$.

A pseudo-code description of this resampling algorithm is given by Algorithm 1 (Fig. 2).

```
Algorithm 2: Resampling Algorithm
```

$[\{\mathbf{x}_k^{j*}, w_k^j, i^j\}_{j=1}^{N_s}] = \text{RESAMPLE } [\{\mathbf{x}_k^i, w_k^i\}_{i=1}^{N_s}]$

- Initialize the CDF: $c_1 = 0$
- FOR $i = 2: N_s$
  - Construct CDF: $c_i = c_{i-1} + w_k^i$
- END FOR
- Start at the bottom of the CDF: $i = 1$
- Draw a starting point: $u_1 \sim \mathrm{U}[0, N_s^{-1}]$
- FOR $j = 1: N_s$
  - Move along the CDF: $u_j = u_1 + N_s^{-1}(j-1)$
  - WHILE $u_j > c_i$
  * $i = i + 1$
  - END WHILE
  - Assign sample: $\mathbf{x}_k^{j*} = \mathbf{x}_k^i$
  - Assign weight: $w_k^j = N_s^{-1}$
  - Assign parent: $i^j = i$
- END FOR

Fig. 2 Algorithm 1

2.3 PSO-SVM

For each newly come frame need to be processed, we use the proposed PSO-SVM to process the previous frames as a regression framework, based on this regression framework, an estimated location of the target will be calculated out.

Support Vector Machines (SVMs) are a set of related supervised learning methods used for classification or regression analysis. Vapnik first proposed a version of SVM for regression (SVR) in 1997 [12]. SVR algorithms have been proven to solve large-scale regression problems efficiently [13] based on studies of approaches such as the smoothing technique and hyperplane thinking and considerations related to kernels. The goal of SVM modeling is to find the optimal hyperplane (classifier boundary) that separates samples and places those in one category on one side of the plane and those in the other category on the other.

Given a training set of labeled pairs(X_i, Y_i), $i = 1, \ldots, n$ where $X_i \in R^d$ and $Y_i \in \{-1, 1\}$ indicate the category information, any point lying on the hyper plane separating the classes satisfies $w\dot{x} + b = 0$, with w being the normal to the hyper plane and $\frac{|b|}{||w||}$ being the perpendicular distance from the hyperplane to the origin. This problem can be transformed using the following quadratic programming optimization problems:

$$\begin{aligned} &\min_{(w,b)} \frac{1}{2}||w||^2 \\ &s.t.\ y_i(w^T \cdot x_i + b) \leq 1 \end{aligned} \tag{11}$$

Particularly, the samples of the data set are mislabeled, not linearly separable, the Slack Variable ξ_i is employed to accommodate the non-linearly separable outliers. Thus the equation above can be rewritten as follows:

$$\begin{aligned} &\min \frac{1}{2} w^T w + C \sum_{i=1}^{n} \xi_i \\ &s.t. \begin{cases} y_i(w^T \cdot \phi(x_i) + b) \geq 1 - \xi_i \\ \xi_i \geq 0 \quad : i = 1, \ldots, n \end{cases} \end{aligned} \tag{12}$$

Suppose training set (X_i, Y_i), $i = 1, \ldots, n$ where $X_i \in R^n$ and $Y_i \in R$ indicate the target value, N is the number of samples, we may define ϵ insensitive loss function as following:

$$|y - f(x)|_\epsilon = \begin{cases} 0, & |y - f(x)| \leq \epsilon \\ |y - f(x)| - \epsilon, & |y - f(x)| > \epsilon \end{cases} \tag{13}$$

Here, $f(x)$ is the regression estimation function, ϵ is the insensitive loss function, The purpose of this learning step is to construct the $f(x)$ to make distance between the target less than ϵ, and at the same time VC dimension of the function is minimum, so that the corresponding target value will be estimated.

Its dual problem is then

$$\min \frac{1}{2}\sum_{i=1}^{n} a_i a_j y_i y_j K(x_i x_j) - \sum_{i=1}^{n} a_i$$
$$s.t. \begin{cases} 0 \le a_i \le C, & : i = 1, \ldots, n; \\ \sum_{i:y_i=+1} a_i = 0; & \\ \sum_{i:y_i=-1} a_i = 0; & \end{cases} \tag{14}$$

Here, C is the penalty factor, and function $K(x_i x_j)$ is the kernel function that describes the behavior of the support vectors. Recently, the most widely used kernel function is the linear kernel function, the polynomial kernel function, and RBF kernel function. The selection of a kernel function influences the results of the SVM regression analysis. This study chooses RBF as its kernel function.

The main difference between operation in linear vector spaces and non-linear vector space by using SVM is that, instead of using the original attributes, the task for non-linear vector space is performed in a transformed attributes. The kernel method is a method for computing similarity in the transformed space using the original attribute set. The main purpose of using kernel function in SVM is making the problem domain into a higher dimension space from the original dimension space, and on the higher dimension space, we use another hyper-plane to finish the classification. When the space rises into an infinite dimension, all instances will be separable. Take formula above as an example, we make the transformed attribute $K(x_i x_j)$ instead of the original attribute. $K(x_i x_j)$ is named as the kernel function. This kernel function makes the problem domain into the dimension space of $K(x_i x_j)$ from the original dimension space, and on the $K(x_i x_j)$ dimension space, another hyper-plane will be calculated out to finish the classification.

$$K(x_i, x_j) = \phi(x_i)^T \phi(x_j) = e^{\left(-\frac{||x_i - x_j||^2}{2\gamma^2}\right)} \tag{15}$$

Here, $\gamma > 0$, C and γ are parameters calculated to make the SVM highly sensitive to training data. They ultimately contribute to reducing the error rate of this classification. We will use particle swarm optimization as a parameter selector for SVM to search the optimized C and γ.

Kennedy and Eberhar proposed Particle Swarm Optimization (PSO) in 1995 which inspired by the behavior of birds. The basic concept of PSO is: each solution to an optimization problem is called a particle and defines a fitness function to measure the degree of superiority of each particle. Each particle is a point in the n-dimensional space, and it closes to two points in the solution space simultaneously, called global optimal solution gbest [14]: the Optimal place it can achieve in the group, and the individual optimal solution pbest: the Optimized place that it ever passed.

In PSO, X_i $[x_{i1}, x_{i2}, \ldots, x_{in}]$ denotes particle i, $gbest = [g_1, g_2, \ldots, g_n]$ indicate the global optimal solution of particle i and $pbest_i[p_{i1}, p_{i2}, \ldots, p_{im}]$ symbolizes the individual optimal solution of particle i. Finally let $v_i^k = [v_{i1}^k, v_{i2}^k, \ldots, v_{in}^k]$ represent the speed of particle i. Each particle updates their speed rate and place based on the following formula:

$$v_{id}^k = w_i v_{id}^{(k-1)} + c_1 r_1 (p_{id}^{(k-1)} - x_{id}^{(k-1)}) + c_2 r_2 (g_d^{(k-1)} - x_{id}^{(k-1)}) \quad (16)$$

$$x_{id}^k = x_{id}^{(k-1)} + v_{id}^k \quad (17)$$

where k is the iteration number; $i = 1, 2, \ldots, m$ and $d = 1, 2, \ldots, n$. m denotes the particle number; r_1 and r_2 are the numbers dotted randomly between [0, 1]; c_1 and c_2 are two regular numbers acted as the acceleration elements; and w_i is the inertia weight, which can change the strength of the search ability [14].

In the SVM part, a PSO is performed to distinguish the parameters C and γ for SVM. The following describes the procedure of this PSO-SVM.

Step 1: Define the stopping condition s, searching space sp.
Step 2: Initialize n (C, γ) randomly.
Step 3: Select the optimal fitness of each individual particle by stochastic/ruled offset, and number each one;
Step 4: Calculate the fitness of particle swarm in the searching space.
Step 5: Select best individuals as particles W.
Step 6: For every fitness of the individual particles in W, classify the differences between each previous fitness/updated fitness pair. This is a comparison process.
Step 7: Update the rule base according to the result of the comparison.
Step 8: If the s is satisfied, terminate the computation; if not, go back to Step 3.

The above procedure details the PSO process of SVM. When a determined computing threshold has been reached, the process terminates. Here, the search space consists of particles and two real values, C and γ. The offset is a selection of random values; these random values change the C and γ after a pretreatment.

Figure 3 shows the main framework of the proposed PSO-SVM.

When the regression step above finished, we can get the estimate location, we do the comparison analysis between basic location (bl) calculated by particle filter and estimate location (el) calculated by PSO-SVM based on the following equation:

$$MachingRate(MR) = \frac{bl \cap el}{bl \cup el} \quad (18)$$

If the MR is larger than 50 %, choose the $bl \cap el$ as the tracked location, otherwise, choose the estimate location as the tracked location.

Figure 4 shows the whole framework of the proposed human tracking system.

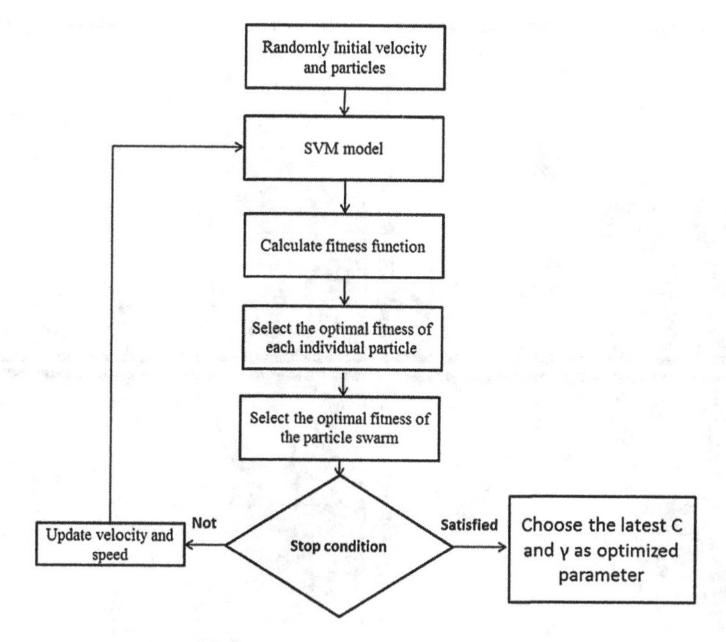

Fig. 3 Framework of PSO-SVM

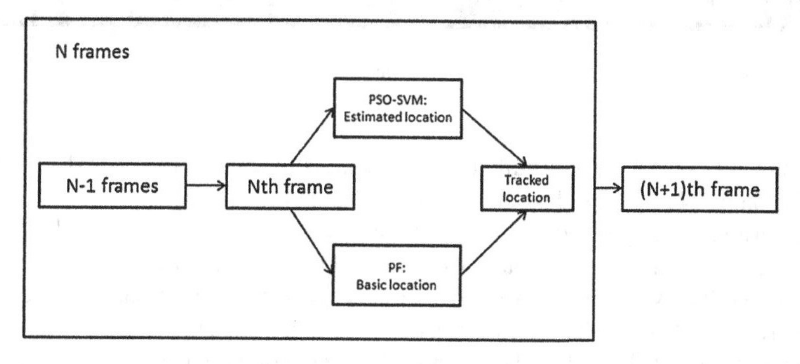

Fig. 4 Framework of proposed human tracking system

3 Experimental Results

To check whether the proposed human tracking system runs well, we runs one simple test video. The result is shown in Fig. 5, we catch 4 frames (435, 443, 451, 459) of the whole frame sequence. It shows that the proposed method can show the target well in the human tracking even occlusion occurred by other object target, in this application which is human.

From the results we can find out that our proposed method can track the target in an accuracy way although occlusion occurred, in each frame, the location of the

Fig. 5 Tracking result in object occlusion situation

human target is marked in a sufficient way. That is to say, the PSO-SVM can solve occlusion problems caused by other object target in human tracking.

To evaluate our method, occlusion caused by firmed obstruction also be tested. We do the experiment of the same video using both conventional particle filter and proposed PSO-SVM. Figure 6 shows the results under the conventional particle filter, we may see that the system missed tracking the target since the target walk through the lamp post and also, when the target walk through other dynamic object, the conventional method also shows the poor result. Figure 7 shows the results under the proposed PSO-SVM, we may see that, to the same frame, our proposed method can track the target even though the target was whole covered by the lamp post, also there is few influence since the target walk through other dynamic object. That is to say, our proposed method can finish the human tracking under occlusion caused by the environment.

In this study, the proposed method and the particle filter method were both applied to the experiments.

As mentioned above, the aim of our method is to solve four problems in a tracking system, namely, to find (1) how to detect the directions of object movement, (2) how to account for changes in the relative sizes of objects in the scene based on their

Fig. 6 Tracking result using conventional particle filter

Fig. 7 Tracking result using proposed PSO-SVM

distance from the camera, (3) how to track objects at a high accuracy and (4) how to minimize processing time.

Results show that our method can track the object more accurately under the complicated environment but the particle filter had difficulty locating a moving target at the same situation, and the tracking result was not good.

To evaluate the proposed method we calculate the accurate rate by the following formulation:

$$AccurateRate(AR) = \frac{tl \cap al}{tl \cup al} \tag{19}$$

Here, tl is the tracked location and al is the actual location.

From the table above we may find out that, in normal situation which means there is no occlusion, both methods can locate target in a accurate way, the conventional particle filter can not track the object when occlusion occurred, the AR in frame 291 and frame 301 is 0 %, which means the system missed tracking the object and shows the wrong tracking result, when the target covered by other dynamic object, the conventional method also failed to track and locate the target which shown in frame 607. The proposed method can locate target in both firmed obstruction occlusion and dynamic object occlusion, but the AR changes when the occlusion rate changes (Table 1).

Table 2 shows that, in any case, the same method takes nearly the same processing time, but the particle filter processes more quickly than the PSO-SVM. However, 41.32 fps almost satisfy the real-time requirement and for tracking system, which requires much greater accuracy, we still use the PSO-SVM to do the tracking step.

From results we found that the proposed method performs well in a more accurately way. This is because as a regression framework, the PSO-SVM can make the estimate location in an accurate way, and by the comparison analysis, we can locate the target by reducing the influence of environment.

Table 1 Result of several frames

Methods	Particle filter (%)	PSOPF-SVM (%)
Frame 276	82.3	84.7
Frame 291	0	77.2
Frame 301	0	76.3
Frame 607	14.3	70.3

Table 2 Processing time

	PSOPF-SVM(s)	Particle filter(s)
Speed	41.32fps	61.27fps

4 Conclusions

A human tracking system has two main issues. The first is the problem of location detection and the second is how to select the targeted human in each frame. In this paper, tracking methods are implemented in a detection module to detect a human location. For the tracking module, we use particle filters to solve the problems we have discussed.

The proposed method provides a more accurate result under occlusion for the systems than conventional methods such as the particle filter, but it requires more processing time although it satisfies the real-time requirement. The surveillance system already takes much time for calculation, so for certain time-sensitive work, this method may not perform well. However, Some methods failed to properly track human locations when an object is under occlusion, and the proposed method can solve this problem.

In the future, we will improve the proposed method into multi object tracking, and also, we will try to raise the processing speed by improving both structure and algorithm.

Acknowledgments This work was supported in part by Nanjing Audit University, Nanjing, Jiangsu, China and "six kinds of peak talents" high level support of Jiangsu.

References

1. Sacchi, C., Regazzoni, C.S.: A distributed surveillance system for detection of abandoned objects in unmanned reailway environments. IEEE Trans. Veh. Technol. **49**, 2013–2026 (2000)
2. Isard, M., Blake, A.: Contour tracking by stochastic propagation of conditional density. Computer Vision-ECCV'96, Lecture Notes in Computer Science, vol. 1064/1996, pp. 343–356 (1996)
3. Repperger, D.W., Ward, S.L., Hartzell, E.J., Glass, B.C., Summers, W.C.: An algorithm to ascertain critical regions of human tracking ability. IEEE Trans. Syst. Man Cybern. **9**, 183–196 (1979)
4. Zhai, Y., Yeary, M.B., Cheng, S., Kehtarnavaz, N.: An object-tracing algorithm based on multiple-model particle filtering with state partitioning. IEEEE Trans. Instrum. Meas. **58**, 1797–1809 (2009)
5. MacCormick, J., Blake, A.: A probabilistic exclusion principle for tracking multiple objects. In: Proceedings of the International Conference on Computer Vision, pp. 572–578 (1999)
6. Carpenter, J., Clifford, P., Fearnhead, P.: Improved particle filter for nonlinear problems. IEE Proc. Radar Sonar Navig. **146**, 2–7 (1999)
7. Ye, L., Zhang, Q., Guan, L.: Use hierarchical genetic particle filter to figure articulated human tracking. In: IEEE International Conference on Multimedia and Expo, pp. 1561–1564 (2008)
8. Crisan, D., Del Moral, P., Lyons, T.J.: Discrete filtering using branching and interacting particle systems. Markov Processes Relat. Fields **5**(3), 293–319 (1999)
9. McKenna, S.J., Nait-Charif, H.: Tracking human motion using auxiliary particle filters and iterated likelihood weighting. Image Vis. Comput. **25**(6), 852–862 (2007)
10. Ristic, B., Arulampalam, S., Gordon, N.: Beyond the Kalman Filter: Particle Filters for Tracking Applications. Artech House (2004)

11. Kitagawa, G.: Monte Carlo filter and smoother for non-Gaussian nonlinear state space models. J. Comput. Geogr. Stat. **5**(1), 1–25 (1996)
12. Smola, A., Vapnik, V.: Support vector regression machines. Advances in Neural Information Processing Systems, vol. 9, pp. 155–161. MIT Press (1997)
13. Collobert, R., Bengio, S.: SVM torch: support vector machines for large-scale regression problems. J. Mach. Learn. Res. **1**(2), 143–160 (2001)
14. Wang, J., Xing, Y., Cheng, L., Qin, F.: The Prediction of Mechanical Properties of Cement Soil Based on PSO-SVM, Computational Intelligence and Software Engineering (CiSE), 2010 International Conference on, pp. 10–12 (2010). doi:10.1109/CISE.2010.5677256

Author Index

A
Abdelmajid, Ben Hamadou, 243
Abelha, António, 115
Adu-Kwarteng, Michael, 255
Afful-Dadzie, Eric, 255
Aknine, Samir, 231
Assylbekov, Zhenisbek, 37

B
Baltabayeva, Assel, 37
Barbucha, Dariusz, 361
Bekishev, Rustam, 37
Bennajeh, Anouer, 231
Ben Said, Lamjed, 231
Biagetti, Giorgio, 413, 425
Bissengaliyeva, Dariya, 37
Bobrowski, Leon, 293
Boryczka, Mariusz, 325
Boryczka, Urszula, 325

C
Chen, Yetian, 141
Coimbra, Ana, 115
Colla, Valentina, 395
Crippa, Paolo, 413, 425
Crispin, Alan, 63
Crockett, Keeley, 63
Czarnowski, Ireneusz, 337

D
Datko, Patrick, 405
Dias, Miguel Sales, 383
Ding, Haochen, 437
Dondi, Riccardo, 27

E
Emna, Hlel, 243

F
Falaschetti, Laura, 413, 425

H
Hamadou, Abdelmajid Ben, 3
Huang, Hung-Hsuan, 203

I
Iwamae, Takayuki, 203

J
Jaffali, Soufiene, 3
Jamoussi, Salma, 3
Janardhanan, Mukund Nilakantan, 77
Jardim, David, 383
Jędrzejowicz, Joanna, 317
Jędrzejowicz, Piotr, 305, 317, 349, 337

K
Kebair, Fahem, 231
Khalfay, Amy, 63
Khosiawan, Yohanes, 77
Komínková Oplatková, Zuzana, 255
Kuwabara, Kazuhiro, 203

L
Lamperti, Gianfranco, 215
Lim, Chee Peng, 447
Lipiński, Zbigniew, 151

M
Machado, José, 115
Madrid, Natividad Martínez, 405
Mamlin, Eldar, 37
Masuya, Satoshi, 129
Mauri, Giancarlo, 27
Mazurek, Jiri, 193

I. Czarnowski et al. (eds.), *Intelligent Decision Technologies 2016*,
Smart Innovation, Systems and Technologies 56,
DOI 10.1007/978-3-319-39630-9

Mbogho, Audrey, 373
Melnykov, Igor, 37
Michna, Zbigniew, 15
Mi, Chuanmin, 141
Mohamed, Turki, 243
Moon, Ilkyeong, 77

N
Nabareseh, Stephen, 255
Neves, João, 115
Neves, José, 115
Nielsen, Izabela, 77, 101
Nielsen, Peter, 15, 101
Nunes, Luís, 383

O
Ochab, Marcin, 165
Ojeme, Blessing, 373
Orcioni, Simone, 413, 425

P
Park, Youngsoo, 77

Q
Quarenghi, Giulio, 215

R
Ratajczak-Ropel, Ewa, 305
Robaszkiewicz, Marek, 281

S
Saastamoinen, Kalle, 51
Sałabun, Wojciech, 181
Salma, Jamoussi, 243
Santos, M. Filipe, 115
Scherz, Wilhelm Daniel, 405
Seepold, Ralf, 405
Sitek, Paweł, 101
Skakovski, Aleksander, 349
Smaili, Kamel, 3
Stańczyk, Urszula, 269

T
Takenaka, Keisuke, 203
Tan, Shing Chiang, 447
Turchetti, Claudio, 413, 425

V
Vannucci, Marco, 395
Velicu, Oana Ramona, 405
Vicente, Henrique, 115

W
Wada, Yudai, 203
Wais, Piotr, 165
Wajs, Wiesław, 165
Wang, Yinchuan, 141
Watada, Junzo, 437, 457, 447
Wątróbski, Jarosław, 89, 181
Wierzbowska, Izabela, 317
Wikarek, Jarosław, 101
Wojtowicz, Hubert, 165

X
Xu, Chao, 457
Xu, Zhenyuan, 457

Z
Zielosko, Beata, 281
Ziemba, Paweł, 181
Zoppis, Italo, 27

Printed in the United States
By Bookmasters